German A. Shneerson, Mikhail I. Dolotenko, Sergey I. Krivosheev

Strong and Superstrong Pulsed Magnetic Fields Generation

De Gruyter Studies in Mathematical Physics

Volume 9

German A. Shneerson, Mikhail I. Dolotenko,
Sergey I. Krivosheev

Strong and Superstrong Pulsed Magnetic Fields Generation

DE GRUYTER

Physics and Astronomy Classification Scheme 2010
07, 41, 52, 85

ISBN 978-3-11-025191-3
e-ISBN 978-3-11-025257-6

Library of Congress Cataloging-in-Publication Data
A CIP catalog record for this book has been applied for at the Library of Congress.

Bibliographic information published by the Deutsche Nationalbibliothek
The Deutsche Nationalbibliothek lists this publication in the Deutsche Nationalbibliografie; detailed bibliographic data are available on the Internet at http://dnb.dnb.de.

Typesetting: PTP-Berlin, Protago TEX-Produktion GmbH, www.ptp-berlin.de
Printing and binding: CPI books GmbH, Leck
♾Printed on acid-free paper
Printed in Germany

www.degruyter.com

Contents

Introduction —— 1

1 Magnetic fields of axially symmetrical magnetic systems used for generation of the strong fields (methods of calculation, assessment of the edge effects) —— 8
1.1 Magnetic field of the systems with the given current distribution —— 8
1.2 The setting of the task for the calculation of a magntic field at a small penetration depth —— 13
1.3 The determination of the parameters of the inductor systems at a strongly-pronounced skin effect according to the simplified field pattern —— 18
1.3.1 The field of single-turn solenoids (flat ring) above the ideally-conducting plane ($h \ll r_1$) —— 18
1.3.2 The field of multiwinding solenoid in the form of a flat spiral above the plane (Figure 1.6b) —— 19
1.3.3 Field of solenoid in off-loading cylindrical screens —— 20
1.4 Edge effects in single-coil magnets. Modelling of problems —— 21
1.5 References —— 27

2 Calculating formulas and the results of numerical estimations of field parameters for typical single-turn magnets —— 29
2.1 The field of the flat ring as an example of the single-turn magnet with sharply pronounced edge effect —— 29
2.2 The coil of a rounded cross-section (an ideally conductive toroid) —— 32
2.3 Thin-wall single-turn magnets —— 35
2.4 The field of rectangular coils with arbitrary ratios of characteristic dimensions —— 39
2.5 Induction of the one-turn magnet placed near the coaxial cylinder or the plane —— 41
2.6 References —— 44

3 Field diffusion into the conductors and their heating —— 45
3.1 Adiabatic heating of conductors at a given current density —— 46
3.2 Linear regime of field diffusion in conductors —— 50

3.3 The surface impedance. Energy losses in the skin layer with sinusoidal current — 52
3.4 The asymptotical values of the magnetic field intensity and current density at the conductor edge under the condition of a pronounced skin effect — 57
3.5 Examples of the diffusion of the uniform pulse electromagnetic field into a medium with constant conductivity — 61
3.6 Energy generation and heating a medium in the case of diffusion of the pulse magnetic field into the conductor — 65
3.7 Heating of a conductor with a current in an external magnetic field — 70
3.8 Minimization of a uniform medium heating under diffusion of the pulse magnetic field — 76
3.9 One-dimensional diffusion of the field into a medium with conductivity depending on the coordinate. Reduction of energy generation in the surface layer — 78
3.10 One-dimensional nonlinear diffusion of the magnetic field into the conductor heated by the eddy current — 81
3.11 Approximate description of the surface effect. "The skin layer method" — 85
3.12 References — 89

4 Matching of the parameters of solenoids and power supply sources — 91
4.1 General requirements to the power supply source — 91
4.2 Optimization of the parameters of the system of solenoids – capacity energy storage — 92
4.3 Optimization of solenoids according to Fabri — 94
4.4 Transformations of energy in a circuit with alternating inductance — 98
4.4.1 Direct current in the element of the electrical circuit with alternating inductance — 99
4.4.2 Energy transformations in the short-circuiting coil with alternating inductance — 99
4.4.3 Railgun powered by energy capacity storage — 100
4.5 On the application of inductive storages for supplying the magnetic systems — 109
4.6 References — 112

5 **Electromagnetic forces and mechanical stresses in multiturn solenoids. The optimization of multilayered windings — 113**
5.1 Asimuthal and axial stresses in the thin-wall turn in the poloidal magnetic field — **114**
5.2 Mechanical stresses in the uniform cylinder with a given current distribution — **117**
5.2.1 A winding with constant current density — **118**
5.2.2 A winding with a current density decreasing inversely with radius (Bitter's solenoid) — **120**
5.3 Mechanical stresses in an equilibrium thin-wall cylinder with current — **121**
5.4 Mechanical stresses in two-component winding — **125**
5.5 Magnets with mechanically separated thin current layer. Series or parallel connection of layers — **129**
5.5.1 A winding with a series connection of current layers — **129**
5.5.2 A winding with parallel-connected layers — **130**
5.6 Multilayer magnet with equally-loaded winding — **131**
5.7 Multilayer magnets with equally-loaded internal reinforcements — **136**
5.8 The plastic deformation and the resource of multiturn magnets — **140**
5.9 References — **145**

6 **Generation of strong magnetic fields in multiturn magnets — 147**
6.1 Traditional constructions of solenoids with spiral multilayer windings — **149**
6.2 Present-day materials used to make windings — **154**
6.3 Special features of constructions of present-day multiturn monolithic magnets with field of 60–80 T — **159**
6.4 The results of tests of multiturn magnets and investigation of their destruction — **164**
6.5 Magnets with record fields — **169**
6.6 Flat helical solenoids — **175**
6.7 References — **178**

7 **Solenoids with quasi-force-free windings — 182**
7.1 Quasi-force-free configurations, an analog of which is a winding of a quasi-force-free magnet — **183**
7.1.1 One-dimensional quasi-force-free magnetic systems: the flat layer and cylinder — **183**

7.1.2 Two-dimensional force-free configurations satisfying the characteristic boundary conditions — 186
7.1.3 Features of current distribution in the face zone of a force-free magnet — 192
7.2 The methods of realization of a quasi-force-free winding. The estimates of residual mechanic stresses in a thin-wall quasi-force-free winding — 194
7.2.1 Quasi-force-free winding with pairs of the equilibrium current layers (number of pairs $N \gg 1$) — 194
7.2.2 Multilayer magnetic systems with variable direction of current in each layer — 196
7.3 Configurations of magnetic systems with equilibrium windings with zero thickness — 201
7.3.1 One-modular configurations — 202
7.3.2 Multimodular systems — 206
7.4 Thin-wall quasi-force-free magnets with current removals — 210
7.4.1 Systems with equally-loaded internal reinforcements — 210
7.5 Comparative estimates of the residual stresses and sizes of magnets with a quasi-force-free winding and loaded outer zone — 213
7.6 Design methods of quasi-force-free magnets — 214
7.7 References — 217

8 Generation of strong pulsed magnetic fields in single-turn magnets. Magnetic systems for the formation of pulsed loads — 219
8.1 Mechanical stresses in a single-turn magnet operating under the condition of a sharply pronounced skin effect — 220
8.2 Assessing the strength of single-turn magnets at short pulses — 222
8.3 Thermoelastic stresses in single-turn magnets — 227
8.4 The destruction of single-turn magnets. The problem of erosion — 231
8.5 Special construction features of single-turn magnets and their power supplies — 237
8.6 Deformed single-turn magnets restored after the discharge — 246
8.7 Magnetic systems used for deformation of solids and the study of their properties — 247
8.8 Magnetic systems for the acceleration of conductors — 259
8.9 References — 266

9 **Generation of ultrahigh magnetic fields in destructive single-turn magnets — 270**

9.1 Physical processes accompanying the generation of megagauss magnetic field in single-turn magnets — **270**

9.2 Modeling problems illustrating the role of different factors leading to the destruction of single-turn magnets — **281**

9.3 Hydrodynamic flows in single-turn solenoids. Application of the model of a noncompressible liquid with ideal conductivity to the description of the deformation of a thick-wall turn — **283**

9.4 Electrical explosion of turns of small thickness. Evaluation of the induction achieved in the destruction of turns with small initial dimensions — **291**

9.5 One-dimensional hydrodynamic flow in the wall of a single-turn magnet. Shock wave in conductors initiated by superstrong magnetic fields — **298**

9.6 General information on the electric explosion of conductors — **301**

9.7 Electric explosion of the skin layer in superhigh magnetic fields. Ideal model — **311**

9.8 The actual processes developing for “slow” and “fast” electric explosions of a conductor surface skin layer in a superhigh magnetic field — **313**

9.9 Computer simulation of a skin layer explosion — **320**

9.10 References — **330**

10 **Magnetic cumulation — 333**

10.1 Initial idea. Brief history. Main trends in development and research — **333**

10.2 MC energy generators — **336**

10.3 Physical processes in magnetic cumulation. Analytical estimates for the MC-1 system — **342**

10.3.1 Induction amplitude and the radius of turnaround for flux compression by an ideal cylindrical shell — **343**

10.3.2 Estimation of the pulse duration of a magnetic field in magnetic cumulation — **344**

10.3.3 The effect of field diffusion on the induction amplitude with magnetic cumulation — **346**

10.3.4 Restrictions on the induction amplitude conditioned by the compressibility of a medium — **349**

10.3.5 Violation of the stability of a liner at flux compression — **352**

10.4 Flux compression systems not using the explosion energy for liner acceleration — **355**
10.4.1 MDC systems with azimuth current in a liner — **357**
10.4.2 Magnetodynamic cumulation in a Z-Θ pinch system — **362**
10.5 Analytical estimations and simulation of magnetodynamic cumulation — **365**
10.6 Explosion devices and solenoids of an initial field used in magnetic cumulation — **379**
10.6.1 Detonation of the explosive charge — **379**
10.6.2 Generation of the initial magnetic field — **382**
10.7 Liners of MC-1 generators — **384**
10.7.1 Commonly used metal liners — **384**
10.7.2 Metal composite liners — **385**
10.7.3 Shock-wave liners with phase transitions — **386**
10.8 Violation of liner stability in flux compression — **390**
10.9 Principle of cascading in MC generators of ultrahigh magnetic fields — **395**
10.10 MC-1 cascade generator. Numerical simulation and experiment — **399**
10.11 Ways of increasing the induction amplitude. Methods for control over the pulse shape. Capabilities of the MC-1 generator — **407**
10.11.1 Methods for control over the shape and amplitude of an induction pulse — **407**
10.11.2 Capabilities of MC-1 generators of ultrahigh magnetic fields — **410**
10.12 Conclusion — **416**
Supplement S10. Calculation of the skin layer thickness and of the parameter q, characterizing the energy in the skin layer — **416**
10.13 References — **418**

Index — 424

Introduction

During the last century, among the interest of physicists and engineers in the generation of more and more strong magnetic fields has strongly increased. This interest is stimulated, firstly, by various applications of strong fields in scientific research and technology. Along with this the effects accompanying the generation of such fields are so versatile and interesting in themselves that their study has allowed us to obtain a variety of new data concerning the field diffusion into conducting media, development of magneto hydrodynamic flows, the appearance of phase transitions in metals, and their properties at high pressures and temperatures. This list is far from complete and does not include the broad scope of the physical effects that are revealed in conductors subjected to the action of strong magnetic fields.

The purpose of this book is to outline the current concepts of the technologies used in the generation of strong and superstrong pulsed magnetic fields and in the description of the physical processes occurring at the action of these fields on the conductors applied in magnetic systems. We suggest that it is necessary to outlay both the qualitative description of these processes and to present their quantitative assessments.

Further, according to the established tradition we shall refer the fields with induction above 10 T as *strong* fields. The fields with induction above 100 T (1 Megagauss) are denoted as *superstrong* or *megagauss* fields. Consequently, the fields with induction above 400–500 T will be denoted as *multimegagauss* fields.

The peculiar feature of superstrong fields is that at their application on the conducting bodies and at the conditions of clearly pronounced skin-effect both the mechanical strength threshold and the threshold of melting of the surface layer are over exceeded. When multimagagauss fields are applied, there occurs the explosion of the skin layer – its evaporation is followed by a loss of conductivity.

The amplitude of the field induction obtained in the magnets of various types is limited by the heating of the coil above the permissible level and its destruction under the action of the electromagnetic forces. The problem of heating is especially significant in magnets designed for the generation of the constant fields without using superconductors. The level of the fields attained in these magnets with complex cooling systems is limited by the values of induction in the scale of 20 T. The application of superconductors exploited at the temperature of liquid helium allows elimination of the problem of heating which permits production of the constant field with induction at approximately 20 T. The higher values of induction could be produced with high-temperature superconductors possessing the high critical field and high critical current. The further progress in the field of application of superconductive magnets is closely connected with the solution of their strength problems.

The heating of the coil could be significantly reduced by the shortening of the applied current pulse, in other words by the formation of the field by the short pulses. Namely, this trend in the technology of the strong magnetic fields is the main focus of this study. It was initiated by Kapitsa [1, 2] and Wall [3] in the 1920s. In experiments with pulse duration of $\leq$ 0.01 s, fields with induction of 30–40 T were obtained. In these experiments the acceptable adiabatic heating of the coil was attained, but the production of even stronger magnetic fields was limited by the insufficient strength of the coils. The further development of the technology of nondestructive magnets is related to the development of more appropriate coil designs and the use of the strongest materials. Nevertheless, in spite of efforts of many laboratories, attempts to construct the magnets for multiple-use purposes with induction exceeding 90–100 T have failed.

Essentially, stronger-pulsed fields could be produced in destructive magnets and by using the method of magnetic cumulation. The first of above approaches was developed in the late 1950s due to significant progress reached at that time in the development of powerful pulsed current generators, using small inductance high voltage condenser banks. In the pioneering work of Furth, Levin, and Waniek [4] fieldd with induction of 160 T were attained. This work provided momentum to the numerous studies in which pulsed fields with small rise times were produced in single turn magnets with characteristic dimensions of the internal radius and length of the order of $\leq$ 1 cm. At present, using this method and with application of generators with relatively moderate energy (of the order of $\leq$ 1 MJ) values of fields with induction more 300 T were obtained [5, 6]. Application of megavolt forming lines with stored energy up to 10 MJ allowed for an increase of the characteristic dimension of the domain where fields are formed and provided the opportunity to obtain multimegagauss fields in spite of explosion of the skin layer [7]

The strongest magnetic fields have presently been attained by the method of magnetic cumulation. This method, based upon the fast compression of the magnetic flux by the walls of the conductive cylinder, was originally realized by the Sakharov group in the USSR and the Fowler group in the USA in the late 1950s. The further development of this technology. based upon the comprehensive study of the physics of the process and due to further improvement of experimental techniques, allowed us to obtain fields with induction up to 2800 T [8]. In these experiments, the conducting cylinder was compressed by explosion products. Another method, namely, the compression of the shell by electromagnetic forces, made it possible to obtain fields with induction up to 720 T [9].

A number of review papers and books have been devoted to the physics and technology of strong magnetic fields. They generally consider nondestructive magnetic systems. In the review by Strahovsky and Kravtsov [10], and also in the books by Parkinson and Mullhall [11], Montgomery [12], Karasik [13], and in the book edited by Komelkov [14], the methods of field calculations with given current distribution and the methods of calculations of mechanical stresses in multiturn magnets of different configuration are described. A detailed description of pulsed fields can be found in the book by Knoepfel [15], in the survey edited by Herlach [16], and in the book by Lagutin and Ozhogin [17]. These books outline the main concepts of the strong field technologies described in the literature before 1990. A rather complete review of the abstracts of the papers is contained in the books [18,19]. The further progress in the technology of pulsed fields has been described in the reviews [20,21] and in the book edited by Herlach and Miura [22]. This book also contains papers devoted to the description of the strong pulsed magnetic fields in various areas of science.

The development of the installation for the obtaining the strong magnetic field generally includes the following stages:
(1) calculation of the solenoid field and selection of its geometrical shape providing the required field configuration and calculation of the current intensity necessary for power supply of magnetic system:
(2) identification of the energy source and its matching with the load;
(3) calculation of the mechanical stresses in the solenoid winding;
(4) analysis of the thermal processes in the coils.

All of these stages, to some extent, are considered in the Chapters 1–4. The first chapter describes the formulation of the task and calculation of the field of magnetic systems, and presents some important formulas. Chapter 2 pays special attention to the problems in which the boundary conditions correspond to the strongly-pronounced skin-effect that is inherently associated with pulsed fields. Chapter 3 presents the problems of field diffusion, with emphasis on the Joule heating and measures taken for its reduction. Nonlinear diffusion is also considered in that chapter. In Chapter 4 we consider some issues of the power supply of magnetic systems, such as: the optimization of the system; magnet-capacity energy storage; and Fabry mode optimization of the magnetic systems with stationary field, power supply of the systems designed for the acceleration of the conductors by electromagnetic forces. Chapters 5 and 6 contain the consideration of nondestructive multiwinding magnetic systems. The calculations of forces and stresses in traditional multiwinding systems with azimuthal current, including windings with equally loaded layers, are also considered. Along with this, in the framework of this book, Chapter 7 also considers the theoretical approaches at the development of quasi-zero-forces magnets, using which a dramatic reduction of mechanical stresses could be obtained. This goal could be achieved by a properly chosen current distribution in the layers of winding and the invention of the configuration providing the equilibrium of the layers. There are reasons to suggest that magnets such as these could allow us overcome the existing strength limitations and attain the megagauss level of fields in nondestructive magnetic systems without an excessive increase of their dimensions and power levels. Therefore, in Chapter 7 attention is given to consideration of the ideas advanced in recent years at the Saint Petersburg State Polytechnical University.

Chapter 8 describes the single-turn magnets designed for repeated use. The assessments of the stresses in the single turn magnets with short pulses when the inertia of material and thermoelastic deformations plays asignificant role are given in Chapter 8. This chapter also presents examples of application of strong pulsed magnetic fields in the various fields of technology where the conductive bodies are subjected to the strong fields to produce their necessary deformations or acceleration.

The last two chapters deali with superstrong fields. Chapter 9 describes the physical effects featuring in destructing single coil magnets at the field generation. Here we present a description of the experiments, a review of experimental data, and the qualitative features of single coil destruction. The character of the coil explosion in Chapter 9 is considered both for the cases of complete destruction (thin winding) and partial destruction (thick wall magnets).

The quantitative description of these processes is made within the frameworks of the simple hydrodynamic model which considers the destruction of the short thick wall coils as two-dimensional flow of ideal noncompressible fluid. The results of the destruction of thin wall coils are obtained using more extensive numerical simulation. Chapter 9 also contains an analysis of the 1D flow in the fields of megagauss range, when the shock wave is formed and the electric explosion of skin layer becomes viable. This process is considered both within the framework of simplest model of flow with ideal conductivity and in a more accurate formulation taking into account the heating of the media in the skin-layer with consequent loss of its conductivity during the expansion. On the basis of the performed analysis, there were obtained an assessments for the parameters of energy sources, used for field generation in destructing single coil magnets.

Chapter 10 of this book contains the description of magnetic cumulation. Here we dwell on the systems in which the magnetic flux is compressed by a cylindrical shell subject to acceleration by explosion products or by electromagnetic forces. The results of the analytical estimation and computer simulation of the process in final stage of which the conversion of kinetic energy to the energy of the magnetic field occurs are presented. Special emphasis is placed on experiments where the loss of instability of the liner were able to be avoided, so that a field with a record induction could be obtained.

The present book is intended both for physicists and engineers who use strong pulsed magnetic fields and are developing the devices for their generation. It also could serve as an important textbook for students studying this field of electrophysics.

Over a period of many years the authors of this book have worked in the field of the generation of superhigh fields and the development of the theory of electromagnetic processes typical for these problems of electrophysics. The questions mentioned are touched upon moreso than the problems of the generation of strong pulse fields in nondestructive multiturn magnets. The corresponding part of the book (Chapters 5 and 6) contains a brief survey which, in general terms, reflects the achievements in the scope of strong magnetic field generation in multiturn magnets.

Chapters 1–7 were written by Prof. Shneerson, Chapters 8 and 9 by Prof. Shneerson and Prof. Krivosheev, and Chapter 10 by Dr. Dolotenko and Prof. Shneerson.

The authors are indebted to Dr. Berezin, who was responsible for the translation of this book. His advice and recommendations cannot be overestimated. We also wish to acknowledge the assistance of Dr. Shedova in the translation.

The authors are grateful to their colleagues G. Kaporskaya and D. Degtev for their great help in compiling the bibliography for the book and design of the manuscript.

Bibliography

[1] P. L. Kapitza, *Proc. Roy. Soc. Ser. A* **105** (1924), 691–710.
[2] P. L. Kapitza, *Proc. Roy. Soc. Ser. A* **115** (1927), 658–683.
[3] T. F. Wall, *J. Inst. Elecrt.Engrs.* **64** (1926), 745–757.
[4] H. P. Furth, M. A. Levine, and R. W. Waniek, *Rev. Sci. Instrum.* **28**(11) (1957), 949–958.
[5] J. W. Shearer, *J. Appl. Phys.* **40**(11) (1969), 4490–4497.
[6] S. I. Krivosheev, V. V. Titkov, and G. A. Shneerson, *J. Tech. Phys.* **42**(4) (1997), 352–366.
[7] R. W. Lemke, M. D. Knudsen, C. H. Harjes, et al., MG-X (2004), pp. 403–404.
[8] S. Takeyama, MG-XIII (2010), pp. 163–169.
[9] B. A. Boyko, A. I. Bykov, M. I. Dolotenko, N. P. Kolokol'chikov, I. M. Markevtsev, O. M. Tatsenko, and A. M. Shuvalov, MG-VIII (1998), pp. 61–70.
[10] G. M. Strahovsky and N. V. Kravtsov, *Silniye magnitniye polya. Uspekhi fizicheskich Nauk.* **70**(4) (1960), 693–714. [*Soviet Phys. Uspekhi* **3** (1960), pp. 260–272].
[11] A. H. Parkinson and B. E. Mullhall, *The generation of high Magnetic Fields*, New York, Plenum Press, 1967.
[12] D. B. Montgomery, *Solenoid Magnet Design*, New York, Wiley Interscience, 1969.
[13] V. R. Karasik, *Fisika i Tehnika silnich magnitnih poley.* Moskow, Nauka, 1964 (in Russian).

[14] P. N. Dashuk, S. L. Zayents, V. S. Komelkov, G. S. Kuchinskiy, N. N. Nikolaevskaya, P. I. Shkuropat, and G. A. Shneerson, in: V. S. Komelkov (ed.),*Tekchnika bolshih impulsnih tokov i magnitnih poley,* Moscow, Atomizdat, 1970 (in Russian).
[15] H. Knoepfel, *Pulse high magnetic fields*, Amsterdam London, North-Holland Publ., 1970.
[16] F. Herlach (ed), *Strong and Ultrastrong magnetic fields and Their Aplications*, Berlin New York Tokyo, Springer-Verlag, 1985.
[17] Lagutin A. S., Ozhogin V.I. *Silniye impulsniye magnitniye polya v fizicheskom experimente*, Moscow. Energoatomizdat, 1988 (in Russian).
[18] S. A. Smirnov, A. V. Georgievskiy, and V. M. Yushnina, *Fisika I tehnika silnih magnitnih poley, Sbornik referatov*, Moscow, Atomizdat, 1970 (in Russian).
[19] G. A. Shvetsov and L. D. Vakulenko (eds), *Ultrahigh Magnetic Fields. 1924–1985 Years Bibliographic index*, Novosibirsk, Hydrodynamics Institute of SB RAN, 1986.
[20] E. I. Bichenkov and G. A. Shvetsov, Megagauss Magnetic Fields. Physics. Technology. Applications, *Zh. Prikl. Mech. Tech. Fiz.* **5** (1997), 90–102. [*J. Appl. Mech. Techn. Phys.* **38**(4) (1997), 578–589].
[21] F. Herlach, Pulsed Magnets, *Rep. Prog. Phys.* 1999, V. 65, pp. 859–920.
[22] F Herlach and N. Miura (eds.), *High Magnetic Fields. Science and Technology*, New Jersey London Singapore Hong Kong Taipei, World Scientific, 2003.

Beginning in 1966 the problems of generation of superhigh magnetic fields were regularly discussed at the international "Megagauss" conferences. Below they are listed along with the bibliography of the conference proceedings. The conference in Santa Fe, which took place in the same year as the regular 11th conference in London, is designated by MG-SF. In the book the each conference will be referred to by its number.

MG-I. H. Knoepfel and F. Herlach (eds.), *Megagauss Magnetic Field Generation by Explosives and Related Experiments*, Brussels, Euratom, 1966.

MG-II. P. J. Turchi (ed.), *Megagauss Physics and Technology*, Proc. of 2nd Intern. Conf. on Megagauss Magnetic Field Generation and Related Topics, Washington, 1979, New York London, Plenum Press, 1980.

MG-III. V. M. Titov and G. A. Shvetsov (eds.), *Ultrahigh Magnetic Fields.Physics. Techniques. Applications*, Proc. of the 3rd Intern. Conf. on Megagauss Magnetic Field Generation and Related Topics, Novosibirsk, June 13–17, 1983, Mockow, Nauka, 1984.

MG-IV. C. M. Fowler, R. S. Caird, and D. J. Erickson (eds.), *Megagauss Technology and Pulse Power Applications*, Proc. of the 4th Intern. Conf. on Megagauss Magnetic Field Generation and Related Topics, Santa Fe, 1986, New York London, Plenum Press, 1987.

MG-V. V. M. Titov and G. A. Shvetsov (eds.), *Megagauss Fields and Pulsed Power System*, Proc. of 5th Intern. Conf. on Megagauss Magnetic Field Generation and Related Topics, Novosibirsk, 1989, New York, Nova Sci. Publ., 1990.

MG-VI. M. Cowan and R. B. Spielman (eds.), *Megagauss Magnetic Field Generation and Pulsed Power Application*, Proc. of the 6th Intern. Conf. on Megagauss Magnetic Field Generation and Related Topics, New York, Nova Sci. Publ., 1994.

MG-VII. V. K. Chernyshev, V. D. Selemir, L. N. Plyashkevitch, (eds.), *Megagauss and Megaamper Pulse Technology and Application*, Proceedings of 7th Intern. Conf. on Megagauss

Magnetic Field Generation and Related Topics, Sarov, Russia, August 5–10, 1996, Sarov, VNIIEF, 1997.

MG-VIII. H. J. Schneider-Muntau (ed.), *Megagauss magnetic field generation, its application to science, and ultra-high pulsed-power technology*, Proc. of 8th Intern. Conf. on Megagauss Magnetic Field Generation and Related Topics, Tallahassee, Florida, USA, 18–23 October 1998, Florida State University, 2004.

MG-IX. Megagauss-IX. V. D. Selemir and L. N. Plyashkevich (eds.), Proc. of 9th Intern. Conf. on Megagauss Magnetic Field Generation and Related Topics, Sarov, Russia, July 7–14, 2002, Sarov, VNIIEF, 2004.

MG-X. Megagauss X. M. von Ortenberg (ed.), Proc. of the 10th Intern. Conf. on Megagauss Magnetic Field Generation and Related Topics, Humboldt University, Berlin, Germany, July 18–23, 2004, Sarov, VNIIEF, 2005.

MG-XI. Megagauss XI. I. Smith and Bu. Novac (eds.), Proc. 11th Intern. Conf. on Megagauss Magnetic Field Generation and Related Topics, London, Great Britain, September 10–14, 2006, London, Loughborough University, 2007.

MG-SF. G. E. Kiuttu, P. J. Turchi, and R. E. Reinovsky (eds.), Proc. of the Intern. Conf. on Megagauss magnetic Fields Generation and Related Topics, Santa Fe, NM, USA, Nov. 5–10, 2006, IEEE Catalog Number: CFP06MEG-CDR, 2008.

MG- XII. G. A. Shvetsov (ed.), Proc. of the 12th Internat. Conf. on Megagauss Magnetic Field Generation and Related Topics, Novosibirsk, Russia, July 13–18, 2008, Novosibirsk Publishing House of the Sibirian Branch Academy of Science, 2010.

MG-XIII. C- Sun, and C. Liu (eds.), L. Yang, W. Zou, W. Liu, Z. Zhang Q. Sun (co-eds.), Megagauss XIII, Proc. of the 13th Intern. Conf. on Megagauss Magnetic Field Generation and Related Topics, Suzhou, China, July 6–10, 2010, Suzhou, 2010.

1 Magnetic fields of axially symmetrical magnetic systems used for generation of the strong fields (methods of calculation, assessment of the edge effects)

1.1 Magnetic field of the systems with the given current distribution

As a rule, the size of the pulsed devices-R_{max} do not exceed a few meters, and this is why at the characteristic frequencies of the order of 1 MHz and lower one can assume that discharge regimes are quasi stationary. This is valid if the case when $\lambda \gg R_{max}$ (λ is the electromagnetic wavelength in a vacuum) is not considered. Therefore the assumption on quasi stationary behavior is applied in all chapters of this book. The vector of induction in the space outside of conductors at these conditions obeys the Maxwell equations [1–3]:

$$\text{rot}\mathbf{B} = 0; \quad \text{rot}\mathbf{E} = -\partial\mathbf{B}/\partial t; \tag{1.1}$$

$$\text{div}\mathbf{B} = 0; \quad \text{div}\mathbf{E} = 0. \tag{1.2}$$

In this chapter we consider the often exploited configurations of solenoid used at the generation of strong and superstrong fields. In the axis symmetrical configurations (Figure 1.1) two types of magnetic structures are possible: poloidal and azimuthal fields. In Table 1.1 we present the components of the vector potential and vectors of induction and current density for both of these field configurations.

At the given current distribution the induction of the toroidal field, having only one component could be calculated from the law of the total current:

$$B_\vartheta(N) = \frac{\mu_0\, i(r)}{2\pi\, r} \tag{1.3}$$

where r isthe radial coordinate of the point N, and $i(r) = \int_0^r \delta_z(r) 2\pi r dr$ is the total current passing through the circle with radius r.

The azimuthal component of the vector potential A_φ, flow function $\psi = rA_\varphi$ and the scalar magnetic potential U_M of the axialy symmetrical poloidal field sat-

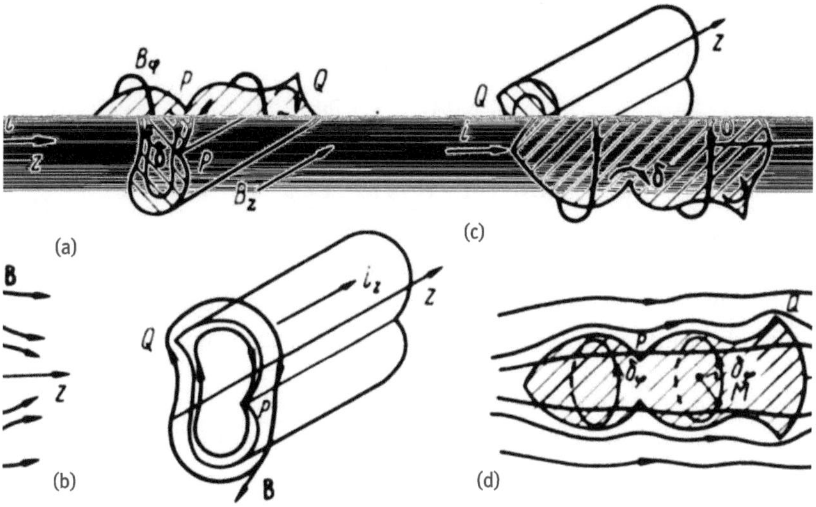

Fig. 1.1: Magnetic systems with the field depending on two coordinates: (a) axis symmetrical toroidal field with $\mathbf{B} = (0, B_\varphi, 0)$, $\boldsymbol{\delta} = (\delta_r, 0, \delta_z)$; (b) axis symmetrical poloidal field with $\mathbf{B} = (B_r, 0, B_z)$, $\boldsymbol{\delta} = (0, \delta_\varphi, 0)$; (c) flat analog to the toroidal field with $\mathbf{B} = (0, 0, B_z)$, $\boldsymbol{\delta} = (\delta_x, \delta_y, 0)$; (d) flat analog to the poloidal field with $\mathbf{B} = (B_x, B_y, 0)$, $\boldsymbol{\delta} = (0, 0, \delta_z)$.

Table 1.1: Two types of magnetic fields: poloidal and toroidal.

Type of the field	Coordinate	**A**	**B**	$\boldsymbol{\delta}$
	r	0	B_r	0
Poloidal field	φ	A_φ	0	δ_φ
	z	0	B_z	0
	r	A_r	0	δ_r
Toroidal field	φ	0	B_φ	0
	z	A_z	0	δ_z

isfy to the following equations reduced from Maxwell equations:

$$\frac{1}{r}\frac{\partial}{\partial r}\left(r\frac{\partial A_\varphi}{\partial r}\right) - \frac{A_\varphi}{r^2} + \frac{\partial^2 A_\varphi}{\partial z^2} = 0 \tag{1.4}$$

$$r\frac{\partial}{\partial r}\left(\frac{1}{r}\frac{\partial \psi}{\partial r}\right) + \frac{\partial^2 \psi}{\partial z^2} = 0 \tag{1.5}$$

$$\frac{1}{r}\frac{\partial}{\partial r}\left(r\frac{\partial U_M}{\partial r}\right) + \frac{\partial^2 U_M}{\partial z^2} = 0. \tag{1.6}$$

The vector of induction of poloidal field outside of conductor is expressed via the functions U_M, A_φ and ψ by the well-known formulas

$$B_r = -\frac{\partial U_M}{\partial r} = -\frac{\partial A_\varphi}{\partial z} = -\frac{1}{r}\frac{\partial \psi}{\partial z} \tag{1.7}$$

$$B_z = -\frac{\partial U_M}{\partial z} = \frac{1}{r}\frac{\partial \left(rA_\varphi\right)}{\partial r} = \frac{1}{r}\frac{\partial \psi}{\partial r}. \tag{1.8}$$

At known axially- symmetrical current distribution on the cross section of solenoid winding, the vector potential of the poloidal field in the arbitrary point N with cylindrical coordinates r, z is expressed in a following way:

$$\begin{aligned} A_\varphi (r, z) &= \frac{\mu_0}{4\pi} \int\limits_T \int\limits_0^{2\pi} \frac{\cos(\varphi - \varphi_t)\delta_\varphi (r_t, z_t) r_t dr_t dz_t}{(r_t^2 + r^2 - 2r_t r \cos(\varphi - \varphi_t) + (z - z_t^2))^{3/2}} \\ &= \frac{\mu_0}{2\pi} \int\limits_T \sqrt{\frac{r_t}{r}} \left[\left(\frac{2}{k} - k\right) K(k) - \frac{2}{k} E(k)\right] \delta_\varphi (r_t, z_t)\, dT (r_t, z_t), \end{aligned} \tag{1.9}$$

where r_t, z_t, φ_t are the coordinates of the point t, on the cross section of the conductor r_t, z_t, φ_t r_t, T and $K(k)$ and $E(k)$ are complete elliptic integrals with module $k = 2(rr_t)^{1/2}[(z - z_t)^2 + (r + r_t)^2]^{-1/2}$.

From here one can obtain the expressions for the induction on the axis of the solenoid:

$$B_z (z, 0) = \left[(1/r)\frac{\partial}{\partial r}(rA_\varphi)\right]_{r=0} = \frac{\mu_0}{2} \int\limits_T \frac{r_t^2 \delta_\varphi (r_t, z_t)\, dT (r_t, z_t)}{\left[r_t^2 + (z - z_t)^2\right]^{3/2}} \tag{1.10}$$

and the important formulas for the large distances from the center of solenoid (r, $z \gg g_{max}$, where g_{max} is the largest axial or radial dimension of the solenoid):

$$A_\varphi = \frac{\mu_0 r M}{4\pi \left(r^2 + z^2\right)^{3/2}}; \tag{1.11}$$

$$\left(B_z\right)_{r=0} = \frac{\mu_0 M}{2\pi z^3}, \tag{1.12}$$

where $M = \pi \int_T \delta_\varphi r_t^2 dT (r_t, z_t)$ – magnetic momentum of the solenoid.

Formulas for induction calculated at the known current distribution on a cross section are given in [4, 5] and others for the windings of different configuration. Let us confine ourselves only by the expression for induction in the coil center of rectangular cross section with the length l, external radius R_2 and internal radius R_1:

$$B(0, 0) = \frac{\mu_0 i w}{2\left(R_2 - R_1\right)} \ln \frac{R_2 + \sqrt{(l/2)^2 + R_2^2}}{R_1 + \sqrt{(l/2)^2 + R_1^2}} \tag{1.13}$$

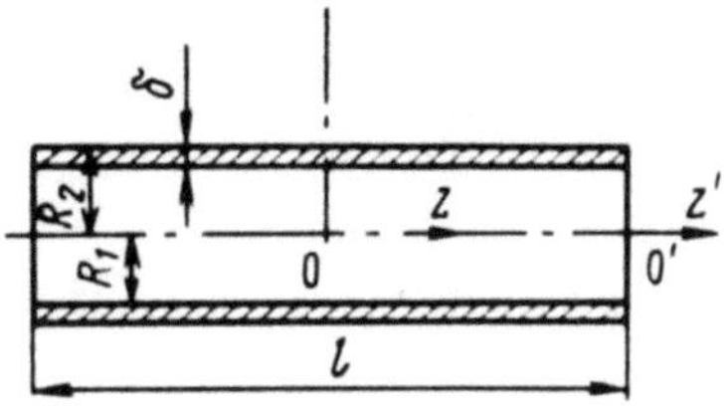

Fig. 1.2: Thin-walled solenoid.

and for the induction on the axis of thin wall solenoid [$R_2 - R_1 = \Delta \ll R_1$ (Fig. 1.2)]:

$$B(0, z) = \frac{\mu_0 J_\varphi}{2}\left[\frac{l/2 - z}{\sqrt{(l/2 - z)^2 + R_1^2}} + \frac{l/2 + z}{\sqrt{(l/2 + z)^2 + R_1^2}}\right], \tag{1.14}$$

where z is the distance from the center of solenoid, and $J_\varphi = iw/l$ is the linear current density. From formula (1.11) follow the expressions

$$\left.\begin{aligned} B_C = B(0, 0) &= \frac{\mu_0 iw}{2}\left[(l/2)^2 + R_1^2\right]^{-1/2} \\ B_T = B(0, l/2) &= \frac{\mu_0 iw}{2}\left(l^2 + R_1^2\right)^{-1/2} \\ \frac{B_T}{B(0, 0)} &= \frac{1}{2}\sqrt{\frac{l^2 + 4R_{12}}{l^2 + R_1^2}}. \end{aligned}\right\}. \tag{1.15}$$

The distribution of induction near the edge of the semiindefinite solenoid with homogeneous current density is expressed as

$$B(0, z_1) = \frac{B_\infty}{2}\left(1 - \frac{z_1}{\sqrt{z_1^2 + R_1^2}}\right), \tag{1.16}$$

where z_1 is the coordinate, counted from the edge of the solenoid (1, 2) and $B_\infty = \mu_0 J_\varphi$ is the induction in the solenoid volume far away from the edge.

In a number of cases it is convenient to use the integral representation of the potentials. If the scalar potential is the absolutely integrated function of the coordinates, it could presented in the form

$$U_M(r, z) = \int_{-\infty}^{\infty} e^{-j\lambda z}\begin{pmatrix} I_0(\lambda r) \\ K_0(\lambda r) \end{pmatrix}\mu(\lambda)\, d\lambda, \tag{1.17}$$

since the integrand satisfies to the Laplace equation. Here the Bessel function of the imaginary argument I is used if the potential is limited at $r = 0$, while the McDonald function K_0 is used if the potential is limited at $r = \infty$. The exponent with imaginary index could be replaced by the function $\sin(\lambda z)$ or $\cos(\lambda z)$

if $U_M(r, z)$ is, consequently, an odd or even function on z. It is also possible to represent $U_M(r, z)$ at $z > 0$ in the form of Fourier–Bessel integral:

$$U_M(r, z) = \int_0^{\infty} e^{-\lambda z} J_0(\lambda r) \chi(\lambda)\, d\lambda. \tag{1.18}$$

For the vector potential, the expressions are valid which are different from (1.17) and (1.18) only by the replacement of the zero-order Bessel functions with the first order Bessel functions.

The functions of the poloidal field flow are described by the same formuls as the functions of the stationary flow at the stationary flux of an ideal noncompressible liquid. In this particular case, at the known induction distribution on the axis of the magnet $B(0, z) = b(z)$ one can to evaluate the field strength near the axis using the well-known in hydrodynamics Wittaker formula [5]:

$$\psi = \frac{1}{\pi} \int_0^{r} r dr \int_0^{\pi} b(z + jr \cos \omega) d\omega, \tag{1.19a}$$

where $j = \sqrt{-1}$.

From the formula (1.19a) one can obtain the representation of induction in the form of a sequence which is valid in the vicinity of the arbitrary point on the axis right up to surface of coils:

$$B_z(r, z) = \sum_{k=0}^{\infty} \frac{(-1)^k}{(k!)^2} \left(\frac{r}{2}\right)^{2k} \frac{\partial^{2k} b(z)}{\partial z^{2k}}; \tag{1.19b}$$

$$B_r(r, z) = \sum_{k=0}^{\infty} \frac{(-1)^{k+1}}{k!\,(k+1)!} \left(\frac{r}{2}\right)^{2k+1} \frac{\partial^{2k+1} b(z)}{\partial z^{2k+1}}. \tag{1.19c}$$

These formuls could be used for the case when the onset of the counting is on the edge of the semiinfinite solenoid and the distribution of the induction is described by formula (1.16). After differentiation of the expression (1.16), it is easy to see that all terms of the sequence for B_z except for the first one ($k = 0$) reduce to zero, if $z_1 = 0$. This results in the constancy of B_z in the plane of the end up of the semiindefinite solenoid:

$$B_z = B_\infty/2 \quad \text{at} \quad z_1 = 0, \ \ 0 \le r \le R_1. \tag{1.20}$$

1.2 The setting of the task for the calculation of a magntic field at a small penetration depth

In spite of the wide range of the parameters of the devices used for the generation of strong-pulsed currents and magnetic fields (current from 1,000 to 10,000,000 A, frequency from 100 to 10,000,000 Hz, and energy from a few Joules to 10,000,000 Joules) there is a general feature distinguishing them from the conventional apparatus operating at the industrial frequency. This is the commonly mentioned skin-effect, appearing presumably in transitional regimes.

It is known that 63 % of the current, varying in the time on sin law with the circular frequency ω, is concentrated in the layer with thickness $\Delta_0 = [2\rho/(\mu\omega)]^{1/2}$, where ρ is the specific resistivity of the medium and 86% in the layer $2\Delta_0$. Parameter Δ_0 is the "classic" penetration depth of the field. For engineering calculations the skin-effect could be assumed to be a strongly pronounced one if the minimal characteristic dimension of the conductor or the system of conductors $R_{\min}$ (thickness of the wall, radius of the surface curvature, distance between the conductors etc.) satisfy the condition $R_{\min} \gg \Delta$. Here Δ is the thickness of the layer (skin-layer) in which is concentrated the major part of the current. This value approximately equals $2\Delta_0$ in the regime of harmonic current vibrations.

At the characteristic frequencies of order of 10^4 Hz and higher $\Delta_0 \le 0,7$ mm (for copper) condition for formation of strongly pronounced skin effect is generally fulfilled. In the case of a transitional regime the notion of penetration depth should be elucidated. As such it could be considered, for instance, one of the values $\Delta' = B_e/[\mu_0\delta(0)]$ or $\Delta'' = \Phi'/B_e$, where B_e and $\delta(0)$ isthe induction and current density on the surface and Φ' the flux penetrated into the conductor (calculating per unit length).

At the strongly-pronounced surface effect one can neglect the influence of the boundary curvature on the current distribution in the skin- layer and assume that the current density $\boldsymbol{\delta}$ changes normal to the boundary in the same way as in the case of conductor with the flat boundary. From the equation $\operatorname{rot}\mathbf{H} = \boldsymbol{\delta}$ follows the known relationship between the field intensity on the boundary of the conductor H_e and the surface current density $\mathbf{J}$ [1]:

$$\mathbf{J} = \int_0^\infty \boldsymbol{\delta}(x)\,dx = [\mathbf{n}_e,\ \mathbf{H}_e], \tag{1.21}$$

$$|\mathbf{J}| = |\mathbf{H}_e|. \tag{1.22}$$

Here the coordinate x is directed normal inside the conductor, the point $x = 0$ lies on the boundary. In formula (1.21) the condition used for current damping is $\delta(\infty) = 0$.

Along with surface current density at the study of current distribution in the thin sheets (the thickness of the sheet is much less than the other characteristic dimensions) we shall use the notion of linear current density

$$\mathbf{J}(N) = \int_{N_1}^{N_2} \boldsymbol{\delta}\, dn, \tag{1.23}$$

where the integral is taken normal to the middle surface of the sheet, and $N_{1,\,2}$ are the points lying on the intersection of the normal with the conductor's boundary.

At the strongly pronounced skin effect

$$\mathbf{J}(N) = \mathbf{J}_1(N_1) + \mathbf{J}_2(N_2), \tag{1.24}$$

where $\mathbf{J}_{1,\,2}$ is the surface current density in the points $N_{1,2}$.

The strongly-pronounced surface effect allows at the calculation of the field outside of conductors and absence of external inductance of contours to exploit, as the first approximation the assumption on the ideal conductivity [6]. The finite penetration depth to take into account in the next approximation at the determination of the losses in conductors and the fields in them [7].

The assumption on the ideal conductivity means equality to the zero of the tangent component of electric field and normal component of the induction:

$$E_\tau = 0, \tag{1.25}$$

$$B_n = 0. \tag{1.26}$$

In the problems with sharply-pronounced skin effect the current distribution on the surface of conductors is not given a priori but is determined in the process of calculations using formula (1.21). In these problems the field calculation is reduced to the solution of some of equations from equations (1.4–1.6). At the approximation of ideal conductivity the boundary condition (1.26) reduces to the conditions for functions U_M, A_φ, and ψ:

$$\frac{\partial U_M}{\partial n}(s) = 0, \tag{1.27}$$

$$r_s A_\varphi(s) = \psi(s) = \text{const}, \tag{1.28}$$

where s is the point, belonging to the contour of the longitudinal (namely passing through the axis z) cross section of the conductor. The condition (1.28) means the constancy of the flux through the any circle of the radius r, lean up on the contour of the coil (Figure 1.3). Besides conditions (1.27), (1.28) there should be given the values of the sought functions on the infinity and total currents in conductors or

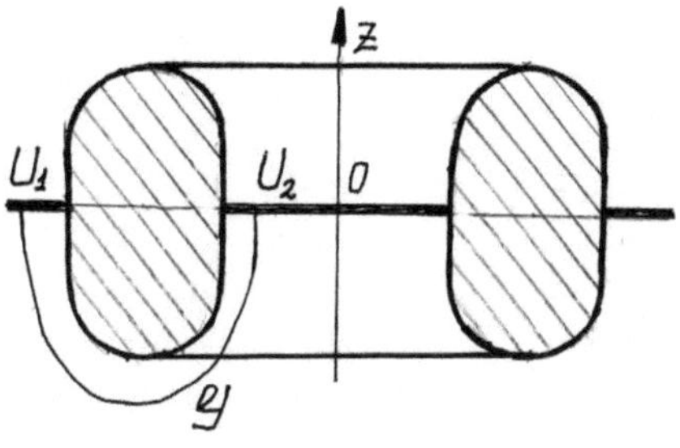

Fig. 1.3: Longitudinal cross section of the solenoid having the body of revolution shape.

should be given additional conditions allowing to determine these currents (for instance the electromotive forces in the magnets power supply network).

The current condition is expressed as

$$\oint_G \vec{H}\vec{dl} = i, \tag{1.29}$$

where integration is taken on the contour G, enveloping the conductor (Figure 1.3).

At the calculation of the field of single-turn magnets in a majority of cases it is necessary to apply the approximate analytical and numerical methods. For direct solution of equation (1.6) one can to use the known finite-difference methods [5] and contemporary programs, using the method of finite elements [6,7]. Application of equation (1.6) for scalar magnetic potential is the most convenient at the numerical calculations when surfaces are known on which the tangent component of induction equals to zero and, therefore, one can assume the boundary condition U_M = const.

As an example we can present the symmetrical configuration (Figure 1.4), where it is sufficient to find the field in half of the region (for instance at $z < 0$).

The boundary conditions in this case will be $U_M = U_1$ at $z = 0$, $r \leq R - a$; and $U_M = U_2$ at $z = 0$, $r \geq R + a$; $\partial U_M / \partial n = 0$ on the coil surface. Current condition $\oint \vec{H}\vec{dl} = i$ in this case acquires the form

$$U_1 - U_2 = -\int_{G_1} \nabla U_M dl = \frac{\mu_0}{2} i, \tag{1.30}$$

where G_1 is the path of integration enveloping the half of the coil.

In the calculation of the magnetic field in axis symmetrical system it is easy to reduce to the solution of the integral equation of the I or II kind; likewise this is done in the electrostatics.

The integral equation of the I kind could be obtained directly from the expression for vector potential (1.9), if we assume that points r, z, and r_t, z_t are placed

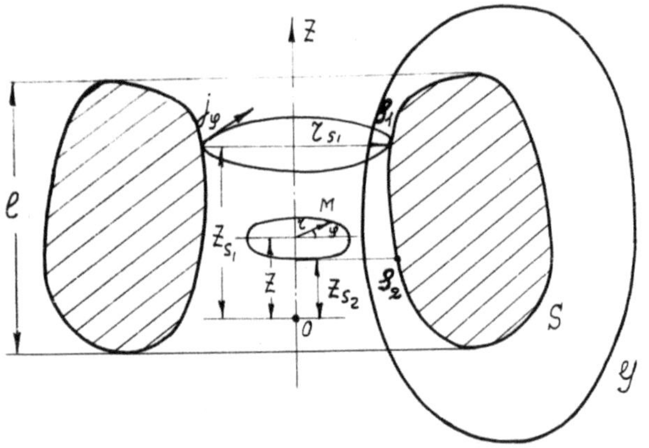

Fig. 1.4: Configuration of the symmetrical case used for the solution of the integral equation for the surface current density J_φ at the condition of the ideal conductivity of the single turn magnet.

on the surface of the coil S (Figure 1.4) [8]. In these points the boundary condition (1.21) is valid, and hence the equation [8]

$$\Phi(s) = \mu_0 \int_S \sqrt{(z_s - z_s')^2 + (r_s + r_s')^2} \times \left[\left(1 - k^2/2\right) K(k) - E(k)\right] J_\varphi\left(s'\right) dS\left(s'\right) + \Phi_e(s) = \text{const} \quad (1.31)$$

is applicable. Here $k = 2(r_s r_s')^{1/2}[(r_s + r_s')^2 + (z_s - z_s')^2]^{-1/2}$, $\Phi_e(s)$ is the component of the flux through the circle of radius r_s produced by the external sources. In [8] the approximate analytical solution of this equation for the case of the short, thin wall cylinder was obtained.

The kernel of the integral equation of the first kind (1.31) is symmetrical, square-summarized, and has logarithmic peculiarity. If the solution of this equation is square-summated, which occurs in the cases when cross section contour is a closed curve, then it is possible for the determination of $J(s)$ to use the iteration process, as proposed in [9]. A solution of the equation (1.31) at the additional condition of (1. 29) could be found by the reduction of it to the system of algebraic equations by splitting Contour S on the fragments and replacing the integration by the summation.

In the ideally conductive magnet with infinitesimally small thickness of the wall, the points s_1' and s_2' (Figure 1.5) merge, and in equation (1.31) we may consider only summary current density $J_\varphi(s') = J_\varphi(s_1') + J_\varphi(s_2') = J_i(s') + J_e(s')$, where J_i and J_e are the current densities from the inner side and the outer side of the magnet wall. Differentiating the expression for vector potential (1.9), one can find the normal component of the conductor's induction. Setting it to the zero we obtain

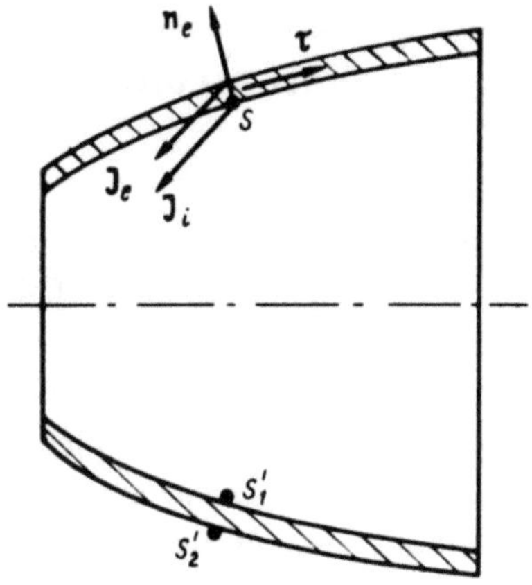

Fig. 1.5: The calculation of current density at the outer end inner surface of the thin-wall magnet.

following equation:

$$\int_S \frac{(z_s' - z_s) f_1(k) \cos(\mathbf{n}, \mathbf{r}) + \left[r_s f_1(k) + r_s' f_2(k) \cos(\mathbf{n}, \mathbf{z})\right]}{2r_s\sqrt{r_s r_s'}} J_\varphi(s')\, dS(s') = 0, \tag{1.32}$$

where $f_1(k) = k[K - \frac{2-k^2}{2(1-k^2)}E]; f_2(k) = \frac{k^3 E}{2(1-k^2)}$, k are modules of the elliptic integrals, the same as in equation (1.31).

In the paper [10], equation (1.32) was used to calculate the current distribution in the thin wall cylinder at the additional condition $\int J_\varphi dS = i$.

At the calculation of the field produced by the conductors with different from zero wall thickness the problem could be reduced to the solution of the integral equation of the second kind for the azimuthal component of the linear current density J_φ [11]:

$$-J_\varphi(s) + \frac{1}{\pi}\int_S J_\varphi(s')\left[Q(s, s') - \pi/\Pi\right] dS(s') = -i/\Pi. \tag{1.33}$$

In the given formulae Π is the perimeter of the boundary of the axial cross section of the coil A, i is the current, τ – the unit vector of tangent to the contour of the cross section in the point s with the coordinates z_s, r_s, and n is the unit vector of the normal,

$$Q(s, s') = \frac{-(z_s - z_s') f_1(k) \cos(\boldsymbol{\tau}, \mathbf{r}) + \left[r_s f_1(k) + r_s' f_2(k)\right] \cos(\boldsymbol{\tau}, \mathbf{z})}{2r_s\sqrt{r_s r_s'}}.$$

1.3 The determination of the parameters of the inductor systems at a strongly-pronounced skin effect according to the simplified field pattern

The rough evaluation of the induction far from the magnet edges and estimation of inductance could be obtained in a number of cases using the simplified field pattern without taking into account the edge effects. The construction of the simplified field pattern is based upon the intuitive considerations, using the condition of nonpermeability of ideally conducting walls for magnetic flux. In this calculation it is assumed that the whole energy of the magnetic field is concentrated in some well-defined region where the induction distribution is supposed as known and outside of which the induction acquires a zero value. The correction for taking into account the edge effects is made in [12]. In the simplest case of the field produced by the long single-coil solenoid the replacement of the true field pattern by the simplified one corresponds to the assumption of zero value of induction outside of cylindrical volume confined by the inner cylindrical surface and planes of magnet butts. It is also assumed that there is constancy of the induction inside this volume (Figure 1.2). Then $B = \mu_0 i/l$, and from here we find the known approximate expression for inductance of long one winding solenoid

$$L = \mu_0 \pi R^2/l. \tag{1.34}$$

Further on we shall present three examples illustrating the method of calculation using a simplified field pattern.

1.3.1 The field of single-turn solenoids (flat ring) above the ideally-conducting plane ($h \ll r_1$)

From the condition of "nonpermeability" of the walls (1.26) and assumption on uniform current distribution on the height of the gap (Figure 1.6a), it follows the expression for the induction in the point A is

$$B(A) = \frac{\Phi}{2\pi h r(A)} = \mu_0 J_\varphi. \tag{1.35}$$

Then one can calculate the current and inductance:

$$i = \int_{r_1}^{r_2} J_\varphi \, dr = \frac{\Phi}{2\pi\mu_0 h} \ln \frac{r_2}{r_1}; \tag{1.36}$$

$$L = \frac{2\pi\mu_0 h}{\ln(r_2/r_1)}. \tag{1.37}$$

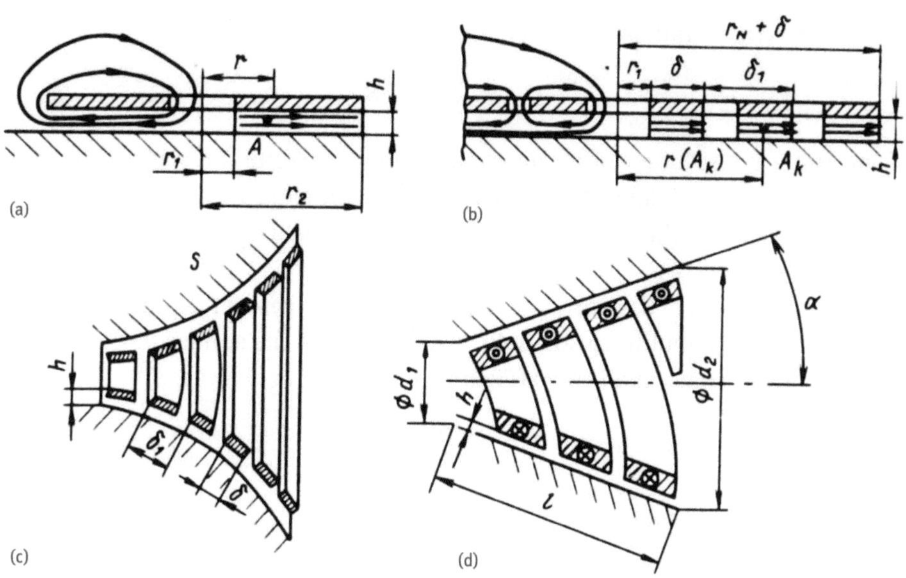

Fig. 1.6: Schemes of solenoids used for calculation of inductance on the simplified field pattern: (a) disk above the surface; (b) flat spiral above the surface; (c) multiturn solenoid, placed near the surface of rotation; (d) solenoid separated by narrow gap from the cone. Examples (a) and (b) show the qualitative field 1 on the left side and the simplified pattern on the right side of the figure.

1.3.2 The field of multiwinding solenoid in the form of a flat spiral above the plane (Figure 1.6b)

This problem could be easily solved at the condition $h \ll \delta_1 - \delta,\ \delta$, when one can assume that the field is concentrated under the windings; in the point A_k under the k-th turn, the induction is determined according to the formula (3.42a): $B(A_k) \approx \Phi_k/[2\pi h r(A_k)]$, from here the inductance of the k-th turn $L_k = 2\pi\mu_0 h/\ln(1 + \delta/r_k)$, and the total inductance

$$L = 2\pi\mu_0 h \sum_{k=1}^{N} \frac{1}{\ln \frac{r_k+\delta}{r_k}}, \tag{1.38}$$

where $r_k = r_1 + (k-1)\delta$.

If the pitch of the spiral $\delta_1 \ll r_1$, then $\ln(1 + \delta/r_k) \approx \delta/r_k$; therefore

$$L \approx \frac{2\pi\mu_0 h}{\delta} \sum_{k=1}^{N} r_k = \frac{2\pi\mu_0 hN}{\delta}\left[r_1 + \frac{(N-1)\delta_1}{2}\right], \tag{1.39}$$

where N is the number of turns.

If the external radius is large enough and $N \gg 1$, then

$$L \approx \pi\mu_0 h N^2 \delta_1/\delta . \tag{1.40}$$

Concerning the accuracy of formulas (1.39) and (1.40) one can suggest, by comparing the results of calculations, using them with more accurate calculations, taking into account the "percolation" of the flux between the coils. These calculations are given in [12].

Formula (1.40) is the particular case of the more general expression for the inductance of the multiturn solenoid, separated from the surface of the body of revolution S by some small gap h (Figure 1.6c):

$$L \approx \frac{\mu_0 V}{\delta\delta_1}, \tag{1.41}$$

where V is the volume confined by the inductor and surface S and δ_1 is the pitch of winding. This result could be obtained assuming that the field is concentrated under the coil ($H = i/\delta$) and suggests that energy of the magnetic field $\mu_0 i^2 V/(2\delta\delta_1)$ equals $Li^2/2$. In the case of the cone-shaped winding, placed near the ideally conduction cone (Figure 1.6d), $V = \pi h (r_1 + r_2) l$; $l = w\delta_1$, therefore

$$L = \mu_0 \pi w^2 h \frac{(d_1 + d_2)\,\delta_1}{2l\delta}. \tag{1.42}$$

This formula turns to the expression for inductance of the flat spiral (1.40) if $\alpha = \pi/2$, $d_1 \ll d_2$, since at this condition $d_2 \approx 2l$.

1.3.3 Field of solenoid in off-loading cylindrical screens

From the simplified field pattern (Figure 1.7) it follows: $B(A_1) = \Phi/F_1$; $B(A_2) = -\Phi/F_2$, where $F_1 = \pi r_i^2$, $F_2 = \pi(R^2 - r_e^2)$.

Summarizing the currents on the outer and inner surfaces of the cylinder we obtain

$$i \approx \frac{\Phi l}{\mu_0 F_1} + \frac{\Phi l}{\mu_0 F_2} = \frac{\Phi l}{\mu_0} \cdot \frac{R^2 - r_e^2 + r_i^2}{r_i^2\,(R^2 - r_e^2)},$$

$$L \approx \frac{\mu_0 r_i^2\,(R^2 - r_e^2)}{l\,(R^2 - r_e^2 + r_i^2)}. \tag{1.43}$$

The condition of solenoid discharging from electromagnetic forces is fulfilled if $F_1 = F_2$ or $r_i^2 = R^2 - r_e^2$.

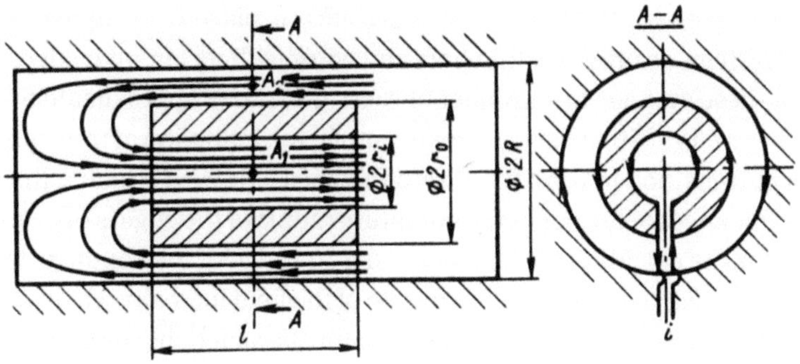

Fig. 1.7: Transition from true to the simplified field pattern at the calculation of the solenoid field in off-loading (discharging) screen.

Knowing the asymptotic dependence of the field strength on the coordinates one can use it for the current calculation. For instance, how it was admitted in Section 2.2, the field of the disk with the hole of the radius R1 on the large distances from the center is characterized by the relation $B\ (M) = \Phi/(4\pi\rho^{\ 2})$.

Assuming that the current density everywhere on the disk (but not only far away from its edge) is determined by formula $J_\varphi = \Phi/(2\pi\mu_0\rho^2)$, then the total current is determined as $i = \Phi/(\mu_0\pi R_1)$. From here we obtain the approximate value of inductance $L = \pi\mu_0 R_1$ instead of the exact value $L = 2\mu_0 R_1$. The error in this case is quite significant, since the major part of the current concentrated in the edge region is calculated inaccurately because the dependence J_φ differs from asymptotic one.

The error could become lower at the calculation of the concentrators of the flux and other devices, where asymptotic dependence is used only for the calculation of the relatively small part of the total current, e. g., the current on the conical parts of concentrator.

1.4 Edge effects in single-coil magnets. Modelling of problems

The magnetic field of the systems, operating at the conditions of strongly pronounced skin-effect is characterized by the probable appearance of the sharp growth of induction in the places where the boundary of conductors is bent. This effect exists in poloidal axially symmetrical fields where, as seen in Figure 1.1, the induction vector lying in the plane *rz* is orthogonal to the circumference, which corresponds to the bending line of the boundary surface. In a toroidal field the induction vector is directed parallel to the edge. In this case the edge does not affect the field outside of the conductor, determined by formula (1.3).

Among the examples considered below, a significant place is occupied by such an idealized systems as infinitely thin bars and wires with rectangular edges. The calculation of these systems in a number of cases gives only approximate expression for the parameters of real devices. For example, the induction of thin bars scarcely changes in transition to the bars of zero thickness. However, taking into account the finite thickness of the bar or the curvature radius of the edge is of principal significance for the calculation of the field near the edge of the conductors. In this case it is insufficient to know the growth rate of the field intensity when approaching the apex of a dihedral angle and the constant characterizing its growth, but it is important to include the real geometry of the region near to the edge of conductor. This should be also taken into account when calculating of the energy released during the heating of the medium in the skin-layer. It is especially important in the case of thin bars, when at a zero thickness approximation not only the local value of power released per unit length increases, but its integral value also increases indefinitely. The problem consists in finding the simple relations allowing to connect the field intensity near to the apex to the field intensity on the real edge. The ratio of above intensities is constant value, if the characteristic dimension (thickness of the bar, radius of curvature) is small enough compared with other dimensions of the magnetic system. For instance when the curvature radius of edge ρ is much less than distance between the conductors d. With this condition one can consider the vicinity of the edge as the solitary region, assuming that the field in this region is flat [13–15]. In the ideal case of sharp angle edge the field in the vicinity of the point O (the apex of dihedral angle) could be described in the local system of coordinates $z_1 = x_1 + jy_1$ (Figure 1.8a). At the condition $\theta < \pi$ the induction of poloidal field in the point M rises unlimited, if the distance from this point to the angle apex tends to zero, this is also the case in the electrostatic field. In real conditions the edge of the conductor is always round shaped, therefore the field induction remains finite.

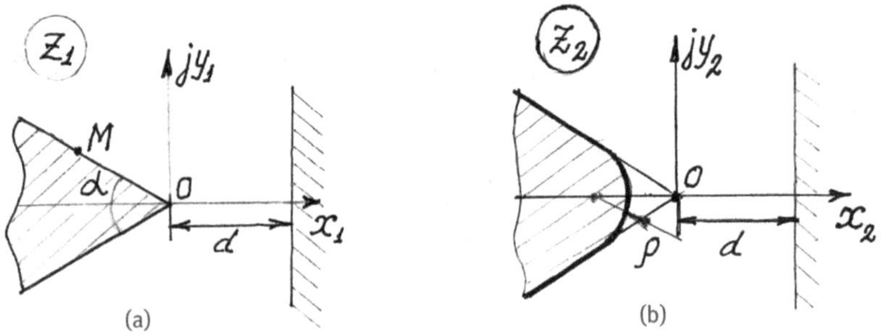

Fig. 1.8: Sharp and round shaped edges of the conductor.

Along with this one can suggest that in the approximation of the ideal conductivity the laws controlling the change of induction and surface current density at the removal from the point O remain the same, as in the case of the sharp bend of the boundary, i.e., the edges were not rounded.

The method of conformal representation allows us to find the low of induction variation in this system. It is represented on the half plane $w > 0$ using the function $w = jCz_1^{\pi/(2\pi-\theta)}$, where C is the real constant. The module of field induction could be calculated using the formula

$$|B| = \left|\frac{dw}{dz_1}\right| = C_1 z_1^{(\theta-\pi)/(2\pi-\theta)}. \tag{1.44}$$

From this formula it is seen, that in the case of $\theta = 0$ the field intensity is increasing according to $|z_1|^{-1/2}$; in the case of $\theta = \pi/2$ it increases proportionally to $|z_1|^{-1/3}$.

Analytical calculation for the ideal system, for instance, in the case of the magnet with zero wall thickness, is much more simple than for magnets with real shape of the boundary. This calculation allows us to find the parameter C_1, which is determined by the general configuration of the magnetic system and given currents. Knowing this parameter one can calculate, as a next step, the field in the vicinity of the real edge, which is presented in the local coordinate system $z_2 = x_2 + jy_2$ (Figure 1.8b). For the calculation of the field in the vicinity of z_2 it is necessary to represent it on the half-plane using the function $w(z_2)$. At the removal from the edge ($|z_2| \gg \rho$) the condition $z_2 \to z_1$ becomes applicable. The function (z_2) should satisfy the condition $w(z_2) = w(z_1)$ if $|z_1| \gg \rho$. If the field in the region z_1 is calculated, then it is possible further on to recalculate the induction on the following formula:

$$|B(z_2)| = |B(z_1)| \left|\frac{dz_1}{dz_2}\right|_{z_1 \to 0}. \tag{1.45}$$

One practically important case is the already mentioned configuration of magnets with thin edges (Figure 1.9). The examples of the axially-symmetrical systems for which, at the first approximation, could be assumed the zero thickness model are considered below.

In all of them is used the similar method of field evaluation near the edge, where the conductor has the shape presented in Figure 1.9a,b. The regions z_1 and z_2 are represented on the upper half-plane $Rew > 0$ using the functions

$$z_1 = \frac{\delta}{\pi} w^2; \quad z_2 = \frac{\delta}{\pi}\left[w\sqrt{w^2-1} - \ln\left(w + \sqrt{w^2-1}\right)\right] + j\frac{\delta}{2}, \tag{1.46a}$$

where δ – the thickness of the flat conductor. Excluding w we find that

$$z_2 = \frac{\delta}{\pi}\left[\sqrt{\frac{\pi z_1}{\delta}\left(\frac{\pi z_1}{\delta} - 1\right)} - \ln\left(\sqrt{\frac{\pi z_1}{\delta}} + \sqrt{\frac{\pi z_1}{\delta} - 1}\right)\right] + j\frac{\delta}{2}. \tag{1.46b}$$

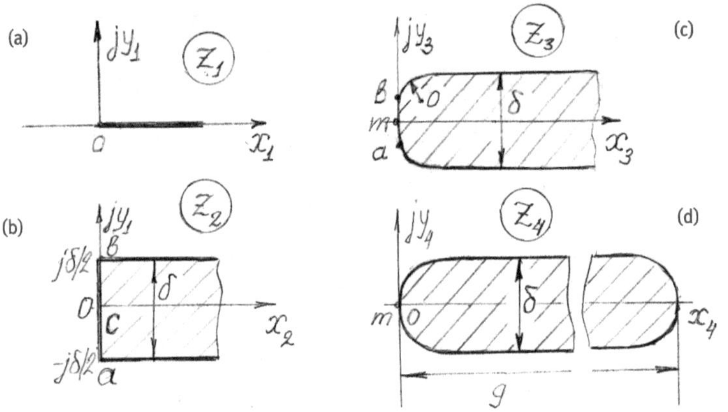

Fig. 1.9: For the calculation of the field and losses near the edge of the flat sheet at the strongly pronounced skin effect.

The induction in the vicinity of the edge of the zero thickness varies as

$$|B(z_1)| = C_1 z_1^{-1/2} \tag{1.46}$$

The values of induction in the point z_2 and point z_1 are coupled by the ratio

$$B(z_2) = B(z_1)\left|\frac{dz_1}{dz_2}\right| = B(z_1)\left|\frac{\pi z_1}{\delta}\right|^{1/2}\left|\frac{\pi z_1}{\delta} - 1\right|^{-1/2}. \tag{1.47}$$

The following expression couples the induction in the point c with ($z_2 = 0$) on the middle of the face of the sheet with finite thickness and the induction in the point z_1 near to the edge of the infinitely thin sheet:

$$B(c) = B(z_1)\sqrt{\frac{|\pi z_1|}{\delta}}. \tag{1.48}$$

From formula (1.49) it follows that for the calculation of the induction on the edge of thin and ideally conducting bars of finite thickness it is sufficient to know the law of induction variation near to the edge of the bars with zero thickness $B(z_1)$.

Using the formula (1.47) we can obtain

$$|B(c)| = C_1\sqrt{\pi/\delta}. \tag{1.49}$$

As an example of the use of his formula we shall consider the flat field of the strip transmission line, namely the conductor with the width g and the thickness $\delta \ll g$ (Figure 1.10).

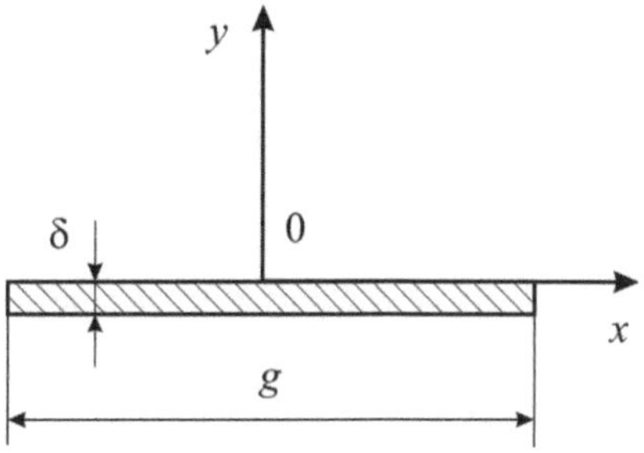

Fig. 1.10: Solitary flat bar.

The surface current density on the each side of ideally conducting sheet in the limit $\delta = 0$ is determined by the dependence of the form

$$J = \frac{B}{\mu_0} = \frac{i}{\pi g\sqrt{1-(2x/g)^2}}. \tag{1.50}$$

Near to the edge, where $z_1 = g/2 - x \ll g/2$, $B(z_1) = \mu_0 i/[2\pi(gz_1)^{1/2}]$; $C_1 = \mu_0 i/(2\pi g^{1/2})$; therefore, using the given formulae one can to calculate the induction in the point c, placed in the middle of the flat edge:

$$B(c) = \frac{\mu_0 i}{2\sqrt{\pi\delta g}}. \tag{1.51}$$

It is quite evident that the calculation method described above also remains significant in the case of thin conductors having the shape different from the one previously considered (for instance, in the case of nonflat sheets). It is only necessary that their thickness δ be much less than other characteristic dimensions, and in particularly the curvature radius of the thin sheet and distance from its edge to the other conductors.

Knowing the induction in the point c, one can calculate the induction distribution in the points of the region z_2 close to the face ab (Figure 1.9b), since the ratio $B(z_2)/B(c)$ is determined by only shape of the edge, if δ is small. Thus, it is easy to find the distribution of the tangent component of the magnetic field induction or the normal component of the electric field intensity along the plane of the face.

Of practical interest is the calculation of the field on the rounded edge (Figure 1.9c, region z_3). The edge of the flat sheet could be rounded only partially or completely. In the first case only direct angles are rounded, and there is a flat fragment ***ab*** (Figure 1.9c). In the second case, the profile of the edge fragment represents itself the smooth curve and the flat part is absent. An example related to the second case is the problem on the field calculation of the thin electrode with length g and thickness δ (at $g \gg \delta$), rounded on the arc with curvature radius $\delta/2$ (Figure 1.9d). Novgorotsev obtained an exact solution of this problem [16]. The maximum of induction takes place in the point m: $B(m) = B_e{\cdot}1.0725\,[\pi g/(2\delta)]^{1/2}$. Here B_e is the induction of the external uniform field, bending the edge of the sheet. Let's construct the ratio $B(m)/B(c)$, where $B(c) = B_e\,[\pi g/(4\delta)]^{1/2}$ – the induction in the point c of the bar with the length g flat edge and introduced in the field B_e. This ratio calculated for the limiting case $g \gg \delta$ does not depend on the specific configuration of the magnetic system and is determined only by the shape of the edge. Thus, for the buss rounded off the arc with a radius $\delta/2$, we have the expression related the induction in the point m to the induction in the middle of a flat edge:

$$B(m)/B(c) = 1.0725\sqrt{2} = 1.517. \tag{1.52}$$

For the complete rounding on the cycloid, calculated in the paper by Cockroft [17], the value of $B(m)/B(c) = \sqrt{1 + \pi/2} = 1.603$ was obtained.

The bars have several advantages, the edges of which are made in such a way that on the rounded rim the induction is constant and equals B_1. At the permanent thickness of the conductor in such a system the edge effect is not strongly pronounced, which is important both with regard to the thermal and mechanical effects and with regard to electrical strength. At this complete round shaping with a permanent induction module [18] the ratio $B_1/B(c)$ acquires the value

$$B_1/B(c) = \sqrt{2}. \tag{1.53}$$

The numerical values of induction, obtained at the complete round shaping, but using various computation technique, do not differ noticeably from each other. It implies that the main influence on the induction is connected with the size of the rounded part, but not with particular geometry of the profile. Along with this, it should be noted the crucially important significance of the rounding on which induction is constant: in this case the induction on the rounded section is minimal.

The broad class of the problems on the design of the electrodes with rounded and sharp angles and with the constant intensity on the rounded section, similar to the problems of the theory of jets in the ideal liquid [19], were solved in the papers by Cockroft [17], Felici [20], Novgorotsev and Shneerson [16], and Novgorotsev and Fatkhiev[18].

The curvilinear sector of the boundary (Figure 1.11) is described by the equation

$$x_1^{3/2} + y_1^{3/2} = r_0^{3/2}. \tag{1.54}$$

The profile of the rounded sector at this consideration is close to the quarter of the circle radius r_0: the difference $\Delta r = r_0 - r$ does not exceed the value $\Delta r = (1.5 - \sqrt{2})r_0 = 0.086r_0$ in the point m. On the sector ab (Figure 1.11) the following

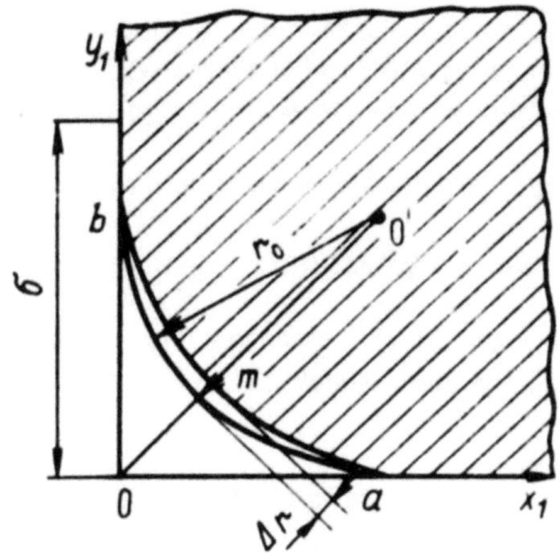

Fig. 1.11: Rounded direct angle.

expression for induction is derived:

$$\frac{B(ab)}{B(\sigma)} = 2^{2/3}\left(\frac{\sigma}{r_0}\right)^{1/3}, \quad (1.55)$$

where $B(\sigma)$ is the induction in the point, situated on the distance σ from the apex of the direct (not rounded) angle. Thus, as in the case of the thin edge, the field near the rounded direct angle (at the small radius of the rounding) is expressed via the solution of the idealized problem which is valid near to the edge.

The similar problem in the case of axial symmetry has been solved by the method described in the papers [21] and in the book [12]. There the profiles of the rounded sector of thick wall single turn magnet at the various ratios of the wall thickness to the radius were calculated.

1.5 References

[1] I. E. Tamm, *Fundamentals of electricity theory*, Moscow, Fismatlit, 2003 (in Russian).

[2] W. R. Smythe, *Static and Dynamic electricity*, 2nd ed., New York Toronto London, 1950.

[3] D. B. Montgomery, *The Magnetic and Mechanical Aspects of Resistive and Superconducting Magnets*, J. Wiley, New York London Sydney Toronto, 1969.

[4] H. E. Knoepfel, Magnetic Fields. *A Comprehensive Theoretical Treatise for Practical Use*, John Wiley, New York Chichester Weinheim Brisbane Singapore Toronto, 2000.

[5] H. Knoepfel, *Magnetic Fields, A Comprehensive Theoretical Treatise for Practical Use*, New York, John Wiley, 2000.

[6] L. G. Loytsansky, *Fluid and gas mechanics*, Moscow, Nauka, 1970 (in Russian).

[7] W. R. Pryor, *Multiphysics Modeling Using Control. A First Principle Approach*, Jones & Bartlett Learning, 2010.

[8] G. A. Shneerson, *J. Tech. Phys.* **31**(1) (1961), 51–54 (in Russian).

[9] Fridman V. M., *Uspekhi matematicheskih Nauk.* **11**(1) (1956), 233–234 (in Russian).
[10] G. Bardotti, B. Berdotti, and L. J. Gianolio, *Math. Phys.* **25**(10) (1964), 1387–1390.
[11] G. N. Kaporskaya, and A. B. Novgorodcev, *Izvestiya Academiyi Nauk SSSR, Energetika I Transport.* **2** (1976), 169–172 (in Russian).
[12] G. A. Shneerson, *Fields and Transiens in Super high Pulse Current Device.* New York, Nova Science Publishers, Inc., 1997.
[13] L. D. Osnovich and V. D. Makelsky, *Izvestiya Vuzov, Energetika* **5** (1970), 16–21; **10** (1971), 24–70 (in Russian).
[14] V. A. Popova and G. A. Shneerson, *J. Tech. Phys.* **47**(10) (1977), 2009–2016 (in Russian).
[15] A. B. Novgorodcev and G. A. Shneerson, *Izvestiya Academiyi Nauk SSSR, Energetika I Transport.* (1983), **6**, 65–74 (in Russian).
[16] J. D. Cockcroft, *J. IEE.* **66**(376) (1928), 385–409.
[17] A. B. Novgorodcev and A. R. Fatkhiev, *Izvestiya Vuzov, Energetika* **2** (1982), 17–21 (in Russian).
[18] M. I. Gurevich, *Theory of ideal liquid Jets*, Moscow, Nauka, 1979. (in Russian).
[19] N. J. Felici, *Revue général de l'électricité* **59**(11) (1950), 479–501.
[20] E. L. Amromin, G. N. Kaporskaya, A. B. Novgorodcev, and G. A. Shneerson, *Elektrichestvo* **3** (1989), 40–46 (in Russian).

2 Calculating formulas and the results of numerical estimations of field parameters for typical single-turn magnets

This chapter presents formulae, graphs, and tables which can be useful to calculate the induction in the center of a single turn magnet and its inductivity.

For some configurations considered here there are solutions in analytical form; however, the main part of presented data has been obtained numerically by the solution of equation (1.6) for a scalar magnetic potential.

2.1 The field of the flat ring as an example of the single-turn magnet with sharply pronounced edge effect

In the previous chapter some methods used in calculations of magnets were described. Thus, it has been shown that at a small penetration depth one can assume, as a first approximation, the ideal conductivity of the conductor. This assumption permits, in a number of cases, the use of a simplified pattern of the field. Along with that, this approximation exhibits the edge effects, leading to the local field enhancement. The example of the system in which these effects become strongly pronounced is the flat ring with a cut (Figure 2.1) [1].

In this case the vector potential of the magnetic field at the condition $R_2 \gg R_1$ could be described using a Fourier–Bessel integral:

$$A_\varphi = \frac{\Phi}{2\pi R_1} \int_0^\infty e^{-\lambda z/R_1} J_1\left(\lambda \frac{r}{R_1}\right) \frac{\sin\lambda}{\lambda} d\lambda . \tag{2.1}$$

One can easily prove that this function satisfies equation (1.4), the condition $2\pi\, rA_\varphi = \Phi$ at $z = 0$ and $r > R_1$, and the condition $\delta A_\phi / \delta r = 0$ at $z = 0$ and $r < R_1$. Further on one can obtain the expressions for the radial component of induction on the plane of disk as well as the axial component in the hole and on the axis of the magnetic system:

$$B_r\big|_{z=0,\, r\geq R_1} = -\left.\frac{\partial A_\varphi}{\partial z}\right|_{z=0} = \frac{\Phi}{2\pi\, r\sqrt{r^2 - R_1^2}} = \mu_0 J_\varphi \tag{2.2a}$$

$$B_z\big|_{z=0,\, r<R_1} = \frac{\Phi}{2\pi R_1 \sqrt{R_1^2 - r^2}} \tag{2.2b}$$

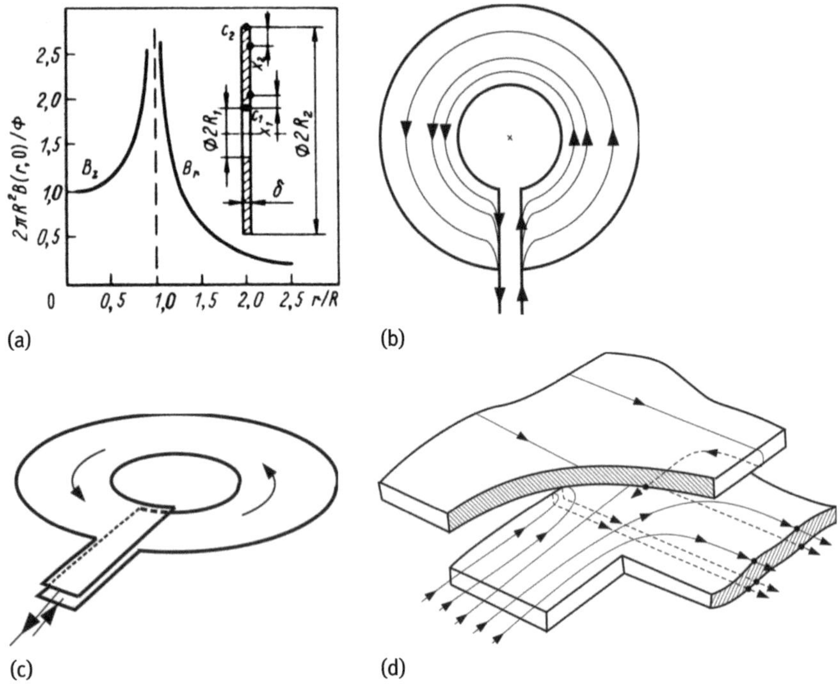

Fig. 2.1: The thin disk with a large ratio of the external radius to internal one.; (a) distribution of the induction in a plane of disk; (b) the magnet with the slot; (c) the magnet with the flat sheets and overlapping of its edges; (d) the current distribution between flat sheets.

$$B_z\,(0,\,z) = \frac{\Phi}{2\pi\,(R_1^2 + z^2)} = \frac{\mu_0 i R_1}{\pi\,(R_1^2 + z^2)}. \tag{2.2c}$$

One can also find the current flowing on both sides of the flat ring along the section from the point with radial coordinate r to $r = \infty$:

$$i\,(r) = \int_r^{\infty} 2J_\varphi\, dr = \frac{\Phi}{\pi\mu_0 R_1} \arcsin \frac{R_1}{r}. \tag{2.3a}$$

The full current is

$$i = \frac{\Phi}{\pi\,\mu_0 R_1}, \tag{2.3b}$$

and the inductivity is expressed by simple formula

$$L = 2\mu_0 R_1. \tag{2.4}$$

Increasing r and z, formulas (2.3) and (2.4) turn into the expression for the induction of the magnetic system in which the flow passes through the orifice of small radius in the flat sheet and is then uniformly distributed over the surface of hemisphere with radius $\rho = (r^2 + z^2)^{1/2}$:

$$B(r, z) = \frac{\Phi}{2\pi (r^2 + z^2)}. \tag{2.5}$$

This system corresponds to the simplified field pattern that is analogous to the configuration under consideration. Using the simplified pattern one can calculate the full current and find the corresponding value of the inductance:

$$\tilde{L} = \pi\mu_0 R_1. \tag{2.6}$$

Unlike the examples considered in Section 1.3, in a given case when the edge effect is quite sharply pronounced, the calculation based on the simplified field pattern leads us to a significant error in the evaluation of the inductance. The formulas (2.2b) and (2.4) give the following expression for the induction in the center of the turn:

$$B_c = B(0, 0) = \mu_0 i/(\pi R_1). \tag{2.7}$$

This value differs from the field value in the center of thin ring coil of the radius R_1 only by the factor $2/\pi$, because the current on the surface of disk is concentrated near the edge of the orifice.

The field near the edge has the “route” behavior $B(R_1 + x_1, 0) = [\Phi/(2\pi R_1)] (2R_1 x_1)^{-1/2}$.

One can find the induction in the point c_1 of the coil with finite thickness of the wall δ using the formulas presented in Section 1.4, since the field in the vicinity of the point c_1 is close to flat. In a particular case of a rectangular edge, the induction in its center equals (Figure 2.1a)

$$B(c_1) = \frac{\Phi}{2R_1\sqrt{2\pi R_1 \delta}}. \tag{2.8}$$

The radial slot of the finite width affects the current distribution over the disk surface. Actually, the current lines tangent to the surface cannot cross the slot. The radial component of the surface current density is not zero at the region with a width comparable with the slot width h (Figure 2.1b) Some part of a current is directed over the disk surface along the slot edge. The passing of a current along the slot leads to some induction decrease in the orifice center.

Note that the influence of the slot can be excluded if the overlapping of its edges takes place, as is shown in Figure 2.1c,d. The current lines on the disk surface remain circular, while the radial current is concentrated between flat sheets.

In a limiting case $R_2 \geq R_1$ one can calculate the field near the outer edge of the ring ($r \approx R_2$), using the mathematical similarity with the problem of the electrostatic field of the point charge, displaced in the center of the round orifice of radius R_2 on the infinite flat the conducting sheet [1].

The analogue of the half charge in our problem will be the flow Φ passing through the orifice of infinitesimal radius R_1. The calculation gives the following expression for the induction which is valid at $r \gg R_1$:

$$B_z = \frac{\Phi R_2}{\pi^2 r^2 \sqrt{r^2 - R^2}} \quad \text{at } r > R_2,\ z = 0; \tag{2.9}$$

$$B_r = \frac{-\Phi}{2\pi r^2} + \frac{\Phi}{\pi^2 r R_2}\left(\frac{R_2^2}{r\sqrt{R_2^2 - r^2}}\operatorname{arctg}\frac{r}{\sqrt{R_2^2 - r^2}} - 1\right) \quad \text{at } r > R_2,\ z = 0. \tag{2.10}$$

From here we can find the field value at the distance x_2 from the edge:

$$B_r(R_2 - x_2,\ 0) \approx \frac{\Phi}{2\pi R_2 \sqrt{2x_2 R_2}}. \tag{2.11}$$

Now, sd in other similar cases, we can find the induction on the edge of the flat ring of final thickness δ, using the fact that the field in the vicinity of the edge is close to flat shape. The induction in the point c_2 of the rectangular shape coil with $R_2 \gg R_1$ and $\delta \ll R_2$ is determined by the formula

$$B(c_2) = \frac{\Phi}{2R_2\sqrt{2\pi\delta R_2}} \tag{2.12}$$

The field of the two counter-switched flat rings separated by a quite thin gap (Figure 2.2) can also be found using Fourier–Bessel transformations:

$$A_\varphi = \frac{\Phi}{2\pi R_1}\int_0^\infty e^{-\lambda z/R_1} J_1\left(\lambda\frac{r}{R_1}\right) J_0(\lambda)\, d\lambda. \tag{2.13}$$

In the plane $z = 0$ the formula (2.13) gives $A_\varphi = \Phi/(2\pi r)$ at $r > R_1$ and $A_\varphi = 0$ at $r < R_1$. The induction on the axis of this system is

$$B_z(0, z) = \frac{\Phi}{2\pi R_1^2}\int_0^\infty \lambda e^{-\lambda z/R_1} J_0(\lambda)\, d\lambda = \frac{\Phi z}{2\pi (R_1^2 + z^2)^{3/2}}. \tag{2.14}$$

In formuls (2.13) and (2.14) Φ is the half of the flow emerging from the gap.

2.2 The coil of a rounded cross-section (an ideally conductive toroid)

The field of the coil having the form of a toroid (Figure 2.3) is calculated by Fock [3]. The current density distribution on the surface of tore of small radius a and

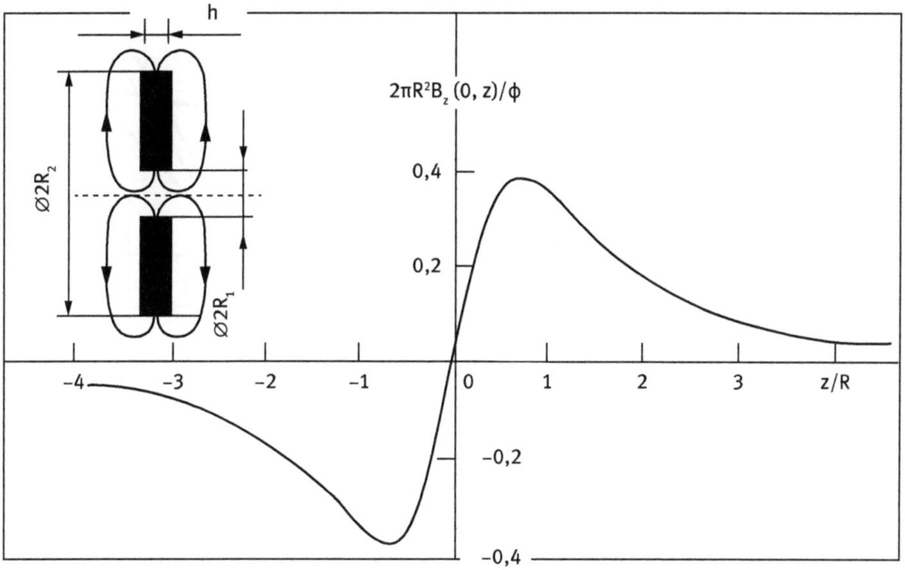

Fig. 2.2: Distribution of the induction along the axis of the magnetic system consisting of two thin disks with currents opposite in direction separated by a narrow insulation gap ($R_2 \gg R_1$).

large R can be calculated by

$$J_\varphi = \frac{iL}{8\sqrt{2\pi^3}\mu_0 a^2}\left(\frac{g-\cos\beta}{g^2-1}\right)^{3/2}\left[\frac{1}{P'_{-1/2}(g)} + 2\sum_{n=1}^{\infty}\frac{\cos(n\beta)}{P'_{n-1/2}(g)}\right], \tag{2.15}$$

where g is R/a, I is the current in a coil, L is the inductance determined by the formula

$$L = \frac{\pi^2\mu_0 a}{2}\sqrt{g^2-1}\left[\frac{Q'_{-1/2}(g)}{P'_{-1/2}(g)} - 2\sum_{n=1}^{\infty}\frac{1}{4n^2-1}\frac{Q'_{n-1/2}(g)}{P'_{n-1/2}(g)}\right]^{-1}. \tag{2.16}$$

As a coordinate of the point on the surface in formula (2.15) and in the graphs the toroidal coordinate β was used, coupled with ϕ (Figure 2.3) by the relation

$$\sin\beta = \left(g^2-1\right)^{1/2}\sin\varphi/(g+\cos\varphi).$$

Obviously, $\beta(0) = 0, \beta(\pi) = \pi\beta \approx \varphi$ if $g \gg 1$.

The values of parameters L and J_ϕ are presented in the graphs of Figure 2.3. There the dependences for the transient surface resistance $Z_i(p)$ are also presented [2].

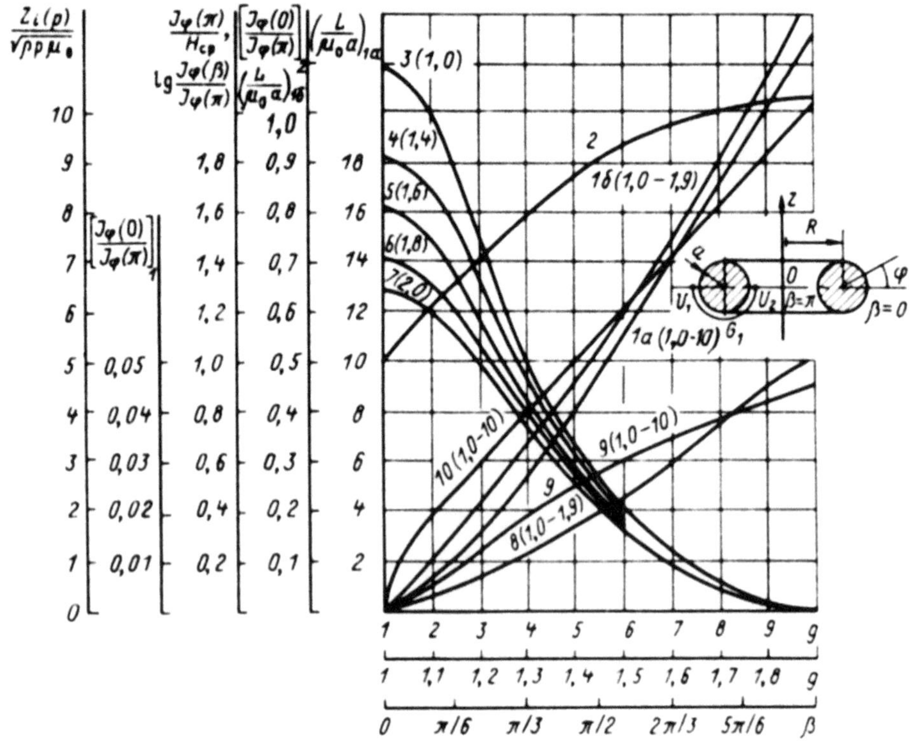

Fig. 2.3: Parameters of the tore coil: **(1a)** $L/(\mu_0 a) = f(g)$, $1 \le g \le 1.9$; **(1b)** $L/(\mu_0 a) = f(g)$, $1 \le g \le 10$; **(2)** $J_\varphi(\pi)/\langle H\rangle$, $1 \le g \le 10$; **(3–7)** $\lg[J_\varphi(\beta)/J_\varphi(\pi)] = f(\beta)$; **(8)** $[J_\varphi(0)/J_\varphi(\pi)]_1 = f(g)$, $1 \le g \le 1.9$; **(9)** $[J_\varphi(0)/J_\varphi(\pi)]_2 = f(g)$, $1 \le g \le 10$; **(10)** $Z_i(p)(\mu_0 p\rho)^{-1/2} = f(g)$, $1 \le g \le 10$.

In formulas (2.15) and (2.16), P_ν' and Q_ν' are Legendre adjunction functions. The curves 3 and 9 in Figure 2.3 show that leveling of the current distribution on the surface with rise of ratio R/a occurs slowly: even for $R/a = 10$ the ratio $J_\varphi(0)/J_\varphi(\pi)$ is equal to 0.45. On the contrary, the inductance (curve 1) agrees with data of [4] and at $R/a = 10$ only differs by 3% from the inductance, calculated under the assumption of the uniform current distribution on the coil surface. As it could be expected the linear current density on the inner coil surface $J_\varphi(\pi)$ at the condition $R/a \approx 1$ approaches the average field intensity in the gap of the coil $\langle H\rangle = \Phi/[\pi\mu_0(R-a)^2]$ (curve 2).

2.3 Thin-wall single-turn magnets

Analytical solutions of the problem of the calculation of the fields of single-turn magnets having the shape of a thin-wall cylinder were obtained for the short coil ($l \leq 2R$) and a "half-infinite" magnet.

Current distribution along the length of a cylinder is described by integral equation (1.31), the core of which, in this case, depends only on differentiating the argument:

$$\frac{L}{\mu_0 R} = \int_{-1}^{1} \sqrt{1 + \frac{g_0^2 (x_1 - x_2)^2}{4}} \left\{K(k)\left(1 - \frac{k^2}{2}\right) - E(k)\right\} y_\varphi (x_1)\, dx_1$$
$$= \int_{-1}^{1} T'(|x_1 - x_2|)\, y_\varphi (x_1)\, dx_1. \qquad (2.17)$$

Here $L = \Phi/i$ is the inductance of the system; $g_0 = l/(2R)$; $k = [1 + g_0^2 (x_1 - x_2)^2/4]^{-1/2}$ the module of the complete elliptic integrals K and E; $y_\varphi = (J_e + J_i)l/i$; J_e; and J_i is the current density on the outer and inner walls of the cylinder; $x_{1,2} = (2z_{1,2} - l)/l$ (Figure 2.4). The approximate solution of the equation (2.17) was obtained in [5].

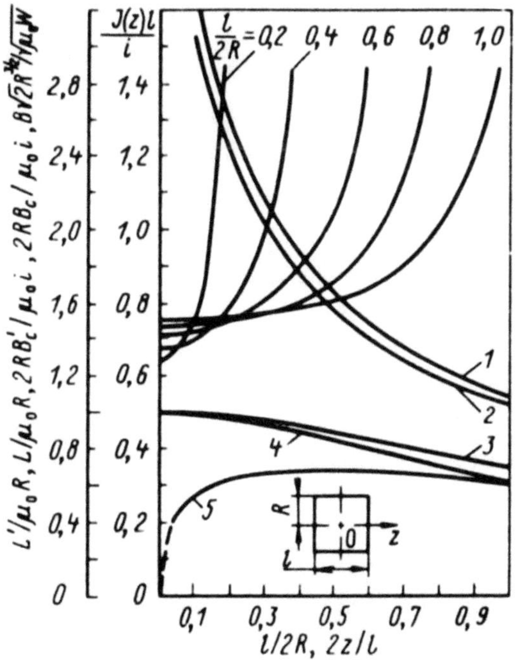

Fig. 2.4: Dependences characterizing the magnetic field in the short thin-wall coil: **(1)** $L'/(\mu_0 R) = f[l/(2R)]$; **(2)** $L/(\mu_0 R) = f[l/(2R)]$; **(3)** $2RB'_C/(\mu_0 i) = f[l/(2R)]$; **(4)** $2RB_C/(\mu_0 i) = f[l/(2R)]$; **(5)** $B_C \sqrt{2}R^{3/2}(\mu_0 w)^{-1/2} = f[l/(2R)]$. The other curves are $J(x)l/i = f(2z/l)$ at different values of $l/(2R)$.

The kernel of the integral equation (2.17) $T'(|x_1 - x_2|) = T(t)$, where $t = |x_1 - x_2|/2$, passes to the function $(1/2)\ln[4/(g_0 t)] - 1$ at $t \to 0$; therefore in the case $g_0 \leq 1$ it is convenient to use its approximate representation on the interval $0 \leq |t| \leq 1$:

$$T(t) \approx T''(t) = \frac{1}{2}\ln\frac{4}{g_0 t} - 1 + a_0 P_0(t) + a_2 P_2(t).$$

The coefficients a_0 and a_2 at the Legendre polynomials P_0 and P_2 are found from the condition of minimum of the mean square deviation of $T''(t)$ from T:

$$a_{0,2} = \int_0^1 \left[T(t) + 1 - \frac{1}{2}\ln\frac{4}{g_0 t}\right] P_{0,2}(t)\, dt,$$

and equation (2.17) acquires the form

$$-\frac{2L}{\mu_0 R} = \int_{-1}^{1} \left[\ln(x-s) + 2 - 3\ln 2 + \ln g_0 - 2a_0 + a_2 - \frac{3}{4}a_2 (x-s)^2\right] y_\varphi(s)\, ds. \tag{2.18}$$

This equation could be solved by the method described in [6]. As a result of these calculations we obtain the expressions for the surface current density and inductance[1]

$$J_\varphi(x) = \frac{2i}{l}\frac{1 + b/2 - bx^2}{\pi\sqrt{1-x^2}}, \tag{2.19}$$

$$L = \mu_0 R\left(\ln\frac{16}{g_0} - 2 + 2a_0 - \frac{a_2}{4} - \frac{9a_2^2}{4}\right) = \mu_0 R\left(\ln\frac{16}{g_0} - 2 + t'\right),$$

where $b = 3a_2/2$.

The parameters, characterizing the field in the short thin-wall single-turn magnet are presented in Table 2.1.

As is seen in Figure 2.4 the linear current density sharply increases near the edge of the coil. On the distance at about $0.6R$ from the edge the current distribution is practically leveling. In [7] the formulas for distribution of the induction on the solenoid axis are given and the results of the measurements which agree well with the calculations are presented. The enhancement of the current near the edges of the solenoid results in a more uniform induction distribution on the axis

1 The calculation error can be estimated by comparing the obtained results with more accurate approximations involving the Legendre polynomials of the 2nd order. Such comparison for the most adverse case $g_0 = 1$ gives the correction for the current density less than 1 %, and for inductance about 0.5 %.

Table 2.1: Parameters of a short single-turn thin-wall magnet.

$l/(2R)$	0	0,1	0,2	0,3	0,4	0,5	0,6	0,7	0,8	0,9	1,0
$L/(2\mu_0 R)$	∞	3,082	2,402	2,020	1,757	1,563	1,412	1,280	1,188	1,101	1,027
t'	0	0,007	0,02	0,043	0,068	0,077	0,129	0,151	0,192	0,223	0,254
b	0	0,014	0,040	0,086	0,125	0,168	0,206	0,247	0,292	0,330	0,372
$2B_cR/(\mu_0 i)$	1,000	0,991	0,963	0,936	0,900	0,848	0,806	0,761	0,713	0,674	0,635
ε'	0	0,410	0,444	0,476	0,495	0,505	0,513	0,519	0,523	0,527	0,530

and in some reduction of the induction in the center, described by

$$B_c = B\left(0, \frac{l}{2}\right) = \frac{\mu_0 i}{\pi R\sqrt{1+g_0^2}}\left(1+\frac{b}{2}\right)E\left(\frac{g_0}{\sqrt{1+g_0^2}}\right) - \frac{b}{g_0^2}\left[K\left(\frac{g_0}{\sqrt{1+g_0^2}}\right) - E\left(\frac{g_0}{\sqrt{1+g_0^2}}\right)\right] \tag{2.20}$$

by comparison with the value B_c' derived at the constant current density. For $g_0 = 1$ the discrepancy consists of about 10%. The discrepancy in the induction is significantly smaller (Figure 2.4). In the limit $l/(2R) \to 0$ we have $L' - L \to \mu_0 R\,(1.5 - 2\ln 2) = 0.114\mu_0 R$.

The solution for the semiinfinite thin-wall magnet obtained using the Wiener–Hopf method [8] is presented in [1]. The final results were obtained in the form of awkward and clumsy formulas. Below we shall present the most significant characteristics of the semiinfinite magnet. In such a magnet with an ideally conductive wall and far from the end the current is concentrated on the inner surface. Near the edge the current is distributed both at the inner and outer surfaces of the cylinder.

Likewise, in the short coil, near the edge of the thin-wall cylinder with zero thickness, the linear current density varies in a following way:

$$J_e = J_i = 0.28\frac{\Phi}{\pi\mu_0 R^2}\sqrt{\frac{R}{z}}, \tag{2.21}$$

where Φ is the flux in the plane of the coil.

Using formulas (1.49) and (1.50) one can calculate the induction at point c, placed in the middle of the flat edge of the magnet, of which the thickness of the wall differs from zero and equals $\delta \ll R$:

$$B\,(c) = \mu_0 J\,(c) = \frac{0.28\Phi}{R\sqrt{\pi\delta R}} \tag{2.22}$$

When removed from the edge the linear current density on the outer side vanishes on the length of order of R. The total current on the outer side is

$$i_e = \frac{0.43\Phi}{\pi\mu_0 R}. \tag{2.23a}$$

On the inner side the linear current density at the removal from the edge tends to the constant value equals $\Phi/(\pi\mu_0 R^2)$.

The total current of a large length magnet could be approximated by

$$i \approx \frac{\Phi l}{\pi\mu_0 R^2} + 2\left(i_e + \Delta i_i\right). \tag{2.23b}$$

This formula besides the current i_e also includes additional current Δi_i, which is concentrated onto the inner cylinder surface. This current appears because of the field rising near the edge

$$\Delta i_i = \frac{0.19\Phi}{\pi\mu_0 R}. \tag{2.24}$$

Thus, the full current in the long coil could be expressed by approximate formula

$$i \approx \frac{\Phi(l + 2\varepsilon' R)}{\pi\mu_0 R^2}. \tag{2.25}$$

The values ε' for short coils ($l/(2R) \leq 1$) are presented in Table 2.1, According to [8],

$$\varepsilon' = \varepsilon'_\infty = 0.62, \quad \text{if} \quad l/R = \infty. \tag{2.26}$$

For longer magnets one can use the results obtained in [9], in which the solenoid field at the fixed values $l/(2R) = 0.2; 0.5; 1.0; 2.0; 2.9; 4.0; 5.0$ was calculated numerically. At $l/(2R) = 5$ we have, according to [9], $\varepsilon'(5) = 0.57$, which is close to the value ε'_∞. The curve $\varepsilon' = f[l/(2R)]$ is presented in Figure 2.5.

From formula (2.25) follows the approximate expression for the induction in the center and for the long single-turn thin-wall magnet:

$$B_c \approx \frac{\Phi}{\pi R^2} = \frac{\mu_0 i}{l + 2\varepsilon' R}. \tag{2.27}$$

In the general case the inductance could be represented in the following form [1]:

$$L = \frac{\pi\mu_0 R^2}{l + 2\varepsilon' R}, \tag{2.28}$$

For approximate calculations one can assume $\varepsilon' = \varepsilon'_\infty = 0.62$ at $l/(2r) \geq 1$, which gives in ε' an error of 17 % at $l/(2R) = 1$ and 9 % at $l/(2R) = 2$. The inductance calculated using the approximate formula (2.28) with $\varepsilon' = \varepsilon'_\infty = 0.62$ exhibits an error of, consequently, 6 % and 2 %.

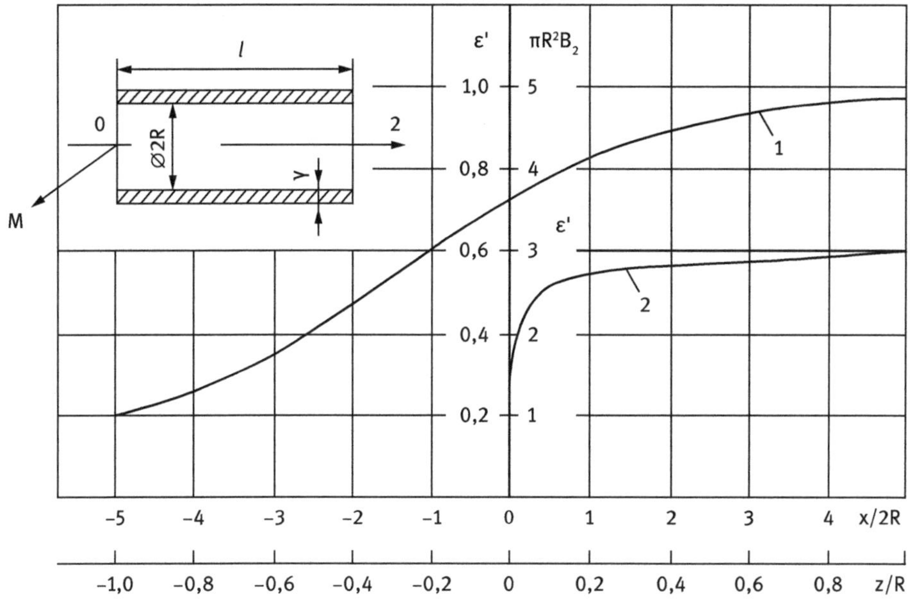

Fig. 2.5: Curves characterizing the field of a thin-wall single-turn solenoid: (1) $\pi R^2 B_z(0, z)/\Phi = f(z/R)$; (2) $\varepsilon' = f[l/(2R)]$.

Figure 2.5 shows the distribution of the induction along the axis near the edge of the semiinfinite solenoid. The induction at point 0 (in a plane of a cylinder cut) $B(0, 0) = B_T = 0.735\Phi/(\pi R^2)$. Let us note that in the coil of the length equal to the diameter ($l/(2R) = 1$) the ratio between the induction at this point and the induction in the center also is 0.735. For comparison note that when the the skin-effect is absent, i.e., at uniform current distribution on the length of multiturn solenoid $B_T/B_c = (1 + 4R^2/l^2)^{1/2}(4 + 4R^2/l^2)^{-1/2}$. The ratio B_T/B_c equals 0.632 at $l/(2R) = 1$ and 0.5 at $l/(2R) = \infty$.

2.4 The field of rectangular coils with arbitrary ratios of characteristic dimensions

In the general case of a rectangular single-turn coil the induction in the characteristic points 1–4 (Figure 2.6) and consequent currents i_1 and i_2 could be expressed in the form

$$[2R_1/(\mu_0 i)]\, B_{1\div 4} = f\,[l/(2R_1),\, R_2/R_1]\,, \quad i_{1,2}/i = f\,[l/(2R_1),\, R_2/R_1]\,.$$

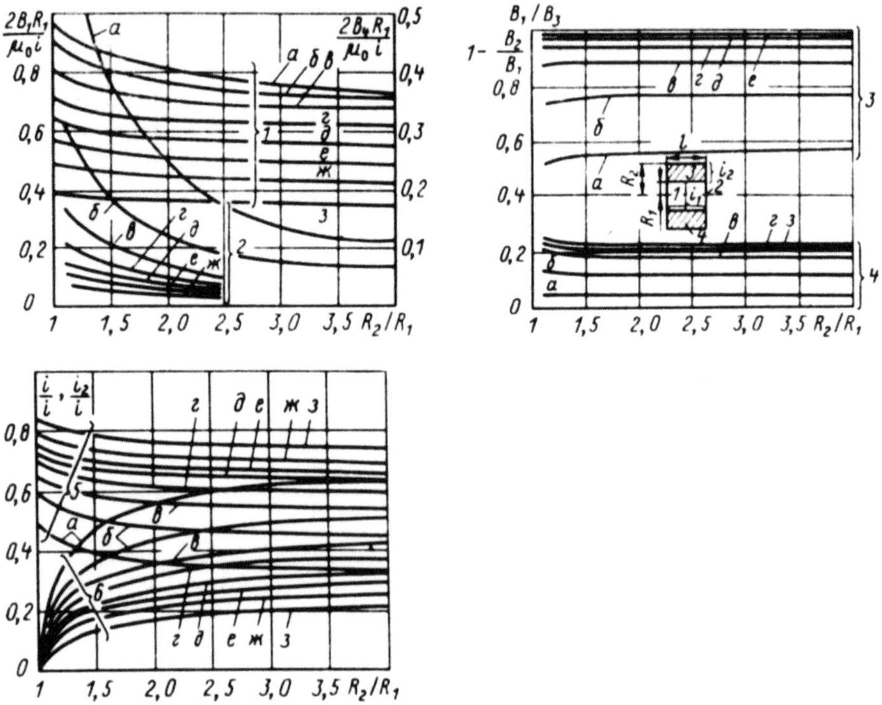

Fig. 2.6: Dependences characterizing the values of the field in rectangular coils: (1) $B_1R_1/(\mu_0 i)$; (2) $2B4R_1/(\mu_0 i)$; (3) B_1/B_3; (4) $1 - B_2/B_1$; (5) i_1/i; (6) i_2/i; the values of the parameter $l/(2R_1)$: (a) 0,2; (b) 0,4; (c) 0,6; (d) 0,8; (e) 1,0; (f) 1,2; (g) 1,5; (h) 2,0.

The results of the calculations, obtained by a net method for solving the Laplace equations for the scalar magnetic potential, are shown in Figure 2.6. They feature a weak dependence of the induction at points 1, 2, 3 upon the ratio R_2/R_1 if it exceeds 1.5. For instance, in the case of solenoid with the ratio of length to inner diameter $l/(2R_1) = 2$ the value of $2B_1R_1/(\mu_0 i)$ varies from 0.390 at $R_2 \approx R_1$ to 0.355 at $R_2/R_1 = \infty$, and in the range $5 \leq R_2/R_1 \leq 13$ from 0.380 to 0.360 [10].

For the generation of the strong pulsed magnetic fields, the thick-wall, single-turn magnets with a rectangular cross section are widely used. In these magnets the ratio R_2/R_1 is much more than 1. It permits for the calculations the assumeption that this ratio equals infinity. The example of this kind of magnet with $l = 0$ is considered in Section 2.1. As follows from formula (2.4), the ratio of the induction in the coil center to the current (geometrical factor) in this magnet could be calculated by the formula

$$\frac{B_c}{i} = \frac{\mu_0}{\pi R_1}. \qquad (2.25)$$

In the more general case the geometric factor depends on the ratio $2R_1/l$.

This and other dependences characterizing the field in the thick-wall coil are presented in Figure 2.6. It should be mentioned that under changes of the ratio $l/(2R_1)$ in the limits from 0 to 1 the ratio $\pi R_1 B_c/(\mu_0 i)$ changes within rather narrow limits from 1 to 0.84. For thick-wall magnets whose length exceed the diameter, the induction in the center could be calculated using the approximate formula

$$B_c \approx \frac{\mu_0 i}{l + \pi R_1/2}, \tag{2.30a)}$$

and the inductance of such a magnet at the arbitrary ratio of a length to diameter could be calculated using the formula

$$L \approx \frac{2\mu_0 R_1}{C' + 2l/(\pi R_1)}. \tag{2.30b)}$$

In this formula the value C' varies within the narrow limits from 1 (at $l = 0$) up to 1.05 (at the condition $l/(2R_1) \gg 1$ [1].

Let us note that induction on the axis of a semiinfinite thick-wall magnet (in the plane of the edge) can be calculated using the formula $B_T = 0.767\Phi/(\pi R_1^2)$. Calculations show that the distribution of the induction out of the cavity of the magnet practically does not depend on its length, if it does not exceed $0.6R_1$.

As was shown above, the overlapping of the slot edges excludes its effect on the current distribution over the plane of the edges. Without overlapping, the current on the ends near the slot edges is directed along the radius rather than the azimuth. This results in some attenuation of the field in the bore. The effect of the slot on the field in a single-turn magnet is characterized by the correction coefficient calculated in [1] for the particular case when the magnet length is much greater than the radius R_1 and $R_1 \ll R_2$. In this case, the induction in the orifice center takes the approximate value

$$B(0) \approx \frac{\mu_0 i}{l}\left(1 - \frac{h}{\pi l}\ln\frac{4R_2}{h}\right). \tag{2.31}$$

The second term is responsible for the role of the slot.

2.5 Induction of the one-turn magnet placed near the coaxial cylinder or the plane

In some magnetic systems operating in conditions with a pronounced skin-effect, conductors are separated by insulating gaps of small thickness. In order to effectively calculate the inductance and the field intensity of such systems, one can use the method of joining. In this case, the field in a narrow gap between conductors

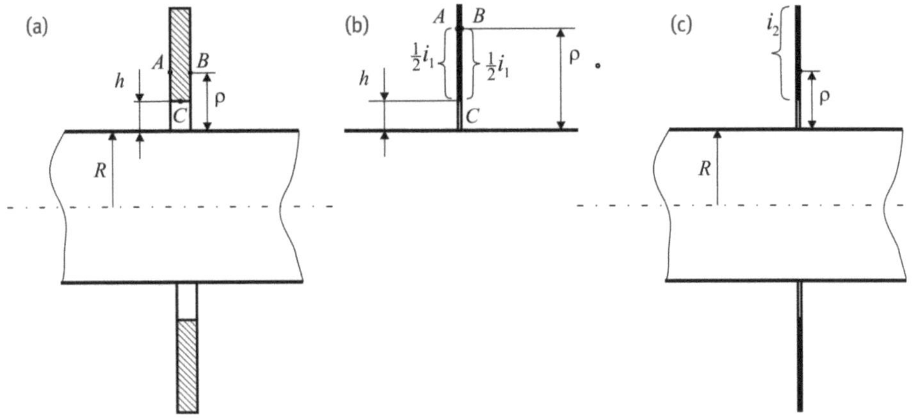

Fig. 2.7: One-turn magnet in the form of a thin flat ring with coaxial cylinder.

is considered as being plane and at the distance from a slot as spatial. The latter can be calculated in approximation of 0 thickness of a gap. For calculation of the total current, the integration is made over the boundary. Up to the definite point of the boundary, the plane problem and on the rest part of the surface the spatial problem is taken into consideration.

An example is illustrated in Figure 2.7a: in the presented system the first of these regions is ***ABC***, and the second occupies the remaining part of the disk surface ($\rho < r < \infty$). The calculated configurations of both regions are shown in Figure 2.7b,c. Let the conditions $h \ll \rho \ll R$ be fulfilled. The current in the region ***ABC*** (Figure 2.7b), calculated with a conformal mapping method, is

$$i_1 \approx \frac{2\Phi}{\pi^2 \mu_0 R} \cdot \ln \frac{2\rho}{h}. \tag{2.32a}$$

The current in the second region (Figure 2.7c) can be presented as

$$i_2 \approx \frac{\Phi}{\pi^2 \mu_0 R} \cdot \left(\ln \frac{R}{\rho} - C_0 + 1 \right), \tag{2.32b}$$

where $C_0 = 0.577$ is the Euler constant. These formulae have been derived in [1, 11]. In the first of them the terms of order h/ρ and in the second of order ρ/R are ignored, and the higher-order small terms are also ignored. Thus, we have the following approximate relation for current i and inductance:

$$i = (i_1 + 2i_2) = \frac{2\Phi}{\pi^2 \mu_0 R} (\ln \frac{2R}{h} - C_0 + 1) \tag{2.33a}$$

$$L = \frac{\Phi}{i} \approx \frac{\pi^2 \mu_0 R}{2\,(\ln (2R/h) - C_0 + 1)}. \tag{2.33b}$$

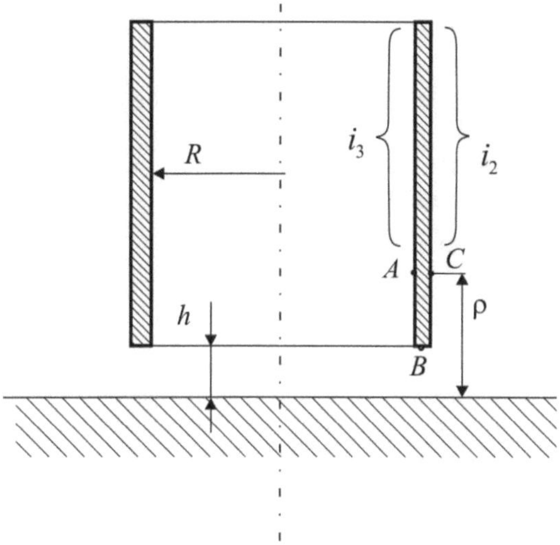

Fig. 2.8: Thin-wall cylinder placed near the plane.

Note that the coordinate of the boundary of the regions do not enter into the ultimate formula for the total current. In some magnetic systems, consideration of the edge effect introduces only an insignificant correction in the inductance, calculated using the simplified picture of the field. In contrast, in the example under consideration the edge effect entirely determines the inductance. It cannot be calculated when considering the simplified picture of the field.

The same approach can be used in the calculation of inductance for the magnetic system shown in Figure 2.8 [1]. In this case, for the current i_1 on the region ABC formula (2.32) is valid, the current $i_2$is determined by formula (2.32b), and the current i_3 on the surface of the long cylinder ($l \gg R$) by

$$i_3 = \frac{\Phi l}{2\pi\mu_0 R^2}\left(1 + \frac{\varepsilon' R}{l}\right) + \frac{\Phi}{\pi^2\mu_0 R}\left(\ln\frac{R}{\rho} - C_0 + \Omega\right), \tag{2.34a}$$

where $\Omega = \int_0^\infty \left[\frac{I_0(x)}{I_1(x)} - \frac{x}{1+x} - -\frac{2}{x}\right]\frac{dx}{x} = -0.893$, $I_0(x)$, $I_1(x)$ are Bessel functions of the imaginary argument.

The approximate formula for inductance has the form

$$L = \frac{\Phi}{2\left(i_1 + i_2 + i_3\right)} = \frac{\pi\mu_0 R^2}{l}\cdot\frac{1}{1 + \frac{2R}{\pi l}\left(\ln\frac{2R}{h} - 0.523\right) + \frac{2\varepsilon' R}{l}}. \tag{2.34b}$$

Here the edge effect has been taken into consideration, including the effect occurring in the vicinity of a slot and near to the end parts of the cylinder as well.

In [1] a great number of configurations have been considered, and the inductance of systems with a narrow gap has been calculated using just the similar method.

2.6 References

[1] G. A. Shneerson, *Fields and Transients in Superhigh Pulse Current Device*, New York, Nova Science Publishers, Inc., 1997.
[2] G. A. Shneerson, *J. Tech. Phys.* **63**(8) (1993), 148–161.
[3] V. A. Fok, *J. Russ. Phys. Chem. Soc., Ser. Phys.* **82**(3) (1930), 281–297.
[4] P. L. Kalantarov and L .A. Zeytlin, *Calculation of inductance*, Reference book. L.: "Energia", Leningrad, 1970 (in Russian).
[5] G. A. Shneerson, *J. Tech. Phys.* **31**(1) (1961), 51–54 (in Russian).
[6] T. Carlemann, *Math. Z.* **15** (1922), 111–120.
[7] P. N., Dashuk, S. L. Zayentz, V. S. Komelkov, N. N. Nikolayevskaya, et al., *High pulsed currents and magnetic fields technique*, Moscow, Atomizdat, 1971 (in Russian).
[8] V. A. Popova and G. A. Shneerson, *J. Tech. Phys.* **47**(10) (1977), 2009–2016 (in Russian).
[9] G. Bardotti, B. Berdotti, and L. Gianolio, *Math. Phys.* **25**(10) (1966) 1387–1390.
[10] V. P. Knyasyev and G.A. Shneerson, *Izvestiya Vusov, Energetika* **4** (1971), 34 (in Russian).
[11] G. A. Shneerson, *Izvtstiya Akademiyi Nauk SSSR, Energetika i Transport* **2** (1969), 85–95 (in Russian).

3 Field diffusion into the conductors and their heating

The heating of conductors significantly affects the exploitation characteristics of magnetic systems. The devices operating at stationary conditions require cooling, and in the design of magnet coils this requirement is considered with great attention [1, 2]. The thermal processes are also significant for the generation of the short pulsed fields [3, 4] Here we turn our attention to the heating of conductors for single pulses with a duration of 10^{-3}–10^{-4} s and less. for this pulse duration the heat exchange between the elements of the winding is becoming a minor factor, which permits us to suppose heating to be an adiabatic process. At nonstationary conditions the current distribution on the cross section of conductors is not uniform due to the skin effect. One should distinguish two limiting regimes presenting the most interesting cases: in the first case the depth of field penetration in conductors is much more than the characteristic dimensions of conductor's cross section, in the second case it is much less than cross section. Therefore, the meaningful criterion is the penetration depth $\Delta = \sqrt{\rho\tau/\mu_0}$, where ρ is the ρ is the specific resistance, and τ the characteristic time (for example the half period of the current pulsation or the duration of the rectangular pulse). If condition $\Delta > G$ is satisfied, where G is the radius or the other characteristic dimension of the conductor, then the skin effect does not affect the current distribution or this effect is pronounced. In this case one can assume the current density to be the same at all points of the cross section. The opposite inequality corresponds to the sharply pronounced skin effect.

The windings of the multiturn coils are generally operating for a long pulse duration, since the inductance of these coils is large enough, likewise the capacity of the feeding low voltage condenser banks. In this type of coils the skin effect is of secondary importance. But, on the contrary, in single-turn solenoids this effect practically is quite strong.

In this context, at the beginning of this chapter we shall consider the adiabatic heating of conductors where there is no skin effect, and further on where it is present. Attention will be focussed both on the linear regimes when heating is small and conductivity is constant and on the regimes where the conductivity drops at the heating which occurs in a sufficiently strong field. The regimes with phase transitions featuring the superhigh fields will be considered in Chapter 9.

3.1 Adiabatic heating of conductors at a given current density

According to the Joule law, the velocity of the changing of the energy volume density q' in the immobile medium of the constant density

$$\frac{dq'}{dt} = C_V \frac{dT}{dt} = \rho\, \delta^2, \tag{3.1}$$

where δ is the current density, C_V is the specific heat per unit volume.

The specific resistance and the energy volume density unambiguously depend on the temperature. Thus, the following dependence takes place:

$$\int_{T_0}^{T} \frac{C_V(T)}{\rho(T)}\, dT = \int_0^t \delta^2 dt = S_I. \tag{3.2}$$

This formula establishes the link between the temperature and parameter S_I, which is referred to as the action integral of current.

The temperature range within which the conductors of magnets with high and ultra high magnetic fields operate extends from the cryogenic (of the order of several °K) up to many thousands.

In nondestroyed multiturn magnets the range of admissible temperatures is restricted by several hundred absolute degrees. The dependences of the action integral on the temperature in this range for copper are shown in Figure 3.1[1]. Commonly, as the initial temperature, whether the normal (293 °K) or the temperature of liquid Nitrogen (77 °K) is chosen, the curves for copper in Figure 3.1 are plotted for these initial temperatures. They show that for windings heated from 77 °K

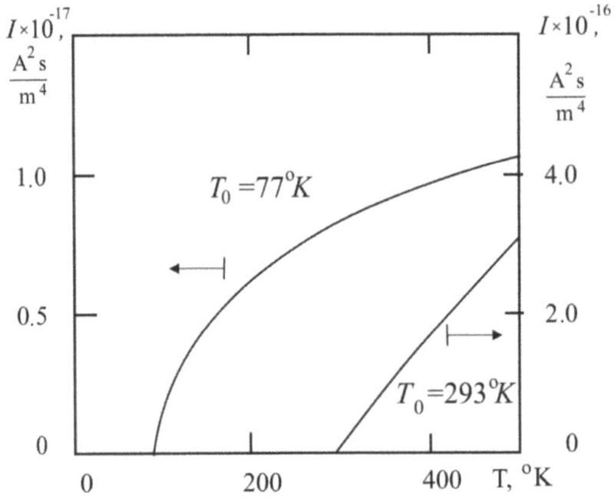

Fig. 3.1: Dependences of the action integral on the temperature increment (copper): (1) an initial temperature 77 °K; (2) initial temperature 293 °K.

1 When building these dependencies used the formulas given in Herlach and Miura [4].

to 293 or 500 °K the admissible values of the action integral are, correspondingly, $7.62 \cdot 10^{16}$ $A^2 \cdot s/m^{-4}$ (point **A**) and $1.07 \cdot 10^{17}$ $A^2 \cdot s/m^{-4}$ (point **B**), while for a winding heated from 293 °K to 500 °K, it is $3.05 \cdot 10^{16}$ $A^2 \cdot s/m^{-4}$ (point **C**). A comparison of the last two values indicates that the initial cooling is rather advantageous. An even greater effect may be obtained by cooling down to the temperature of the liquid Helium. Von Ortenberg and Mueller [5] have shown that the action integral is three times greater for the range (4,4–77) ^{0}K than for the range (77–300) °K. This effect is caused by the sharp increase of the ratio C_V/ρ at the temperature of liquid Helium.

In the range of cryogenic temperatures the specific resistance increases in the presence of magnetic field (magnetoresistive effect). The correspondence dependence for copper is described by the formula given in the review [4]:

$$\rho\,(B, T) = \rho\,(0, T)\left[1 + 10^{-3}\left(B\frac{\rho_0}{\rho\,(0, T)}\right)^{1,1}\right], \tag{3.3}$$

where ρ_0 is the specific resistance at the temperature 273 °K , and $\rho(0, T)$ is the specific resistance at the temperature T in the absence of a field. At the normal temperature the field influence on the conductivity is negligible. At the temperature of liquid Nitrogen in the field with induction 100 T the specific resistance increases by 16 %. With further heating, the field influence decreases. This gives grounds for ignoring the effect of magnetoresistivity in calculations of coils with the initial temperature 77 °K and above [3]. The detailed data on the temperature dependence of the resistance and on its relation to the integral of action can be found in the paper by Grőssinger [31].

A great many magnetic systems operate at high temperatures. These are multiturn magnets designed for both multiple and single use. In the first case the admissible temperature should not exceed the threshold value at which the surface erosion appears. For a small number of pulses this threshold is close to the melting temperature and becomes essentially lower as the number of pulses increases (see Chapter 8).

The destroyed single-turn magnets and conductors applied in a magnetic cumulation may heat up to and higher than the energy of sublimation. After the phase transition the conductivity is described by a rather complicated dependence on time and concentration, which will considered in Chapters 9 and 10, devoted to the generation of ultrahigh magnetic fields. It is convenient to describe the specific resistance of solid and liquid conductors operating at high temperatures in the form of dependence on the increment of the volume density of the thermal energy $\rho/\rho_0 = f(\Delta q')$ (Figure 3.2). Here $\Delta q'$ is the indicated increment in the range from the normal to the current temperature.

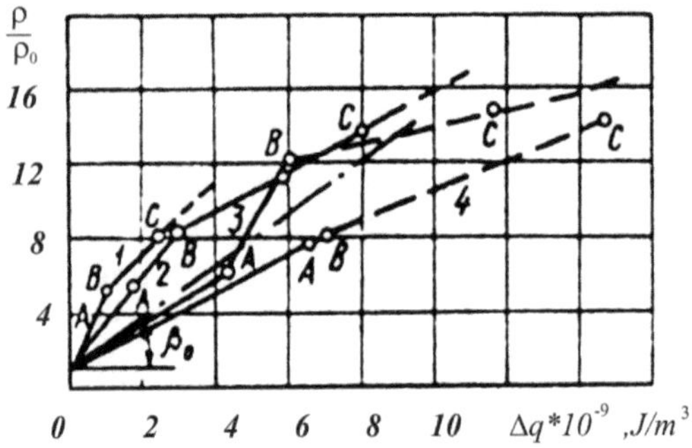

Fig. 3.2: Dependences of the specific resistance on the energy generated in a unit volume at heating of the conductor; (A) beginning of melting; (B) the melt; (C) beginning of evaporation; (1) lead, (2) aluminum; (3) copper; (4) tantalum. The dot-dashed line corresponds to the averaged value of parameter β in the range $\Delta q' \le Q_2$ (for copper).

For estimation the function $\rho(q')$ could be conveniently approximated by a linear dependence:

$$\rho \approx \rho_0 \left[1 + \left(q' - q'(0)\right)\right] = \rho_0 \left(1 + \beta \cdot \Delta q'\right), \tag{3.4a}$$

where ρ_0 is the initial value of the specific resistivity, $\Delta q' = q' - q'(0)$ the volume energy increment. In this formula β is the thermal coefficient of resistivity, values of which are given for some metals in the Table 3.1. This table presents three values of the coefficient β: β_1 corresponds to variations of $\Delta q'$ from T = 20 °C to completion of the melting, β_2 is determined by a slope of the line $\rho/\rho_0 = f(\Delta q')$ behind the boiling point, β_3 by a slope of the straight line drawn between the points of melting termination and beginning of the evaporation. For copper $\beta_1 \approx 1{,}7 \cdot 10^{-9}$, $\beta_2 \approx 1.4 \cdot 10^{-9}$ m^3/J. Depending on the range of values $\Delta q'$ one should use in analytical assessments the corresponding values of parameter β.

Substituting (3.3) in formula (3.1) after integration we obtain

$$\ln \frac{\rho}{\rho_0} = \rho_0 \beta\, S_I; \rho = \rho_0 \exp\left(\rho_0 \beta\, S_I\right). \tag{3.4b}$$

In the coil where the skin effect is absent the current density is i/F, where i is a current in the separate conductor with cross section F. In this simple case the heating of conductors in adiabatic approximation is easily calculated by the given formulas. Since $\Delta q' \approx C_V \Delta T$, (where C_V is the heat capacity at the constant volume, ΔT is the temperature increment) one can find, knowing $\Delta q'$, the increment in temperature.

The coefficient of the volume specific heat as a function of temperature is presented in Table 3.2. In approximate calculations for the specific heat in the range from the room temperature to the melting point one can take the averaged value $\langle C_V \rangle$. For copper $\langle C_V \rangle \approx 3.6 \cdot 10^6$ J/(m^3· K).

Table 3.1: $Q_{1,2,3} = \gamma_0 \varepsilon_{1,2,3}$, where ε_1 is the density of the thermal energy (J/kg) of the melt at the melting temperature; ε_2 the same at the boiling point, at normal pressure; $\varepsilon_3 = \varepsilon_2 + \Delta\varepsilon$, where $\Delta\varepsilon$ is the latent heat of evaporation at the normal pressure; $B_{S1,2,3} = \sqrt{2\mu_0 Q_{1,2,3}}$, $\beta_1 = (\rho(Q_1) - \rho_0)/Q_1$, $\beta_2 = (\rho(Q_2) - \rho(Q_1))/(Q_2 - Q_1)$, $\beta_3 = (\rho(Q_2) + \rho(Q_1))/(Q_1 + Q_2)$, $B_{01} = \sqrt{2\mu_0/\beta_1}$, $B_{02} = \sqrt{2\mu_0/\beta_2}$, $B_{03} = \sqrt{2\mu_0/\beta_3}$.

Parameters		Copper	Iron	Aluminum	Tantalum	Lead
γ_0	$\mathrm{Kg/m^3} \cdot 10^{-3}$	8.9	7.8	2.7	16.5	11.3
ρ_0	$\Omega \cdot \mathrm{m} \cdot 10^8$	1.7	10	2.9	15.5	20.6
C_1	Km /s	3.96	3.80	5.25	3.37	2.03
λ		1.50	1.58	1.39	1.16	1.52
Q_1	$\mathrm{J/m^3} \cdot 10^{-9}$	6.0	7.3	2.9	7.7	0.78
B_{S1}	T	123	135	85	138	44
Q_2	$\mathrm{J/m^3} \cdot 10^{-9}$	11.6	17.2	8	14.7	2.3
B_{S2}	T	170	210	100	190	75
Q_3	$\mathrm{J/m^3} \cdot 10^{-9}$	54	67	33	22	3.1
B_{S3}	T	370	410	320	235	88
β_1	$\mathrm{m^3/J} \cdot 10^9$	1.7	1.35	2.5	1.04	4.9
B_{01}	T	38	43	32	49	23
β_2	$\mathrm{m^3/J} \cdot 10^9$	0.48		1.1	0.86	2.24
B_{02}	T	73		48	54	34
β_3	$\mathrm{m^3/J} \cdot 10^9$	1.4		2	0.98	4.2
B_{03}	T	43		36	50.5	26.5

Table 3.2: The coefficient of the volume-specific heat.

T, K	20	40	80	150	250	293	400	600	1000
C_V, $10^6 \mathrm{J/m^3K}$	0.064	0.466	1.81	2.87	3.35	3.43	3.52	3.70	4.05

As an example let us calculate the increment in temperature per pulse in conductors comprising the single layer solenoid winding with a number of turns $w' = 1000\,1/m$ that generates the pulse of magnetic field with the shape of decaying sinusoid: $B = 20\exp(-10^3 t)\sin(6 \cdot 10^4 t)$. The copper conductors have a rectangular shape with the sides 2 and 1 mm. The amplitude of the current density is $\delta_m = B_m/(\mu_0 w' F) = 7.95 \cdot 10^9\ \mathrm{A/m^2}$; and the current action integral $S_I = \int_0^\infty \delta^2 dt \approx \frac{1}{4} T_0 \delta_m^2$, where $1/T_0 = 10^3\ \mathrm{s^{-1}}$ is constant in the exponential factor. Further on we find: $S_I = 1.6 \cdot 10^{16}\ \mathrm{A^2 m^{-4}\ s}$; $\rho/\rho_0 = \exp(\rho_0 \beta\ S_I) = \exp(1.7 \cdot 10^{-8} \cdot 1.7 \cdot 10^{-9} \cdot 1.6 \cdot 10^{16}) \approx 1.59$; $\Delta q' = (\rho/\rho_0 - 1)/\beta = 3.48 \cdot 10^7\ \mathrm{J/m^3}$; $\Delta T = 94\,°\mathrm{K}$.

3.2 Linear regime of field diffusion in conductors

As a sampling problem we shall consider the penetration of the pulsed magnetic field in the conductive half-space. Let us examine the diffusion into the conductor of fairly weak magnetic fields, when heating is not so strong that it would be necessary to take into account the variation of the resistance with growth of the temperature. The Maxwell equation for the conductive media with penetrating magnetic field [6, 8] is

$$\frac{\partial B}{\partial x} = -\mu_0 \delta_y = -\mu_0 \frac{E}{\rho}, \tag{3.5}$$

$$\frac{\partial B}{\partial t} = -\frac{\partial E}{\partial x}, \tag{3.6}$$

where we see that the induction component $B = B_z$, electric field intensity $E = E_y$ and specific resistivity ρ, could be transformed into the equation

$$\frac{\partial B}{\partial t} = \frac{\partial}{\partial x}\left(D_M \frac{\partial B}{\partial x}\right), \tag{3.7}$$

where $D_M = \rho/\mu_0$ is the coefficient of the magnetic field diffusion.

To evaluate the penetration depth one can use the formula

$$\Delta \approx \sqrt{D_M t} = \sqrt{\rho\, t/\mu_0}\,. \tag{3.8}$$

Similar evaluations are made in the theories of gas diffusion and thermal conductivity, where the processes are also described by parabolic equations [7].

Cases of nonstationary diffusion of the magnetic field in a medium with ρ_0 = const have been examined in a number of publications (for instance, [3, 4, 8, 9]).

Let us consider at the beginning the simple solution in the form of wave propagating with the constant velocity. On this assumption the quantities would depend on the argument $x - u\,t$, where u is the phase velocity. Then for an arbitrary function $f(x,t)$ the following equalities are valid:

$$\frac{\partial f}{\partial x} = \frac{df}{dz} \cdot \frac{\partial z}{\partial x} = \frac{df}{dz}, \tag{3.9a}$$

$$\frac{\partial f}{\partial t} = \frac{df}{dz} \cdot \frac{\partial z}{\partial t} = -u\frac{df}{dz}. \tag{3.9b}$$

The differential equation in partial derivatives (3.7) transforms to the standard differential equation:

$$D_m B_z'' + u B_z' = 0, \tag{3.9c}$$

which has the solution

$$B = C \exp\left(-uz/D_m\right) = C \exp\left(-ux/D_m\right) \exp\left(u^2 t/D_m\right). \tag{3.9}$$

This solution exists only if on the boundary ($x = 0$) the field induction increases exponentially: $B(0) = B_e = B_0 \exp(t/t_0)$. In this case $C = B_0$, $D_m/v^2 = t_0$ or $v = \sqrt{D_m/t_0}$.

Thus, the ultimate solution acquires the form

$$B = B_0 \exp(t/t_0) \exp\left(-x/\sqrt{Dt_0}\right). \tag{3.10a}$$

The current in this case is also distributed exponentially:

$$\delta = -\frac{1}{\mu_0}\frac{\partial B}{\partial x} = \frac{B_0}{\mu_0\sqrt{Dt_0}} \exp(t/t_0) \exp\left(-x/\sqrt{Dt_0}\right). \tag{3.10b}$$

The distance at which the current decreases **e** times compared to its boundary value, in this particular case, does not change, but the total current grows exponentially.

In those cases when the field does not grow infinitely on the boundary but reaches some fixed value, the solution has an entirely different character. Because the field in the depth of the conductor is absent ($B(\infty) = 0$), increasing the induction value B_e on the boundary to the given one B_0 resulted in the linear current density rising up to the value $j_0 = B_0/\mu_0$. In this case the region which experiences a current flow broads with time according to formula (3.8), and hence the current density in the surface layer decreases.

The classical example of this situation is the regime of steady state oscillations, when the induction on the boundary changes, obeying the sin law. This can be represented in the exponential form $B_e = B_{e,m} \exp(j\omega t)$. Further on we can use the formula (3.9) and substitute the parameter t_0 with the quantity $(-j/\omega)$, where $j = \sqrt{-1}$. The following rearrangements give the familiar dependence for induction and current density:

$$B(x, t) = B_{em} \exp\left(-\frac{x}{\Delta_0}\right) \sin\left(\omega t - \frac{x}{\Delta_0}\right), \tag{3.11a}$$

$$\delta\,(x, t) = -\frac{1}{\mu_0}\frac{\partial B}{\partial x} = \frac{B_{em}\sqrt{2}}{\mu_0 \Delta_0} \exp\left(-\frac{x}{\Delta_0}\right) \sin\left(\omega t - \frac{x}{\Delta_0} + \frac{\pi}{4}\right). \tag{3.11b}$$

In the above formulas the parameter $\Delta_0 = (2\rho_0/\omega\mu_0)^{1/2}$ is the classic penetration depth, that is equal to the distance on which the amplitudes of the induction and current density decrease **e** times.

In the general case, the solution of equation (3.7) could be obtained using Laplace transformation. At the zero initial condition ($B(x,\ 0) = 0$) we obtain the equation for the transformed values:

$$\frac{d^2\bar{B}(x,p)}{dx^2} - \frac{p}{D_M}\bar{B}(x,p) = 0. \tag{3.12a}$$

Using the boundary conditions $B(0, t) = B_e(t) = \bar{B}_e(p)$, $B(\infty, t) = 0$, we obtain the dependence for the transformed value of induction:

$$\bar{B}(x, p) = \bar{B}_e(p) \exp\left(-\sqrt{\mu_0 p/\rho_0}\, x\right). \tag{3.12b}$$

With this result we have the expressions for the intensity of an electric field and the current density:

$$\bar{E}(x, p) = \sqrt{p\rho_0/\mu_0}\, \bar{B}(x, p); \quad \bar{\delta}(x, p) = \sqrt{p/(\rho_0\mu_0)}\, \bar{B}(x, p). \tag{3.12c}$$

3.3 The surface impedance. Energy losses in the skin layer with sinusoidal current

In considering the current flow on the boundary of the conducting half-space one can introduce the concept of *surface impedance* (transitional resistance) of the skin layer. This is the resistance of the square part of the surface, oriented in such a way that current flow lines are parallel to the side of the square. The expression for such a resistance has the form

$$Z_0(p) = \frac{\bar{E}_\tau(p)}{\bar{J}(p)}, \tag{3.13}$$

where $\bar{E}_\tau(p) = \bar{\delta}(0, p)$ is the representation of the electric field intensity on the boundary, $\bar{J}(p)$ the representation of the current linear density. From formula (3.13) follows

$$Z_0(p) = \sqrt{p\mu_0\rho_0}, \tag{3.14}$$

and at sinusoidal current

$$Z_0(p) = \sqrt{j\omega\mu_0\rho_0} = R_0 + j\omega L_0, \tag{3.15}$$

where $R_0 = \sqrt{\omega\mu\rho_0/2}$ is the active resistance of the square part of the surface layer of conductor at the alternative current, $L_0 = \sqrt{\mu\rho_0/(2\omega)}$ its inductance.

Taking into consideration that $J(p) = \bar{H}_e(p)$, we obtain the relationship between the tangent components of the electric and magnetic field intensities which is valid at the conditions of a sharply pronounced surface effect:

$$\bar{E}_\tau(p) = \bar{H}_\tau(p) \cdot Z_0(p). \tag{3.16a}$$

This condition could be used as a boundary condition for the calculation of a field outside of the conductor. For a sinusoidal field it acquires the form (the boundary condition of Leontovich [10])

$$\bar{E}_\tau = \bar{H}_\tau \sqrt{j\omega\mu\rho_0}. \tag{3.16b}$$

From these expressions follows the boundary condition of the third kind for the vector potential on the surface of the conductor:

$$\frac{d\bar{A}}{dx} + \sqrt{\frac{\mu_0 p}{\rho_0}}\,\bar{A}. \tag{3.16c}$$

In the case of ferromagnetic media and at high frequencies the tangent component of the field intensity may acquire noticeable values. For instance, in a case of the iron with $\mu \approx 1000\mu_0$ and $\rho_0 \approx 10^{-7}\,\Omega\cdot$m at $H \approx 10^3$ A/m (in such a field the iron is not yet saturated) we have $E_\tau \approx 10^2$ V/m at $\omega = 10^6$ Hz and $E_\tau = 10^3$ V/m at $\omega = 10^8$ Hz. In the absence of the current, $E_\tau = 0$. In the case of nonferromagnetic conductors, E_τ is usually small, in spite of the fact that the field intensity H_τ may assume higher values. In order to evaluate the utmost practically achievable values of E_τ in conductors where current density is limited by the conditions of the heating, let us take into account that copper conductor heated by the current without the heat removal is melted, if condition $\delta^2\tau \approx 10^{16}$ A^{2}m^{-4}s is present, where τ is the time of the current passage. At $\tau = 10^{-6}$ s the limiting permissible current density has the value of order of 10^{11} A/m^2, which corresponds to $E_\tau \approx 10^3$ V/m = 10 V/cm, since ρ is of order $10^{-8}\,\Omega\cdot$m. The normal components of intensity could be many orders of magnitude higher than this small value. Therefore in a great number of the calculations it is possible to assume $E_\tau = 0$, which means considering the conductor as being ideal when evaluating the electric field outside conductor. In other words, it is permissible to assume on its boundaries the condition

$$E_\tau = -\frac{\partial A_\tau}{\partial t} - \frac{\partial U}{\partial \tau} = 0, \tag{3.17}$$

where A_τ is the tangent to the boundary component of the vector potential and U the scalar potential of the electromagnetic field. The condition $E_\tau = 0$ is approximate for the conductors with current and exact in the case when current is absent.

Knowing the distribution of the surface current density, taken as the approximation of ideal conductivity, one can calculate the energy losses in the skin layer. In the case of sinusoidal current, the average power, released in the skin layer of the conductor and limited by the surface S, could be calculated using the following formula, which includes the active component of the surface impedance and integration is carried out on the surface:

$$\langle P \rangle = \frac{1}{2}\sqrt{\frac{\omega\mu_0\rho}{2}}\int_S J_m^2\,dS. \tag{3.18}$$

Here J_m is the amplitude value of the linear current density. This formula could be used for the calculation of the losses on the total surface or on its part, if the corrections, taking into account the two-dimensional character of the field diffusion, could be neglected. The sharpest deviation from the one-dimensional character of the diffusion is revealed near the edges of electrodes. This effect will be considered at the point 3.4. It has a local character and weakly affects the total energy release in the conductor at the small penetration depth. Due to this consideration, in calculations by formula (3.18) one can use the values of the surface current density, obtained in the approximation of the ideal conductivity.

The unlimited growth of the field intensity near the edge of the conductor does not lead to the divergence of the integral (3.31) if the angle $\alpha > 0$. This special case corresponds to the edge of the thin conductor, e. g., the edge of the strip-line or the single turn magnet, having the shape of a thin disk with an orifice. If, in the first approximation, instead of the real edge of finite thickness δ we assume its thickness to be equal to zero, we obtain the result in the configuration presented in Figure 1.9a (region z_1), in which the square of the linear current density, calculated in the approximation of ideal conductivity, rises by $1/x_1$. In this case the integral (3.18) diverges. The initial configuration z_2 is characterized by the complex potential $W = u + jv$ mapping the calculated region on the canonical one (Figure 1.9b). In this representation the equalities $J^2 = |dW/dz_2|^2$ and $\int J^2 dS = \int |dW/dz_2| du$ are valid. In this example the function $W(z_2) = Aw(z_2)$, where $w(z_2)$ is expressed by the formula (1.45). The value of the constant A should be chosen by such a way that the linear current density with distance from the edge does not differ from its value in the system with a sheet of zero thickness. This value is given by the general configuration of the magnetic system and i currents and described by the dependence of the type $|J| = C/x_1$, where the coordinate x_1 is measured from the edge of sheet. Along with that, the values of the current linear density should coincide with the distance from the edge both in the system with a zero thickness sheet and in the initial system, where the module of the linear current density at $x_2 \gg \delta$ takes the value $|J| \approx A\sqrt{\pi/(4\delta\, x_2)}$. Far from the edge, $x_2 = x_1$. Equating two expressions for $|J|$ we find the value of the constant $A = 2C\sqrt{\delta/\pi}$. Further on we find the value of the integral in the formula for the power of the losses on the conductor section from the point **c**, located in the middle of the edge, to the point with the coordinate $j\delta/2 + x_0$ where $x_0 \gg \delta$:

$$\int_c^n |J^2| ds = A^2 \int_{w(c)}^{w(n)} |dw/dz_2| dw = 2C^2 \left(\int_0^1 (1 - w^2)^{-1/2} dw + \int_1^{w(n)} (w^2 - 1)^{-1/2} dw \right). \tag{3.19a}$$

The first integral is the number $\pi/2$, but in the second one the value $w(n) \approx (\pi x_0/\delta)^{1/2} \gg 1$ it should be taken as an upper limit. The second integral is $\ln(w(n) + \sqrt{w(n)^2 + 1}) \approx \ln(2\sqrt{\pi x_0/\delta})$. Taking into account all the above mentioned we obtain

$$\int_{c}^{n} |J^2| dS \approx C^2(\pi + \ln(4\pi x_0/\delta)). \tag{3.19b}$$

Instead of using this formula, one can carry out the integration of the square of linear current density on the surface of zero thickness sheet to the point x_0. The lower integration limit θ should be chosen in such a way that the integral

$$\int_{\theta}^{x_0} (C^2/x)dx = C^2 \ln(x_0/\theta_0), \tag{3.20a}$$

calculated for the zero thickness sheet, equals the integral determined by formula (3.18). As a result we arrive at the equation for estimating the number θ_0:

$$\theta_0 = \frac{e^{-\pi}}{4\pi}\delta = 3.49 \cdot 10^{-3}\delta. \tag{3.20b}$$

It is worth noting that the obtained solution does not depend on the parameter x_0. Thus, the calculation of the losses is reduced to the integration of the current density square on the whole surface of the zero thickness sheet, excluding the section of the width θ_0 adjacent to the edge of the sheet. This calculation method of loses offered in [11] allows us to take into the account the finite sheet thickness and could be used if it essentially exceeds the thickness of the skin layer.

As an example, let us use the obtained formulas to calculate the losses in the strip line with thickness g with a given current $i = i_m \sin(\omega t)$, for the condition that the conductor thickness equals δ. We use formula (1.51) for the linear current density. To take into account the finite conductor's thickness, in the calculations in formula (3.18) one should deviate from the conductor edge by the distance θ, instead of integrating over the whole surface. As a result, we obtain the following expression for the average value of losses per unit length:

$$\begin{aligned} \langle P' \rangle &= 2\sqrt{\frac{\omega\mu_0\rho}{2}} \int_{0}^{g/2-\theta} \frac{i_m^2 dx}{\pi^2 g^2 (1 - (2x/g)^2)} \\ &\approx \frac{1}{2}\sqrt{\frac{\omega\mu_0\rho}{2}} \frac{i_m^2}{\pi^2 g}\left[\ln\frac{g}{\delta} + \pi + \ln(4\pi)\right]. \end{aligned} \tag{3.21}$$

Here the condition $\theta_0 \ll g/2$ is taken into account.

This process of loss calculation could be used for the thin disk with the orifice. The dependence of the surface current density on the radius in this case is described by formula (2.2). In formula (3.18), at the integration, one should exclude the part of the surface on which the radius obeys the condition $R_1 < r < R_1 + \theta_0$.

The average power loss in the skin layer at the alternative current is determined by the following formula:

$$\langle P\rangle = 2\pi\sqrt{\frac{\omega\mu_0\rho}{2}}R^2 i_m^2 \int\limits_{R_1+\theta}^{\infty} \frac{dr}{r(r^2-R_1^2)}$$

$$\approx \pi\sqrt{\frac{\omega\mu_0\rho}{2}} i_m^2 \left[\ln\frac{R_1}{\delta} + \pi + \ln(2\pi)\right]. \tag{3.22}$$

When calculating the impedance of the conductor and the energy losses in a skin layer, one can avoid the integration over the surface if the procedure of differentiation of inductance on the normal to the boundary is used. The surface impedance of the linear conductor could be calculated using following formula:

$$Z_0'(p) = \sqrt{\frac{p\rho}{\mu_0}}\frac{\partial L'}{\partial n_i}, \tag{3.23a}$$

where n is the internal normal to the boundary. The calculations using this formula can be made most simply in the case where the boundary of the cross section of the conductors is a circle or regular polygon. Then the derivative over the internal normal takes the form

$$\frac{\partial L'}{\partial n_i} = -\frac{\partial L'}{\partial R}, \tag{3.23b}$$

where R is the radius of the inscribed circle or the radius of the conductor of the round cross section. An example is calculation for the two-conductor line with the conductors with a round cross-section with radii R_1 and R_2 and distance D between them. In this case the induction per unit length is expressed by the known formula [9]

$$L' = \frac{\mu_0}{2\pi}\operatorname{Arch}\frac{D^2-R_1^2-R_2^2}{2R_1R_2}. \tag{3.24a}$$

The surface impedance of each conductor can be calculated by (3.23b):

$$Z_0'(p)_{1,2} = -\sqrt{\frac{p\rho_{1,2}}{\mu_0}}\left(\frac{\partial L_{1,2}'}{\partial R_{1,2}}\right) = \frac{\sqrt{\mu_0 p\rho_{1,2}}}{2\pi}\frac{D^2+R_{1,2}^2-R_{2,1}^2}{R_{1,2}\sqrt{(D^2-R_1^2-R_2^2)^2-4R_1^2R_2^2}} \tag{3.24b}$$

3.4 The asymptotical values of the magnetic field intensity and current density at the conductor edge under the condition of a pronounced skin effect

The edge effect is the characteristic feature of conductors with a polygon cross section. To evaluate the field diffusion in the vicinity of the edge, it is necessary to solve the two-dimensional problem. Here we shall restrict our considerations to the specific features of two-dimensional linear diffusion of poloidal and toroidal fields, which allows us to evaluate the characteristic parameters for conductors with an acute angle rim. The peculiarities of nonlinear two-dimensional diffusion will be considered in Chapter 9, when we describe the experiments for obtaining superstrong magnetic fields.

In Chapter 1 we outlined 1 how under the conditions of a pronounced skin effect the poloidal magnetic field intensity rises in the vicinity of the apex of acute angles. Assuming an ideal conductivity, the dependence of poloidal field intensity in the point M on distance s between this point and the top of angle (Figure 3.3) is of the form

$$H(M, t) = C(t)\, s^{\alpha}, \tag{3.25}$$

where $\alpha = (\theta - \pi)/(2\pi - \theta)$. Here, the coefficient $C(t)$ depending on time t and angle θ appears. This coefficient is determined by the configuration of the magnetic system and the currents in the conductors. When approaching the apex of the angle the intensity rises infinitely if the condition $\theta \leq \pi$ holds.

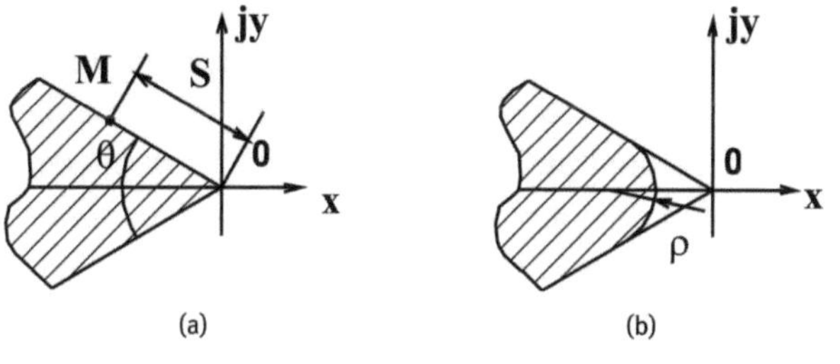

Fig. 3.3: The conductors (a) with an acute angle rim and (b) with a rounded rim.

Let us consider the practical case when the penetration depth of the field varies according to the sinusoidal law and differs from zero, but is small by comparison with all the other dimensions of the magnetic system. In this case the intensity in a point ***O*** is restricted, as for a rounded rim of an ideal conductor. It was shown in Chapter 1 that in the conductor with a circular rim the intensity amplitude in the point M at the rim is defined by

$$H_m(M) = \lambda C_m \rho^{\alpha}, \tag{3.26}$$

where C is the amplitude value of the coefficient $C(t)$, ρ is the geometry parameter of a round segment of a rim which has the dimension of length (for example, ρ is the radius of curvature), λ is a dimensionless factor depending on the angle θ, configuration of the segment and location of a point $\boldsymbol{M}$ on this segment.

Based on considerations of the dimensions, it is reasonable to suppose that when the skin layer has a small depth, the dimension ρ in the expression for the intensity in the point $\boldsymbol{O}$ can be replaced by the penetration depth Δ, whereby the dimension of the length and the amplitude of intensity will be determined by the formula

$$H_m(O) = \gamma C_m \Delta^{\alpha}. \tag{3.27}$$

In this formula we have γ instead of the number λ, the other dimensionless factor depending only on the angle θ.

As in magnet systems with a rounded rim, the number γ is only defined by a field configuration nearest to a vertex of angle. Hence, this number does not depend both on the specific configuration of the magnet system as a whole and currents in it. These factors affect only the value of the coefficient C_m, which, as we mentioned above, can be calculated for the specific system at in approximation of an ideal conductivity without regard for the rounding. It is evident that the number γ is constant only in the case when the depth penetration is much less than the other dimensions, e. g., the distance to other conductors.

To define the number γ it is necessary to calculate the field by solving the Maxwell's equations in the conductor and outside, taking into consideration the continuity of the normal and tangent intensity components at the interface. The analytical solution of this two-dimensional problem of diffusion theory presents severe difficulties and does not exist in the literature. In addition, in the problem under consideration, it is possible for a specific system and each angle θ, using the numerical methods, to evaluate the field near the vertex of the angle, depending on nondimensional coordinates x/Δ and y/Δ. The results (for instance, the field intensity and current density at point $\boldsymbol{O}$) will be independent of the chosen configuration of the magnetic system considered in numerical calculations.

As an example we consider the field of the conductor with current located at the distance d from the vertex of dihedral angle (Figure 3.4).

At ideal conductivity, the intensity amplitude of the field in the point located at distance $s \ll d$ from the vertex of angle can be calculated by

$$H_m(M) = \frac{i_m}{\pi} \frac{\beta}{d^{\beta} s^{1-\beta}}, \tag{3.28}$$

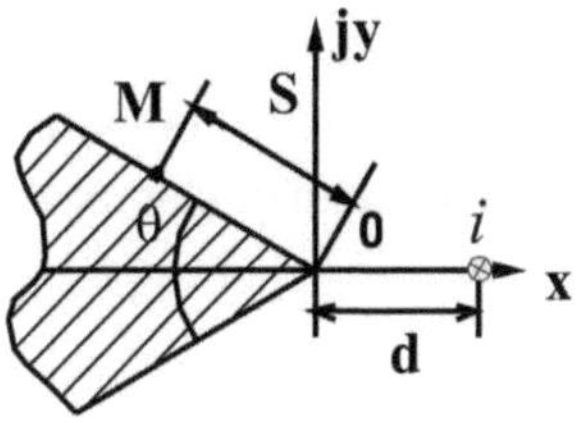

Fig. 3.4: Conductor with a current located near the conductor with an acute-angled rim.

where $\beta = 1 + \alpha = \pi/(2\pi - \theta)$, i_m is the current amplitude. In this example the number C_m takes the following value:

$$C_m = \frac{i_m}{\pi}\beta\, d^{-\beta}. \tag{3.29}$$

The numerical estimation allows us to determine the magnetic field intensity at point ***O*** at a given distance to the conductor and the current in it along with a given circular frequency ω and material conductivity ρ.

Further, the number γ can be determined using formulas (3.27) and (3.29):

$$\gamma = \frac{\pi}{\beta}\left(\frac{H(O)d}{i}\right)\left(\frac{\Delta}{d}\right)^{1-\beta}, \tag{3.30}$$

where $\Delta = \sqrt{2\rho/(\mu_0\omega)}$.

Table 3.3 presents the results of estimation made for the configuration shown in Figure 3.4. In this estimation we take the angle $\theta = \pi/2$, the size $d = 0.2 \cdot \sqrt{2} = 0.28$ m, the current density $3.14 \cdot 10^3$ A, and the intensity in the vertex of angle was calculated for four frequency values, the penetration depth calculating for a material with specific resistance $10^{-6}\Omega \cdot$ m.

Table 3.3: Dependence of γ and ξ on the frequency.

f	Δ/d	γ	ξ
500	$7.92\ 10^{-2}$	0.943	2.2
1000	$5.58\ 10^{-2}$	0.947	2.19
2000	$3.95\ 10^{-2}$	0.959	2.19
4000	$2.79\ 10^{-2}$	0.966	2.185
8000	$1.97\ 10^{-2}$	0.968	2.18

The data presented show that with an error less than 1 % the number γ remains constant when $\Delta/d \leq 0,1$. Thus, it is plausible that the calculated approximated value $\gamma = 0.97$ correlated with the asymptotic dependence for the intensity in the vertex of rectangular angle. Checking the calculations have shown that errors of the numerical estimation do not affect the value of the number γ, determined with the mentioned accuracy. Thus the approximate value of the magnetic field intensity in the apex of rectangular angle at a small penetration depth can be calculated with help of the following approximation formula:

$$H_m(O) \approx 0.97\, C_m \Delta^{-1/3}. \tag{3.31}$$

The amplitude of current density in the point $\boldsymbol{O}$ can be calculated based on considerations of dimension:

$$\delta_m(O) = \xi\, H_m(O) / \Delta, \tag{3.32}$$

where the number ξ depends on the angle θ. In a case $\theta = \pi/2$ according to data listed in Table 3.2 we have $\xi \approx 2.18$.

Thus the amplitude of current density at point $\boldsymbol{O}$ can be calculated by a formula containing the constant appearing in the expression for the field intensity of ideal conductor in vicinity of this point:

$$\delta_m(O) = \eta\, C_m \Delta^{\alpha-1} = \eta\, C_m \left(\frac{\mu_0\, \omega}{2\rho_0}\right)^{\frac{1-\alpha}{2}}, \tag{3.33}$$

where $\eta = \xi\,\gamma$.

In contrast to the poloidal field, in which the current is parallel to the edge of dihedral angle with vertex in the point $\boldsymbol{O}$ (Figure 3.5), in the toroidal field the vector of the magnetic induction is parallel to the edge. Along with this, the current density at point $\boldsymbol{O}$ is equal to zero if the condition $\theta < \pi$ is satisfied. At this point the electromagnetic force becomes zero, and the Joule heating is absent. As an example to illustrate the preceding, consider the result of calculating diffusion into the rectangular edge ($\theta = \pi/2$) with an instantaneous switching of the external field, wherebx induction $B_z = B_e$ [9]. The induction distribution in the region $x > 0$, $y > 0$ (Figure 3.5) is described by the following dependence on the coordinates:

$$B_z(x, y, t) = B_e \left[1 - \mathrm{erf}\left(\frac{x}{2d_0}\right) \mathrm{erf}\left(\frac{y}{2d_0}\right)\right], \tag{3.34a}$$

where erf $[x/(2d_0)]$ is the probability integral, $d_0 = (\rho_0 t/\mu_0)^{1/2}$. The current density varies lengthwise the bisector of an angle according to the following law:

$$\delta(M) = \frac{B_e\sqrt{2}}{\mu_0\sqrt{\pi}\, d_0}\, \mathrm{erf}\left(\frac{r}{2\sqrt{2}\, d_0}\right) \mathrm{erf}\left(-\frac{r^2}{8\, d_0^2}\right). \tag{3.34b}$$

Here r is the distance from the vertex of angle to point $\boldsymbol{M}$ (Figure 3.5). The current density equals zero in the vertex of angle and decreases with $r \to \infty$. The maximum value of the current $\delta_m \approx 0.335\, B_e/(\mu_0 d_0)$ density is reached at the point with coordinate $r_m \approx 1.75\, d_0$.

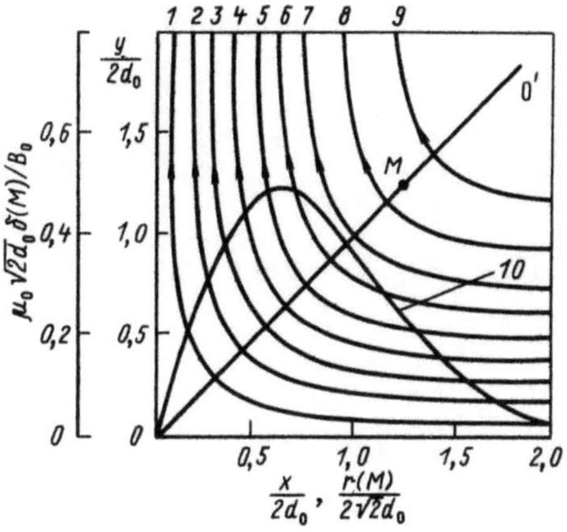

Fig. 3.5: The current density in the vicinity of the rectangular edge of the conductor with the instant switching-on of the external field of the induction $B_e = B_z$: (1–9) the lines of current (between the adjacent lines the current $\Delta J = 0.1B_e/\mu_0$); is passing); (10) the dependence for the current density on the line 00″.

3.5 Examples of the diffusion of the uniform pulse electromagnetic field into a medium with constant conductivity

Formulas (3.12) can be used to calculate the current distribution in a skin layer with different regimes of induction at the boundary. The general expressions for the induction and current density can be obtained using the Duhamel integral:

$$B(x, t) = \frac{2}{\sqrt{\pi}} \int_0^t \left[\frac{\partial}{\partial \tau} B(0, \tau)\right] \left(\int_\theta^\infty e^{-y^2} dy\right) d\tau, \tag{3.35a}$$

where $\theta = (x/2)[\rho_0(t-\tau)/\mu_0]^{-1/2}$,

$$\delta(x, t) = \frac{1}{\sqrt{\pi \mu_0 \rho_0}} \int_0^t \frac{\partial B(0, \tau)}{\partial \tau} \exp\left(-\frac{x}{2}\sqrt{\frac{\mu_0}{\rho_0 (t-\tau)}}\right) \frac{d\tau}{\sqrt{t-\tau}}, \tag{3.35b}$$

$$\delta(0, t) = \frac{1}{\sqrt{\pi \mu_0 \rho_0}} \int_0^t \frac{\partial B(0, \tau)}{\partial \tau} \frac{d\tau}{\sqrt{t-\tau}}. \tag{3.35c}$$

In the case of the step pulse ($B_e = 0$ at $t < 0$, $B_e = B_1 = \text{const}$ at $t > 0$) we have

$$\bar{B}(x, p) = \frac{B_1}{p} \exp\left(-\sqrt{\mu_0 p/\rho_0}\, x\right), \tag{3.36a}$$

$$B(x, t) = B_1\left[1 - \frac{2}{\sqrt{\pi}} \int_0^\chi e^{-y^2} dy\right] = B_1 \operatorname{erfc}(\chi), \tag{3.36b}$$

where $\chi = x\sqrt{\mu_0/(4\rho_0 t)}$. Deep in the conductor ($x \to \infty$) the induction damps according the law

$$B(x, t) = \frac{2B_1}{x}\sqrt{\frac{\rho_0 t}{\pi\mu_0}}\exp\left(\frac{-\mu_0 x^2}{2\rho_0 t}\right). \tag{3.36c}$$

For the boundary condition $B_e = B_1$ at $0 < t < t_0$, $B_e = 0$ at $t > t_0$ the solution takes the simple form

$$B(x, t) = B_1\left[\operatorname{erfc}\left(\frac{x}{2}\sqrt{\frac{\mu_0}{\rho_0 t}}\right) - \operatorname{erfc}\left(\frac{x}{2}\sqrt{\frac{\mu_0}{\rho_0(t - t_0)}}\right)\right]. \tag{3.37}$$

In the limiting case where $t_0 \ll t$, one can assume that the induction has the form of a δ-function: $B_e = B_1 t_0 \delta(t)$, $\bar{B}_e(p) = B_1 t_0$. Then

$$\bar{B}(x, p) = B_0 t_0 \exp\left(-\sqrt{\frac{\mu_0 p}{\rho_0}}x\right) = \frac{B_1 t_0 x}{2t}\sqrt{\frac{\mu_0}{\pi\rho_0 t}}\exp\left(-\mu_0\frac{x^2}{4\rho_0 t}\right). \tag{3.38}$$

In this example the induction distribution deep inside the conductor is non-monotonous: the induction reaches the maximum value $B_m = B_1(t_0/t)(2\pi e)^{-1/2}$ at point $x_m = \sqrt{2\rho_0 t/\mu_0}$.

For the boundary condition $B_e = B_1 (t/t_0)^\alpha$, where $\alpha > 0$, the current density at the surface is described by the formula [3]

$$B(0, t) = \frac{B_1}{\sqrt{\rho_0\mu_0 t}}\cdot\frac{\Gamma(1+\alpha)}{\Gamma(\alpha + 1/2)}. \tag{3.39}$$

When the external field increases exponentially, $B_e = B_1\exp(t/t_0)$, the solution obtained above by the other method (see equation (3.9))

$$B(x, t) = B_1\exp\left(\frac{t}{t_0} - \frac{x}{x_0}\right) = B_e\exp\left(-\frac{x}{x_0}\right),$$

is valid, where $x_0 = (\rho_0 t_0/\mu_0)^{1/2}$. In this example the depth of the field penetration, which is defined as the ratio of the flux in the conductor to the induction at the surface B_e, is a constant value.

Let us next consider a case of practical importance, where the switching field varies according to the sinusoidal law $B_e(t) = B_{em}\sin\omega t$, $t \geq 0$. In this case

$$\bar{B}_e(p) = B_{em}\frac{\omega}{p^2 + \omega^2},$$

$$B(x, t) = B_{em}\exp\left(-\frac{x}{\Delta_0}\right)\sin\left(\omega t - \frac{x}{\Delta_0}\right) + \Psi_B, \tag{3.40a}$$

where Ψ_B can be described in two equivalent forms [12–14]:

$$\Psi_B = -\frac{2B_{em}}{\pi}\int_0^{\gamma_0}\sin\left(\omega t - \frac{x^2}{2\Delta_0^2\gamma^2}\right)\exp(-\gamma^2)d\gamma, \quad \gamma_0 = \frac{x}{\Delta_0\sqrt{2\omega t}}, \tag{3.40b}$$

$$\Psi_B = \frac{B_{em}\omega}{\pi}\int_0^{\infty}\exp(-\lambda t)(\lambda^2 + \omega^2)^{-1}\sin\left(x\sqrt{\frac{\mu_0\lambda}{\rho_0}}\right)d\lambda. \tag{3.40c}$$

In formula (3.40a) the first term corresponds to the regime of stationary sinusoidal oscillations. The term Ψ_B characterizes the regime of switching of sinusoidal field from "zero" in contrast to the regime of stationary oscillations.

Figure 3.6 shows the dependences $B(x, t)$ plotted according to [13, 14].

In the early stage of the process ($\omega t < \pi/2$), the induction distribution in the nonstationary regime differs noticeably from that in the regime of stationary sinusoidal oscillations. In later stages ($\omega t > \pi/2$) the difference is pronounced

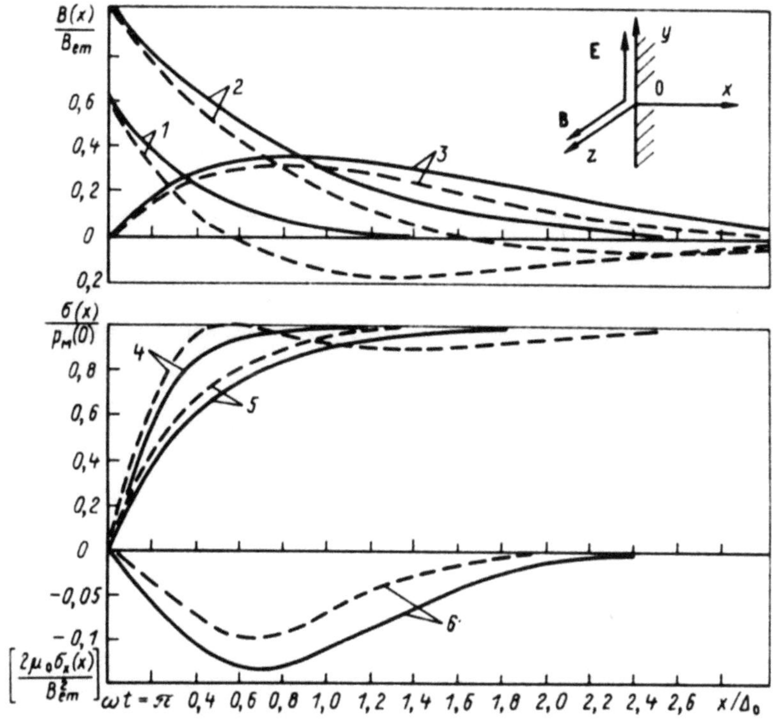

Fig. 3.6: The penetration of a sinusoidal field into the conductive half-space. *Solid lines*: switching of the field from zero, $B_e = 0$ $(t < 0)$; $B_e = B_{em}\sin(\omega t)$ $(t > 0)$. *Dotted lines*: the stationary regime: (1) $B(x)$, $\omega t = 0,6$; (2) $B(x)$, $\omega t = \pi/2$; (3) $B(x)$, $\omega t = \pi$; (4) $\sigma_x(x)/p_M(0)$, $\omega t = 0,6$; (5) $\sigma_x(x)/p_M(0)$, $\omega t = \pi/2$; (6) $2\mu_0\sigma_x(x)/B_{em}^2$, $\omega t = \pi$.

only for points where $x > \Delta_0$. It is of interest to note that in the nonstationary regime (switching of the sinusoidal field from zero) the monotone damping of $B(x,t)$ at $x \to \infty$ occurs, in contrast to the oscillating behavior of damping in a a stationary case. The asymptotic behavior of the function Ψ_B was studied in [15], where it was shown that at the condition $x \gg \Delta_0$ and $\omega t \gg x/\Delta_0$

$$\Psi_B \approx \frac{-x}{2\sqrt{\pi}\Delta_0}(\omega t)^{-3/2}\exp\left(-\frac{x^2}{2\Delta_0^2\omega t}\right). \tag{3.41}$$

Initially with increasing ωt, the function $\Psi_B(x, t)$ increases to the maximum $0.46B_{em}(\Delta_0/x)^2$ and then decreases as $(\omega t)^{-3/2}$. Thus, the induction amplitude at $x \to \infty$ decreases in a nonstationary solution as $(\Delta_0/x)^2$ and not as $\exp(-x/\Delta_0)$. The nonstationary behavior of the diffusion process revealed itself vividly in the behavior of the current density in the initial stage of the process. During the first half-period, the current density can be calculated using a formula from [16]:

$$\delta\,(0, t) = \frac{\sqrt{2}B_{em}}{\mu_0\Delta_0}\left[\cos(\omega t)\cdot C(\omega t) + \sin(\omega t)\cdot S(\omega t)\right], \tag{3.42}$$

where *C and S* are Frenel integrals. The curve ***1*** in Figure 3.7 shows that initially the function $\delta\,(0,t)$ differs noticeably from the stationary value, which can be obtained in a limiting case $\omega t \gg 1$, when $C(\omega t) = S(\omega t) = 1/2$. For instance, at the moment of maximum induction of the external field we have $\delta\,(0, t) = 0,86B_{em}/(\mu_0\Delta_0)$ instead of $B_{em}/(\mu_0\Delta_0)$ corresponding to a regime of stationary oscillations.

In the process of diffusion of the pulse field in the skin layer, the sign of the volume electromagnetic forces can change. This occurs after the maximum of the external field is reached, when the induction inside the conductor can exceed by

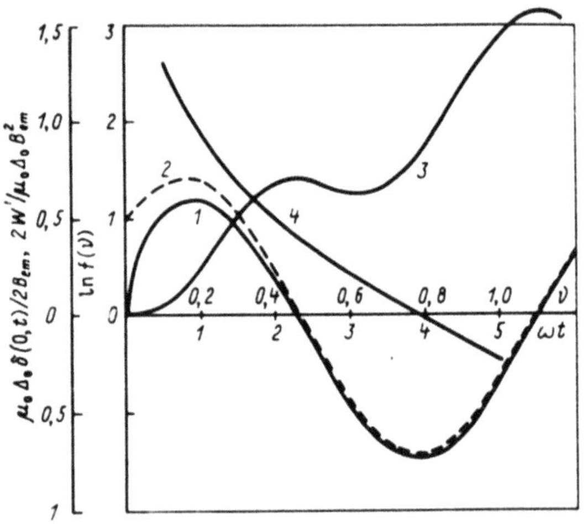

Fig. 3.7: The current density at the conductor surface and the energy characteristics of diffusion of the sinusoidal field into the conductor: (1) $\mu_0\Delta_0\delta\,(0, t)/(2B_{em})$; (2) the same in the stationary regime; (3) $2W'/(\mu_0\Delta_0B_{em}^2)$; 4 – $\ln f(\nu)$.

absolute value the induction of the external field. The characteristic parameter of this process is the difference of the magnetic pressures $\sigma_x = P_M(0) - P_M(x)$, where $P_M(0) = B_e^2/(2\mu_0)$, $P_M(x) = B^2(x)/(2\mu_0)$. The curves in Figure 3.6 show in particular that at the moment when the induction of the external field takes the value of zero ($\omega t = \pi$), σ_x is negative, and the maximum of ratio $|\sigma_x \cdot 2\mu_0/B_{em}^2|$ becomes close to 0.15.

The diffusion of fields, varying by damping sinusoidal law $B_e = B_{em}\exp(-\beta\ t)$ $\sin(\omega\ t)$, was considered by Bronshtein [13] and Bondaletov [14]. The quite cumbrous calculation formula taken from their works can be found in [9]

3.6 Energy generation and heating a medium in the case of diffusion of the pulse magnetic field into the conductor

In the following we shall consider the energy generation in a skin layer und the condition that the boundary of the conductor is a plane, or curved so insignificantly that we can assume the diffusion process of the pulse magnetic field to be one-dimensional.

The power flow through the boundary of the conductor can be calculated using Poynting's formula $P' = E(0,t)B(0,t)/\mu_0$, and all the energy generated in the conductor can be further calculated by integrating over time. A characteristic example is diffusion of a field with induction which varies according to the sinusoidal law without damping. This process is described by formula (3.40). The time dependence of the energy supplied per unit surface of the conductor in this regime is presented in Figure 3.7. In this case all the energy is built up in time and increases indefinitely at $t \to \infty$. This dependence can be used when calculating the energy balance in the condenser discharge circuit containing the conductors with the pronounced skin effect. At the moment of the first maximum of the external field ($\omega t = \pi/2$) we have $W' = 0.54 B_{em}^2 \Delta_0/\mu_0$. Included here is the energy of losses $W_R' = 0,31 B_{em}^2 \Delta_0/\mu_0$ and the energy of magnetic field $W_M' = 0,23 B_{em}^2 \Delta_0/\mu_0$. For comparison note that in a stationary regime the enhancement of energy of losses over the time of changing of the external field induction from zero to maximum consists of $\Delta W_R' = (\pi/8) B_{em}^2 \Delta_0/\mu_0 = 0.39 B_{em}^2 \Delta_0/\mu_0$. The maximum value of the magnetic field energy in the skin layer is $W_M' = (1 + 1/\sqrt{2}) B_{em}^2 \Delta_0/(8\mu_0) = 0.21 B_{em}^2 \Delta_0/(\mu_0)$. Knowing the equation for the current density transformed after Laplace (see formulas (3.12)) it is possible, using the Parceval's formula, to calculate the entire amount of energy released in the conductor over the time of the existence of the external field. Using the same formula one can also calculate the energy generated in the surface element of the conductor [9].

In the case of the field changing according to the law $B_e = B_{em}\exp(-\beta t)$ $\sin(\omega t)$ we have

$$W_R' = \int_0^\infty \int_0^\infty \rho_0 \delta^2(x, t)\, dt\, dx = \frac{\rho_0}{2\pi} \int_0^\infty dx \int_{-\infty}^\infty |\delta(js)|^2\, ds, \qquad (3.43a)$$

where

$$\bar{\delta}(js) = \frac{B_{em}\omega}{\sqrt{\mu_0\rho_0}} \left[\frac{p^{1/2}\exp\left(-x\sqrt{\mu_0 p/\rho_0}\right)}{(p+\beta)^2 + \omega^2} \right]_{p=js}. \qquad (3.43b)$$

As a result of transformation we obtain

$$W_R' = \frac{B_{em}^2 \Delta_0}{8\sqrt{2}\mu_0} \frac{1}{\nu\sqrt{1+\nu^2}} \left[\left(\sqrt{1+\nu^2}+1\right)^{1/2} - \left(\sqrt{1+\nu^2}-1\right)^{1/2} \right] = \frac{B_{em}^2 \Delta_0}{8\sqrt{2}\mu_0} f(\nu),$$

where $\nu = \beta/\omega$. The function $\ln f(\nu)$ is shown in Figure 3.7.

We already mentioned that the energy released in the conductor due to switching of sinusoidal nondamping field increases indefinitely at $t \to \infty$. However, the quantity $\Delta W'_R$ remains finite, and it is equal to the difference of energies released in the conductor in the transient regime and in the regime of stationary oscillations at $t > 0$. The calculations with the help of Parceval's formula give the next equation for $\Delta W_R'$:

$$\Delta W_R' = -\frac{B_{em}^2 \Delta_0}{2\pi\mu_0} \int_0^\infty \left(\sqrt{y} - \frac{y^2 - 3}{4} \right) \frac{dy}{(y^2-1)^2} = -\frac{B_{em}^2 \Delta_0}{16\mu_0}. \qquad (3.44)$$

In the case where the unipolar pulse has the form of the one-half period of sinusoid ($B_e = B_m \sin(\omega t)$ at $0 < \omega t < \pi$; $B_e = 0$ at $t < 0$,one can calculate the total energy generated in a skin layer, using Parceval's formula. The form of the external field induction, according to Laplace, is $B_e(p) = \omega B_m(\omega^2 + p^2)^{-1}$ $[1 + \exp(-p\pi/\omega)]$. Further, using formula (3.43a) we obtain after transformations the next equations for the energy of Joule losses:

$$W_R' = \frac{B_{em}^2}{\mu_0 \pi} \Delta_0 \int_0^\infty \frac{\sqrt{x}(1 + \cos(\pi x))\, dx}{(1 - x^2)^2} = 1.997 \frac{B_{em}^2}{\mu_0 \pi} \Delta_0. \qquad (3.45)$$

Along with the total energy, the volume energy density released in different parts of the conductor during the field diffusion is also of considerable interest. In particular, for the surface element of the medium ($x = 0$) and sinusoidal field rise we have

$$\Delta q'(0, t) = \int_0^t \delta^2(0, t)\, \rho_0 dt = \frac{B_{em}^2}{\mu_0} \int_0^{\omega t} f^2(\omega t)\, d(\omega t). \qquad (3.46)$$

As seen from this formula, the value $\Delta q'(0,t)$ depends only on the phase of sinusoid ωt, and it does not depend either on the frequency or on the specific resistivity.

The lack of dependence for $\Delta q'(0,t)$ on ρ_0 is determined by the fact that the power of losses is $\delta^2\rho_0$, where the first factor is proportional to $\rho_0^{-1/2}$ because of an increase in the penetration depth with increasing ρ_0. Similarly, the energy generation $(1/\omega)\int_0^{\omega t}\rho_0\delta^2(\omega\, t)d(\omega\, t)$ over the phase interval ωt does not depend on ω, since δ is proportional to $\omega^{1/2}$. Thus, irrespective of the frequency and conductivity, for the moments of maximum induction and its transition through zero ($\omega t = \pi/2, \pi, 2\pi$) we have

$$\begin{aligned} \Delta q'(0,\ \pi/2) &= 1,63B_{em}^2/(2\mu_0), \\ \Delta q'(0,\ \pi) &= 2,18B_{em}^2/(2\mu_0), \\ \Delta q'(0,\ 2\pi) &= 5,51B_{em}^2/(2\mu_0). \end{aligned} \tag{3.47}$$

Here the values of the energy released in a unit volume have been calculated using formulas (3.42). They are compared to the maximal value of the external field energy density. For the adiabatic heating of the medium the corresponding temperature increments $\Delta T = \Delta q'(0,t)/C_V$ for copper (the volume specific heat $C_V = 3,7\cdot 10^6\ \mathrm{J\cdot m^{-3}K^{-1}}$) are $\Delta T(0,\ \pi/2) = 0,185B_{em}^2$; $\Delta T(0,\ \pi) = 0,234B_{em}^2$; $\Delta T(0,\ 2\pi) = 0,595B_{em}^2$ [°K, T]. In particular, the surface temperature in the field of induction 74 T rises by 1000 °K to the moment of the first induction maximum at the boundary.

For a pulse having the "standard" form of a half-period sinusoidal, the Laplace transformed dependence for B_e has a form

$$\bar{B}_e(p) = B_{em}\frac{(1+\exp(-p\pi/\omega))\,\omega}{p^2+\omega^2}, \tag{3.48a}$$

and the Fourier transforms $B(x,s)$and $\delta(0,s)$ are

$$B(x,s) = B_{em}\frac{[1+\exp(-js\pi/\omega)]\ \omega}{\omega^2-s^2}\exp\left(-x\sqrt{j\mu_0 s/\rho_0}\right) \tag{3.48b}$$

$$\delta(0,s) = B_{em}\sqrt{j\mu_0 s/\rho_0}\,\frac{[1-\exp(-js\pi/\omega)]}{\omega^2-s^2}\omega \tag{3.48c}$$

We can evaluate the volume energy density using Parceval's formula:

$$\Delta q'\ (0,\ \infty) = \rho_0\int_0^\infty \delta^2\ (0,\ t)\ dt = \frac{\omega^2 B_m^2}{2\pi\mu_0}\int_{-\infty}^{\infty}\frac{\left|s\,[1+\exp(-js\pi/\omega)]^2\right|ds}{(\omega^2-s^2)^2},$$

or

$$\Delta q'(0,\infty) = \frac{2B_m^2}{\pi\mu_0}\int\limits_0^\infty \frac{x(1+\cos(\pi x))\,dx}{(1-x^2)^2} = \frac{B_m^2}{\mu_0}\left[Si(\pi) - \frac{\pi}{2}\right] = 2.42\frac{B_m^2}{2\mu_0}. \quad (3.48d)$$

$$\Delta T(0,\infty) = 6\cdot 10^{-7}\,B_m^2/(2\mu_0) = 3.8\cdot 10^{-13}J_m^2, \quad (3.38e)$$

where J_m is the amplitude of the linear current density.

A comparison of the obtained result with the above data for $\Delta q'(0,\pi)$ shows that after external field extinction the extra heating of the conductor caused by damping of eddy currents is $0.24B_m^2/2\,\mu_0$.

If the external field is the exponentially damped sinusoid $B_e = B_{em}\exp(-\beta t)\sin(\omega t)$ we have the following expression for $\Delta q'(0,\infty)$ derived by Karpova and Titkov [17] with the help of Parceval's formula:

$$\Delta q'(0,\infty) = \frac{B_{em}^2}{2\mu_0}\frac{\omega}{2\beta}\left[\frac{1}{2} - \frac{1}{\pi}arctg\left(\frac{\beta^2-\omega^2}{2\beta\omega}\right)\right]. \quad (3.49)$$

The above formuls can be used to evaluate conductor heating if we ignore thermal conductivity. The validity of this assumption is revealed by comparing the penetration depths of electromagnetic and thermal waves. The equation for the temperature can be obtained by changing B by T and D_M by $D_T = \lambda/C_V$ in equation (3.7), where D_T is "the coefficient of the temperature-conductivity" (by analogy with the coefficient of diffusion), λ is the coefficient of thermal conductivity, and C_V is the volume specific heat. For similar behavior of the temperature and induction at the conductor surface, the ratio of the penetration depths of the field and temperature is, according to equation (3.7),

$$\frac{\Delta_B}{\Delta_T} = \sqrt{\frac{D_B}{D_T}} = \sqrt{\frac{\lambda\,\mu_0}{C_V\rho_0}} \approx 10 \quad \text{(for copper)}. \quad (3.50)$$

From the above estimation it follows that in approximate calculations we can consider the conductor heating to be adiabatic.

Earlier, for the poloidal field, we derived relations allowing the calculation of the amplitude of the high-frequency current density in the vertex of a dihedral angle limited the conductor (point ***O***, Figure 3.3a).

Further calculations make it possible to determine the power of Joule losses at point ***O*** and estimate the energy generated in the unit volume when switching the pulse magnetic field, without regard for conductivity. In this way the estimated value of temperature in the angular point can be obtained.

The volume density of the energy generated at point ***O*** is given by the integral

$$\Delta q'(O) = \int\limits_0^\infty \delta^2(O,t)\rho_0 dt \quad (3.51)$$

If the Fourier transform of $C_m(\omega)$ is known, for further calculations one can use Parcevale's formula, similar to the procedure of calculations of the energy generation in the skin layer of the conductor with a flat boundary:

$$\Delta q'(O) = \frac{1}{2\pi}\int_{-\infty}^{\infty} |\delta_m(O,\omega)|^2 \rho_0 d\omega, \tag{3.52}$$

where $\delta_m(O,\omega)$ is determined by equation (3.33). Using the previous formulas we have

$$\Delta q'(O) = \frac{1}{2\pi}\eta^2\rho_0\left(\frac{\mu_0}{2\rho_0}\right)^{1-\alpha}\int_{-\infty}^{\infty}\left|C_m^2(\omega)\,\omega^{1-\alpha}\right| d\omega, \tag{3.53}$$

Consider now a pulse in a form of a single half-wave sinusoid with circular frequency ω_0. In this case $C(t) = D\sin(\omega_0 t)$ at $t < \pi/\omega$, and $C(t) = 0$ at $t > \pi/\omega$. The Laplace transformed dependence for $C(t)$ has the form in (3.48a).

The Fourier transform is

$$C_m(\omega) = D\frac{[1+\exp(-j\omega\pi/\omega_0)]\,\omega_0}{\omega_0^2-\omega^2}. \tag{3.54}$$

Thus we obtain the following equation for the energy density:

$$\begin{aligned}\Delta q'(O) &= \frac{D^2\eta^2}{2\pi}\left(\frac{\mu_0}{2}\right)^{1-\alpha}\rho_0^\alpha\omega_0^2\int_{-\infty}^{\infty}\frac{[1+\cos(\pi\,\omega/\omega_0)]\,s^{1-\alpha}ds}{(\omega_0^2-\omega^2)^2}\\ &= \frac{1}{\pi}D^2\eta^2\mu_0\left(\frac{2\rho_0}{\mu_0\omega_0}\right)^\alpha\int_0^{\infty}\frac{(1+\cos(\pi x))\,x^{1-\alpha}dx}{(1-x^2)^2}.\end{aligned} \tag{3.55}$$

In the particular case of the rectangular angle, when $\theta = \pi/2$, we have

$$\int_0^{\infty}\frac{(1+\cos(\pi x))\,x^{4/3}dx}{(1-x^2)^2} \approx 1.98. \tag{3.56}$$

In this case $\eta = 2.11$, and

$$\Delta q'(O) \approx 2.82\,D^2\mu_0\left(\frac{2\rho_0}{\mu_0\omega_0}\right)^{-1/3}. \tag{3.57}$$

Let consider further the field near the boundary of the two flat buses with gap h (Figure 3.8). Let the bus thickness g be much more than the gap between them. The heating of the bus rim can be calculated by the above mentioned formula, using expressions describing the field intensity in the vicinity of point $\boldsymbol{A}$ (the vertex of angle, Figure 3.8).

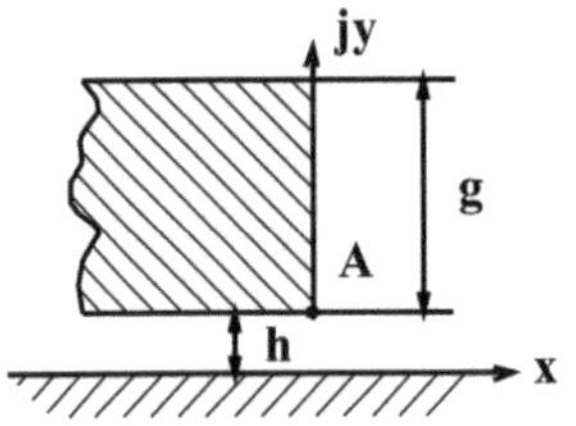

Fig. 3.8: For calculation of the energy generation on the conductor rim (at point A).

One can find these expressions using the method of conformal mapping with the help of the Kristoffel–Schwarz's formula [11]:

$$D = H_{\infty}\left(\frac{2h}{3\pi}\right)^{1/3}, \tag{3.58}$$

where H_{∞} is the field amplitude between buses far from the boundary. Further, we obtain the following expressions for the intensity amplitude and current density at point $\boldsymbol{A}$ in the field varying by the sinusoidal law with circular frequency, and for the energy density of Joule heating in the field varying as a half-wave sinusoid with the same frequency:

$$\begin{aligned} H_m(A) &\approx 0.58\,H_{\infty}\left(\frac{h}{\Delta_0}\right)^{1/3}, \\ \delta_m(A) &\approx 1.27\,\frac{H_{\infty}}{\Delta_0}\left(\frac{h}{\Delta_0}\right)^{1/3}, \\ \Delta q'(A) &\approx 2.03\,q'_M\left(\frac{h}{\Delta_0}\right)^{2/3}, \end{aligned} \tag{3.59}$$

where $q'_M = \mu_0 H_{\infty}^2/2$ is the maximal value of the energy density of the magnetic field far from the bus edge, and $\Delta_0 = (2\rho_0/\mu_0\omega_0)$ is the penetration depth of a sinusoidal field. In the particular case at $h/\Delta_0 = 10$ we have: $H_m(A) \approx 1.24\,H_{\infty}$, $\delta_m(A) \approx 2.73\,(H_{\infty}/\Delta_0)$, $\Delta q'(A) \approx 9.28\,q'_M$.

For comparison note that far from the edge of the buses the volume density of energy generated in the surface layer of the conductor is $2.42q'_M$. These calculations confirm that the field enhancement near the rim results in more heating at point A than in the region of uniform field and allow estimating this effect.

3.7 Heating of a conductor with a current in an external magnetic field

Let us consider the plane conducting layer of thickness d with the current, of which the linear density is $\boldsymbol{J}$ (Figure 3.9). Let us also assume that this layer is in the external field with induction $\boldsymbol{B}_1$ directed in parallel with $\boldsymbol{J}$. On the external

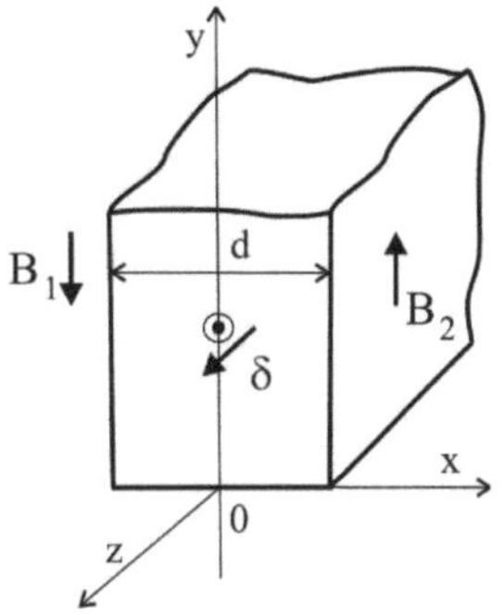

Fig. 3.9: For the calculation of heating of the conductor placed in the external field, directed along the conductor current.

boundary of the layer (in points $x = \pm d/2$) the induction takes the value

$$B_z\left(\frac{d}{2}\right) = B_z\left(-\frac{d}{2}\right) = B_1, \tag{3.60a}$$

$$B_y\left(-\frac{d}{2}\right) = B_2 = -\mu_0 J/2, \tag{3.60b}$$

$$B_y\left(\frac{d}{2}\right) = \mu_0 J/2 = -B_2. \tag{3.60c}$$

The boundary values of induction are the given temporal functions. The initial value of induction in the conductor is assumed to be zero. Here we shall consider the linear problem of the field diffusion into the layer d, i.e., the conductivity is believed to be constant.

The induction components, subjected to Laplace transformation, satisfy the diffusion equation (3.12a), the solution of which enables us to find for the each field the current density on the layer boundaries, where it achieves a maximal value:

$$\bar{\delta}_y\left(-\frac{d}{2}, p\right) = \frac{\bar{B}_1(p)}{\mu_0}\sqrt{\frac{p}{D_M}}\,th\left(\frac{d}{2}\sqrt{\frac{p}{D_M}}\right), \tag{3.61a}$$

$$\bar{\delta}_z\left(-\frac{d}{2}, p\right) = \frac{\bar{B}_2(p)}{\mu_0}\sqrt{\frac{p}{D_M}}\,cth\left(\frac{d}{2}\sqrt{\frac{p}{D_M}}\right), \tag{3.61b}$$

where $D_M = \rho/\mu_0$ is the coefficient of the magnetic field diffusion.

The action integral determining the heating in time of pulse duration takes a maximal value on the boundaries:

$$S\left(\frac{d}{2}\right) = S\left(-\frac{d}{2}\right) = \int_0^\infty \left[\delta_z^2\left(\frac{d}{2}, t\right) + \delta_y^2\left(\frac{d}{2}, t\right)\right] dt. \tag{3.62a}$$

Further on the Parseval formula is useful:

$$S\left(\pm\frac{d}{2}\right) = \frac{1}{2\pi}\int_{-\infty}^{\infty}\left[\left|\bar{\delta}_z\left(\frac{d}{2}, js\right)\right|^2 + \left|\bar{\delta}_y\left(\frac{d}{2}, js\right)\right|^2\right] ds, \tag{3.62b}$$

where

$$\bar{\delta}_z\left(\frac{d}{2}, js\right) = \frac{1}{\mu_0}\left(\frac{d\bar{B}_y(x, js)}{dx}\right)_{d/2}, \quad \bar{\delta}_y\left(\frac{d}{2}, js\right) = -\frac{1}{\mu_0}\left(\frac{d\bar{B}_x(x, js)}{dx}\right)_{d/2}, \quad j = \sqrt{-1}.$$

Let us consider the process of switching of the field when the temporal dependence of boundary values B_1 and B_2 is a sine half-wave with a circular frequency ω. In this case

$$\bar{B}_{1,2}\left(\frac{d}{2}, p\right) = \omega\, B_{m1,2}\left(\omega^2 + p^2\right)^{-1}\left[1 + \exp\left(-\frac{p\pi}{\omega}\right)\right], \tag{3.63}$$

where $B_{m1,2}$ are the induction amplitudes on the boundaries. On rearrangement, we come to the following expressions for the increment of the volume density of the energy on the layer boundary [19]:

$$\begin{aligned}\Delta q_1 &= \frac{B_{m1}^2\rho}{\pi\mu_0^2 D_M}\int_0^{\infty}\left[\frac{\omega\sqrt{js}}{\omega^2 - s^2}\left(1 + \exp\left(-js\pi/\omega\right)\right) th\left(\frac{d}{2}\sqrt{\frac{js}{D_M}}\right)\right]^2 ds \\ &= \frac{B_{m1}^2}{\pi\mu_0}\int_0^{\infty}\frac{g\left(1 + \cos\left(\pi g\right)\right)}{\left(1 - g^2\right)^2}\left[1 - \frac{2\cos\left(k\sqrt{g}\right)}{ch\left(k\sqrt{g}\right) + \cos\left(k\sqrt{g}\right)}\right] dg;\end{aligned} \tag{3.64a}$$

$$\Delta q_2 = \frac{B_{m2}^2}{\pi\mu_0}\int_0^{\infty}\frac{g\left(1 + \cos\left(\pi g\right)\right)}{\left(1 - g^2\right)^2}\left[1 + \frac{2\cos\left(k\sqrt{g}\right)}{ch\left(k\sqrt{g}\right) - \cos\left(k\sqrt{g}\right)}\right] dg. \tag{3.64b}$$

Here the characteristic parameter $k = d/\Delta_0$ is present ($\Delta_0 = \sqrt{2\rho/(\mu_0\omega)}$ is the classical depth of the field penetration). The corresponding dependences are shown in Figure 3.10. In the limit $k \gg 1$ these curves have the asymptote $\Delta q \cdot 2\mu_0/B_{m1,2}^2 = 2.42$. This case is described in Section 3.4. It corresponds much more to the field diffusion into the layer with a thickness than with the depth of field penetration.

For a weakly-pronounced skin effect ($k \ll 1$) it is possible to use the approximate formulas

$$1 - \frac{2\cos\left(k\sqrt{g}\right)}{ch\left(k\sqrt{g}\right) + \cos\left(k\sqrt{g}\right)} \to \frac{k^2 g}{2}, \quad 1 + \frac{2\cos\left(k\sqrt{g}\right)}{ch\left(k\sqrt{g}\right) - \cos\left(k\sqrt{g}\right)} \to \frac{2}{k^2 g}.$$

Using the values of the integrals

$$\int_0^{\infty}\frac{g^2\cdot(1 + \cos(\pi g))}{(1 - g^2)^2} dg = \int_0^{\infty}\frac{(1 + \cos(\pi g))}{(1 - g^2)^2} dg = \frac{\pi^2}{4}$$

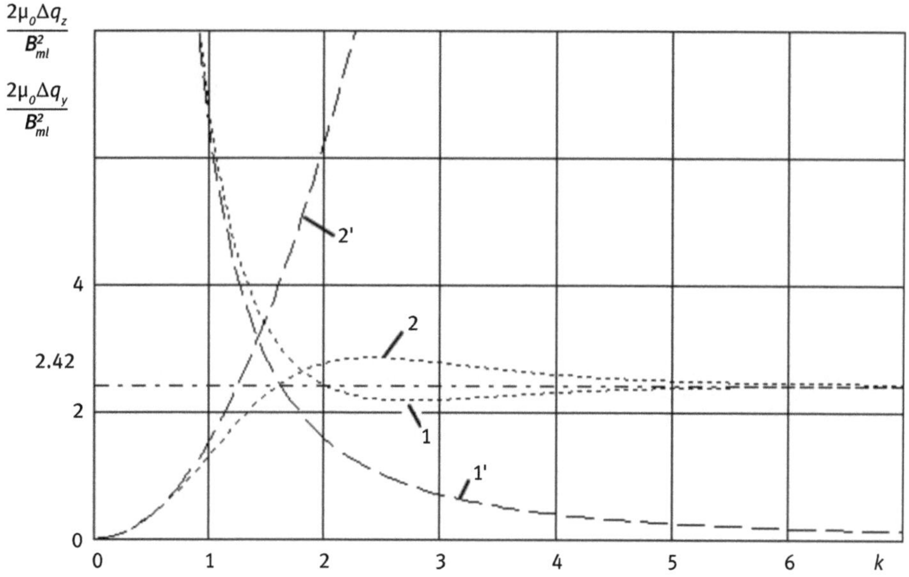

Fig. 3.10: The energy release in the conductor placed in the external field, as a function of the characteristic parameter $k = d/\Delta_0$: (1) $\frac{2\mu_0\Delta q_z}{B^2_{m2}} = f(k)$; (2) $\frac{2\mu_0\Delta q_y}{B^2_{m1}} = f(k)$; (1′) $\frac{2\mu_0\Delta q_z}{B^2_{m2}} = \frac{2\pi}{k^2}$; (2′) $\frac{2\mu_0\Delta q_y}{B^2_{m1}} = \frac{\pi}{2}k^2$.

we obtain the following dependences:

$$\frac{2\mu_0\Delta q_y}{B^2_{m1}} = \frac{\pi}{2}k^2; \quad \frac{2\mu_0\Delta q_z}{B^2_{m2}} = \frac{2\pi}{k^2}. \tag{3.65}$$

These curves describe the layer of small thickness in the case where the heating by induced currents is low, but the heating by the longitudinal current increases, due to reduction of the cross section of the conductive layer at the constant linear current density.

The obtained data can be used for calculating the heating of the multilayered winding with an inner radius R_0 and at a given shape of the current pulse. This task was solved numerically by Li and Herlach in [20], who examined the conductivity variation caused by heating. They noted a nonuniformity in a winding heating due to the influence of eddy currents. The same effect can be described analytically within the framework of the constant conductivity model. In a long winding with N equidistant layers with the same rectangular turns (d is a thickness), the induction in the gap from the left side of the n-layer obtains the value

$$B(n) = B_0\left(1 - \frac{n-1}{N}\right) \tag{3.66a}$$

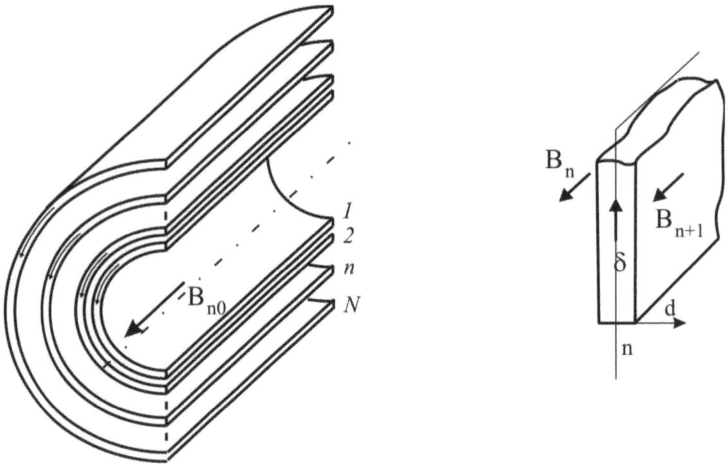

Fig. 3.11: For the calculation of heating of multilayer winding in the pulse magnetic field.

(Figure 3.11). Here B_0 is the induction on the magnet axis. In the derived formulae one should take

$$B_1 = [B(n) + B(n+1)]/2 = B_0[1 - (2n-1)/(2N)], \tag{3.66b}$$

$$B_2 = [B(n) - B(n-1)]/2 = B_0/(2N). \tag{3.66c}$$

Here the condition $Nd \ll R_0$ is taken. In the first layer (at $n = 1$)

$$(B_1)_1 = B_0\,(1 - 1/(2N))\,;(B_2)_1 = B_0/(2N). \tag{3.66d}$$

In the last layer (at $n = N$)

$$(B_1)_N = B_0/(2N) = (B_2)_N\,. \tag{3.66e}$$

The characteristic feature is the nonuniformity of heating. The current density takes the largest values on the left boundary of each layer. Its Laplace transform has the form

$$\left[\bar{\delta}_y\left(-\frac{d}{2}\right), p\right]_n = \frac{1}{\mu_0} B_{0m}\sqrt{\frac{p}{D_M}}\frac{\omega(1+\exp(\omega\, p))}{\omega^2+p^2}\left[\xi_n\, th\left(d\sqrt{\frac{p}{D_M}}\right) + \eta_n cth\left(d\sqrt{\frac{p}{D_M}}\right)\right],$$

where $\xi_n = (1 - (2n-1)/N)$, $\eta_n = 1/(2N)$.

With a weakly pronounced skin effect it is possible to use the expansions $th(d\sqrt{p/D_M}) \approx d\sqrt{p/D_M}$, $cth(d\sqrt{p/D_M} \approx (1/d)\sqrt{D_M/p}$ and calculate the energy volume density by the Parseval formula:

$$\left[\Delta q\left(-\frac{d}{2}\right)\right]_n = \frac{2B_{0m}^2}{\pi\mu_0}\int_0^{\infty}\frac{g(1+\cos(\pi g))}{(1-g^2)^2}\left|\xi_n(1+j)\frac{k}{2}\sqrt{g}+\eta_n(1-j)\frac{1}{k\sqrt{g}}\right|^2 dg$$
$$= \frac{B_{0m}^2}{2\mu_0}\left(\xi_n^2\frac{\pi k^2}{2}+\eta_n^2\frac{2\pi}{k^2}\right).$$

With the weakly-pronounced skin effect the relations for the energy increment in the first and last layers follow from the derived formulas.

Most heat release takes place in the first layer:

$$(\Delta q)_1 = (\Delta q_1)_1 + (\Delta q_2)_1 = \frac{B_0^2}{2\mu_0}\left[\left(1-\frac{1}{2N}\right)^2\frac{\pi k^2}{2}+\frac{1}{(2N)^2}\frac{2\pi}{k^2}\right], \tag{3.67a}$$

$$(\Delta q)_N = (\Delta q_1)_N + (\Delta q_2)_N = \frac{B_0^2}{2\mu_0}\frac{1}{(2N)^2}\cdot\left(\frac{\pi k^2}{2}+\frac{2\pi}{k^2}\right). \tag{3.67b}$$

An excess of energy release in the first layer over the one in the last layer is

$$(\Delta q)_1 - (\Delta q)_N = \frac{B_0^2}{2\mu_0}\left[\left(1-\frac{1}{2N}\right)^2-\frac{1}{(2N)^2}\right]\frac{\pi k^2}{2}. \tag{3.68a}$$

At the condition $N \gg 1$ we have

$$\Delta(q)_1 - (\Delta q)_N = \frac{B_0^2}{2\mu_0}\frac{\pi d^2\omega\mu_0}{4\rho}. \tag{3.68b}$$

The characteristic feature is that at $N \gg 1$ the dependence $(\Delta q)_1 = f(k)$ has a minimum in the region of the weakly pronounced skin effect, where the formula (3.67a) is valid (Figure 3.10). This enables us to find the numbers $k = k_{opt}$ and $(\Delta q)_{1,\min}$, corresponding to the minimal heating of the first layer:

$$k_{opt} = \left[N\left(1-\frac{1}{2N}\right)\right]^{-1/2}, \tag{3.69a}$$

$$(\Delta q)_{1,\min} = \frac{B_0^2}{2\mu_0}\left(1-\frac{1}{2N}\right)\cdot\frac{\pi}{N}. \tag{3.69b}$$

For the field with induction of 100 T at $N = 10$ the minimal increment of temperature of the first layer is

$$(\Delta T)_1 = (1/C_V)\,(\Delta q)_{1,\min} \approx 350\,\mathrm{K}. \tag{3.69c}$$

Here $C_V \approx 3.4\cdot 10^6$ (J/m$^3\cdot$ K) is the copper volume specific heat.

3.8 Minimization of a uniform medium heating under diffusion of the pulse magnetic field

The problem of the minimization of thermal energy generated in the surface layer by the choice of the appropriate pulse shape of external field is of considerable practical interest. The solution of the problem depends on imposed complimentary conditions [9]. Rosenbluth, Furth, and Case [21] calculated the optimal shape of the front and tail of the pulse having flat peak which provides a minimum total energy generation at the surface. Figure 3.12 shows the shape of the pulse and examples of dependences $\Delta q'(0, \infty)$ on the characteristic parameters t_0/t_1 and t_1/t_2 along with the example from [21] of a pulse with an infinite tail. In this case the condition $t_2 = \infty$ is satisfied, and the total energy generated in the surface layer is close to energy generated by a pulse having linear front of induction ($B_e = B_m t/t_0$, $t < t_0$), flat the top ($B_e = B_m$, $t_0 < t < t_1$) and decaying according to $B_e = B_m t_1/t$, $t > t_1$ (Figure 3.12).

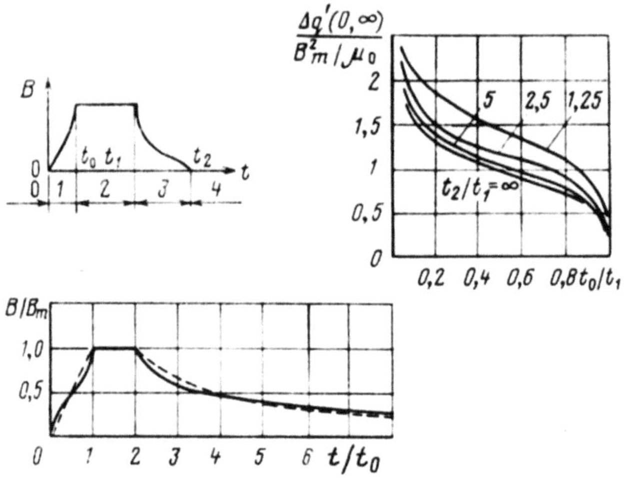

Fig. 3.12: Presented is a pulse with flat top providing a minimum of energy generation in the conductor surface layer. Here are: the pulse shape; dependences $\Delta q'(0, \infty)/(B_m^2/\mu_0)$ at different values t_1/t_2; the dependence $B/B_m = f(t/t_0)$ for $t_2 = \infty$, $t_0 = t_1/2$ (*solid line*), and dependences $B/B_m = t/t_0$ at $t < t_0$, $B/B_m = t_2/t$ at $t > t_2$ (*dotted line*).

In Figure 3.13 a set of dependences $B_e/B_m = f(t/t_2)$ is shown for symmetrical pulses ($t_0 = t_2 - t_1$) which characterize parameter $\beta = t_0/t_1$. For such pulses, $\Delta q'(0, \infty) = (B_m^2/\mu_0)[K'(\beta)/K(\beta)]$. Here $\mathbf{K}(\beta)$ and $\mathbf{K}'(\beta)$ are complete elliptic integrals.

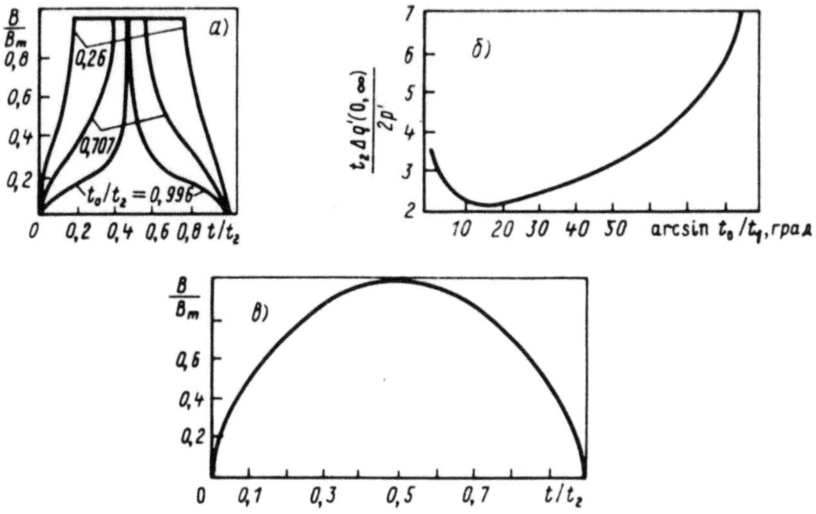

Fig. 3.13: Pulses providing a minimal energy generation: (a) examples of symmetrical pulses with a flat top; (b) the energy generation depending on ratio t_0/t_1 at fixed value of $p' = \int_0^{t_2} [B^2/(2\mu_0)]dt$ for pulses with a flat top; (c) the pulse shape providing minimal energy generation at given p'.

The conditions can be stated at which the energy generated in the surface conductor element assumes a minimal value at the given integral $1/(2\mu_0) \int_0^{\infty} B_e^2(t)dt = p'$. This integral defines the total impulse of the electromagnetic force applied to the unit surface. For the symmetric pulse with a flat top this condition is satisfied at the minimal value of the function $\Delta q' t_2/(2p')$, which is close to two and reached at $\beta \approx 0.26$ (Figure 3.13b).

The problem can be stated of defining of induction pulse shape in the interval $0 < t < t_2$ which provides a minimal energy generation for the fixed value p' without resorting to the complementary assumption for the constant induction in the interval $t_0 < t < t_1$. This problem was been investigated by Titkov [9].

The dependence $B/B_m = f(t/t_2)$ is shown in Fig 3.13c. The volume energy density related to this pulse, is $\Delta q'(0, \infty) = 2.76B_m^2/(2\mu_0)$, the parameter p' being $p' = 1.18B_m^2 t_2/(2\mu_0)$. By comparison, for the sinusoidal single pulse t_2 we have, according to (3.48b), $\Delta q'(0, \infty) = 2.42B_m^2/(2\mu_0)$, and $p' = B_m^2 t_2/(2\mu_0)$. The difference in energy generation at equal parameters p' and t_2 for both compared pulses is only about 3.5 %.

Thus, the single sinusoidal pulse differs little from the optimal one, providing the minimal energy generation at the given external field amplitude and the pulse duration.

3.9 One-dimensional diffusion of the field into a medium with conductivity depending on the coordinate. Reduction of energy generation in the surface layer

The study of the surface effect in a medium with conductivity that does not depend on time, but varies in space, reduces to the solution of equation (3.7), where $D = \rho/\mu_0 = f(x)$, with boundary conditions $B(0, t) = B_e(t)$, $B(\infty, t) = 0$. Here we shall consider the simplest examples, showing the qualitative peculiarities of the field diffusion into the nonuniformed medium. In the simplest case of two-layer medium, where $\rho = \rho_1$ at $0 \le x \le d$, $\rho = \rho_2$ at $x > d$, the induction acquires the form

$$\bar{B} = \bar{B}_1 = \bar{B}_e(p) \frac{ch[a_1(d-x)] + \sqrt{\rho_2/\rho_1} sh[a_1(d-x)]}{ch(a_1 d) + \sqrt{\rho_2/\rho_1} sh(a_2 d)}, \quad 0 \le x \le d, \tag{3.70a}$$

$$\bar{B} = \bar{B}_2 = \bar{B}_e(p) \frac{\exp[-a_2(x-d)]}{ch(a_1 d) + \sqrt{\rho_2/\rho_1} sh(a_2 d)}, \quad x > d, \tag{3.70b}$$

where $a_{1,2} = (\mu_0 p/\rho_{1,2})^{1/2}$. This solution satisfies the continuity condition of the induction of the magnetic field and electric field intensity on the boundary of the separation. The further calculations allow us to choose the thickness of the layer with higher conductivity ($\rho_1 < \rho_2$) which is deposited on the poorly conducting substrate for enhancement of the surface effect, (for a more detailed description see [22]). The transitional resistance in this case is

$$z(p) = z_1(p) \frac{\sqrt{\rho_2/\rho_1} + th(a_1 d)}{1 + \sqrt{\rho_2/\rho_1} th(a_1 d)}, \tag{3.70c}$$

where $z_1(p) = (\mu_0 \rho_1 p)^{1/2}$. In a limit $d \to \infty$ $z(p) = z_1(p)$, and at $d \to 0$ $z(p) = z_2(p) = (\mu_0 \rho_2 p)^{1/2}$.

With continuous dependence $\rho(x)$, the exact solution of the equation of diffusion is only possible in some particular cases. The qualitative peculiarities of the surface effect in a medium with monotonously decreasing specific resistance can be seen in the case when $\rho = \rho_0 (x/x_0)^{-\alpha}$, where $\alpha \ge 0$. After the Laplace transformation, the expression for induction at zero initial conditions acquires the form

$$\bar{B}(x, p) = 2\bar{B}_e(p) \left(\frac{1}{2+\alpha}\right)^{\nu} p_1^{\nu/2} x_1^{(1+\alpha)/2} \times K_\nu \left[\sqrt{p_1} \frac{2x_1^{1+\alpha/2}}{2+\alpha}\right], \tag{3.71}$$

where $p_1 = (\mu_0 x_0^2/\rho_0)p$; $x_1 = x/x_0$; $\nu = (1+\alpha)/(2+\alpha)$ and K_ν is the McDonald function. At the instantaneous increase of the external field up to the value B_0 the induction and the current density are described by the following formulas:

$$B(x,t) = \frac{2B_0}{\Gamma(\nu)} \int_u^\infty e^{-u^2} u^{2\nu-1}\,du, \tag{3.72a}$$

$$\delta(x,t) = \frac{B_0}{\mu_0 x_0 \Gamma(\nu)} t_1^{-\nu} \left(\frac{1}{2+\alpha}\right)^{\frac{\alpha}{2+\alpha}} x_1^{\alpha} e^{-u^2}, \tag{3.72b}$$

where $u = x_1^{(2+\alpha)/2}/[(2+\alpha)\sqrt{t_1}]$, $t_1 = t\rho_0/(\mu_0 x_0^2)$.

The case $\alpha = 0$ is considered above [formulas (3.36)]. As it follows from formula (3.72b), near the surface ($x \to 0$) the current density changes as x^α, i.e., proportionally to the conductivity. At large values of the argument u the exponential factor plays a main part, which defines the damping of the current density when $x \to 0$. With increasing conductivity ($\alpha > 0$), the current density has a maximum iat the point $x_{1m} = [t_1\alpha(2+\alpha)]^{1/(2+\alpha)}$. The coordinate of the maximum $x = x_0 x_{1m}$ increases monotonically with time. The amplitude of the current density drops proportionally to $t^{-1/(2+\alpha)}$. This behavior is confirmed by the examples of current distribution given in Figure 3.14 for the values $\alpha = 2$ and 1.

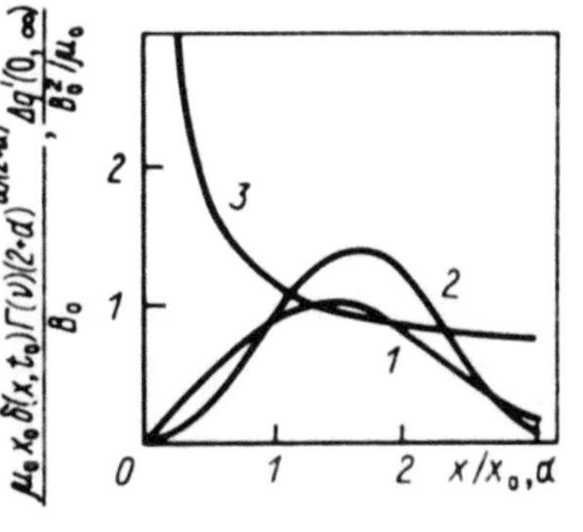

Fig. 3.14: The field diffusion into a medium with conductivity rising according to the power law: (1, 2) distribution of the current density at the instant $t_0 = \mu_0 x_0^2/\rho_0$ ($1 - \alpha = 1$; $2 - \alpha = 2$); (3) dependence $\mu_0 \Delta q'(0,\infty)/B_0^2 = f(\alpha)$.

The shift of the current density maximum in the medium with increasing towards the depth of electroconductivity leads to a more uniform energy release and a reduction of the general level of Joule losses. In this case the total energy generation for unlimited time does not depend on the coordinate and is determined at the instant rise of the external field by the formula

$$\Delta q'(x,\infty) = \frac{B_0^2}{\mu_0} \frac{\Gamma\left(\frac{\alpha}{2+\alpha}\right)}{\Gamma^2\left(\frac{1+\alpha}{2+\alpha}\right)} 2^{-\frac{\alpha}{2+\alpha}}. \tag{3.73c}$$

In a medium with the constant conductivity $\Delta q'(0,\infty) = \infty$, and with growth α, the value $\Delta q'(0,\infty)$ decreases, approaching to a limiting value $\Delta q'(0,\infty) = B_0^2/(2\mu_0)$ at $\alpha \to \infty$ (Figure 3.14).

An analytical solution could also be obtained at the exponential dependence of specific resistance on the coordinate: $\rho = \rho_0 \exp(-x/x_0)$. If $B(0, t) = B_0$ at $t > 0$, then

$$B(x, t) = B_0 \exp\left[-\frac{\mu_0 x_0^2}{\rho_0 t} \exp\left(\frac{x}{x_0}\right)\right], \tag{3.74a}$$

$$\delta(x, t) = \frac{B_0 x_0}{\rho_0 t_2} \exp\left[\frac{x}{x_0} - \frac{\mu_0 x_0^2}{\rho_0 t} \exp\left(\frac{x}{x_0}\right)\right], \tag{3.74b}$$

$$\Delta q'(x, t) = \frac{B_0^2}{2\mu_0} \exp\left[-\frac{2\mu_0 x_0^2}{\rho_0 t} \exp\left(\frac{x}{x_0}\right)\right]. \tag{3.74c}$$

At $t \to \infty$ $\Delta q'(x, t) \to B_0^2/(2\mu_0)$.

The opportunity to reduce the heating in a nonuniform medium at the *sin*-variation of the external field ($B_e = B_m \sin(2\pi t/T)$) has been studied in the papers [23–25]. Numerical calculations were made for a medium with the dependence for the specific resistance of the form of $\rho(x) = \rho_\infty + \Delta\rho \exp(-x/d)$, and also for a two- and a three-layer medium. In a first case, the maximal energy generation $\Delta q_m'$ to the instant $t = T/2$ takes place in the depth of the conductor. The ratio $\Theta_q = \mu_0 \Delta q_m'/B_m^2$ is the function of two dimensionless parameters $P_1 = \rho_0 T/(2\mu_0 d^2)$ and $P_2 = \rho_\infty/\Delta\rho$. As it is seen from curves in Figure 3.15a, the minimal value of the ratio could be decreased from the value 1.09, corresponding to a uniform medium ($P_2 \gg 1$) close to 0.3 at $P_2 \approx 10^{-2}$, i.e., more than threefold.

Let us now consider the two-layer medium in which, unlike the previous example, the conductivity of the external layer is lower than in the substrate. In such a medium the spatial distribution of the volume energy density for a fixed phase of sinusoid is determined by two parameters: $P_1 = \rho_1 T/(2\mu_0 d^2)$ and $P_2 = \rho_2/\rho_1 < 1$. At given P_2 in two limiting cases $P_1 = 0$ and $P_1 = \infty$, the energy generation will be the same as in the uniform medium, i.e., not depending on ρ_1 and ρ_2. In the first of these limiting cases the current practically does not penetrate into the substrate and is concentrated in the surface layer with specific resistivity ρ_1 and the maximum of energy generation occurs on the boundary of the conductor. In the second case the current is distributed in the substrate and the maximum of the energy generation is on the boundary of the media. In the general case the function $\Delta q'(x, T/2)$ has two maximums on both boundaries. The lowest energy generation at given P_2 occurs at some value of P_1, when both maximums become equal to each other (Figure 3.15b). The simulation shows that an absolute minimum of energy generation takes place at the condition $P_2 \approx 0,1 \div 0,2$ and $P_1 \approx 10$, and its value is approximately $0.67 B_m^2/\mu_0$. This is approximately 40 % less than for the uniform conductor, but essentially more than at the exponential conductivity distribution.

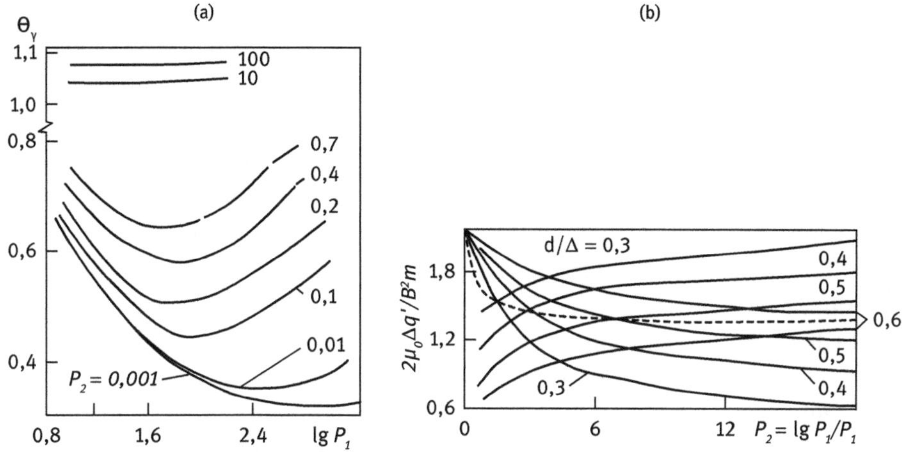

Fig. 3.15: The energy generation in a two-layer medium: (a) the dependence of the maximal energy generation on the parameter P_1 at different $P_2 = \rho_\infty/\Delta\rho$ for a medium in which $\rho(x) = \rho(\infty) + \Delta\rho \exp(-x/d)$; (b) energy generation at the instant $t = T/2$ on the boundary of layers [rising of $\Delta q'(0, \infty) = f(\rho_1/\rho_2)$] and on the surface [decreasing of $\Delta q'$] depending on P_2 at different $d/\Delta = \sqrt{1/P_1}$ (the *dashed line* is the line of optimum).

The numerical calculations for the three-layer medium which were done by Karpova and Titkov [26] showed that in this system the lowest energy generation also takes place at the condition of its mutual equality on all interfaces and equals $0.54B_m^2/\mu_0$, at $\rho_2/\rho_1 \approx 0.35$, $\rho_3/\rho_1 \approx 0.1$; $d_1/d_2 \approx 0.5$, and $\rho_1 T(\mu_0 d_1^2) \approx 9$. Here $\rho_{1,2,3}$ are consequently the specific conductivities of the first and second layers and substrate, and $d_{1,2}$ the thickness of the first and second layer.

3.10 One-dimensional nonlinear diffusion of the magnetic field into the conductor heated by the eddy current

As we have seen in Section 3.1, one can approximate the dependence of the specific conductivity on the energy enhancement in the unit volume by the straight line. Let us introduce the characteristic magnetic field B_0, in which the energy density equals $1/\beta$:

$$B_0 = \sqrt{\frac{2\mu_0}{\beta}}. \tag{3.75}$$

For copper the characteristic induction equals 42 T, if parameter β presented in Table 3.1 is given the value $\beta = \beta_3$. Based on qualitative considerations, one can expect that magnetic field, rising in a time monotonously, would not result in the essential heating of the conductor, unless induction is less than B_0. In such a field the resistivity will be close to ρ_0. However, in field, induction much more than B_0 the change of resistivity will necessarily occur. Therefore the parameter B_0 is a characteristic one from the point of view of the influence of nonlinear effects on the diffusion of a magnetic field. The nonlinear diffusion of the strong magnetic field has been extensively studied in many works [3, 8, 27–29]. On order to avoid significant mathematical difficulties which appear at the solution of this problem, we will try to find it in the form of a running wave and assume that all characteristic physical values which are the functions of the spatial coordinate x and time, would depend only on the argument $z = x - ut$ [27]. We assume that the wave propagates in a semiinfinite medium, and at $z \to \infty$ the conditions $B \to 0$ and $E \to 0$ are fulfilled. The solution would not be of a general nature, but corresponds to a particular form of the boundary conditions. In this particular case the partial derivatives on time and the coordinate are replaced by the derivatives on the variable z:

$$\frac{\partial}{\partial t} = -u\frac{d}{dz}, \quad \frac{\partial}{\partial x} = \frac{d}{dz}.$$

Then equations (3.5) and (3. 6) acquire the form

$$\frac{dB}{dz} = -\mu_0 \delta\ (z) = -\frac{\mu_0 E\,(z)}{\rho\ (z)}; \tag{3.76}$$

$$u\frac{dB}{dz} = \frac{dE}{dz} \tag{3.77}$$

From equation (3.77) and the above-mentioned conditions at $z \to \infty$ follows the simple relation between electric field intensity and induction of the magnetic field:

$$E = uB. \tag{3.78a}$$

The heating of the conductor is described by the equation

$$\frac{\partial}{\partial t}(\Delta q') = -u\frac{d(\Delta q')}{dz} = E\delta = -\frac{uB}{\mu_0}\cdot\frac{dB}{dz} \tag{3.78b}$$

whence it follows that

$$\Delta q' = \frac{B^2}{2\mu_0}. \tag{3.79}$$

In the field of a running wave the density of the energy generated in the element of the conductor at the adiabatic heating equals the energy density of the magnetic field. From here, according to equation (3.4a), we find

$$\rho = \rho_0\left(1 + \beta\frac{B^2}{2\mu_0}\right) = \rho_0\left(1 + \frac{B^2}{B_0^2}\right). \tag{3.80}$$

On substitution of E and ρ in the eq. (3. 76) we obtain the equation with separable variables for B

$$\frac{1}{B}\left(1+\frac{B^2}{B_0^2}\right)dB=-\frac{\mu_0 u}{\rho_0}dz. \tag{3.81a}$$

Its solution has the form

$$\ln B+\frac{B^2}{2B_0^2}=-\frac{\mu_0 u z}{\rho_0}+C. \tag{3.81b}$$

On the surface of the conductor ($x_0 = 0$ and $z = -ut$) at the initial moment we have $B_e(t) = B_e(0)$. Assuming $x = 0$ and $t = 0$ we find the constant of integration: $C = \ln B_e(0) + B_e^2(0)/2B_0^2$. Further, we obtain the law for the change of induction sustaining the regime of the wave propagating with the constant velocity:

$$\ln\frac{B_e(t)}{B_e(0)}+\frac{B_e^2(t)-B_e^2(0)}{2B_0^2}=\frac{\mu_0 u^2 t}{\rho_0}. \tag{3.82}$$

At the moment $t = 0$ the induction jumps up to the value $B_e(0)$. If $B_e(0) \ll B_0$, then after the jump the induction rises exponentially and at $B_e > B_0$ changes proportionally to $t^{1/2}$. This part of the process is the most interesting for the study of the diffusion in a strong field. Thus, it could be stated that in the field with the induction growing according to the law

$$B_e(t)=B_0\,(t/t_0)^{1/2}, \tag{3.83}$$

and the instantaneous value of induction satisfies to the condition $B_e > B_0$, the current wave propagates in the conductor with constant velocity. It could be found from the eq. (3.82), neglecting all the terms besides $B_e^2/2B_0^2$ in its left part:

$$u\approx\sqrt{\frac{\rho_0}{2\mu_0 t_0}}. \tag{3.84}$$

The induction distribution on depth can be obtained by eliminating the term $\mu_0 u^2 t/\rho_0$ from (3.81) and (3.82):

$$\ln\frac{B(x,t)}{B_e(t)}+\frac{B^2(x,t)-B_e^2(t)}{2B_0^2}=-\frac{\mu_0 u x}{\rho_0}. \tag{3.85}$$

As it follows from the given formula the induction near the surface vanishes linearly while at the $x \to \infty$ it vanishes exponentially $B \approx B_e \exp(-\mu_0 u x/\rho_0)$.

If we plot the dependence $B(x, t)$ for a chosen instant of time t, then for smaller times it could be easily obtained by shifting of the plot by length $-ut$. It

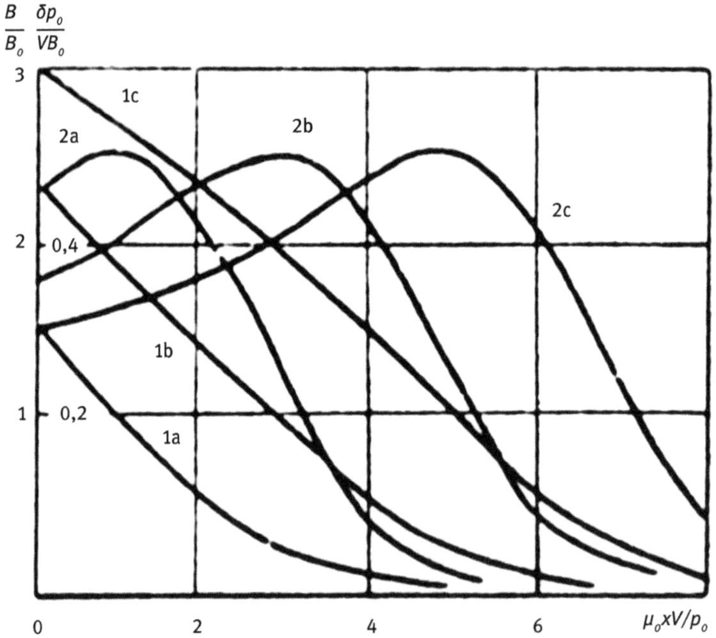

Fig. 3.16: Distribution of the induction and current density at the diffusion of the strong magnetic field in the conductor (simulation of the wave with constant phase velocity u): 1. $B/B_0 = f(\mu_0 xu/\rho_0)$; 2. $\rho_0 \delta/(uB) = f(\mu_0 xu/\rho_0)$. The curves (a), (b), and (c) correspond to $B_e(t)/B_0 = 1,5$; 2,3 and 3,0.

follows from the form of the obtained solution, in which all the sought-for values depend on the argument $z = x - ut$.

The dependence for the current density $\delta(x, t)$ is not monotonous (Figure 3.16).

It follows from this expression for δ that

$$\delta = \frac{E}{\rho} = \frac{uB}{\rho_0 (1 + B^2/B_0^2)}. \tag{3.86}$$

From equation (3.86) it follows that at the condition $B_e > B_0$ the current density has the maximum in the point where the equality $B(x, t) = B_0$ is valid. The maximal value of current density is

$$\delta_m = \frac{u B_0}{2\rho_0}. \tag{3.87}$$

The physical meaning of the nonmonotonous dependence of $\delta(x)$ consists in the fact that when in a strong field $B_e > B_0$, the layers of the conductor adjacent to the surface are strongly heated, their conductivity drops, and part of the current penetrates in deeper cold layers with high conductivity. This process competes with a skin effect – i.e., the displacement of the current affected by the eddy electric field towards the surface. As a result of the joined action of above factors the main part of the current will be concentrated in a region, where $B \approx B_0$, and on the boundary of conductor the current density drops as $t^{-1/2}$:

$$\delta\ (0,\,t) \approx \frac{1}{\sqrt{\rho_0 \beta\, t}}. \tag{3.88}$$

The self-similar solutions obtained in the study [28] show that this law remains valid at other values of the exponent α, characterizing the process of external field increasing. Numerical calculations [29] show, that at the power law growth of external field ($B_e = \text{const} \cdot t^{\alpha}$) and at the change of the power index in the range $0,5 < \alpha < 2$ the density of energy generated on the surface is close to the value

$$\Delta q'\ (0,\,t) = \chi B_e^2/2\mu_0, \tag{3.89}$$

where the number χ varies in a narrow region $1 > \chi > 0,83$ for $B_e \geq 2B_0$.

3.11 Approximate description of the surface effect. “The skin layer method”

In many publications the integral characteristic of the diffusion of the field are described using the simplified approach, the skin layer method. In calculations the notion of the skin layer is helpful, which is determined differently, depending on the authors’ preferences and purposes of the stated task.

The surface density can be connected with the volume current density at the boundary by the relation

$$|J| = |H_e| = \Delta' \delta\ (0,\,t)\,. \tag{3.90}$$

The value Δ' can be interpreted as a layer thickness with the constant current density $\delta\,(0,\,t)$, in which the whole surface current is concentrated. In order to estimate the flux in the conductor per unit length, one can use the formula

$$\Phi' = \Delta'' B_e. \tag{3.91}$$

Such an estimate corresponds to the assumption that the field induction remains equal to B_e up to the conventional field–conductor boundary, and beyond the boundary vanishes. In both formulas there are the thicknesses of the skin layers Δ' and Δ'', which have the dimension of length. They can be naturally considered as coordinates of the conventional field–conductor boundary. The thicknesses of the skin layer can be estimated by

$$\Delta', \Delta'' = b_{1,2}\sqrt{\rho\, t/\mu_0}, \tag{3.92}$$

where the numbers $b_{1,2}$ could be obtained by solving the modeling problem on the field diffusion in the medium at the given law of the rise of the external-field induction $B_e(t)$. Evidently the different dependences $B_e(t)$ correspond to the different numbers $b_{1,2}$; however, when estimating, some fixed values are ascribed to these numbers.

By using the "skin layer method", one can avoid the solution of the differential equations in partial derivatives at the description of the field diffusion. It has been effectively used, for instance, in the studies by Sakharov [30], Lyudaev [31], Shearer, Abracham at al, [32], Kalitkin and Serova [33], and in some other studies. As an example, in the paper by Shearer, Abraham et al. [32], for the calculation of the penetration depth the equation, based upon considerations of dimensionality, they used

$$\Delta\frac{d\Delta}{dt} = D_0\frac{\rho}{\mu_0}, \tag{3.93}$$

where D_0 is the dimensionless constant.

The equation describing one-dimensional field diffusion into a conductor could be rearranged to the form containing the parameters $\Delta'$and Δ''. In the general case we assume that the boundary of the conducting medium (the field-conductor boundary) shifts in respect to the medium. In this case the medium itself could be immobile or moving in the reference frame of observer.

The equation of induction in the system connected to the medium has the form (3.7). The boundary conditions are

$$B_z(x_1) = B_e = \mu_0\int_{x_1}^{\infty}\delta_y(x)\,dx; \quad B_z(\infty) = 0,$$

where δ_y is the current density. Integrating the right part of equation (3.7) from x_1 to ∞, we obtain

$$\int_{x_1}^{\infty}\frac{\partial}{\partial x}\left(\frac{\rho}{\mu_0}\frac{\partial B_z}{\partial x}\right)dx = -\frac{\rho\,(x,\,t)}{\mu_0}\frac{\partial B_z}{\partial x}\bigg|_{x_1} = \rho\,(x_1, t)\,\delta_y(x_1, t) = \frac{\rho_1 B_e}{\mu_0\Delta'}, \tag{3.94a}$$

where $\rho_1 = \rho\,(x_1,\,t)$ is the value of specific resistance on the boundary of the conducting medium. Integration of the left part of equation (3.7) gives

$$\int_{x_1}^{\infty}\frac{\partial B_z}{\partial t}dx = \frac{d}{dt}\left(\int_{x_1}^{\infty}B_z dx\right) + B_z(x_1)\frac{dx_1}{dt} = \frac{d}{dt}\left(B_e\Delta''\right) + B_e\Omega, \tag{3.94b}$$

where $\Omega = dx\ /dt$ is the velocity of the boundary displacement.

Thus, we come to the equation

$$\frac{1}{B_e}\frac{d}{dt}(B_e\Delta'') - \frac{\rho_1}{\mu_0\Delta'} + \Omega = 0. \tag{3.95}$$

In Chapter 9 this generalized equation will be used for the description of the model of skin layer explosion, during which the surface elements of the medium lose their conductivity due to evaporation. Here we consider the case of the medium with an immobile boundary. In relatively weak field, when $\rho \approx \rho_0 = \text{const}$, the above given exact solution of the differential equation (3.7) offers a means of calculating $\Delta'$and Δ''. For example, with the exponential rise of the external field ($B_e = B_1 \exp(t/t_1)$) we have $\Delta' = \Delta'' = (\rho_0 t_1/\mu_0)^{1/2}$and the equality $p' = \Delta''/\Delta' = 1$ is maintained. In the case of the exponential rise of the external field ($B_e = B_1(t/t_1)^\alpha$) the numberp' takes the value [29]

$$p' = \Gamma^2(1+\alpha)/[\Gamma(\alpha+1/2)\Gamma(\alpha+3/2)]. \tag{3.96}$$

This value lies in the range $2/\pi < p' < 1$ when α changes in the range $0 < \alpha < \infty$. The relative error is not beyond the limits ± 15 %, if for all values of α we take the value $p' = 0.85$, corresponding to the case $\alpha = 1$.

With this assumption equation (3.95) is rearranged to the form

$$\Delta''\frac{d}{dt}\left(B_e\Delta''\right) = \frac{\rho_1}{\mu_0}p'B_e. \tag{3.97}$$

Its solution gives the following dependence for a thickness of the skin layer Δ'':

$$\Delta'' = \frac{1}{B_e}\sqrt{\frac{2\rho_0 p'}{\mu_0}\int_0^t B_e^2\,dt}. \tag{3.98a}$$

For the above-mentioned particular case of the exponential induction growth we obtain

$$\Delta'' = \sqrt{\frac{2\rho_0 t\, p'}{\mu_0(1+2\alpha)}}. \tag{3.98b}$$

From the preceding, the calculation by the approximate formula (3.98b) gives a result close to the exact solution. Since the value p' weakly depends on the increased rate of induction, it can be assumed to be constant for many of the problems of linear diffusion, where the boundary induction increases monotonically.

It is reasonable to also use the approximate equation (3.97) in calculations of nonlinear diffusion of the strong field. In particular, it allows the estimation of the diffusion influence on the current variations in a circuit with a conductor in which the strong magnetic field penetrates.

Table 3.4: Parameters of the skin layer model.

α	$p' = \Delta''/\Delta'$	$\vartheta = 2\mu_0 q'(0)/B_e^2$	$p'\vartheta$
0.5	0.33	1.0	0.33
1.0	0.34	0.89	0.30
1.5	0.35	0.85	0.30
2.0	0.38	0.84	0.32

The study [29] presents the results of numerical calculations for nonlinear diffusion of the field increasing according to a power law. The data obtained show that with changing of the exponent α in the limits from 0.5 to 2 and under condition $B_e \gg B_0$, the ratio Δ''/Δ'changes in a rather narrow range. This data is presented in Table 3.4. The table contains the values of the parameter $\vartheta = 2\mu_0 q'(0)/B_e^2$, where $q'(0)$ is the volume density of the energy in the boundary element of the immobile medium ($x_1 = 0$). The table also presents the values of the product $p'\vartheta$. The latter varies insignificantly in the indicated range of the exponent α. On this basis, for the case of nonlinear diffusion equation (3.95) can be written in the form

$$\Delta'' \frac{d}{dt}\left(B_e \Delta''\right) \approx \rho_0 \frac{p'\,\vartheta\, B_e^3}{\mu_0 B_0^2}, \tag{3.99a}$$

where the characteristic induction $B_0 = \sqrt{2\mu_0/\beta}$ is present, and for the parameter ρ_1 the approximate formula $\rho_1 \approx \rho_0 B_e^2/B_0^2$ was used.

From equation (3.99a) it follows that

$$\Delta'' = \frac{1}{B_e}\left(\frac{\rho_0 p'\vartheta}{\mu_0 B_0^2}\int_0^t B_e^4 dt\right)^{1/2}. \tag{3.99b}$$

The solution of this equation and the equation of the contour of the capacitive storage discharging onto the slot with an immobile boundary are considered in [29]. In this case the equation of the discharge circuit has the form

$$\frac{d}{dt}\left(2gB_e\Delta'' + \frac{L_0 B_e b}{\mu_0}\right) + \frac{b}{\mu_0 C_0}\int_0^t \frac{B_e}{\mu_0} dt = U_0, \tag{3.100a}$$

where g is the slot length; b, the slot width; L_0, the induction of the external circuit; C_0, the battery capacity; U_0, the initial voltage.

The limiting case where the condition $L_0 \gg \Delta''\mu_0 g/b$ is fulfilled corresponds to the regime when the induction amplitude B_m is close to the calculated value for the discharge in the contour $\mathbf{L_0 - C}$: $B_m \approx B_{m0} = (U_0\mu_0/l)(C_0/L_0)^{1/2}$. In the general case the solution is determined by the dimensionless parameters

$$a_1 = \rho_0 \beta\, \pi\, U_0^2 \rho\, \vartheta\, C^{3/2} / (2b^4 \sqrt{L_0});$$

$$a_2 = 2p' g \mu_0 / L_0.$$

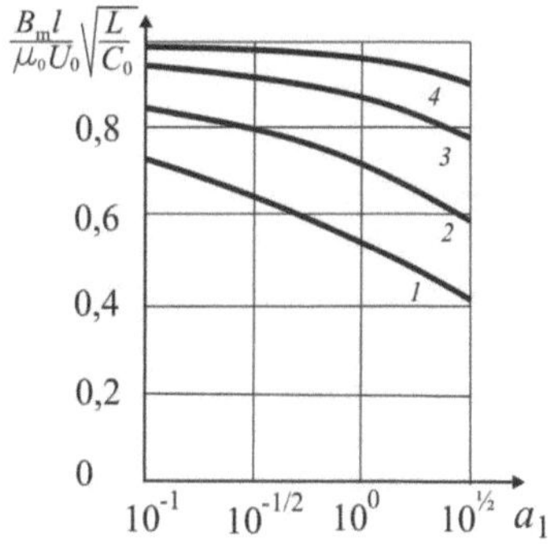

Fig. 3.17: The induction amplitude in a slot included in the oscillatory circuit. The value of the parameter a_2: (1) $a_2 = 2$; (2) $a_2 = 1.5$; (3) $a_2 = 1$; (4) $a_2 = 0.5$.

The curves in Figure 3.17 show that in the strong field (at the condition $a_{1,2} > 1$) the induction amplitude decreases compared to B_{m0}. In the limiting case $a_{1,2} \gg 1$, the current rise in the circuit is described by the following dependence:

$$i(t) = \frac{bB_e(t)}{\mu_0} = b\left(\frac{B_0 U_0}{2\mu_0 g}\right)^{1/2} \left(\frac{t}{\vartheta\, p' \rho_0 \mu_0}\right)^{1/4}. \tag{3.100b}$$

In the analogous conditions in the weak field, at $\rho = \rho_0$, the induction increases proportionally to $t^{1/2}$. Faster diffusion into the medium, subjected to the heating by eddy currents, results in retardation of the current rise in the discharge circuit.

3.12 References

[1] D. N. Parkinson and B. E. Mulhall, *The Generation of High Magnetic Fields*, New York, Premium Press, 1967.

[2] D. B. Montgomery, *The Magnetic and Mechanical Aspects of Resistive and Superconducting Magnets*, New York London Sydney Toronto, J. wiley, 1969.

[3] H. E. Knoepfel, *Pulsed High Magnetic Fields*, Amersterdam London, North-Holland Publishing, 1970.

[4] F. Herlach and N. Miura (eds.), *High Magnetic Fields. Science and Technology*, Vol. 1, New Jersey London Singapore Hong Kong Taipei Bangalor, World Scientific, 2003.

[5] M. Von Ortenberg, H.-U. Mueller, **MG-8**, 1998, pp. 171–174.

[6] I. E. Tamm, *Fundamentals of electrisity theory* Moskow, Fismatlit, 2003 (in Russian).

[7] A. N. Tihonov and A. A. Samarskiy, *Equations of mathematical physics* 5th edn., Mowcow, GITTL, 1977 (in Russian).

[8] H. E. Knoepfel, *Magnetic Fields. A Comprehensive Theoretical Treatise for Practical Use*, New York Chichester Weinheim Brisbane Singapore Toronto, John Wiley & Sons Inc., 2000.

[9] G. A. Shneerson, *Fields and Transients in Super high Pulse Current Device*, New York, Nova Science Publishers, Inc., 1997.

[10] M. A. Leontovith, in: *Issledovaniya po rasprostraneniyu radiovoln*, pp. 5–12, Moscow Leningrad, AN USSR, 1948 (in Russian).

[11] L. A. Vainstain and S. M. Zhurav, *Tech. Phys. Letters* **12**(12) (1986), 723–727 (in Russian).

[12] H. S. Carslaw and I. C. Jaeger, *Conduction of heat in solids*, 2nd edn., Oxford, Clarendon Press, 1959.

[13] L. V. Bronstein, *Elektrichestvo* **11** (1968), 67–71 (in Russian).

[14] B. N. Bondaletov, *Elektrichestvo* **8** (1975), 55–58 (in Russian).

[15] V. I. Volosov and B. V. Chirikov, *Zhurnal Tech. Phys.* **30**(5) (1960), 508–511 (in Russian).

[16] W. R. Smythe, *Static and Dynamic electricity*, 2nd edn., New York Toronto London, 1950.

[17] I. M. Karpova and V. V. Titkov, *Izvtstiya Akademiyi Nauk SSSR, Energetika I Transport* **5** (1988), 83–90 (in Russian).

[18] M. A. Lavrentyev and B. B. Shabat, *Methods of complex variable functions theory*, Moskow, Nauka, 1985 (in Russian).

[19] G. A. Shneerson, O. S. Koltunov, and S. I. Krivosheev, Adamian Yu. E., et al., *IEEE Trans. on Plasma Science* **38**(8) (2010), 1731–1737.

[20] L. Li and F. Herlach, *J. Phys. D Appl. Phys.* **31** (1998), 1320.

[21] M. N. Rosenbluth, H. P. Furth, and K. M. Case, *J. Appl. Phys.* **45** (1974), 1097–1099.

[22] V. M. Mikhailov, *Pulsed electromagnetic fields*, Kharkov, Vyssha Shkola, 1979 (in Russian).

[23] A. Farinsky and L. Karpinski, Nowak, *J. Techn. Phys.* **20**(2) (1979), 265–280.

[24] Yu. E. Adamyan, V. V. Titkov, and G. A. Shneerson, *Izvetstiya Akademiyi Nauk SSSR, Energetika I Transport* **5** (1984), 105–107 (in Russian).

[25] I. M. Karpova, V. V. Titkov, and G. A. Shneerson, *Izvestiya Akademiyi Nauk SSSR, Energetika I Transport* **3** (1988), 122–127 (im Russian).

[26] I. M. Karpova and V. V., *Titkov, Electrichestvo* **12** (1999), 20–26 (in Russian).

[27] A. R. Bryant, **MG-I**, 1966, p. 183.

[28] G. A. Shneerson, *J. Tech. Phys. (J. Techn. Phys.)* **37** (1967), 513–519 (in Russian).

[29] G. A. Shneerson, *J. Tech. Phys. (J. Techn. Phys.)* **43** (1973), 419–428 (in Russian).

[30] A. D. Saharov, *Uspehi Fisicheskich Nauk* **88**(4))1966), 725–734 (in Russian).

[31] R. Z. Lyudaev, **MG-VII**, 1996, pp. 86–114.

[32] R. Grossinger, *J. Phys. D* **17**(5) (1982), 1541–1550.

4 Matching of the parameters of solenoids and power supply sources

4.1 General requirements to the power supply source

The choice of the power supply source is a necessary stage in the development of general purpose magnetic systems. With the great diversity of the solenoids configurations used in practice, most commonly we must deal with the problem of inducing of the field in the finite volume V_0, where the induction amplitude in some characteristic point attains the value B_m. If the field in the volume V_0 were strictly uniform and absent outside of it, the magnetic field energy confined in the volume V_0 would be equal to

$$W_1 = \frac{B_m^2}{2\mu_0} V_0. \tag{4.1}$$

However in practical conditions the field is also distributed beyond the boundaries of the volume V_0 and, in addition, it is nonuniform within the limits of the volume. The total magnetic field energy of the solenoid at the current peak t_m is given by

$$W_m(t_m) = \int_V \frac{B_m^2(t_m)}{2\mu_0} dV. \tag{4.2}$$

It iss possible to express the relationship between W_m and W_1 with the help of the nondimensional coefficient k_1:

$$W_1 = k_1^2 W_m(t_m), \tag{4.3}$$

where in a general case k_1 is determined by the coil geometry and depends on time due to the fact that in the thickness of a winding in the alternating field the eddy currents are induced, the spatial distribution of which changes in time. In the coils with windings of thickness much more than the penetration depth and in solenoids operating under conditions of the pronounced skin effect, we can assume that the coefficient k_1 is constant, determined only by the magnetic field configuration. The energy at the current peak is

$$W_m = L_c i_m^2/2, \tag{4.4}$$

were L_C is the solenoid inductance. Then we have the expression for the induction amplitude at the instant t_m:

$$B_m = k_1 \cdot i_m \sqrt{\frac{\mu_0 L_c}{V_0}}. \tag{4.5}$$

The source used for the magnet supply must provide the transfer of energy $W_m(t_m)$ to the solenoid, accounting for the losses in the circuit. In the devices where the short single pulses of a field are produced, it is impossible, over the discharge time (10^{-4} s and less) to remove the energy dissipated in the resistances of the winding. The role of the resistance mainly manifests itself in a decrease of the current amplitude and whence a decrease of the induction amplitude.

In stationary systems where it is necessary to use devices for removal of the energy generated in the winding resistance, the main problem is the reducing of the loss of power at the unchangeable value of induction.

Below we consider the simplified problems which illustrate the approach to the choice of optimal parameters of both types of systems. In addition, we consider the problem of the conversion of the energy supply source in the kinetic energy of a projectile accelerated by electromagnetic forces.

4.2 Optimization of the parameters of the system of solenoids – capacity energy storage

In an ideal case the energy stored in the electrical field of the capacity storage $W_E = C U_0^2/2$ (C is the bank capacity, U_0 is the initial power of storage) is converged into the magnetic energy of the uniform field of induction B_m, confined in the volume V_0. Then based on the condition of the energy balance one can determine B_m:

$$B_m = U_0 \sqrt{\frac{\mu_0 C}{V_0}}. \tag{4.6}$$

However, in practice the other dependence holds:

$$B_m = k U_0 \sqrt{\frac{\mu_0 C}{V_0}}, \tag{4.7}$$

where $k = k_1 k_2$. The coefficient k_1, depending only on the solenoid geometry, was introduced previously. Using equation (4.1) we find

$$k_2 = \frac{i_m}{U_0} \sqrt{\frac{L_C}{C}} \tag{4.8}$$

Thus the coefficient k_2 is the ratio of the current amplitude to its calculated value. The latest corresponds to the ideal conditions of the discharge, when the inductance of the source and connecting elements equals zero and the energy losses are absent in a whole circuit.

The optimization problem consists of finding from the solenoids parameters the one which has the largest value of the coefficient k_2.

Let us show such an optimization by the simple example assuming that the solenoid winding has given dimensions of the longitudinal section and the number of coils w only changes due to the change of the cross section of the winding wire. When simplified estimates cannot take into account the heating conductors and skin effect, then resistance is constant. The inductance of the solenoid is $L_C \approx L_1 w^2$ and resistance $R_C \approx R_1 w^2$ (on the assumption that the current distribution over the cross section is uniform), where L_1 and R_1 are constant. Taking into account the intrinsic inductance of the capacity bank, which together with the inductance of the connecting conductors is L_0, we can write k_2 using the known expression for the current amplitude in an L–R–C-circuit in the oscillatory discharge regime:

$$k_2 = \sqrt{\frac{L_1 w^2}{L_1 w^2 + L_0}} \exp\left[-\frac{\xi}{\sqrt{1-\xi^2}} \arcsin\sqrt{1-\xi^2}\right], \tag{4.9}$$

where $\xi = R_1 w^2 [C/L_1 w^2 + L_0)^{1/2}]/2 < 1$. Here we suggest that the winding resistance R_C is much more than the resistance of other elements of the circuit. The first factor in equation (4.9) (the coefficient of utility η) characterizes the transfer of energy from the storage into the load, where the active resistance of the circuit is not present. This factor increases monotonously with increasing w. It has the minimal value $\eta = \sqrt{L_1/(L_1 + L_0)}$ when $w = 1$. With increasing w the exp. index will grow proportionally to w. Therefore in a circuit with $L_1 \ll L_0$ one should increase the number of windings for increasing k_2 but because of the increasing resistance, only until the exponential factor begins to have an effect. For solenoids with $L_1 \ll L_0$ there are the optimal number of turns w_m at which k_2 attains the maximum value [1] (Figure 4.1)

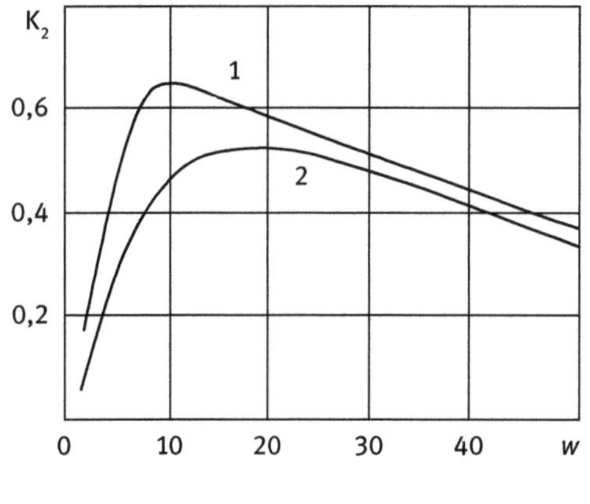

Fig. 4.1: The dependences of the coefficient of utility k_2 on the number of turns: (1) $L_0 = 1\,\mu H$, $L_1 = 0.02\,\mu H$, $C = 3 \cdot 10^{-3} \Phi$, $R_1 = 10^{-4}\,\Omega$; (2) $L_0 = 5\,\mu H$, the rest parameters as in (1).

The single turn coils ($w = 1$) also can exhibit the high efficiency of utilizing the storage energy, which, however only takes place if the powered source inductance is less than the inductance of the magnet. Therefore, in the case where single turn coils are used, we have to meet the rigid requirements for the construction of the capacity storage; otherwise reducing transformers are needed to enhance the efficiency of transfer of the storage energy into the load. In systems with transformers the first factor in equation (4.9) for the single turn magnet with a capacity takes the form $\eta = \sqrt{L_1 n^2/(L_1 n^2 + L_0)}$. The application of reducing transformers, in some sense, is equivalent to the using of the multiple turn coil with the winding number n, which results in an increase of the utility coefficient. However, the transformer adds both the resistance and inductance in the discharge circuit. In addition, the induction amplitude in the load reduces, due to the nonideal coupling between windings. Therefore the system with transformer calls for the optimization in every specific case.

The problem of matching solenoid parameters and capacity storage is different when it is needed to provide the largest rate of the induction rise in the solenoid $(dB/dt)_0$. If, neglecting the edge effect, consider the field to be uniform over the solenoid cross section S_C, it is easy to connect the initial power of the capacity bank with the initial rate of the induction rise:

$$\frac{U_0 L_C}{L_C + L_0} = \left(\frac{d\Psi}{dt}\right)_0 = wS_C\left(\frac{dB}{dt}\right)_0,$$

where Ψ is the flux-coupling, and w is the number of turns. Assuming as previously $L_C = w^2 L_1$, we find

$$\left(\frac{dB}{dt}\right)_0 = \frac{U_0}{S_C} \cdot \frac{L_1 w}{L_0 + L_1 w^2}. \tag{4.10}$$

Under condition $L_1 \ll L_0$ the initial rate of the induction rise $(dB/dt)_0$ is maximum if $w = \sqrt{L_0/L_1}$, while if $L_1 \leq L_0$, the largest rate of the induction rise occurs in a single turn magnet. This circumstance resulted in an application of single turn magnets and low-inductance capacity banks in experiments on plasma heating with powerful azimuth discharges (θ-pinchs) and in experiments on the generation of ultrahigh magnetic fields.

4.3 Optimization of solenoids according to Fabri

The classic problem is the optimization of the multiple turn solenoid operating in stationary regime under the conditions of minimal losses in the coil at given induction (Fabri's problem) [2, 3]. Our base is the known expression for the induction in the center of the multiple turn solenoid:

$$B_C = \frac{\mu_0}{2} \int_S \frac{\delta\, r^2 dS\,(r,\ z)}{\left(r^2 + z^2\right)^{3/2}}, \tag{4.11}$$

where r and z are cylindrical coordinates of the point in which the current density equals δ, and S is the longitudinal section of the coil. We assume that the current is uniformly distributed over the coil section. In practice, however, the conductors occupy only a part of the section, since there are inserts between them. This fact should be taken into account in what follows, with a help of the filling factor $\lambda = d_0/dS$, where dS_0 is the part of the cross section element occupied by conductors. An average current density δ in equation(4.11) is related to the current density in conductors δ_0 by the formula $\delta_0 = \delta\, dS/dS_0$. Let us go in (4.11) to the nondimensional parameters dependent only on the coil geometry and relative distribution current in it:

$$B_C = \frac{\mu_0 \lambda\ r_0 \delta_{0m}}{2} \int_S \frac{\left(r'\right)^2 \delta_0' dr' dz'}{\left[\left(r'\right)^2 + \left(z'\right)^2\right]^{3/2}}, \tag{4.12}$$

where the minimal inner radius of the solenoid r_0, the maximal current density in the conductor δ_{0m}, and the relative values $r' = r/r_0$, $z' = z/z_0$, $\delta' = \delta_0/\delta_{0m}$, are introduced assuming, for simplicity, the parameter λ to be constant for all cross sections of the winding. Further let us write the expression for the power which is generated in the winding resistance and removed when cooling:

$$P = \int_S R\,(r,\ z)\,(di)^2 = \int_S 2\pi\ r\rho\delta_0^2 dS_0\,(r, z) = 2\pi\,(\delta_{0m})^2\,\rho\lambda r_0^2 \int_{S'} \left(\delta_0'\right)^2 r'\ dr'\ dz'. \tag{4.13}$$

Excluding δ_{0m} we obtain the expression for the induction:

$$B_C = \mu_0 G \sqrt{\frac{\lambda\, P}{\rho\, r_0}}, \tag{4.14}$$

which contains the nondimensional parameter, Fabri's factor:

$$G = \int_{S'} \frac{\left(r'\right)^2 \delta_0' dr' dz'}{\left[\left(r'\right)^2 + \left(z'\right)^2\right]^{3/2}} \left[2\pi \int_{S'} \left(\delta_0'\right) r' dr' dz'\right]^{-1/2}. \tag{4.15}$$

Its numerical value depends only upon the form of the solenoid cross section and relative current distribution in it. With unchangeable power P and steady values of the parameters ρ, r_0, and, λ, the larger G is, the higher the magnetic field. Hence, the optimization of the magnetic system in the context of the considered problem is to find such a solenoid form and current distribution that the parameter L attains the maximal value. For this purpose we can use numerical simulation. The results of calculations for a set of configurations are given in [3]. In the solenoid of the rectangular cross section at the constant current density ($\delta_0' = 1$) the value G depends on the parameters $\alpha = R_2/R_1$ and $\beta = l/2R_1$, where l is the length, $R_{1,2}$ are the inner and outer radii of the solenoid (Figure 4.2). The maximal value $G_m \approx 0.142$ takes place when $\alpha \approx 3$ and $\beta \approx 2$. The current distribution of the form $\delta' = (1/r')$ is more preferable and then at $\alpha \approx 6$ and $\beta \approx 2$, $G_m = 0.166$. If in the solenoid of the rectangular cross section we assume for the nondimensional current density the dependence of the form $\delta_0' = (1/r')\,[(1+\beta)/(r'^2+\beta^2)]^{-1/2}$, the value of G_m can be even larger.

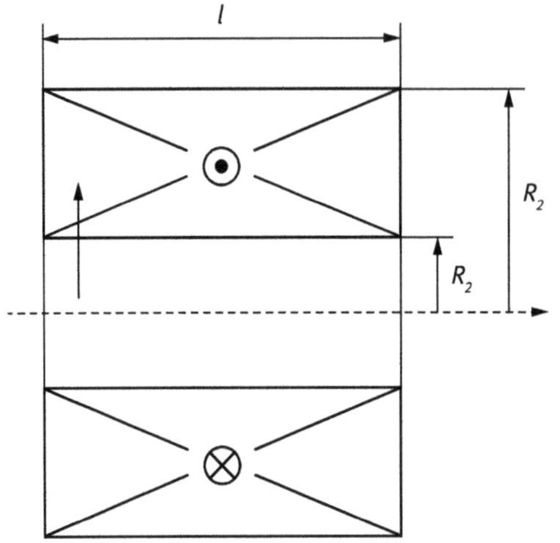

Fig. 4.2: Solenoid of rectangular cross-section.

In this case the current density at small values r' drops as $1/r'$ and at large values r' as $(r')^{-2}$. Then the maximal value of Fabri's factor is $G_m = 0.180$, which is reached in the solenoid under condition $R_2 \gg R_1$ ($\alpha = \infty$) and $\beta \approx 2$.

The analogous optimization can be performed for the solenoid of a trapezoidal or other form of the cross section. Along with this, it is interestzing to find the answer to the question on how the maximal induction can be attained at the prescribed loss of power, steady values of λ, r_0, δ_{0m}, and related current distribution, determined on the assumption that the current is continuously distributed in an unlimited region (Kelvin's problem [3]). Otherwise, one should solve the variation problem on determining the function $\delta_0'(r', z')$ such that it provides the minimum of the functional (4.13) on the additional condition (4.12). Solving the problem by the Lagrange method, we combine the functional

$$F = \int_{S'} \left[\xi \, \frac{\delta_0' r'^2}{\left[(r')^2 + (z')^2\right]^{3/2}} + \left(\delta_0'\right)^2 r' \right] \cdot dS', \tag{4.16a}$$

where ξ is the indefinite factor. Further, we find the condition when the partial derivative of the integrand turns out to be zero:

$$\frac{\xi \, (r')^2}{\left[(r')^2 + (z')^2\right]^{3/2}} + 2\delta_0' r' = 0. \tag{4.16b}$$

From this follows the dependence for the current density:

$$\delta_0' = \frac{-\xi \, r'}{2\left[(r')^2 + (z')^2\right]^{3/2}}. \tag{4.16c}$$

The constant $\xi = -2$ if we assume that in the point $r' = 1$, $z' = 0, \delta_0' = 0$. Hence the current relative distribution has the form

$$\delta_0' = \frac{r'}{\left[(r')^2 + (z')^2\right]^{3/2}} \tag{4.16d}$$

with $G_m = 0.217$ [3]. The Kelvin solenoid with the current distribution of the form (4.16) provides the highest field at the given dissipated power. However, in practice obtaining such a current distribution is troublesome: the winding must be continuously distributed over the whole space; in this case the current density increases indefinitely at point $r' = 0$, $z' = 0$. It is possible to simplify the solenoid construction by decreasing the Fabri's factor. Then, going from the "ideal" construction to the solenoid of the rectangular cross section with $\delta_0' = \text{const}$, the factor G decreases from 0.217 to 0.142, which is approximately 1.5 times.

The problems of optimization of multiturn magnets designed to generate pulsed magnetic fields should be posed in another way. The energy supply of such magnets is determined, to a considerable extent, by the sizes of the magnetic system, and, thus, the question of the strength of the winding takes on prime significance. The temperature restrictions also retain their importance. In many publications, for example in [4], the approaches to the problem of optimization are considered with regard to the mentioned factors.

4.4 Transformations of energy in a circuit with alternating inductance

Along with the problem of the transfer of the storage energy into the energy of the magnetic field of the coil, the problem of the transformation of the field energy into the kinetic energy of a body accelerated by electromagnetic forces is also quite important. Examples are the constructions for launching conductive projectiles and for their deformations, as well as the systems used for the plasma acceleration. Here we analyze only the energetic aspects of the acceleration process, using in our consideration the simplified "electrotechnical" model of the accelerator (rail gun). In this model the projectile of mass m is propelled by electromagnetic forces. In the general case the accelerator can be considered as an element of the electrical circuit with alternating inductance connected to the energy source (Figure 4.3).

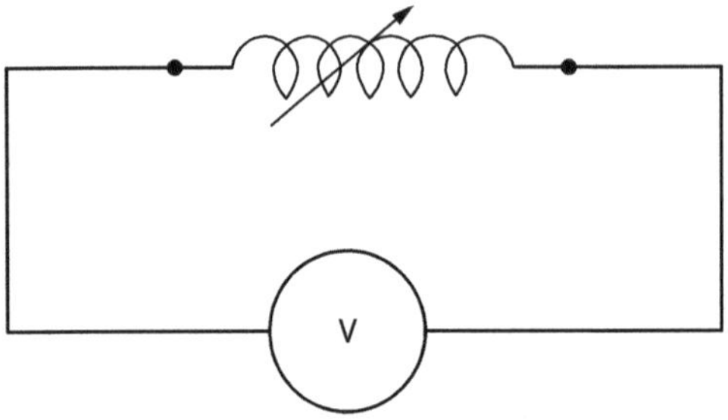

Fig. 4.3: Element of the electrical circuit with alternating inductance used for the analysis of the energetic balance of magnetic systems.

The power transferred from the source is

$$P = iU = i\,(d\psi/dt), \tag{4.17}$$

where U is the power at the entrance, i is the current in circuit, Ψ is the flux-coupling; $\Psi = Li$, L is inductance of the given circuit element. It is possible to rearrange this formula to the form

$$P = \frac{d}{dt}\left(\frac{Li^2}{2}\right) + \frac{i^2}{2}\frac{dL}{dt}. \tag{4.18}$$

Here $Li^2/2$ is the magnetic field energy. Thus we can consider that the supplied energy is converted into the magnetic field energy and other energy forms, for instance, the kinetic energy of a projectile.

In fact, the term $(i^2/2)(dL/dt)$ can be represented as the product of the force $(i^2/2)(dL/dx)$ by projectile velocity $u = dx/dt$. Then their product is the power expended in acceleration.

Let us consider some particular cases of the energy transformation for the condition of changing inductance.

4.4.1 Direct current in the element of the electrical circuit with alternating inductance

In this example $i = i_0$ = const and the both terms in equation (4.18) are equal.

The work produced over the finite time is equal to the increment of the magnetic field:

$$\Delta A = \frac{i_0^2}{2}\Delta L = \Delta W_{\mathrm{M}} = \frac{1}{2}\Delta W. \tag{4.19}$$

Here $\Delta W = \Delta A + \Delta W_M$ is the total energy transferred to the device used for projectile accelerating. With direct current the efficiency $\eta = \Delta A/\Delta W$ equals 0.5.

4.4.2 Energy transformations in the short-circuiting coil with alternating inductance

At the absence of losses in the short-circuiting coil with alternating inductance the energy of the magnetic field is conserved. One such system is the inductive energy storage. In our case the condition $U = 0$ holds. Then in the system with alternating inductance at the absence of losses the energy is also conserved, but the magnetic field energy can be converted into the projectile kinetic energy and vice versa.

The first scheme describes the operation of the inductive storage as a supply source in the system of the projectile acceleration. The second one describes the energy transfer to the storage when the conductive projectile is decelerated by electromagnetic forces. In such a manner the explosion-magnetic generators operate [5]. The explosion in them is used for accelerating the conductor which transfers the kinetic energy to the field.

From the condition $U = 0$ follows that both the total magnetic flux and energy are constant ($\psi = \psi_0$ = const, $W = W_0$). The instantaneous current value is defined by $i(t) = \psi_0/L(t)$, where $L(t)$ is the instantaneous inductance value. Let the inductance be changed from the initial value $L(t_1)$ to its terminal one $L(t_2)$. Then the change of the kinetic energy, according to equation (4.19), is

$$\Delta A = \int_{t_1}^{t_2} \frac{i^2}{2} dL = \int_{t_1}^{t_2} \frac{\psi_0^2 dt}{2L^2(t)} = \frac{\psi_0^2}{2}\left(\frac{1}{L(t_1)} - \frac{1}{L(t_2)}\right). \tag{4.20}$$

The total energy is conserved, so that the change of the magnetic field energy is $\Delta W_M = -\Delta A$. The value ΔA is positive if the inductance increases $[L(t_2) > L(t_1)]$. In this case the field energy is converted to the kinetic energy. With decreasing inductance the magnetic energy increases, the projectile is decelerated, and its kinetic energy drops.

The efficiency of the accelerating system is defined by

$$\eta = \frac{\Delta A}{W_0} = 1 - \frac{L(t_1)}{L(t_2)}, \tag{4.21}$$

assuming that at the initial instant the projectile was at rest, and all the energy was confined to the magnetic field of the storage $W_0 = \psi_0^2/(2L(t_1))$.

From formula (4.21) it can be seen that with the condition $L(t_2) \gg L(t_1)$ the efficiency of conversion of the magnetic field energy into the kinetic energy can be quite high in the circuits without losses.

4.4.3 Railgun powered by energy capacity storage

In this section we consider the conversion processes in a rail gun, which is the most simple device used for the acceleration of conductors by electromagnetic forces. The following analysis remains within the scope of the electrotechnical model, which describes the movement of a projectile with constant mass along parallel conductors. For considering of the acceleration process the scheme presented in Figure 4.4 is commonly used.

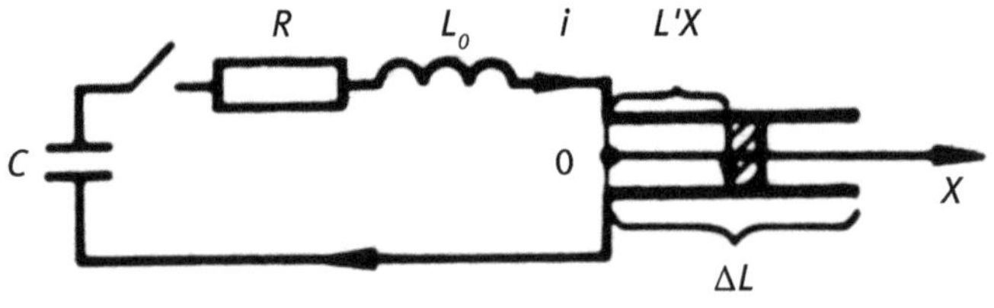

Fig. 4.4: Schematic for acceleration of the conductive projectile in a circuit with alternating inductance (electrotechnical model of a rail gun).

The major practical issue when considering the operation of the circuit in the device for acceleration of the conductive projectile, is the choice of the parameters (capacity C, initial inductance L_0, active circuit resistance R, mass of the accelerated projectile m, inductance $L(x)$, linear inductance L') yielding the efficient conversion of the capacitor bank energy into the kinetic energy of the accelerated projectile. A similar problem was considered by Artsimovich et al. in [6]. In succeeding years the electrotechnical model of a railgun was widely used in describing the processes in devices for acceleration of plasmoids and solid conductors by electromagnetic forces [7–12]. Linhart [7] studied the limit regime of slow varying inductance in the oscillating contour with $R = 0$ when the relative velocity of variations of the current oscillations frequency is much less than the frequency itself (adiabatic regime). In this case the work of the electromagnetic forces on the length where the inductance increment is δL, expressed by

$$\delta A = \frac{1}{2} \int_{L}^{L+\delta L} i^2(t)\, dL \approx \frac{1}{4} i_m^2(L)\, \delta L. \tag{4.22}$$

Here L is the instantaneous inductance value, and i is the circuit current. When integrating we take into account only the contribution from the slow varying component of i^2 and ignore the sign-changing component. The starting energy of the capacity bank is the sum of the work of the electromagnetic forces and energy of the magnetic field at peak current:

$$W_0 = A + \frac{1}{2} L i_m^2, \tag{4.23}$$

From this equality we can find the current amplitude. Further, within the limit of small increments we can replace δL by dL and then, using (4.22), obtain the approximate equation for A:

$$\frac{dA}{dL} = \frac{W_0 - A}{2L}. \tag{4.24}$$

Further we find the accelerator's efficiency:

$$\eta = \frac{A}{W_0} = 1 - \sqrt{\frac{L_0}{L_0 + \Delta L}}. \tag{4.25}$$

In the general case of the simple R–L–C-contour, the movement of the projectile of mass m, accelerated by electromagnetic forces, along the coordinate x, is defined by the equation

$$m \frac{d^2 x}{dt^2} = \frac{i^2}{2} \frac{dL}{dx}, \tag{4.26}$$

where i is the circuit current, and L is the inductance of the system.

We present the equation of the electrical circuit with resistor R and inductance C in the form

$$\frac{d^2}{dt^2}(Li) + R\frac{di}{dt} + Ci = 0, \tag{4.27}$$

where $L = L_0 + L(x)$ is the circuit inductance. It includes the starting inductance L_0 and the inductance increment $L(x)$ induced by a displacement of the projectile by distance x. In the following, we restrict ourselves to the simplest case when $L(x)$ changes proportionately to the projectile displacement: $L = L'x$, where $L' = \text{const}$. Then equation (4.27) takes the form

$$\frac{d^2}{dt^2}\left(\left(L_0 + L'x\right) i\right) + R\frac{di}{dt} + Ci = 0. \tag{4.28}$$

The initial conditions for the system of equations (4.26) and (4.28) are: $i(0) = 0$, $di/dt(0) = U_0/L_0$, $x(0) = 0$, $dx/dt(0) = 0$. Here U_0 is the initial voltage of the capacity bank.

The last value means that the projectile at the initial instant is immobile. In the considered system of the two second-order equations there are six parameters: U_0, C, L_0, L', R, m. One other parameter, ΔL, should be added to this list, which is the inductance increment at the length of acceleration: $\Delta L = l/L'$, where l is the length of the accelerator. The acceleration ends at the instant t_0 when the inductance attains the value $L = L_0 + \Delta L$. In some cases the condition $\Delta L = \infty$ can be assumed. This is helpful if the capacity bank entirely discharges earlier than the projectile escapes the acceleration length l.

The simulation of the discussed problem for each set of the parameters presents no difficulties if conventional computer programs are used. More complicated are the calculations for the parameters which should be varied for the purpose of the finding the optimal functioning regime. One example of this would be a search for obtaining the maximal efficiency of the accelerator. A detailed investigation of this problem is contained in [9, 10, 13]. Numerical modeling of a great number of variants for a wide range of changing parameters in combination with the compact representation of the results is a sound theory, as well as the analytical study of the asymptotical regimes. For this purpose it is reasonable to move to the nondimensional variables and thus decrease the number of the variable parameters.

Let us introduce the basic inductance L_B, the basic time $t_B = \sqrt{L_B C}$, and the basic length $x_B = L/L'$, and the new variables $\tau = t/t_B$, $y = x/x_B$ and $z = i\sqrt{L}/(U_0\sqrt{C})$. Then the rearrangement of the equation (4.28) gives

$$\frac{d^2}{d\tau^2}\left(\left(\frac{L_0}{L_B} + y\right)z\right) + \rho\frac{dz}{d\tau} + z = 0, \tag{4.29a}$$

where $\rho = R\sqrt{C/L_B}$. The initial conditions for the current take the form $z(0) = 0$, $(dz/d\tau)_0 = L_B/L_0$. The equation of movement is also rearranged to

$$\frac{d^2y}{d\tau^2} = gz^2, \tag{4.29b}$$

where $g = (CU_0L')^2/(2mL_B)$. As a result, we have two nonlinear second-order differential equations, the solution of which is determined by three parameters: L_0/L_B, ρ, g. The basic inductance can be taken arbitrarily. At first it makes sense to consider such regimes of acceleration when the energy losses in the resistor R are insignificant, and it is possible to put $\rho = 0$. Then the solution is determined only by two parameters. Numerical modeling of a great number of variants results in plotting of the parameter $\eta = m(dx/dt)^2/(CU_0^2)$. This parameter (the efficiency of the accelerator) is the ratio of the kinetic energy at the end of the acceleration length to the initial energy of the capacity storage. The results of modeling for a great number of variants are shown in Figure 4.5. As a parameter of L_B we take the value $L_B = \Delta L$, then g takes the value $g = (CU_0L')^2/(2m\Delta L) = k$.

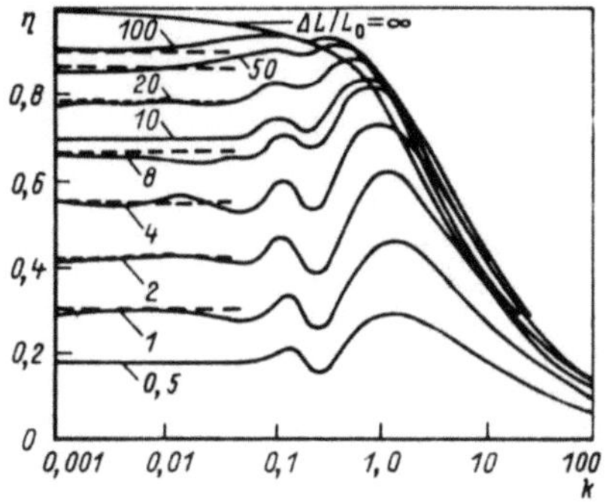

Fig. 4.5: Efficiency of the acceleration in the contour without losses.

According to equation (4.25), at small k the efficiency practically is independent on k (dotted lines in Figure 4.5). In this range of parameters the discharge current appears as oscillations with damping amplitude, and the displacement of the projectile occurs over time of few periods of the current (Figure 4.6).

Characteristically the growth of η in a range $k \ll 1$ is nonmonotonic: there are small local extremes on the curve $\eta(k)$. This is due to the fact that with the oscillating discharge the velocity builds up by pushes, as a result of which the heavier projectile can be acted upon by more half-periods of the current to the end of the acceleration and thus obtains somewhat more energy, although the main tendency in a range $k < 1$ is the growth of efficiency with decreasr of the initial inductance.

With the increase of k (decreasing of projectile mass) the speed-up at length Δx occurs for less and less number of periods. In the range of parameters of k of order 1 the efficiency reaches the maximum and then decreases. For the optimal conversion of the capacity bank energy into the energy of the accelerated projectile in the system with the linear changing inductance, the condition needs to be satisfied [10, 13]:

$$k = \left(L'CU_0\right)^2/(2m\Delta L) = 1 \div 2. \tag{4.30}$$

In this case the projectile accelerates mainly during the first have-wave of the current in the discharge contour, as it seen in Figure 4.7.

The range of large values of k corresponds to the projectiles of small mass. In this range, efficiency decreases because the light projectiles attain the predetermined inductance change early than the capacitor discharge happens. Therefore

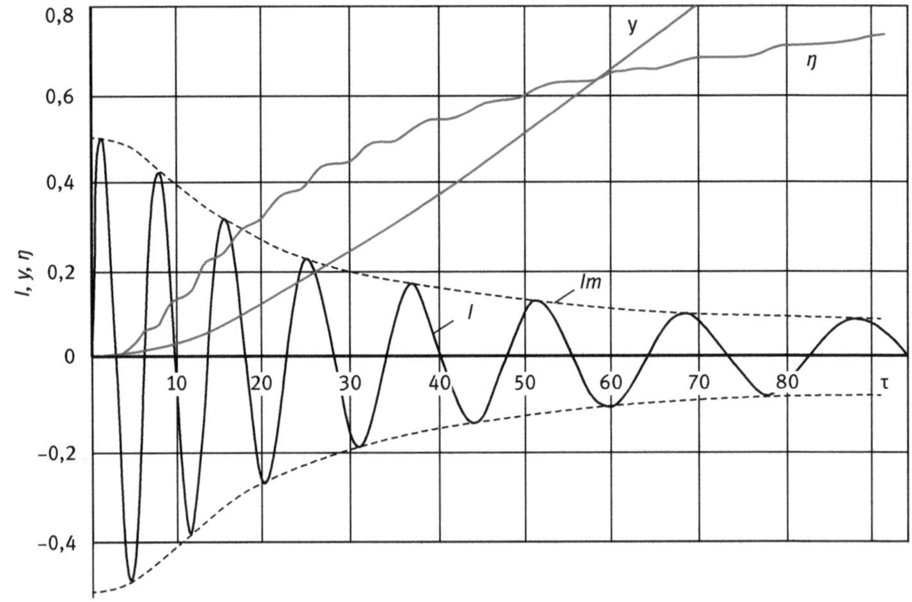

Fig. 4.6: Nondimensional coordinates of the projectile $y = xL'/L_0$; the kinetic energy $\eta = m(dx/dt)^2/CU_0^2$ and the current $I = (i\sqrt{C/L_0})/U_0$ as functions of the non dimensional time $\tau = t/\sqrt{L_0C}$ for the acceleration of the projectile of large mass, $[(CU_0L')^2/(2mL_0) = 0.016]$.

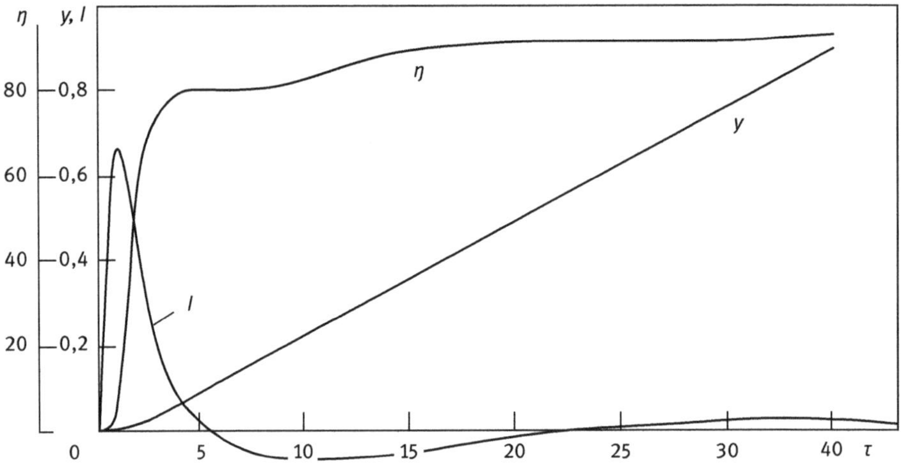

Fig. 4.7: Plots characterizing the acceleration of a light projectile: $(CU_0L')^2/(2mL_0) = 4.1$. The notations are as in Figure 4.6.

the significant part of its energy over the acceleration time does not manage to convert into the magnetic field energy and then to the kinetic energy of the accelerated body.

Maximal efficiency depends upon the ratio of the initial inductance L_0 to its variation on the acceleration length ΔL. This dependence is quite insignificant. Thus, at $k = 2$ with decreasing of $L_0/\Delta L$ from 0.3 to 0, η changes from 0.66 to 0.75. Therefore for development of the generators designed for the acceleration of conductors there is no need to tend to decrease their intrinsic (parasitic) inductance.

Figure 4.5 shows that in the range $k \gg 1$ the curves $\eta = f(k, \Delta L/L)_0$ come closer at $\Delta L/L_0 \gg 1$. The limiting case $\Delta L/L_0 = \infty$ at condition $k \gg 1$ can be simulated, assuming that the source power is constant during the movement of the light projectile. In this case the current changes according the dependence $i(t) = U_0 t/L = U_0 t/(L' x)$. Using (4.26) we obtain

$$x(t) = \left[\frac{9U_0^2 t^4}{8mL'}\right]^{1.3}, \quad i(t) = \left[\frac{8U_0 m}{9L'^2 t}\right]^{1.3} \tag{4.31}$$

and find the asymptotic value of efficiency

$$\lim \eta\big|_{\Delta L/L_0 \to \infty,\, k \to \infty} = \frac{4}{3\sqrt{k}}. \tag{4.32}$$

For other limiting regimes of the movement of the light projectile ($k \gg 1$), when $\Delta L \ll L_0$, we can find η assuming that the current is linearly growing $i = (U_0/L_0)t$:

$$\lim \eta\big|_{\Delta L/L_0 \to 0,\, k \to \infty} = \frac{4\Delta L}{\sqrt{3k} L_0}. \tag{4.33}$$

In the case of the acceleration of heavy projectiles the active resistance causes a sharp drop in efficiency. When considering the regimes of the projectile acceleration in the circuit with losses it is reasonable to choose $L_B = L_0$. Then the parameter g takes the value $g = q = (CU_0 L')^2/(2mL_0)$. An analysis of the calculations carried out for a wide range of nondimensional parameters $\rho = R(C/L_0')^{1/2}$ shows that the maximal value of efficiency and behavior of the function $\eta = f(q, \Delta L/L_0)$ in a region of the large values q are not very sensitive to the losses in the contour if $\rho \leq 0.1$. However, in the region of “slow regimes” ($q \leq 1$) the effect of the parameter ρ is rather significant. From the curves in Figure 4.8 it can be seen that at a fixed value of ρ the functions plotted for different values of $\Delta L/L_0$ merge with decreasing q. The physical meaning of this is quite apparent: when the heavy body is speeding up, the current in the circuit, due to the losses in the resistance, damps earlier than the projectile quits the length of the acceleration. This is why the kinetic energy is independent on ΔL. Evidently, the whole process is determined by

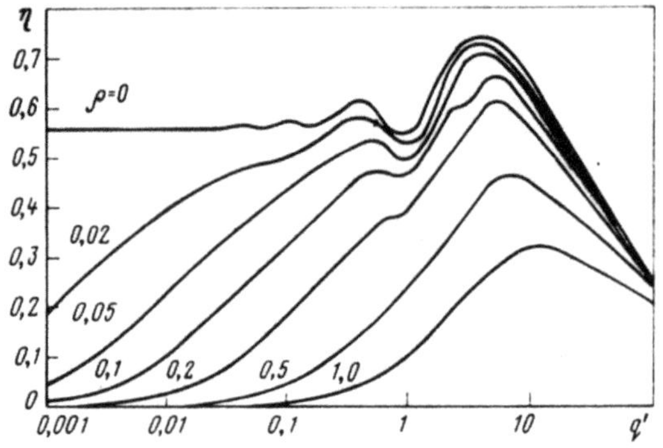

Fig. 4.8: Efficiency of the acceleration in the contour with losses, $\Delta L/L_0 = 0.2$.

the same relations as in the case of the body acceleration in a rail gun of infinite length ($\Delta L/L_0 \to \infty$).

In absence of active resistors, the efficiency of such a rail gun equals 1. In the circuit with resistance R its losses W_R and the kinetic energy at the end of the acceleration W_k depends upon the argument $\int_0^\infty i^2 dt$:

$$W_k = \left(L' \int_0^\infty i^2 dt\right)^2 /(8m); \quad W_R = R \int_0^\infty i^2 dt. \tag{4.34}$$

Since $W_0 = W_k + W_R$, we obtain the following equation for η (in nondimensional form): $\eta + 2(2\eta/q')^{1/2} - 1 = 0$. From here, as Belyaeva has shown in [1],

$$\eta = \frac{2\rho^2}{q}\left(\sqrt{1 + \frac{q}{2\rho^2}} - 1\right)^2 . \tag{4.35}$$

At $q \to 0$ the limit of this expression is $\eta = q/(8\rho^2)$, which is the asymptote of the curves $\eta = f(\rho, \Delta L/L_0, q)$ at $q \to 0$ (fig. 4.8).

General relations derived for the rail gun model can be used not only for systems with linear dependence of induction on coordinates, but in more sophisticated cases with other dependences $L(x)$ as well. Taking into account that the $\eta(\kappa)$ in the vicinity of maximum changes only slightly, when choosing the parameters providing the highest efficiency of conversion of capacity energy into kinetic energy, it is possible to use the approximate relation $L' = (L(\Delta x) - L(0))/\Delta x$.

As an example we consider the system used for the induction acceleration of a disk in the field of the multiturn magnet (Figure 4.9a). The dimensions shown in this figure are $R_1 = 0.01$ m, $R_2 = 0.06$ m, $a = 0.003$ m, $l = 0.02$ m. The mass of a copper disk of thickness $h = 0.01$ m is 1.13 kg. It has to be accelerated on the path $\Delta x = S = 0.03$ m. Consider the case where the capacity bank of 200 μF and voltage 10 kV is used as the energy supply. We have to determine such a number of the solenoid turns which provides the maximum velocity at the end of the run.

In the case of a sharply pronounced skin effect, the conductivity of the projectile can be assumed to be perfect, and the solenoid inductance can be calculated with the aid of the formula

$$L = L_0 - M, \tag{4.36}$$

where L_0 is the inductance of a solenoid, and M is the mutual inductance of the system of two solenoids with the opposite currents, shown in Figure 4.9b. Both parameters can be calculated with the help of the tables given in the book of Kalantarov and Tseitlin, *Calculation of inductances* [15]. The inductance L_0 can be determined using the formula

$$L_0 = 10^{-7} w^2 (R_1 + R_2)\Phi, \tag{4.37}$$

where the dimensionless number Φ is the function of two parameters $\rho' = (R_2 - R_1)/(R_1 + R_2)$ and $\alpha' = l/(R_1 + R_2)$. In the example under consideration, $\rho' = 0.86$, $\alpha' = 0.29$. The values of the parameter Φ are given as graphs in [15]. In the given case $\Phi = 5.5$, $L_0 = 38.5 \cdot 10^{-9} w^2$ Hn, where w is the number of turns.

The mutual inductance of the solenoid and its mirror reflection can be calculated by

$$M = 0.5(L_{123} + L_2 - 2L_{12}), \tag{4.38}$$

where L_{123}, L_{12}, L_2 are the inductances of the solenoids of length $2(a + x + l)$, $2(a + x) + l$ and $2(a + x)$, correspondingly. In this example the coefficients Φ in formulas for the initial inductance have the following values: $\Phi_{123} = 4.43$, $\Phi_{12} = 5.40$, $\Phi_2 = 6.90$. In calculations of each of the inductances entering in the formula for M, one should take into account that the number of turns in the solenoids is proportional to their lengths. For instance, in the calculation of L_{123} the coefficient 4.43 has to be additionally multiplied by 2.3^2, since the solenoid length of inductance L_{123} is 4.6 cm, whereas the length of the initial solenoid with the number of turns w is 2 cm. In such a calculation all inductances found contain the same product w^2. Finally, we find the mutual inductance corresponding to the initial position of the projectile: $M(0) = 20.3 \cdot 10^{-9} w^2$ Hn. At the end of the run one should take $x = S$. In the considered example where $S = 0.03$ m, we have $\Phi_{123} = 3.84$, $\Phi_{12} = 4,86$, $\Phi_2 = 6.62$ and $M(S) = 2.6 \cdot 10^{-9} w^2$. The increase of the inductance on the whole path of acceleration is equal to the change in the mutual inductance. In the given case this increase is $\Delta L(S) = 17.7 \cdot 10^{-9} w^2$ Hn. For the optimization calculations one can take the following approximated value of the parameter $L' = \partial L/\partial x$:

$$L' \approx \Delta L(S)/S. \tag{4.39}$$

Thus, in the considered example we have the following set of parameters for the magnetic system necessary for the further calculations: $L(0) = L_0 - M(0) =$

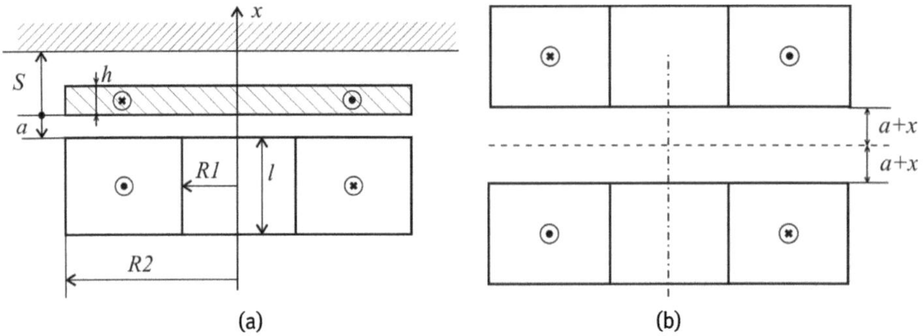

Fig. 4.9: The model of the device for electromagnetic acceleration of a flat disk in the field of the multi-turn magnet: (a) the initial configuration of the magnetic system; (b) the system of two magnets with opposite currents which allows taking into account the skin effect.

$18.2 \cdot 10^{-9} w^2$ Hn, $L' = 5.9 \cdot 10^{-7} w^2$ Hn/m, $\Delta L/L(0) = 0.97$. Ignoring the losses in the winding resistance, we can evaluate the optimal value of the parameter $k = (CU_0 L')^2/(2m\Delta L)$. According to Figure 4.6, for the found ratio $\Delta L/L(0)$ the maximum of the kinetic energy corresponds to $k \approx 1$. In this case the efficiency is about 40 %. The dimensionless coefficient k is proportional to the square of the number of turns. Here $k = 3.5 \cdot 10^{-5} w^2$. The optimal regime of acceleration is realized when the winding has 170 turns.

Some quite different experiment conditions correspond to the problem of acceleration of conductive cylinder in the devices designed for generation of x-ray pulses in inertial thermonuclear fusion, for plasma heating and producing ultrahigh magnetic fields through compression of trapped magnetic flux. These systems will be considered in Chapter 10. Here we deal only with only energetic facet of the problem for two versions of magnetic systems: Z- and θ-pinches (Figure 4.10).

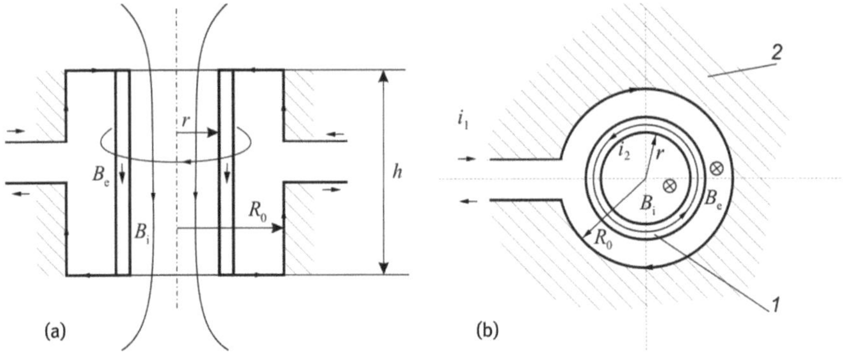

Fig. 4.10: (a) Z-pinch; (b) θ-pinch.

In the former case $L = \mu_0 h/(2\pi)\ln(R_0/r)$. As an estimating value of induction we can assume $\Delta L = \mu_0 h/(2\pi)\ln 2$, which corresponds to the compression of the cylinder up to half its original radius R_0. In this case $L' \approx -(\partial L/\partial r)_{r=R_0} = \mu_0 h/(2\pi R_0)$. In the case of θ-pinch when estimating energy parameters, we can assume $L = \mu_0\pi(R_0^2 - r^2)/l$, $\Delta L \approx \mu_0\pi R_0^2/l$, $L' = -(\partial L/\partial r)_{r=R_0} = 2\pi\mu_0 R_0/l$.

Using the above formula we can calculate, for instance, the strength of the generator applied for compression of a Z-pinch of mass 10^{-4} kg, length $2\cdot 10^{-2}$ m, and original radius $2\cdot 10^{-2}$ m. At the given capacity energy its voltage can be evaluated based on formula $U = WL'\sqrt{2/(m\Delta L)}$, followed from the condition $k \approx 1$.

If $W = 10^6$ J, then, using the above mentioned formula, we have $U \approx 3$ MV. Such high voltages are inherent in installations where, due to compression of the light liners, the powerful pulses of x-ray radiation can be obtained for studies of an inertial thermonuclear fusion.

The cited cases show that the choice of optimal parameters based on the condition $k \approx 1$ embraces the devices of various types with radically distinct projectile mass and electric parameters.

4.5 On the application of inductive storages for supplying the magnetic systems

Figure 4.11 shows the simplest array of the inductive storage in which the energy is stored in the inductance L_0 when the switch K is closed, and passes in the inductive load L_1 when it opens. In this section we consider the systems in which the inductance of a load does not change.

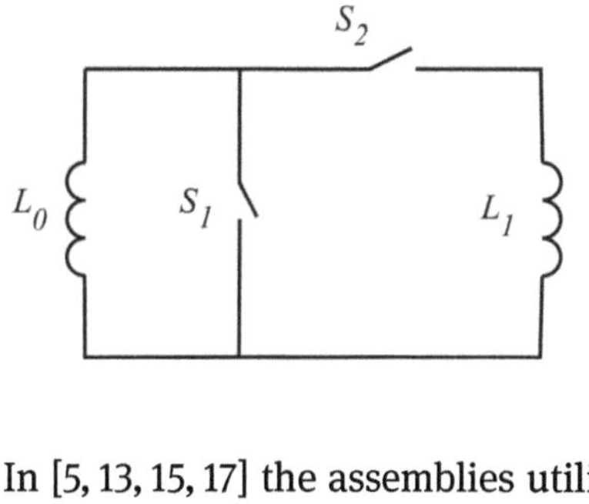

Fig. 4.11: A schematic drawing of the energy output from an inductive storage.

In [5, 13, 15, 17] the assemblies utilizing the inductive storages of energy for supplying the high field magnets are described. In this assembly first the energy is relatively slowly delivered into storage. For energy transfers in the load, the switch $\boldsymbol{S}_1$ opens the circuit and switch $\boldsymbol{S}_2$ closes it.

In the simplest circuit the initial current in the load equals zero. The energy delivered to the load is determined by the formula

$$W_1(t) = W_0(0)\,L_0^2/(L_0 + L_1)^2, \tag{4.40}$$

where $W_0(0) = L_0 i_0^2(0)/2$ is the stored energy. The energy transfer coefficient $\eta = W_1(t)/W_0(0)$ has its maximum at condition $\eta_{max} = 0,25$ [5]. In this case half the energy is dissipated in the switch, and the rest is equally distributed between the inductances L_0 and L_1.

The efficiency of the energy transfer can be increased if the initial current in the load differs from zero [13]. For instance, in the circuit presented in Figure 4.11 the η shows a maximum when

$$\frac{i_1(0)}{i_0(0)} = \frac{L_1}{L_0} \cdot \frac{1}{1 + 2L_1/L_0}. \tag{4.41}$$

In this circuit the lowest switching losses take place if $L_0 \ll L_1$ and $i_1(0) = 0.5 i_0(0)$. Under these conditions half the energy $W_0(0)$ transfers into the load, 1/4 is left in the L_0, and 1/4 is lost when switching.

The system designed for conversion of the capacitor energy to kinetic energy has certain peculiarities. In this case formula (4.21) characterizes the transfer of energy to a projectile after the switching of the inductive capacitor is finished. Then the energy W_0 is the initial energy of the inductive capacitor minus the switching losses. If the capacitor with initial inductance L_0 and current $i_0(0)$ at instant t_1 is switched to the load with initial inductance L_1, then, as was shown above, with ideal switching the energy $\Delta W = L_0 L_1 i_0^2(0)/(2L(t_1))$ is lost, where $L(t_1) = L_0 + L_1$. Thus

$$W_0 = \frac{L_0 i_0^2(0)}{2} - \Delta W = \frac{L_0 i_0^2(0)}{2} \cdot \frac{L_0}{L_0 + L_1}. \tag{42a}$$

Taking into account the switching losses, the efficiency is [13]

$$\eta = \frac{\Delta A}{L_0 i_0^2/2} = \frac{\Delta L\, L_0}{(L_0 + L_1 + \Delta L)(L_0 + L_1)}, \tag{42b}$$

where $\Delta L = L(t_2) - L(t_1)$. It turns to zero at condition $L_0 \ll L_1$ or $L_0 \gg \Delta L, L_1$. In the former case the main amount of energy is expended for switching losses. In the latter the efficiency drops because the change of relative inductance becomes small with displacement of a projectile. The efficiency is maximal at condition

$$L_0 = L_1 \sqrt{\frac{L_1 + \Delta L}{L_1}}, \tag{42c}$$

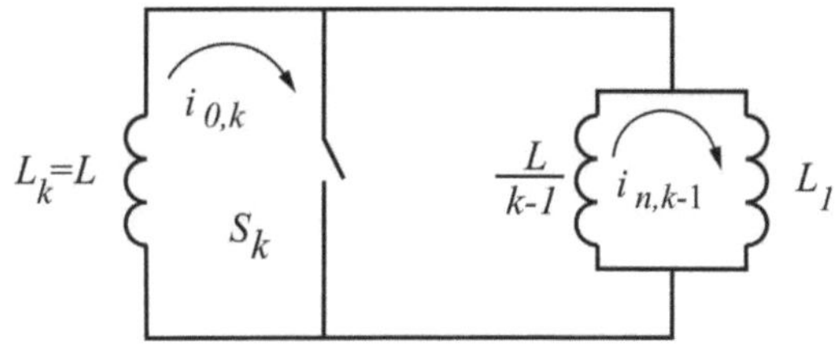

Fig. 4.12: A schematic drawing of "dissipationless" energy output in multisection inductive storage

when the capacitor inductance is the geometric mean of initial and final values of the load inductance:

$$\eta_{\max} = \frac{\Delta L}{L_1\left(1 + \sqrt{1 + \Delta L/L_1}\right)^2}. \tag{42d}$$

Circuits of "dissipationless" switching have been described in papers by Larionov, Spevakova, Stolov, and Azizov [17] and Lototsky [18, 19]. For instance, in the circuit shown in Figure 4.12, the next section of the capacitor with an inductance L_k and current $i_{0,k}$ is connected to the load L_1 by a switch P_k after the previous $(k-1)$ sections are already connected to the load carrying a current $i_{1,k-1}$. In the considered circuit all inductances are equal to L_1 and all currents $i_{0,k} = i_0(0)$. When the next section of the capacitor is switched on, the current in the load differs from zero. It results in a decrease of switching losses much as in the discussed simplest scheme of the one-step switching at the presence of initial current in the load. The analysis and optimization of the multistep storage are given in [13], where it is shown that at similar initial currents $i_{0,k}$, when the initial current in the load is lacking $i_1(0) = 0$ and a number of sections $N \gg 1$, the losses in the switches can be as small as one wishes. The energy transferred into the load is expressed by the relation

$$W_1(t) = W_0(0)\,\frac{\ln^2(1+\lambda)}{\lambda}, \tag{43a}$$

where $W_0(0) = NLi_0^2(0)$ is the total initial energy of all sections of the capacitor with inductance L apiece, and $\lambda = L_1N/L$. The maximal efficiency $\eta = W_1(t)/W_0(0)$ is achieved at condition $\lambda \approx 0.39$ and is $\eta_{\max} \approx 0.65$.

Energy transfer efficiency can be increased to a greater extent if inductances of the alternatively connected sections L_k are different and governed by the law

$$L_k \approx \sqrt{\frac{2L_1W_0}{L_0^2}}\exp\left(-i_0k\sqrt{\frac{L_{\mathrm{H}}}{2W_0}}\right). \tag{43b}$$

This formula has been derived through the solution of the optimization problem in the procedure of which the function $L_k(k)$ was calculated by Euler's method at given equal initial currents $i_k = i_0$.

The cited formula holds if the condition $i_0[L_1/(2W_0)]^{1/2} \ll 1$ is fulfilled. This means that the energy of a magnetic field in a load with current i_0 is much less than the energy of the capacitor. Energy transfer efficiency can be as close to unity as one wishes if the number of sections is sufficiently large. This is seen from the formula for η:

$$\eta \approx \left[1 - \exp\left(-Ni_0\sqrt{\frac{L_{\mathrm{H}}}{2W_0}}\right)\right]^2 . \tag{43c}$$

Other more complicated schemes of "dissipationless" energy transfer have been proposed in [19].

4.6 References

[1] K. S. W. Champion, *Proc. Phys. Soc. (London)* B **53** (1950), 795–806.
[2] C. Fabri, *Eclairage Electrique* **17** (1898) 133–141.
[3] D. B. Montgomery, *Solenoid Magnet Design*, New York, Wiley Interscience, 1969.
[4] L. J. Campbell and J. Schilling, in: F. Herlach and N.. Miura (eds.), *High Magnetic Fields. Science and Technology*, pp. 153–203, New Jersey London Hong Kong Taipei Singapore, 2003.
[5] H. Knoepfel, *Pulsed High Magnetic Fields*, Amsterdam, London, North-Holland Publishing Company, 1970.
[6] L. A. Arzimowitch, C. Yu. Lukyanov, I. M. Podgorniy, et al., *J. Exp. Theor. Phys.* **33**(7) (1957), 3–10 (in Russian).
[7] J. G. Linhart, *Nucl. Fusion* **1**(1) (1960), 78–81.
[8] L. V. Dubovoy and Y. A. Beresin, *J. Tech. Phys.* **34**(10) (1964), 1867–1870 (in Russian).
[9] S. G. Alihanov, G. I. Budker, G. N. Kichigin, et al., *J. Appl. Mech. Tech. Phys.* **4** (1966), 38–41(in Russian).
[10] A. B. Novgorodcev and G. A. Shneerson, *Izvestiya Akademii Nauk, Energetika i Transport.* **2** (1970), 154–161 (in Russian).
[11] G. A. Shvetsov, V. M. Titov, Y. L. Bashkatov, et al., **MG-III**, pp. 177–182.
[12] V. N. Bondaletov, E. N. Ivanov, S. A. Kalichman, and Y. P. Pitchugin, **MG-III**, pp. 234–238.
[13] G. A. Shneerson, *Fields and Transients in Superhigh Pulse Current Device*, New York, Nova Science Publishers, Inc., 1997.
[14] E. I. Belayeva, *Issledovaniya deformirovaniya tonkich obolochek*, Dissertaziya. Moskowskiy Energeticheskiy Institut, 1969.
[15] P. L. Kalantarov and L. A. Tseytlin, *Rschet Induktivnostey.Spravochnaya kniga*, Leningrad, Energiya, 1970 (in Russian).
[16] R. Carruthers, Energy storage for thermonuclear research, in Proceedings of the Inst. of El. Eng. Convention on Thermonuclear Processes. 29–30 April. 1959.
[17] B. A. Larionov, F. M. Spevakova, A. M. Stolov, and E. A. Azizov, in: E. P. Velikhov (ed.), Fizika i Technika moschnih impulsnih system, Energoatomizdat, pp. 66–105, 1987 (in Russian).pp.66–105
[18] A. P. Lototskiy, Preprint of Kurchatov Institut, No. 3714/14, Moscow, 1982.
[19] A. P. Lototskiy, *Elektrichestvo* **6** (1985), 64–66 (in Russian).

5 Electromagnetic forces and mechanical stresses in multiturn solenoids. The optimization of multilayered windings

Electromagnetic forces affecting the conductors in magnetic systems with high and ultrahigh fields may lead to the destruction of these systems if mechanical stress in the elements of a winding exceed the acceptable values. In the well-known experiments by Kapitza the strength constraints were already the main obstacle limiting the amplitude of the induction. The calculation of forces and stresses is a necessary stage in the development of magnetic systems. The initiation of these calculations is attributed to the work of Kapitza and Cockroft [1, 2]. A necessary stage in the development of the high field magnets is the calculation of the forces and strength. In the widely known books by Montgomery [3], Knoepfel [4], and others the calculation formulae for stresses are presented, obtained by the solution of the appropriate problems in elasticity theory. In this chapter we give the formulas allowing us, with a definite idealization, to carry out the evaluation of magnet strength. We restrict ourselves to simple configurations allowing for analytical solutions with the help of which we can estimate mechanical forces and the qualitative characteristics of their distribution. A more detailed analysis must consider the real configuration of the magnet, the nonuniformity of a winding (consisted of conductors and insulating interlayers), the presence of reinforcements, the action of axial forces, etc. In this chapter we focus on calculations of forces and strengths for the case of a quasistatic regime, when the characteristic time interval of the process is much larger than the period of the elastic eigen oscillations of the magnetic system. The role of inertia effects will be considered in Chapter 8. In recent years a more comprehensive analysis is carried out using the available programs for the solution of the complex problem of the estimation of parameters of fieldd, forces, and stresses, using the method of finite elements. Some results of such calculations are given in this chapter as well as in Chapter 6, in which nondestructive magnets are in example.

Further on we consider axial-symmetric magnetic systems subjected to radial and axial electromagnetic forces. The components of the elastic stress tensor $\sigma_{\varphi\varphi}$, σ_{rr}, σ_{zz}, which are the solutions of the linear problem of the elasticity theory, are expressed as a sum of two constituents: the former corresponds to radial electromagnetic forces, and the second to axial ones (for instance, $\sigma_{\varphi\varphi} = \sigma'_{\varphi\varphi} + \sigma''_{\varphi\varphi}$). In notations of the tensor components one lower index is used for instance, σ_{φ} instead of $\sigma_{\varphi\varphi}$.

5.1 Asimuthal and axial stresses in the thin-wall turn in the poloidal magnetic field

The simplest approach to the calculation of stresses in a magnet winding is to break down the winding into its elements in the form of conducting thin cylindrical layers carrying current, which mechanically do not interact with each other (the model of separated layers). For example, this model has been described by Montgomery [3].

Let us consider the element of the solenoid winding with azimuth current as a turn of length l and of thickness h which is much less than the radius r (Figure 5.1). The azimuth stress in this element could be approximately evaluated without resorting to the solution of differential equations of the elasticity theory, but assuming that the turn is in the equilibrium due to the joint action of electromagnetic forces and azimuth stretching stresses. The effect of the radial stress can be neglected because the turn thickness is small, and it does not interact mechanically with other layers of the winding. This case may occur when layers of the winding are separated by gaps or in the monolithic winding there are absent the radial stresses due to proper distribution of the electromagnetic forces.

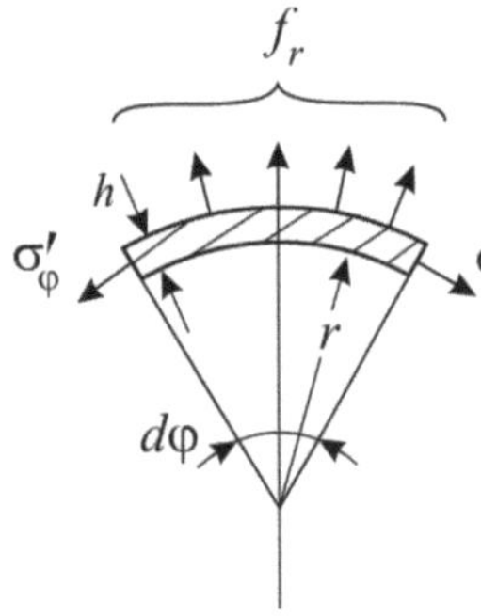

Fig. 5.1: Schematic for calculation of equilibrium of a thin winding element.

Let us consider the action of radial forces. Let us express the equilibrium conditions of the winding element in the shape of a closed uniform ring with evenly distributed azimuth current. Each unit surface is affected by the radial force f_r which is the result of electromagnetic forces, while in azimuth direction each of cross-sections is affected by forces $\sigma_\phi hl$. Projecting forces on the symmetry axis we arrive at the equilibrium equation:

$$lf_r r d\varphi - \sigma'_\varphi hl d\varphi = 0.$$

From this it follows that

$$\sigma'_\varphi = \frac{r}{h} f_r. \tag{5.1a}$$

It should be noted that in this approach the inertial forces are not taken into account, which means that we are considering the static problem. It is valid for the pulsed magnetic fields if the characteristic pulse duration much more than the period of the natural mechanical oscillations of the magnetic system.

If the axial components of the magnetic field induction inside and outside of the layer are B_1 and B_2, then

$$f_r = \frac{B_1^2 - B_2^2}{2\mu_0} = \frac{B_1 + B_2}{2} j_\varphi \tag{5.1b}$$

where j_φ is the azimuthal current per unit length of the system. In a case of a large length single cylinder we have $B_2 = 0$. Then

$$\sigma'_\varphi = \frac{B_1^2 r}{2\mu_0 h} = \frac{B_1 j_\varphi r}{2h}. \tag{5.1c}$$

This stress differs from the magnetic pressure of the internal field $P_M = B_i^2/(2\mu_0)$ by the factor R/h. The action of the field on a single-layer winding is analogous to that of a gas with the pressure P_M on the wall of the tube.

The strength of a material may be characterized by a permissible stress $[\sigma]$. In order to evaluate the strength characteristics of materials with relation to the generation of strong magnetic fields, it is convenient to replace $[\sigma]$ by a conventional magnetic ultimate stress B_M connected with $[\sigma]$ by the relation

$$B_M = \sqrt{2\mu_0 [\sigma]}. \tag{5.2}$$

The values of the parameter B_M for some materials are given on the diagram in the paper by Furth, Levine, and Waniek [5] (Figure 5.2). The diagram represents the materials whose ultimate strength attains 50–60 T. These materials were applied until the advent of new composite wires of the type of Cu–Nb, Cu–Ag, and others.

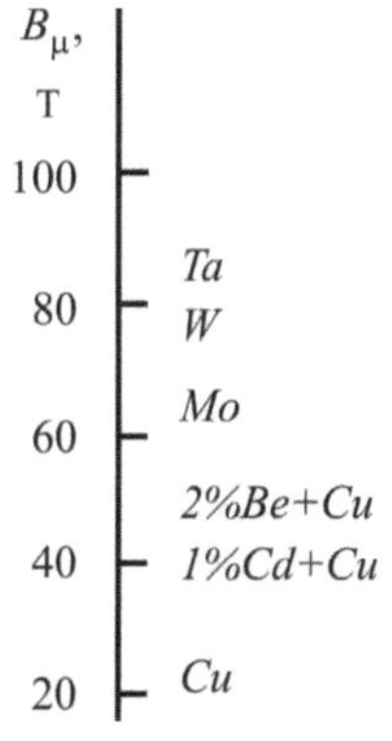

Fig. 5.2: "Magnetic limit of strength" for various metals corresponding to the static loading regime at 20 °C.

These diagrams do not reflect the fact that the ultimate strength depends on the number of acting pulses. In solenoids designed for prolonged operation (hundreds of pulses and more), the ultimate strength decreases, e.g., for copper from 26 to 19 T. The ultimate strength depends on the pulse length and increases when the pulse length decreases. This question is considered in Chapter 8, where the more elaborated table is given, containing also the other ultimate values of the induction which are determined by the yield point for stresses produced by electromagnetic forces and for thermoelastic stresses.

For a winding of finite length the simplest case is a winding with free ends. The axial strength has maximum in the middle of a winding ($z = 0$). In the wall of the solenoid of length l and radius R it can be calculated by

$$\sigma_z'' = \int_{-l/2}^{0} \delta_\varphi B_r dz, \tag{5.3a}$$

where B_r is the induction radial component in the wall [6, 7]. Further, we consider a case when the winding length is much larger than its diameter. With this assumption we can consider that the induction distribution on the axis of the magnet with distance from its end is identical to that of half-infinite solenoid:

$$B_z(0, z') = \mu_0 h \delta_\varphi \left(1 + \frac{z'}{\sqrt{(z')^2 + R^2}}\right), \tag{5.3b}$$

where $z' = z + l/2,\ h \ll R$.

It is convenient to continue the calculations using the Witteker's formula [8] to construct the flux function based on the given axial induction distribution:

$$\psi(r, z') = \frac{\mu_0 \delta_\varphi h}{2\pi} \int_0^r r\, dr \int_0^\pi B_z(0, (z' + jr\cos\omega))\, d\omega. \tag{5.3c}$$

Further on we can define the radial induction component in a winding, $B_r(R, z') = (-1/R)(\partial\psi/dz')$ and calculate the axial strength using equation (5.3a):

$$\sigma_z'' = -\frac{\mu_0 h \delta_\varphi^2}{2\pi R} \int_0^R r dr \int_0^\pi \int_0^\infty \frac{R^2 dz' d\omega}{\left[(z' + jr\cos\omega)^2 + R^2\right]^{3/2}} = -\frac{B_1^2}{4\mu_0} \frac{R}{h}. \tag{5.3d}$$

Here B_1 is the induction in a middle part of the large length solenoid.

Note that in the considered example the axial component of the tensor of elastic stresses is half of the azimuth component and has the opposite sign.

Formula (5.3d) overestimates the axial strength of the medium plane of a one-layer winding. The more accurate value accounting for the first terms of the strength expansion have been calculated in the study of Ivanchik and Sannikov [9]:

$$\sigma_z'' = -\frac{B_1^2}{4\mu_0}\frac{R}{h}\left(1 - \frac{7R^2}{2l^2}\right). \quad (5.3e)$$

5.2 Mechanical stresses in the uniform cylinder with a given current distribution

The magnet with the monolithic winding and continuous current distribution is the simplest model of the actual two-component system including conductors and insolating interlayers clamped by a special compound. Using this model, one can approximately estimate stresses and obtain the starting data for more accurate calculations accounting for the media inhomogeneity.

For monolithic magnets with a rectangular winding the analytical formulae cited in [10–12] give us the opportunity to calculate to some accuracy the stresses in a winding. Mechanical stresses arising under the action of radial forces will be calculated here analytically for a uniform cylinder of infinite length and supplemented with results of numerical calculations of the various parameters $\alpha = R_2/R_1$ and $\beta = l(2R_1)$.

In the accepted analytical model the components of the stress and strain tensors are functions of one coordinate r and satisfied known equations of the elasticity theory [13, 14]:

$$\varepsilon_r' = \frac{1}{E}\left(\sigma_r' - \mu\left(\sigma_\varphi' + \sigma_r'\right)\right), \quad (5.4a)$$

$$\varepsilon_\varphi' = \frac{1}{E}\left(\sigma_\varphi' - \mu\left(\sigma_r' + \sigma_z'\right)\right), \quad (5.4b)$$

$$\varepsilon_z' = \frac{1}{E}\left(\sigma_z' - \mu\left(\sigma_r' + \sigma_\varphi'\right)\right) = 0, \quad (5.4c)$$

$$\frac{d}{dr}\left(\varepsilon_\varphi' r\right) - \varepsilon_r' = 0, \quad (5.4d)$$

$$\frac{d}{dr}\left(\sigma_r' r\right) - \sigma_\varphi' = -f_r \cdot r, \quad (5.4i)$$

$$\sigma_z' = \mu\left(\sigma_r' + \sigma_\varphi'\right), \quad (5.4j)$$

where E is the elasticity module, μ is the Poisson coefficient, and $f_r = \delta_\varphi B_z$ is the volume electromagnetic force.

In many constructions the ends of the magnets are fastened by external clamping devices. It allows exclusion of the axial deformations and the assuming that $\varepsilon_z' = 0$. Then further transformations lead us to the following system of differential equations:

$$\frac{d}{dr}\left(r\left(\sigma'_\varphi + \sigma'_r\right)\right) - \left(\sigma'_\varphi + \sigma'_r\right) = -\frac{rf(r)}{1-\mu}, \tag{5.5a}$$

$$\frac{d}{dr}\left(r\left(\sigma'_\varphi - \sigma'_r\right)\right) + \left(\sigma'_\varphi - \sigma'_r\right) = \frac{rf(r)}{1-\mu}\cdot(1-2\mu). \tag{5.5b}$$

This has the solution

$$\sigma'_{r,\varphi} = -\frac{1}{2(1-\mu)}\int_{R_1}^{r} f(r)\,dr - \frac{1-2\mu}{2(1-\mu)}\frac{1}{r^2}\int_{R_1}^{r} r^2 f(r)\,dr + C_1 \mp \frac{C_2}{r^2}, \tag{5.5c}$$

where R_1 is the inner radius of the cylinder, $C_{1,2}$ are constants defined by boundary conditions. In a case of free boundaries, the conditions $\sigma'_r(R_1) = \sigma'_r(R_2) = 0$ are fulfilled. The cylinder loaded by electromagnetic forces is in equilibrium without external reinforcements. Then the constants $C_{1,2}$ are given by the formulas

$$C_1 = \frac{R_2^2}{R_2^2 - R_1^2}[\cdot\frac{1}{2(1-\mu)}\int_{R_1}^{R_2} f(r)dr + \frac{1-2\mu}{2(1-\mu)}\frac{1}{R_2^2}\int_{R_1}^{R_2} r^2 f(r)dr, \tag{5.6a}$$

$$C_2 = C_1 \cdot R_1^2. \tag{5.6b}$$

Further we consider as an example the analytical estimation of stresses in a solid winding with the given current distribution and some results of numerical calculations as well.

5.2.1 A winding with constant current density

The induction distribution in this case for a uniform cylinder of infinite length is described by

$$B = B_i \frac{R_2 - r}{R_2 - R_1} \tag{5.7a}$$

and

$$f_r(r) = \frac{B_i^2}{\mu_0}\cdot\frac{R_2 - r}{(R_2 - R_1)^2}. \tag{5.7b}$$

Calculations using formulas (5.5) lead to the following dependence for σ_r' è σ_φ', in which the nondimensional magnitudes $\alpha = R_2/R_1, \rho = r/R_1$ are used:

$$\sigma'_{r,\varphi} = \frac{B_i^2}{2\mu_0}\frac{1}{(\alpha-1)^2}\left\{\frac{1}{1-\mu}\left(\alpha-\frac{1}{2}-\alpha\rho+\frac{\rho^2}{2}\right)\right. \quad (5.8a)$$

$$\mp\frac{1-2\mu}{1-\mu}\left(\frac{\alpha\rho}{3}-\frac{\rho^2}{4}+\frac{1}{4\rho^2}-\frac{\alpha}{3\rho^2}\right)+\frac{\alpha^2}{\alpha^2-1}$$

$$\left.\cdot\left[\frac{-1}{1-\mu}\left(\alpha-\frac{\alpha^2}{2}-\frac{1}{2}\right)+\frac{1-2\mu}{1-\mu}\left(\frac{\alpha^2}{12}+\frac{1}{4\alpha^2}-\frac{1}{3\alpha}\right)\right]\left(1\mp\frac{1}{\rho^2}\right)\right\}.$$

The stress σ_z' can be calculated using (5.4j). On the internal boundary the azimuth stress is a maximum:

$$\sigma_\varphi'(R_1) = \frac{B_i^2\alpha^2}{\mu_0(\alpha-1)^2(\alpha^2-1)}$$
$$\cdot\left[\frac{1}{1-\mu}\left(\frac{\alpha^2}{2}+\frac{1}{2}-\alpha\right)+\frac{1-2\mu}{1-\mu}\left(\frac{\alpha^2}{12}+\frac{1}{4\alpha^2}-\frac{1}{3\alpha}\right)\right]\Bigg\}. \quad (5.8b)$$

It takes the value

$$\sigma_\varphi'(R_1) = \frac{B_i^2}{12\mu_0}\frac{(7-2\mu)}{(1-\mu)}, \quad (5.8c)$$

when $R_2 \gg R_1$.

The strength of the winding can be characterized by von Mises criteria [15]:

$$\sigma_M = (1/\sqrt{2})\sqrt{(\sigma_\varphi-\sigma_z)^2+(\sigma_\varphi-\sigma_r)^2+(\sigma_z-\sigma_r)^2}. \quad (5.9)$$

The numerical results obtained for the dependence of maximal values of the $2\mu_0\sigma_M/B_i^2$ on the aspect ratio α and parameter $\beta = 2R_1/l$ are shown in Figure 5.3[1].

The calculations have been performed for two types of sealing of the winding end parts: for free ends and for a rigid sealing. The characteristic feature is that in the considered range of the parameters the maximal value of the von Mises stresses differs from the maximal azimuth stress by less than 20 %. This is clear from the values indicated by points in Figure 5.3. The maximal stresses σ_M and σ_φ do not depend anymore on the length of the magnet, provided that the length is larger than the internal diameter by a factor of 3–4. Note that the sealed ends weakly affect the maximal value of the von Mises stress; however, the relation between the latter and the maximal azimuth stresses changes. In the magnet with free ends we have $\sigma_{M,\max} > \sigma_{\varphi,\max}$, while for the sealed ends the opposite inequality takes place.

1 Numerical calculatons have been made by a method of finite elements [16].

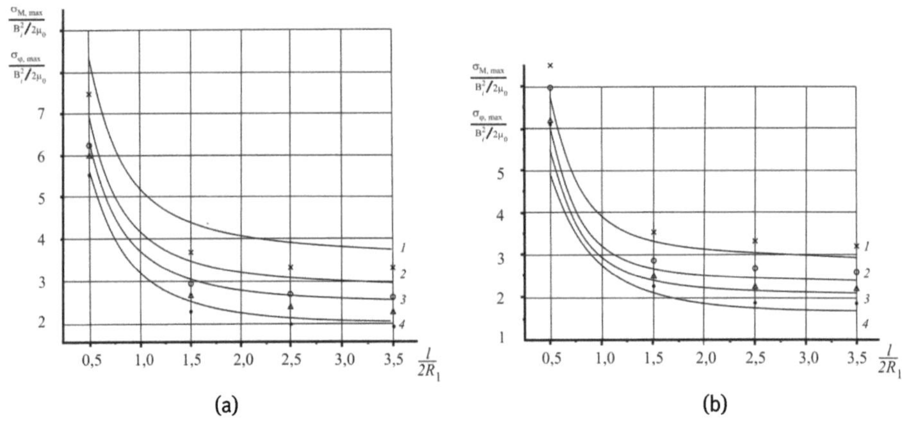

Fig. 5.3: Mechanical stresses $2\mu_0\sigma_{M,\max}/B_i^2$ (*lines*) and $2\mu_0\sigma_{\phi,\max}/B_i^2$ (*points*) caused by radial forces in a monolithic winding with the constant current density. (a) Free ends of a winding. (b) A rigid sealing of ends. The values of the aspect ratio $\alpha = R_2/R_1$ and the notations of the points: (1) $\alpha = 1.5$ (×); (2) $\alpha = 1.75$ (∘); (3) $\alpha = 2$ (Δ); (4) $\alpha = 3$ (•).

5.2.2 A winding with a current density decreasing inversely with radius (Bitter's solenoid)

For this winding the change of the current and induction density are described by the following formulas for a magnet with $l/(2R_1) \gg 1$:

$$\delta_\varphi = \frac{B_i}{\mu_0 r \ln\alpha}, \tag{5.10a}$$

$$B = B_i\left(1 - \frac{\ln(r/R_1)}{\ln\alpha}\right). \tag{5.10b}$$

The dependence of this type occurs in a winding made from the disks (Bitter's solenoids) or thick-wall cylinder at the absence of a skin effect [3, 11]. Calculations using the formulas above result in the following dependence for the stresses caused by radial forces in a winding with fastened ends:

$$\begin{aligned}\sigma'_{r,\varphi} = \frac{B_i^2}{2\mu_0\ln\alpha}\Bigg\{&-\frac{1}{1-\mu}\left(\ln\rho - \frac{\ln^2\rho}{2\ln\alpha}\right)\\ &\mp\frac{1-2\mu}{1-\mu}\left[\frac{1}{2}\left(1-\frac{1}{\rho^2}\right)\left(1-\frac{1}{2\ln\alpha}\right) - \frac{\ln\rho}{2\ln\alpha}\right]\\ &-\frac{\alpha^2}{1-\alpha^2}\left[\frac{1}{2(1-\mu)}\ln\alpha + \frac{1-2\mu}{1-\mu}\left(-\frac{1}{2\alpha^2} - \frac{1}{4\ln\alpha}\left(1-\frac{1}{\alpha^2}\right)\right)\right]\left(1\mp\frac{1}{\rho^2}\right)\Bigg\}.\end{aligned} \tag{5.11a}$$

This formula does not take into account the end effects, that is, refers to magnets of a length far exceeding its external radius. Figure 5.4 shows the numerically calcu-

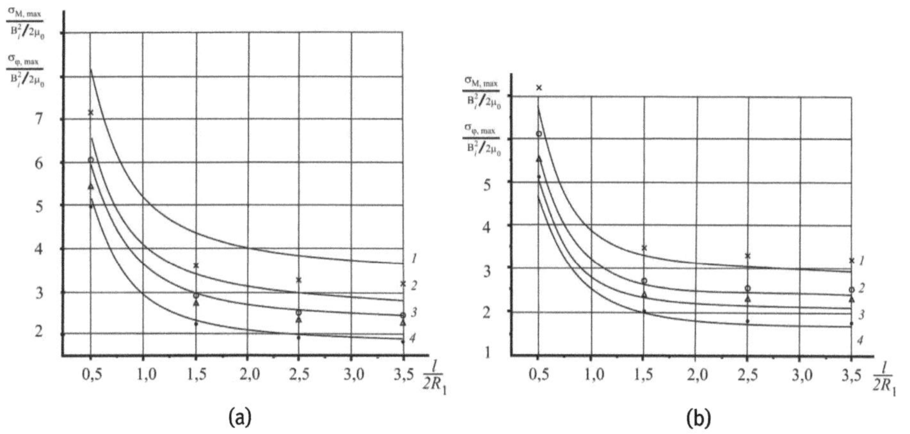

Fig. 5.4: Mechanical stresses $2\mu_0\sigma_{M,\max}/B_i^2$ (*lines*) and $2\mu_0\sigma_{\phi,\max}/B_i^2$ (*points*) caused by radial forces in a monolithic winding with the current density decreasing inversely with radius.
(a) Free ends of a winding. (b) A rigid sealing of ends. The values of the aspect ratio $\alpha = R_2/R_1$ and the notations of the points: (1) $\alpha = 1.5$ (×); (2) $\alpha = 1.75$ (∘); (3) $\alpha = 2$ (Δ); (4) $\alpha = 3$ (•).

lated maximal value of the stress obtained by the von Mises formula for solenoids of the different length and aspect ratio.

It has its maximum at the internal boundary ($\rho = 1$):

$$\sigma'_\varphi(R_1) = \frac{B_i^2}{2\mu_0 \ln\alpha} \cdot \frac{\alpha^2}{\alpha^2 - 1}\left[\frac{1}{1-\mu}\ln\alpha - \frac{1-2\mu}{(1-\mu)\alpha^2}\left(1 + \frac{1-\alpha^2}{2\ln\alpha}\right)\right]. \tag{5.11b}$$

In the limit $\alpha \gg 1$ we have

$$\sigma'_\varphi(R_1) = \frac{B_i^2}{2\mu_0(1-\mu)}. \tag{5.11c}$$

Note that, in spite of the fact that the current distribution for monolithic magnets of these two types are not the same, the maximal values of the von Mises stress differ insignificantly. The distinctions, which are beyond the scope of the monolithic magnet model, can take place in real windings with the different configuration of conductors and insulating interlayers between them.

5.3 Mechanical stresses in an equilibrium thin-wall cylinder with current

Mechanical stresses in a winding can be decreased if it is subjected to the opposite directed electromagnetic forces which can be partially compensated. In Chapter 7 we consider the construction of the quasi force-free multilayer windings in which the compensation results in adequate decreasing of stresses. Here we consider the equilibrium of a thin-wall, one-layer winding with the axial and azimuth currents 1 installed in the crossed magnetic field produced by the azimuth current distributed in a conductive cylinder 2, and in the azimuth field produced by axial current in conductor 3 (Figure 5.5).

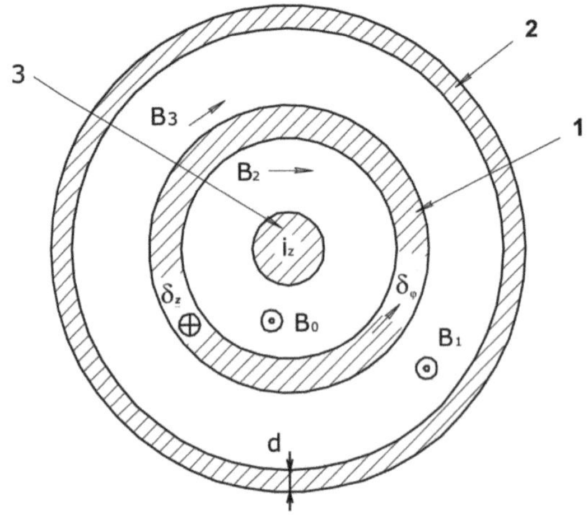

Fig. 5.5: Solenoid in an external field. (1) Thin-wall winding with axial and azimuth currents in an external magnetic field. (2) Magnet generating an external axial field. (3) Conductor with axial current.

In the general case the field in solenoid ***1*** has axial and azimuth components. The axial field with induction $\mathbf{B}_1$ is produced by t magnet ***2*** installed outside. For generating of the azimuth field $\mathbf{B}_2$ it is necessary to accommodate in a cavity of the magnet the conductor ***3*** with the axial current i. When the current in solenoid ***1*** is switched on, the field is produced in it by the help of the winding stacked at the angle γ to the axis. As a result, the azimuth field with the induction $\mathbf{B}_3$ is generated in a gap between magnets ***1*** and ***2***, and the axial field with induction B_0 is generated in a gap between solenoid ***1*** and conductor ***3*** (Figure 5.5). The radial component of the current density in the winding equals zero. Neglecting the discrepancy of the winding we assume that both the azimuth component of the current density δ_φ and axial one δ_z are constant in the winding thickness. Due to the superposition of the intrinsic field of the solenoid 1 and external field of solenoid ***2*** the following induction distribution is formed; see Figure 5.6.

In a layer of small thickness, both components exhibit linear dependence on the coordinate $x = r - R$:

$$B_z(x) = B_0 + (B_1 - B_0)\xi, \tag{5.12a}$$

$$B_\varphi(x) = B_2 + (B_3 - B_2)\xi, \tag{5.12b}$$

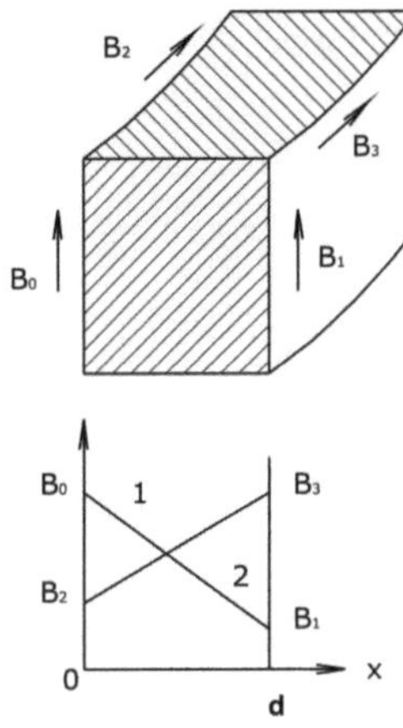

Fig. 5.6: Induction distribution in a winding of a thin-wall magnet.

where $\xi = x/d$ and

$$\delta_\varphi = (B_0 - B_1)/(\mu_0 d), \quad \delta_z = (B_3 - B_2)/(\mu_0 d). \tag{5.12c}$$

Let the characteristic dependence be the total force acting on the winding element of thickness x:

$$F(x) = \int_R^{R+x} f_r(x)\, dr, \tag{5.13}$$

where $f_r(x) = \delta_\varphi B_z - \delta_z B_\varphi = -\frac{1}{\mu_0}[\frac{\partial}{\partial r}(B_z^2 + B_\varphi^2) + \frac{B_\varphi^2}{r}]$.

Assume that $d \ll R$; then we can ignore the last term in the given formula. The current in a winding can be chosen in such a way that the equilibrium condition is fulfilled:

$$B_0^2 + B_2^2 = B_1^2 + B_3^2. \tag{5.14}$$

In view of condition (5.14), for the force $F(x)$ we obtain

$$F(x) = (1/2\mu_0)\left[B_0^2 + B_2^2 - \left(B_z^2(x) + B_\varphi^2(x)\right)\right]. \tag{5.15}$$

This functions reduces to zero at $x = 0$ and $x = d$. After substitution of equations (5.12a) and (5.12b) in (5.15) we have

$$F(x) = (1/2\mu_0)\left(\xi - \xi^2\right)\left[(B_0 - B_1)^2 + (B_2 - B_3)^2\right]. \tag{5.16}$$

The function $F(x)$ attains maximum F_m in a middle plane of a winding ($\xi = 1/2$):

$$F_m = (1/8\mu_0)\left[(B_0 - B_1)^2 + (B_2 - B_3)^2\right]. \tag{5.17}$$

The equations of the elasticity theory (5.5) allow us to estimate the azimuth and radial stresses in the equilibrium layer of small thickness. For a winding with fixed ends ($\varepsilon_z = 0$), the stress modules peak in a middle of a layer ($\xi = 1/2$) and have the values[2]

$$\sigma'_{\varphi\,\max} = -\frac{2-\mu}{3(1-\mu)}F_m, \quad \sigma'_{r\,\max} = -F_m,$$

$$\sigma'_{z\,\max} = \mu(\sigma'_{r\,\max} + \sigma'_{\varphi\,\max}) = \frac{\mu(5-4\mu)}{3(1-\mu)}F_m, \tag{5.18}$$

where μ is the Poisson coefficient. Further on we can calculate the equivalent stress (according to von Messes) for the equilibrium winding:

$$\sigma'_M = \frac{1}{\sqrt{2}}\left((\sigma'_r - \sigma'_\varphi)^2 + (\sigma'_\varphi - \sigma'_r)^2 + (\sigma'_z - \sigma'_r)^2\right)^{1/2} = bF_m. \tag{5.19}$$

In the particular case when $\mu = 0.3$ we have $b \approx 0.4$.

The ratio $2\mu_0\sigma'_M/B_0^2$ can be represented as a function of two nondimensional parameters $g = B_1/B_0$ and $t = B_2/B_0$. With regard to the equilibrium condition (5.14) we have

$$\frac{2\mu_0\sigma'_M}{B_0^2} = \frac{b}{4}\left[(1-g)^2 + \left(t - \sqrt{1+t^2-g^2}\right)^2\right] \tag{5.20}$$

From the given formulae it follows that the turns in the equilibrium winding should be stacked to the angle to the axis, which is

$$\gamma = \mathrm{arctg}\frac{B_0 - B_1}{B_3 - B_2} = \mathrm{arctg}\frac{1-g}{\sqrt{1+t^2-g^2} - t} \tag{5.21}$$

In this case the vector of the current linear density is in parallel with the vector $\mathbf{B}(0) - \mathbf{B}(d)$, rotated by angle $\pi/2$.

In a case when the azimuth field B_2 is not switched on ($t = 0$), the stress in a winding of the solenoid 1 can be determined by

$$\frac{2\mu_0\sigma'_M}{B_0^2} = \frac{b}{2}(1-g). \tag{5.22}$$

For instance, at the condition $g = 0.3$ we have $(2\mu_0\sigma'_M/B_0^2) = 0.35b \approx 0.14$. Here the angle is $\gamma \approx 0.64 \approx 36^\circ$.

Let us consider the case when the induction of the azimuth field on the inner side of the solenoid installed in the external field is equal to the induction of the

2 As in Section 5.2, in calculations of stresses only the radial component of the electromagnetic force is taken into account.

poloidal field on its external side ($B_2 = B_1, t = g$). For the winding to be in an equilibrium, the condition $B_3 = B_0$ should be fulfilled. Then

$$\frac{2\mu_0\sigma'_M}{B_0^2} = \frac{b}{2}(1-g)^2, \quad \gamma = \pi/4. \tag{5.23}$$

When $g = 0.3$ we have $(2\mu_0\sigma_M/B_0^2) = 0.245\, b \approx 0.1$.

Thus, the added axially located solenoid 1 permits us, with unchangeable induction B_0 at the axis, to decrease the field generated by the second magnet 2 to the value gB_0, as well as the magnetic pressure acted on the external magnet to the value $g^2B_0^2/(2\mu_0)$. Besides, as in formulas (5.21), (5.22) and considered examples demonstrate, the stresses in the equilibrium winding of the added solenoid can be significantly less, as compared to the magnetic pressure of the estimated field $B_0^2/(2\mu_0)$ [17].

Using the slightly loaded added solenoid 1 one can attain the higher field in comparison with magnetic system 2 operating near the strength limit. For instance, in order to increase the induction near the axis by 30 %, without increasing the load on the initial magnet system, one can position in the cavity of the magnet the solenoid with equilibrium winding characterized by parameters $t = 0$, $g = 0.77$, $\gamma \approx 20°$.

5.4 Mechanical stresses in two-component winding

In an operating winding, the gaps between the circular or rectangular conductors are packed with isolation. The ratio of conductive volume to the total volume of a winding (the filling factor) λ is usually about 0.5–0.9. As early as 60–70 years ago, glass-epoxy compound and other strong materials were widely used as insulating materials[3]. The strength characteristic of the winding conductors can greatly differ from those of the insulating layers. The elasticity module and strength limit of such materials as the glass–epoxy compound is lower in comparison with the winding's conductors. On the contrary, some current insulating materials (for instance, Zylon [18] and carbon fibre [25]) exceed the conductors in strength characteristics.

The calculation of stresses in the two-component winding is rather troublesome. In the next chapter the results of the computer calculations for certain windings with sharply different mechanical characteristics of the conductors and insulating layers will be presented.

3 The examples of windings with insulation of different type are considered in the next chapter.

The study [19] gives the analytical estimation of stresses in a winding consisting of the alternate cylindrical conductive layers and dielectric. The results are shown in Figures 5.7 and 5.8. Figure 5.7 corresponds to the case when the elasticity module of the winding material is $E_1 = 100$ GPa_(copper), and for the dielectric it is assumed to be equal to $E_2 = 60$ GPa (glass–epoxy compound). In this example the main part of the load is taken by the conductor. The azimuth stresses in conductors here are higher in comparison with a case of a homogeneous media with the elasticity module E_1 and uniform radial distribution of the current. In Figurre 5.8 the dielectric (graphite fiber) has the higher elasticity module compared to copper: $E_2 = 1.9$ GPa. Here the main load is taken by insulating interlayers.

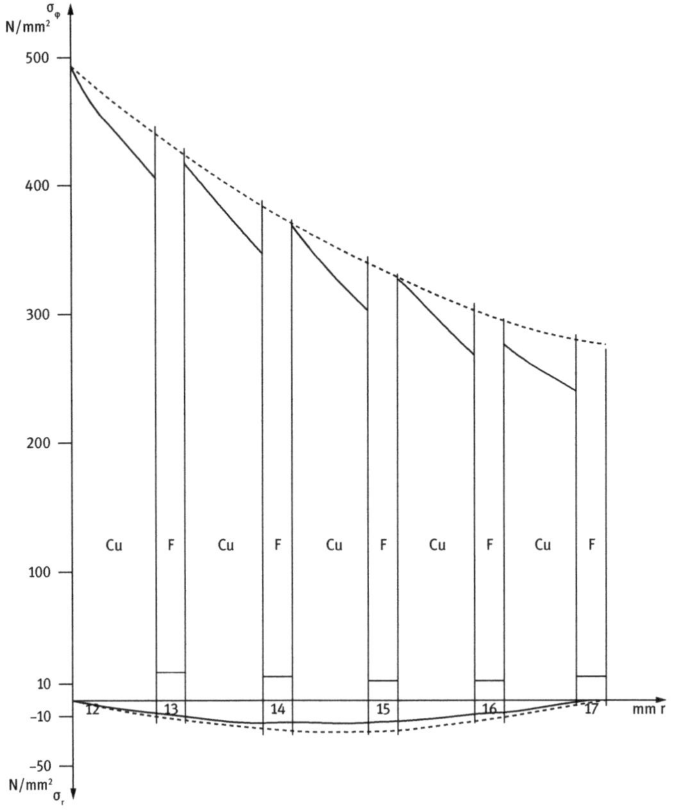

Fig. 5.7: orrAzimuth and radial stresses in the multilayer magnet with the constant current density. The induction on the axis is 18.75 T. M = copper conductors; I = insulating interlayers. The *dotted line* corresponds to calculations according the model of hollow homogenous cylinder [19].

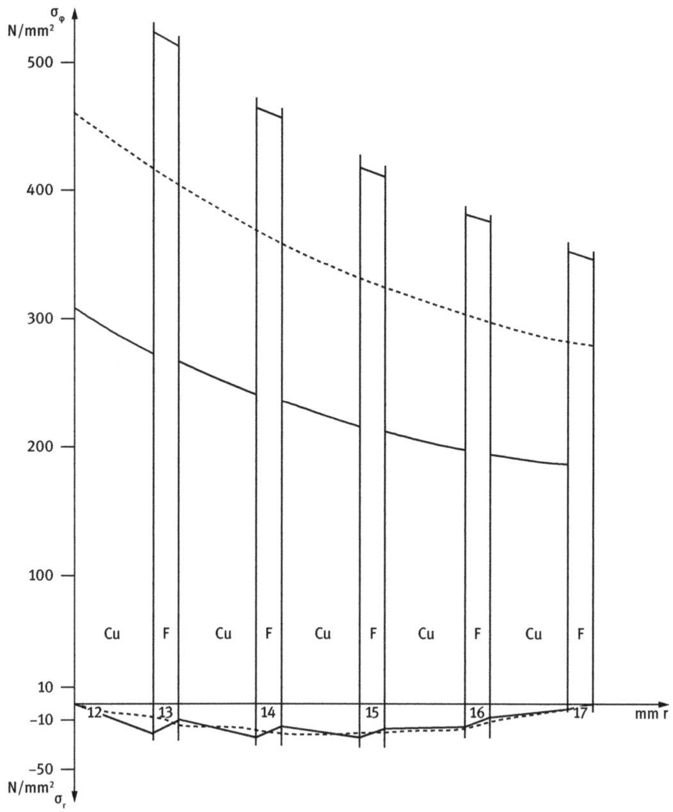

Fig. 5.8: Distribution of stresses in the solenoid as shown in Figure 5.7 but the carbon fiber is used as insulation (E_2 = 190 GPa).

It should be noted that the strength characteristics of materials, especially those of the dielectric, could only be specified with some uncertainty. Therefore when calculating, it is reasonable to use the simplified estimates. One can consider individually the response of the two-component system to the radial and axial forces. This method is used when calculating the strength of reinforced elastic material, the analog of which is the winding (there the conductors are considered to be "the reinforcement" and insulation to be "the binder").

These forces act differently on the winding. The radial forces result in stretching of turns. For calculations of stresses in both materials in the case of the winding soaked with binding compound, one can use "the law of mechanical mixing" [20]. In the simplified form it leads to the following dependences for the azimuth stresses $\sigma'_{\varphi 1}$ in conductors (the elasticity module is E_1) and those $\sigma'_{\varphi 2}$ in the dielectric (the elasticity module is E_2) [21, 22]

$$\sigma'_{\varphi 1} = \frac{\sigma'_{\varphi 0}}{\lambda + (1-\lambda)(E_2/E_1)}, \quad \sigma'_{\varphi 2} = \frac{\sigma'_{\varphi 0}}{(1-\lambda) + \lambda(E_1/E_2)}, \tag{5.24a}$$

where λ is a filling factor, $\sigma'_{\varphi 0}$ is equivalent stress:

$$\sigma'_{\varphi 0} = \lambda \sigma'_{\varphi 1} + (1-\lambda)\sigma'_{\varphi 2}. \tag{5.24b}$$

The fact that these formulas can be used for evaluation of the strength of the coils with insulating interlayers was proved in the work of Katrukhin and Doroshenko [23].

These formulas follow from the assumption that the media are bonded together and the azimuth deformation ε'_{φ} is the same. The magnitude $\sigma'_{\varphi 0}$ is the stress calculated for the monolithic media with elasticity module $E_0 = \lambda E_1 + (1-\lambda)E_2$.

For multilayer media these formulas give the same qualitative result as shown in Figures 5.7 and 5.8.

In magnets in which the conductors are held in position by the cylinders made from dielectrics with high elasticity module the insulation cylinders experience the main load, but the conductors do not. The examples of such constructions are given in the next chapter.

In multiturn magnets with a spiral winding the conductors in each layer are separated by insulating interlayers not only in a radial direction but in an axial direction as well. These interlayers are subjected not only to azimuth stresses considered above, but axial and radial stresses just take place in the monolithic media. In two-component media, the normal-to-boundary components of the elastic stress tensor are the same both in the conductor and in insulation. For instance, the axial stress σ_z in a winding with rectangular conductors does not change as one goes from conductor to dielectric in the axial direction. It was noted above that in a two-component winding subjected to radial forces the conditions could be possible when one layer is more loaded than the other. On the contrary, under action of axial forces, both layers are equally loaded. Therefore, when using relatively "soft" and low-strength insulation, the application of the hard conductors does not involve a decrease of stresses in the insulation arising under axial forces. All one has to do is to decrease the compressing axial forces and firmly fasten the faces of the magnet.

5.5 Magnets with mechanically separated thin current layer. Series or parallel connection of layers

5.5.1 A winding with a series connection of current layers

When a large number of current layers of equal thickness are connected in a series one can consider the average current density to be the same in all layers. The study [24] deals with strengths involved in such a solenoid. In this problem with a great number of mechanically separated and electrically isolated layers, one can consider the current distribution to be continuous and use equation (5.1), taking the condition $\delta_\varphi = B_i/\mu_0(R_2 - R_1) = \text{const}$. The induction in the magnet linearly drops from B_i at $r = R_1$ to zero at $r = R_2$:

$$B = B_i \frac{R_2 - r}{R_2 - R_1} \tag{5.25}$$

Further using equation (5.2b) we come to the next expression for the azimuth stress:

$$\sigma'_\varphi = \frac{B_i^2}{\mu_0} \frac{r(R_2 - r)}{(R_2 - R_1)^2}. \tag{5.26}$$

Azimuth stress in the layers changes nonmonotonically if condition $\alpha > 2$ holds. Figure 5.9 shows ia example of the azimuth stress versus radius.

On condition $\alpha \leq 2$, the greatest stress occurs at the inner boundary of the coil:

$$(\sigma'_{\varphi,\max})_{\alpha \leq 2} = \frac{B_i^2}{\mu_0(\alpha - 1)}. \tag{5.27a}$$

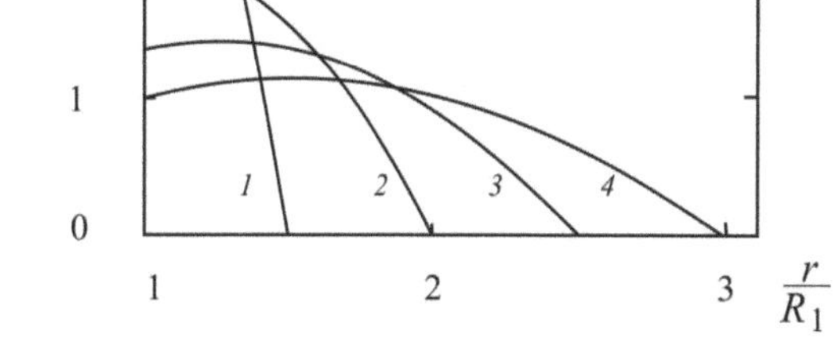

Fig. 5.9: The distribution of the azimuth stresses in a long multi-layer magnet connected in series with the thin layers of the winding. Values of the parameter $\alpha = R_2/R_1$: (1) 1.5; (2) 2.0; (3) 2.5; (4) 3.0.

On condition $\alpha \geq 2$ the maximum of the σ'_ϕ appears inside the winding in the point with the coordinate $r = R_2/2$. It takes the value

$$(\sigma'_{\varphi,\max})_{\alpha\geq 2} = \frac{B_i^2}{4\mu_0}\frac{\alpha^2}{(\alpha-1)^2} \tag{5.27b}$$

and lies in the range $(2-0.5)B_i^2/(2\mu_0)$ at the change of the aspect ratio in the limits $2 < \alpha < \infty$ [26, 31].

Ivanchik and Sannikov [9] have calculated the axial force acting on the current layer with radius r in a winding with rectangular cross-section having the inner radius R_1 and external radius R_2 when the current density is constant. Using the results of this study, one can calculate the axial stress in the middle plane of the magnet of length l and write it as the sum of the two terms of expansion:

$$\sigma''_z = -\frac{B_i^2}{12\mu_0(R_2-R_1)^2}\left[\left(-2r^2+3rR_2-\frac{R_1^3}{r}\right)-\frac{7r(R_2^3-R_1^3)}{2l^2}\right]. \tag{5.27c}$$

Here the second term in relation to the former is a magnitude of the order $(R_2/l)^2$.

5.5.2 A winding with parallel-connected layers

Another dependence for the stress will be at the parallel connection of separated layers of equal thickness. In this case, in a long solenoid we have the following expressions for the current density, induction, and azimuth stress [3, 26]

$$\delta_\varphi = \frac{B_i}{\mu_0 r\ln\alpha}, \tag{5.28a}$$

$$B = B_i\left(1-\frac{\ln(r/R_1)}{\ln\alpha}\right), \tag{5.28b}$$

$$\sigma'_\varphi = \frac{B_i^2}{\mu_0\ln\alpha}\left(1-\frac{\ln(r/R_1)}{\ln\alpha}\right). \tag{5.28c}$$

Azimuth stress in this case acquires the highest value on the boundary of the inner winding's and then decreases monotonously (Figure 5.10).

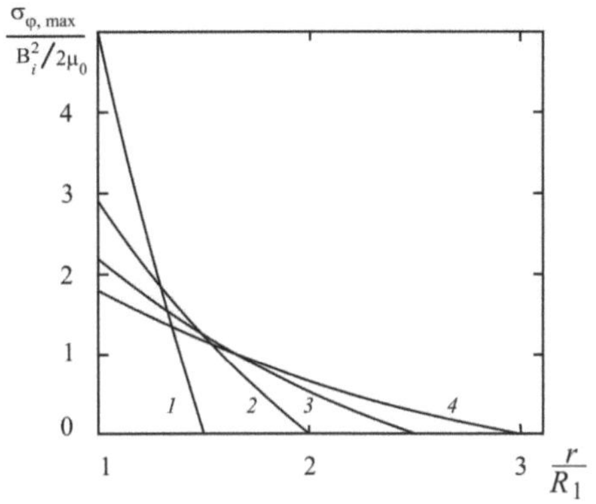

Fig. 5.10: Azimuth stresses arising in the magnet winding with parallel-connected separated layers under the action of radial electromagnetic forces. Values of the parameter $\alpha = R_2/R_1$: (1) 1.5; (2) 2.0; (3) 2.5; (4) 3.0.

5.6 Multilayer magnet with equally-loaded winding

A multilayer winding consisting of noninteracting mechanically thin layers possesses the greatest strength if all layers are equally loaded. Proposals to make the multilayer solenoids with equally loaded winding were considered in papers [26–32]. Let the layers be of the same thickness; then the condition of equal strength for the layers subjected to radial forces has the form

$$\sigma'_\varphi = [\sigma] = rB_z\delta_\varphi = \text{const.} \tag{5.29}$$

The actual solenoid represents a design consisting of several winding layers, reinforced by reinforcements. The current distribution in a real system is discrete. In [29] the analytical calculations were carried out for the multilayer equally-loaded winding. To simplify the research of the optimal solenoid design, one can replace the discrete current distribution by a continuous one [26–28, 30–32]. Then the average current density in layer δ_ϕ which could be changed step-wise from layer to layer is replaced by continuous function $\delta_\phi(r)$ related to induction by the equation

$$\delta_\varphi = \frac{-1}{\mu_0}\left(\frac{\partial B_z}{\partial r} - \frac{\partial B_r}{\partial z}\right), \tag{5.30}$$

where B_z and B_r are axial and radial components of induction. These components are also the continuous functions of the coordinate r. In what follows we restrict ourselves to the calculation of stresses in the middle part of the solenoid with a length significantly exceeding the external radius. In this case we can ignore the second term in equation (5.30). The system of equations (5.29) and (5.30) is useful to find the field distribution in the equally-loaded long coil with the internal radius R_1 and external radius R_2:

$$B_z = \sqrt{B_i^2 - 2\mu_0[\sigma]\ln(r/R_1)}, \tag{5.31a}$$

where B_i is the field inside of the solenoid.

Outside of the coil $B = B_e$, and hence it follows that

$$[\sigma] = \frac{B_i^2 - B_e^2}{2\mu_0 \ln(R_2/R_1)}, \tag{5.31b}$$

$$B_z = B_i \sqrt{\frac{\ln(R_2/r) + \frac{B_e^2}{B_i^2} \ln(r/R_1)}{\ln(R_2/R_1)}}, \tag{5.31c}$$

$$\delta_\varphi(r) = \frac{B_i}{2\mu_0 r \sqrt{\ln \frac{R_2}{R_1}}} \cdot \frac{1 - \frac{B_e^2}{B_i^2}}{\sqrt{\ln \frac{R_2}{r} + \frac{B_e^2}{B_i^2} \ln \frac{r}{R_1}}}. \tag{5.31d}$$

Figure 5.11 shows the distribution of induction and of current density in the winding in the absence of the external field. In this limiting case the current density increases infinitely when approaching the external boundary. The current density at the external boundary remains finite if the magnet is placed in the external field:

$$\delta_\varphi(R_2) = \frac{B_e}{2\mu_0 R_2} \cdot \frac{\frac{B_i^2}{B_e^2} - 1}{\ln \frac{R_2}{R_1}}. \tag{5.31e}$$

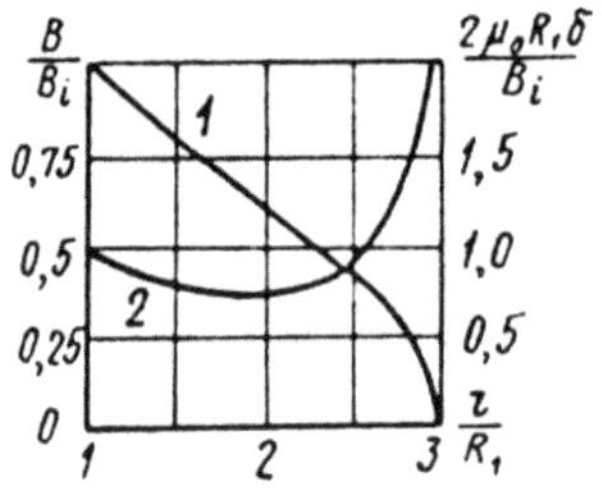

Fig. 5.11: Distribution of the induction and currcurrent density in idealized equally-loaded solesolenoid with ratio $R_2/R_3 = 3$: (1) B/B_i; (2) $2\mu_0 R_1 \delta/B_i$.

It should be noted that the assumption on the absence of mechanical interaction between the layers of the winding agrees with the condition of the constancy of the azimuth stress along the radius: in this case the relative deformation of all layers is equal, and hence, the radial forces between the adjacent layers are absent. Actually, if the two adjacent thin layers do not interact before the deformation (for instance, were separated by an infinitesimal small gap), then at equal σ_ϕ all layers will be exposed to the equal relative deformation. The similarity transformation occurs and the gap is maintained, which means that mechanical interaction between the layers do not result.

In the above idealized equally-loaded solenoid without an external field, the current density is distributed nonmonotonously (Figure 5.11). It shows minimum at $r = R_2/\sqrt{\mathbf{e}} \approx 0.6R_2$ and grows indefinitely at $R = R_2$.

It is interesting to compare the conditionally allowed stress $[\sigma]$ with a magnetic limit of strength B_M. In the coil with soft insulation stress in the conductor is $\sigma'_\varphi \approx [\sigma]/\lambda$. Therefore B_M coupled with $[\sigma]$ by the ratio $B_M = \sqrt{2\mu_0[\sigma]/\lambda}$. It is easy to see that the induction obtained in the solenoid with equally-loaded winding exceeds B_M as $[\ln\alpha]^{1/2}$, where the aspect ratio $\alpha = R_2/R_1$. Theoretically the field in the solenoid could be indefinitely higher than B_M, but in practice very high aspect ratios should be applied. For instance, In order to obtain the field with induction $B_i = 50$ T in the solenoid with a copper winding (without additional reinforcements) which has $B_M = 26$ T, it is necessary to have $\ln\alpha = 3.7$ or $\alpha = 40$, which is difficult to fulfill.

As was noted, in the idealized equally-loaded winding of a large length the current density at the external boundary is infinitely high. It was shown in [32] that along with this at the external boundary the logarithmic divergence takes place for the axial stress. The authors of this work propose excluding the mentioned effects by terminating the winding at radius $R^* < R_2$. Using this technique it is possible to avoid excessively high stresses and overheating of external layers of the winding. The calculation have shown that in the optimized magnet of a large length at the cutoff of external part, and taking into account the action of axial forces, it is possible to ensure the constancy of von Mises stress. For such a magnet the aspect ratio is larger compared to the aspect ratio calculated without regard to the axial forces. For example, when $B_i = 2B_M$, the aspect ratio increases approximately twofold.

Formula (5.31d) shows that the current density and the axial stress remain finite at the external boundary in the magnet placed in the external field. Another method to obtain the equally-loaded winding without a sharp increase of the current density is the use of an external reinforcement. This follows from calculations carried out in [35]

Along with the growth of the aspect ratio as the induction increases, the magnet length, characterized by the parameter $\beta = l/(2R_1)$, should somewhat increase as well. This leads to an additional increase of energy. Therefore, the corrections, taking into account the magnet finite length, should be introduced in the relations describing the magnet of infinite length. In [28, 33] the current density distribution in a mid-plane of the equally-loaded magnet was calculated based on the approximating formula for induction B_z in the point with coordinate r:

$$B_z(r,0) \approx \mu_0 \int_r^{R_2} \frac{\delta_\varphi(r)\cdot dr}{\sqrt{1+(2r/l)^2}}. \qquad (5.32a)$$

Here the contribution in the axial induction component induced by currents inside the cylinder with radius r is not taken into consideration. Besides, it is assumed that a field of currents outside of this cylinder is homogeneous inside it. Both assumptions are valid for the mid-part of a magnet where the condition $2r/l \ll 1$ is fulfilled. In order to calculate the admissible azimuth stress in the mid-plane of a magnet, caused by radial forces, formula (5.2b), $[\sigma'_\varphi] = rB_z(r,0)\langle\delta_\varphi(r)\rangle$, is used, where $\langle\delta_\varphi\rangle$ can be found with the help of (5.32a):

$$\langle\delta_\varphi\rangle = -\frac{2}{\mu_0 l}\sqrt{(l/2)^2 + r^2}\cdot\frac{dB_z}{dr}. \tag{5.32b}$$

Thus, for induction we have the equation

$$\frac{rB_z(r,0)}{\mu_0}\cdot\frac{dB_z(r,0)}{dr}\cdot\sqrt{1+\left(\frac{2r}{l}\right)^2} = -[\sigma]. \tag{5.32c}$$

Its solution has the form

$$B(r,0) = \left(2\mu_0[\sigma]\cdot\ln\left[\frac{R_2}{r}\cdot\frac{1+\sqrt{1+(2r/l)^2}}{1+\sqrt{1+(2R_2/l)^2}}\right] + B_e^2\right)^{1/2}. \tag{5.32d}$$

In contrast to (5.31c), here the additional factor is present: $(1+\sqrt{1+(2r/l)^2})/(1+\sqrt{1+(2R_2/l)^2})$. It is less than unity, i.e., the magnetic energy density in the magnet of the finite length is less than in the solenoid of the infinite length with an equally-loaded winding. In the center of the magnet the induction is [28, 33]

$$B_i \approx B(R_1,0) = [2\mu_0[\sigma]\ln\left(\alpha\cdot\frac{\beta+\sqrt{1+\beta^2}}{\beta+\sqrt{\alpha^2+\beta^2}}\right) + B_e^2]^{1/2}. \tag{5.32e}$$

Hence, instead of equation (5.31b), the approximate relation for the aspect ratio of the magnet of finite length is

$$\alpha = \frac{2\beta\left(\beta+\sqrt{1+\beta^2}\right)\alpha_0}{\left(\beta+\sqrt{1+\beta^2}\right)^2 - \alpha_0^2}, \tag{5.33}$$

where $\alpha_0 = \exp[(B_i - B_e)/B_M]^2$ is the aspect ratio for a magnet of infinite length. From the expression above it is seen that at a constant ultimate strength in the equally loaded winding of finite length the aspect ratio is higher compared to a magnet of infinite length. According to these estimations, the magnet with an equally-loaded winding can be achieved only under the condition

$$\beta + \sqrt{1+\beta^2} \geq \alpha_0. \tag{5.34a}$$

In the case $\beta \gg 1$) this condition is

$$l/(R_1) > \alpha_0. \tag{5.34b}$$

More accurate calculation of an equally-loaded winding in the external field was made by Doroshenko [34]. In this paper a field of parameters α and β is given, in which the stress deviations from the constant value in the mid-plane of the magnet are rather small, and thus formula (5.33) can be used for estimation.

Although the given estimation follows from the approximate relation (5.32a), it is reasonable to assume that the equally-loaded winding with sufficiently small ratio of the stress to the magnetic pressure cannot be developed if its length is commensurable with the inner radius. For example, in order to have the equally-loaded magnet with an aspect ratio $\alpha = 20$, its length, according to inequality (5.34b), must satisfy the condition $l/(R_1) \geq 20$. In the absence of the external field the induction on the axis of such a magnet satisfies the condition $B_i \leq (1/2) \ln \alpha \cdot B_M \approx 1.5 B_M$.

The effect of the length on the aspect ratio becomes insignificant at the condition $\beta \gg \alpha_0 \gg 1$ as is seen from the expansion into a series of the relation (5.33):

$$\alpha = \alpha_0 \left(1 + \frac{\alpha_0^2}{4\beta^2} + \cdots \right). \tag{5.34c}$$

A more complete computer analysis of the relation between the length and aspect ratio of the equally loaded magnet is made in [35]. This work presents the results of numerical calculations of effective stresses in an equally-loaded winding of finite length, according to von Mises, and accounting for axial forces. For the model with continuous distribution of the azimuth current the following expression was derived:

$$\ln \alpha = p + q \ln \alpha_0, \tag{5.35}$$

where $p = 0.22$ $q = 1.15$. As is seen from this work, the basic contribution to stresses, as calculated with the von Mises formula, is determined by axial forces. According to (5.35), at the condition $\ln \alpha_0 \approx 3$, the aspect ratio takes the value $\alpha = \exp(3.67) = 39.25$ instead of $\alpha_0 = 20$. The additional contribution, related to the effect of the winding finite length, is not significant. This correlates with the above estimations for long systems.

Some restrictions are inherent in the analysis, based on assumptions that the longitudinal section of the winding is rectangular in shape and the current distribution is continuous. The latter assumption results in the limitless growth of the current density on the external boundary of the winding. In an actual winding with discrete layers, the current density does not increase indefinitely at $r = R_2$. As is shown, this effect is also absent in the case of a winding fixed by an external reinforcement, and of a winding which is in an external field. These factors are taken into consideration in the numerical modeling of actual magnetic systems.

However, the model of the equally-loaded winding with continuous current distribution allows the magnetic energy of the solenoid to be generally characterized together with the requirements for the power supply system, while taking into consideration the admissible level of the adiabatic heating of conductors.

The magnetic energy per unit length of the solenoid with an equally-loading winding calculated without accounting for the axial forces and in the absence of the external field can be found using the relations [30]:

$$W'_M \approx \frac{\pi\lambda B_M^2 R_1^2}{4\mu_0}\left(\left(\frac{R_2}{R_1}\right)^2 - 1\right) = 0.5\,\pi\lambda\,(R_2^2 - R_1^2)\,[\sigma]. \tag{5.36a}$$

For a magnet of finite length the formula for the energy has the form

$$W_M = f\pi R_1^2(\alpha^2 - 1)V\lambda\,[\sigma], \tag{5.36b}$$

where V is the volume of the magnet, including the volume of the internal cavity. Here the coefficient f takes a value 0.37 with regard for axial forces, and 0.35 if the dependence of the aspect ratio on the magnet length is taken into account [35].

The given estimations show that in a case of magnets with induction level close to 100 T and above $\alpha \gg 1$, the size and energy of such magnets sharply increase compared to magnets operating in the field of 70–80 T. This is confirmed by data on the contemporary developments of big magnets used for the generation of fields of the megagauss level (see below).

5.7 Multilayer magnets with equally-loaded internal reinforcements

Here we present the solutions of the modeling problems, describing equally-loaded magnets with internal reinforcements [36, 37]. Such magnets have recently found wide use, when insulating materials stronger than conductors appeared. The winding of such a magnet consists of two-component layers, the properties of which were described in Section 5.4. We will take into consideration not only the problems of strength, but also the restrictions imposed by the heating of conductors.

Let us consider a long winding consisting of several layers of equal thickness d_1 with reinforcements of thickness d_2. It is possible to assume that at the condition (5.34b) the derived solution rather accurately describes the field in a mid-plane of the magnet. This follows from the similarity to the system of equally loaded conductors. Further we shall assume that the current density is equal over the cross section in all conductors and varies according to the same law $\delta_\varphi = \delta_{\varphi,m} f(t)$. Here, $f(t)$, which is a dimensionless function of time ($f(t) \le 1$). On the axis of the magnet the induction varies as $B_i = B_{i,m} f(t)$. The induction in each layer of the winding is proportional to the current: $B(r,t) = B_m(r) \cdot f(t)$. Here $\delta_{\varphi,m}$, $B_m(r)$ are maximal values of the corresponding parameters.

The current density is chosen according to the conditions of heating at a given pulse duration. As in [36–39], we assume that all layers hold by reinforcements and are mechanically independent. Let us further assume that the azimuth stresses are equal in all conductors and in reinforcements. In the first they are equal to σ_c, and in the latter to σ_R. At the current peak the stresses take the admissible values $[\sigma_C]$ and $[\sigma_R]$ As previously, at a sufficiently large number of layers it is possible to use the model of continuous current distribution. The averaged current density determined in the framework of this model is connected with the actual current density by a relation $\langle \delta_\varphi \rangle = \lambda \delta_\varphi$, where the filling factor $\lambda = d_1/(d_1 + d_2) = -(dB/dr)/(\mu_0 \delta_\varphi)$ is a function of radius r. The equivalent stress in an equally-loaded winding, according to (5.24b), (5.29), satisfies the equation

$$[\sigma_{\varphi 0}] = [\sigma_C]\lambda + [\sigma_R](1 - \lambda) = -(Br \cdot dB/dr)/\mu_0. \tag{5.37}$$

Thus, the following differential equation, determining the induction distribution at the peak current, takes place:

$$[\sigma_R] - ([\sigma_C] - [\sigma_R]) \frac{1}{\mu_0 \, \delta_\varphi} \frac{dB_m(r)}{dr} = -\frac{rB_m(r)}{\mu_0} \frac{dB_m(r)}{dr}, \tag{5.38a}$$

or

$$\frac{dr}{dB_m(r)} + \frac{rB_m(r)}{\mu_0 \, [\sigma_R]} - \frac{[\sigma_R] - [\sigma_C]}{\delta_\varphi \, \mu_0 \, [\sigma_R]} = 0. \tag{5.38b}$$

If the solenoid is in the external field with induction B_e, the solution of this linear equation should be sought at the following boundary conditions: $B_m(R_1) = B_{m,i}$, $B_m(R_2) = B_{e,m}$. We shall give a solution similar to the one derived by Askenazy [39]:

$$y = \frac{r}{R_1} = \exp\left(-x^2\right) \left[\exp\left(x_i^2\right) + \gamma \cdot \int_x^{x_i} \exp\left(t^2\right) dt\right], \tag{5.40a}$$

where $x = B_m(r)/B_R$, $x_i = B_{i,m}/B_R$, $x_e = B_{e,m}/B_R$, $B_R = \sqrt{2\mu_0 [\sigma_R]}$ is the magnetic ultimate strength of the reinforcement material.

The condition on the external boundary can be used to calculate the aspect ratio:

$$\alpha = \frac{R_2}{R_1} = \exp\left(-x_e^2\right) \left[\exp\left(x_i^2\right) + \gamma \cdot \int_{x_e}^{x_i} \exp\left(t^2\right) dt\right]. \tag{5.40b}$$

We introduced the notation

$$\gamma = \frac{\sqrt{2}\,([\sigma_R]-[\sigma_C])}{\delta_{\varphi,m}\,R_1\sqrt{[\sigma_R]\,\mu_0}} = \left(1-\frac{[\sigma_C]}{[\sigma_R]}\right)\frac{\delta_0}{\delta_{\varphi,m}} \tag{5.40c}$$

in these relations. Here $\delta_0 = (2\sigma_R/\mu_0)^{1/2}\cdot(1/R_1)$ is a characteristic current density. It can be presented as $\delta_0 = B_R/(\mu_0 R_1)$. For example, for the magnet with an internal radius $R_1 = 10^{-2}$ m and $\sigma_R = 2.5$ GPa, we have $B_R = 79$ T, $\delta_0 \approx 6,3\cdot 10^9$ A/m^2.

We can mentiont here two limiting cases. In the first, both materials have the same mechanical properties: $[\sigma_C] = [\sigma_R] = [\sigma]$. Then the azimuth stress is the same in the whole layer, the coefficient $\gamma = 0$, and the aspect ratio is determined by the formula following from (5.31b):

$$\alpha_C = \exp\left[\left(B_{i,m}^2 - B_{e,m}^2\right)/(2\mu_0\,[\sigma_C])\right]. \tag{5.41a}$$

In the second limiting case, the condition of small thickness of the conductor ($d_1 \ll d_2, \lambda \ll 1$) is assumed. Besides, the ratio $\delta_0/\delta_{\varphi,m}$ and coefficient γ are equal zero, therefore for the aspect ratio we have

$$\alpha_R = \exp\left[\left(B_{i,m}^2 - B_{e,m}^2\right)/(2\mu_0\,[\sigma_R])\right], \tag{5.41b}$$

In fact, in the first case the whole load falls on the conductors, in the second on the dielectric reinforcements. Numbers α_C and α_R can be significantly different. For example, at conditions $B_i = 80$ T, $B_e = 0$, $[\sigma_C] = 10^9$ Pa, $[\sigma_R] = 2.5\cdot 10^9$ Pa we have $\alpha_C = 12.9$; $\alpha_R = 2.8$.

In the general case the aspect ratio, according to formula (5.40b), depends on dimensionless parameters $B_{i,m}/\sqrt{2\mu_0\sigma_R}$, $B_{e,m}/\sqrt{2\mu_0\sigma_R}$, the ratio $[\sigma_R]/[\sigma_C]$, and parameter γ.

At the equal current density in the layers of equal thickness we obtain

$$\delta_{\varphi,m} = (B_{i,m} - B_{e,m})/(\mu_0 n d_1), \tag{5.41c}$$

$$\frac{\delta_0}{\delta_{\varphi,m}} = \frac{B_R}{B_{i,m} - B_{e,m}}\,\frac{n d_1}{R_1}. \tag{5.41d}$$

For the magnet with above-given strength parameters and such values as $d_1 = 2$ mm, $n = 8$, $R_1 = 10$ mm at condition $B_e = 0$ the current density $\delta_{\varphi,m}$ is $3.97\cdot 10^9$ A/m^2 for $B_{i,m} = 80$ T and $4.96\cdot 10^9$ A/m^2 for $B_{i,m} = 100$ T, and the parameter γ is 0.95 and 0.76, respectively. The aspect ratio takes values $\alpha = 4.21$ and $\alpha = 6,82$, respectively. The significant increase of the aspect ratio is shown at relatively small increase of the induction.

Similar to the system with equally-loaded conductors, under condition $B_{e,m} = 0$ a sharp growth of the current density at the external boundary takes place. In the rather weak external field this effect is absent. Figure 5.12 demonstrates the induction, and average current density as the functions of radius for the example where $B_{i,m} = 100$ T, $B_{e,m} = 15$ T. In this case the average current density differs little at the external and inner boundaries. With the external field switched-on, the aspect ratio slightly changes: it is equal to 6.69 instead of 6.82 for $B_{e,m} = 0$. The values of the dimensionless average current density $\langle\delta\rangle_m/\delta_0$ and of the filling factor differ only by numerical factor $\delta_0/\delta_{\varphi,m}$, which is equal to 1.42 in the considered example with the field 100 T. Thus, according to the data in Figure 5.12, in a magnet with an external field the number λ is close to 0.23 in the middle part of the winding and close to 0.4 near its boundaries.

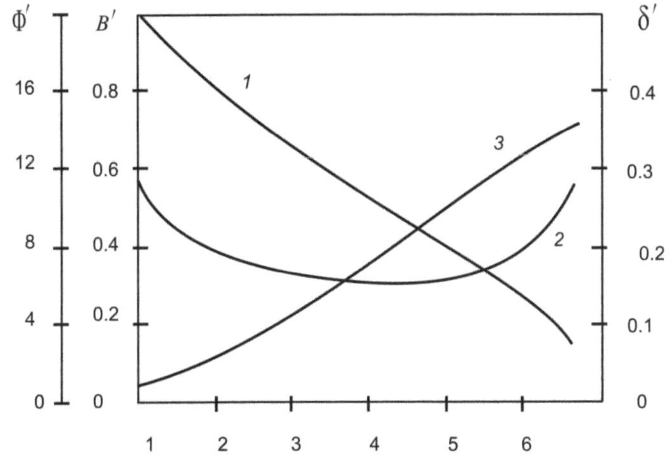

Fig. 5.12: The induction average current density and the dependence characterizing the relative distribution of the magnetic flux in the middle plane of the long magnet with equally loaded internal reinforcements. $B_{i,m} = 100$ T, $B_{e,m} = 15$ T. (1) B_m/B_R. (2) $\langle\delta_{\varphi,m}\rangle/\delta_0$. (3) $\Phi_m/(\pi R_1^2) = \Gamma(y) + B_{i,m}/(2B_R)$.

In the model of continuous current distribution it is possible to consider the voltage induced in the element of winding of the thickness dr, carrying the current with linear density $\langle\delta_\varphi\rangle dr = (-1/\mu_0)dB_z$.

The voltage is

$$U_L(r) = N(r) \cdot \frac{\partial\Phi(r,t)}{\partial t}, \tag{5.42a}$$

where $N(r)$ is a number of turns in the elementary layer of the winding and Φ is the flux:

$$\Phi = \int_{R_1}^{r} 2\pi r\, dr \cdot B(r) + \pi R_1^2 B_i. \tag{5.42b}$$

Further we use the solution derived above:

$$\Phi = 2\pi f(t)\, R_1^2\, B_R\, (\Gamma(y) + B_{i,m}/2\, B_R)\,, \tag{5.42c}$$

where $\Gamma(y) = \int_1^y x(y)y \cdot dy$. Figure 5.12 shows the numerically calculated dependence $\Phi_m/(\pi R_1^2) = \Gamma(y) + B_{i,m}/(2B_R)$ for the considered example at $B_{i,m} = 100$ T, $B_{e,m} = 15$ T. Thus, the voltage induced in the winding layer is

$$U_L = 2\pi N \frac{\partial f}{\partial t} R_1^2 B_R \left(\Gamma(y) + \frac{B_{i,m}}{2B_R} \right). \tag{5.43a}$$

The voltage drop across the resistance of the layer is determined by

$$U_R = 2\pi r\rho\delta_\varphi N = 2y\rho\, N\, (B_{i,m} - B_{e,m}) \frac{f(t)\, R_1}{\mu_0 n d_1}. \tag{5.43b}$$

On the internal layer of the winding let the number of turns be $N(R_1) = 20$. For further estimations we assume that the current pulse can be approximated by one half-wave of the sinusoid with a rise time $\tau = 2m$s. Thus, $f(t) = \sin(\omega\, t)$ at $t \le \pi/\omega$, where $\omega = \pi/(2\tau) = 0.79 \cdot 10^3$ 1/s. For these conditions $\delta_m \approx 5 \cdot 10^9$ A/m^2, and the action integral is about 510^{16} A^{2}m^{-4}s, which does not exceed the admissible values for such material as Cu–Nb composite ($\rho \approx 2.6 \cdot 10^{-8}$ Ohm.m).

The amplitude of the total induced voltage in the example above is $U_{L,m} =$ 38 kV. If the layers are connected in series, the contribution in $U_{L,m}$ of the inner part of winding (first four layers) is 7.2 kV, and of the outer part 30.8 kV. The amplitude of the total voltage drop in the resistance is substantially less. It is 4.86 kV, and the contribution of the parts is 1.04 kV and 2.74 kV. The change of the number of turns affects on the voltage distribution. Particularly, the decrease of N from 20 to 10 in the outer part of the winding allows the induced voltage to be decreased from 30.8 kV to 15.4 kV.

Thus, the increase of the width of conductors with retention of their thickness, leading to the change of number of turns, opens the additional opportunities for optimization of the magnetic system. In this case, evidently, the different parts of the magnet should be energized separately.

5.8 The plastic deformation and the resource of multiturn magnets

When the stress in a magnet exceeds the yield point of the material, additional analysis is needed to take into account the plastic deformation.

An example would be the uniform thin layer of the winding of the magnet with the mechanically separated thin current layers. Consider the diagram of loading of the material (the dependence $\sigma = f(\varepsilon)$), assuming that the conducting layer is not confined by the reinforcements subjected to the elastic deformation. Figure 5.13a shows schematically the simplified deformation process in the course of the first cycle. It is based on the assumption that when the maximum stress σ_m is attained and the magnetic pressure is dropping, the dependence of the stress on the relative deformation corresponds to the line ***AB***. This line is parallel to the line $\sigma = E\varepsilon$, following from the Hooke law. We consider, for example, the microcomposite Cu + 24 % Ag, whose properties are described in the work by Han, Walsh, Toplosky, and Lu [41]. The dependence of the azimuth stress on the relative deformation (see Figure 5.13a) is plotted according to the data from their work. The region near the yield point can be presented by the formula which was used in the work of Eyssa, Markiewitz, and Pernambuco–Wise [42]:

$$\sigma_\varphi = E\varepsilon_\varphi \quad \text{when } \sigma_\varphi \le \sigma_S,$$

and

$$\sigma_\varphi = \sigma_1 - (\sigma_1 - \sigma_S)\exp\left(-E'\Delta\varepsilon_\varphi/(\sigma_1 - \sigma_S)\right) \quad \text{when } \sigma_\varphi > \sigma_S, \tag{5.44}$$

where $\Delta\varepsilon_\varphi(\sigma_\varphi) = \varepsilon_\varphi - \sigma_S/E$. The points in Figure 5.13a are the values of σ_φ, calculated by this formula, in which $\sigma_S = 600$ MPa, $E' = 1.55\cdot10^5$ MPa, $\sigma_1 = 1120$ MPa.

It follows from the given formula that the residual azimuth deformation after the first pulse is

$$\Delta\varepsilon_1(0) = \frac{\sigma_1 - \sigma_S}{E'}\,\mathrm{Ln}\frac{\sigma_1 - \sigma_S}{\sigma_1 - \sigma_m}. \tag{5.45}$$

In the general case the layers of the winding have mechanical contact with other elements of the magnetic system. This results in the complication of the loading diagram. The analytical evaluation of the plastic deformation was carried out by Gersdorf, Müller, and Roeland [6] and Melville and Mattocks [43].

The authors of [6] calculated the stresses occurring in the course of the first pulse, taking into account that the layers are deformed in a different manner. Some of the layers experience plastic deformation, and the stress in other layers does not reach the yield point. They showed that after the first pulse, when switching the field, in the system with reinforcements the plastically-deformed winding may be compressed by the external reinforcement, which is deformed elastically. In this case the negative (compressing) residual stresses appear in the winding. Such a process is shown schematically in Figure 5.13b [43]. Before a new cycle initial stress exists in the system, which is oppositely directed in relation to the stress produced by the magnetic field. This raises the stress of the winding. Li and Herlach [44], by means of computer analysis, managed to describe the stress state of the multilayer winding with external and internal reinforcements.

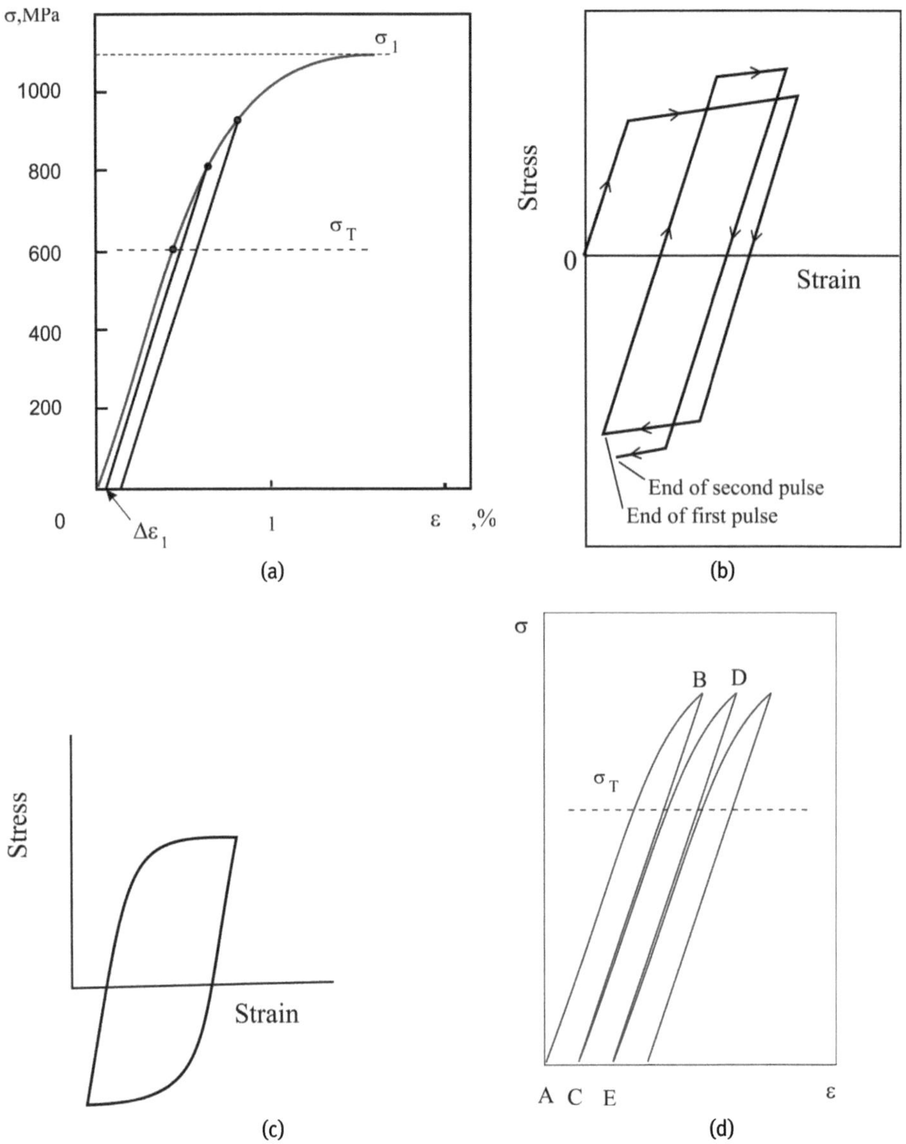

Fig. 5.13: Examples of the diagrams of loading in the case of plastic deformation of the winding. (a) Dependence $\sigma_\varphi = f(\varepsilon_\varphi)$ taken from the work [41], and the diagram describing the plastic deformation of the solitary layer induced by a single unipolar pulse. (b) The schematic drawing of loading of the magnet with reinforcements, according to Melville and Mattocks [43]. (c) A diagram in the form of the hysteresis loop. (d) Schematic drawing of the accumulation of deformation in the solitary layer for the repeated pulses and for the material without the reinforcements (1) the beginning of the next pulse in turn, (2) the end of the pulse.

The magnet in initial state was at the temperature of cryogenic nitrogen and compressed by an external reinforcement. They studied the history of the formation of the stress state of the magnet and pointed out that the stabilization of the process takes place after several cycles. This effect is peculiar to mechanical systems subjected to the action of cyclic loads. The diagram of loading of such systems after stabilization acquires the form of the hysteresis loop (Figure 5.13c).

Analytical estimates and calculations allow us to find the stresses in the winding and reinforcements, and to determine the conditions under which the stress does not exceed the admissible value. However, in order to evaluate the resource of the magnet, i.e., the number of pulses preceded the destruction of the magnet, these data are not sufficient. There are no such estimates in the publications dealing with the calculations of strengths in the windings of multiturn magnets.

In fracture mechanics the methods for evaluation of the service life of mechanical systems subjected to the action of cyclic loads are developed [45]. There are two kinds of processes which lead to the endurance failure: multi- and few-cycle processes. The the former corresponds to the functioning at stresses lower than the yield point. Under this condition the number of pulses before the destruction, for the particular form and duration of the pulse, depends on the ratio of the maximum stress to the yield point σ_m/σ_S. In the installations loaded conservatively the condition $\sigma_m/\sigma_S < 1$ is satisfied, and the resource can be rather significant (about 10^4–10^5 cycles and higher). In this regime in the course of elastic deformation in the micro-volumes (weak seeds, structure imperfections) the local plastic flow takes place, resulting in incipient microcracks. Then the cracks develop gradually, eventually leading to the destruction of the winding. The empirical formulas for evaluation of the resource can be found in many publications where the problem of the multicycle fatigue has been examined.

For generation of strong magnetic fields it is more interesting to give an estimate of the resource under the condition when stresses exceed the yield point. Such an approach is justified when it is desirable to obtain the field as strong as possible, and, along with this, to provide the acceptable resource of a magnet. The destruction of a magnet, in this case, is caused by so-called few-cycle fatigue, when the service life may number in the hundreds or even tens of pulses. In such regimes the plastic deformation can develop in several ways. The regime with the reinforcement is possible for the model of the solitary layer, and at the second and next pulses the dependence $\sigma(\varepsilon)$ corresponds to the straight line ***BC*** which is parallel to the line $\sigma = E\varepsilon$ in Figure 5.13d. In this case elastic deformation, characterized by the higher yield point, takes place. The regime with no reinforcement is the least favorable, and at the second pulse the process is described by the curve ***CDE***. In this case, after the second pulse an increase of the residual deformation occurs (Figure 5.13d). At the next pulses the residual deformation is accumulated.

The resource is determined by the acceptable value of the residual deformation. In this case for the lower estimate of the solitary layer of the winding one can assume that in each next pulse the increase of the deformation is the same as in the first pulse. Then the number of pulses N responsible for the destruction of the winding (the resource of the system) is determined by the acceptable deformation ε_{max}:

$$N = \frac{\varepsilon_{max}}{\Delta\varepsilon_1}. \tag{5.46a}$$

In the above model

$$N = \frac{\varepsilon_{max} \cdot E'}{(\sigma_1 - \sigma_S) \ln \frac{\sigma_1 - \sigma_S}{\sigma_1 - \sigma_m}}. \tag{5.46b}$$

For the material described in [41], the maximal acceptable deformation is $\varepsilon_{max} = 3 \cdot 10^{-2}$. Then, for the pulse of the load with the stress amplitude $\sigma_m = 700$ MPa we find $N \approx 42$. The last quantity becomes equal to 10 at $\sigma_m = 800$ MPa.

The model considered above has illustrative value. It is analogous to the system in which the layers of the winding do not interact mechanically and the reinforcement of the material is absent. This model demonstrates the effect of the stress on the service life under conditions when the accumulation of deformation takes place.

In the presence of the mechanical contact with the neighboring parts of the magnetic system, the layers of the winding during the cycle are subjected to the action of the sign-changing load and, as mentioned above, the dependence of the stress on the deformation may have the hysteresis loops (Figure 5.13c). In fracture mechanics, for the estimation of the few-cycle resource of such systems the Manson and Koffin empirical formula is used [45]:

$$N \approx \left(\frac{M}{\Delta\varepsilon}\right)^2. \tag{5.47}$$

In this formula N is a number of pulses prior to the failure of the mechanical system, $\Delta\varepsilon$ is a change of the residual deformation during a cycle, M is the parameter of a metal. For metals this parameter is of the order of unity.

This formula can be used to evaluate the service-time of the most loaded parts of the magnetic system which have the calculated dependence $\sigma = f(\varepsilon)$ in the form of a hysteresis loop.

5.9 References

[1] P. L., Kapitza, *Proc. Roy. Soc. (London)* **A115** (1927), 658–683.
[2] J. D. Cockroft, *Phil. Trans. Roy. Soc. (London)* **A227** (1928), 317–343.
[3] R. B. Montgomery, *Solenoid Magnet Design*, New York London Sydney Toronto, John Wiley and Sons, 1969.
[4] H. Knoepfel, *Pulsed High Magnetic Fields*, Amsterdam, London, North-Holland Publishing, 1970.
[5] H. P. Furth, M. A. Levine, and R. W. Waniek, *Rev. Sci. Instr.* **28** (1957), 949–958.
[6] R. Gersdorf,. F. A. Müller, and L. W. Roeland, *Rev. Sci. Inst.* **36**(8) (1965), 1100–1109.
[7] S. Askenazy, *Physica* **B177** (1992), 36–40.
[8] L. G. Loytsansky, *Mechanics of Fluid and Gas*, Moscow, Wysshaya Shkola, 1985 (in Russian).
[9] I. I. Ivanchik and D. G. Sannikov, *J. Appl. Mech. and Tech. Phys.* **5** (1966), 77–83 (in Russian).
[10] A. A. Kuznetsov, *Sov. Phys.-Tech. Phys.* **30** (1960), 555–561.
[11] A. A. Kuznetsov, *J. Tech. Phys.* **31** (1961), 944–947 (in Russian).
[12] A. A. Kuznetsov, *J. Tech. Phys.* **31** (1961), 1657–1662 (in Russian).
[13] L. D. Landau and E. M. Lifshitz, *Teoriya uprugosti*, Moscow, Nauka, 1987 (in Russian)
[14] S. P. Timoshenko and J. Gudier, *Theoriya Uprugosti*, Moskow, Nauka, 1979 (in Russian).
[15] H. Liebowitz, *Fracture, an advanced treatise*, Academic Press, 1969.
[16] W. R. Pryor, *Multiphysics Modeling Using Control. A First Principle Approach*, Jones & Bartlett Learning, 2010.
[17] G. A. Shneerson, *Tech. Phys. Letters* **37**(19) (2011), 51–56 (in Russian).
[18] Y. K. Huang, P. H. Frings, and E. Hennes, *Composites* **B33** (2002) 109–115.
[19] J. Liedl, W. F. Gauster, H. Haslacher, and R. Grossinger, *IEEE Trans. on Magnetics* **MAG-17**(6) (1981), 3286–3288.
[20] Z. Hashin and B. W. Rosen, *Trans. of ASME, Ser. E, Journ. of Appl. Mechanics* **31**(2) (1964), 71–82.
[21] D. S. Abolinsh, *The Mechanics of Polymers* **4** (1965), 52–59 (in Russian).
[22] D. S. Abolinsh, *The Mechanics of Polymers* **3** (1966), 372–379 (in Russian).
[23] Y. K. Katrukhin and A. P. Doroshenko, *Pribory I Technika Experimenta* **6** (1985), 162–165.
[24] H.-J. Shneider–Muntau, Polyhelix Magnets *IEEE Trans. on Magnetics* **MAG-17**(5) (1981), 1775–1778.
[25] H. Witte, H. Jones, V. Galanov, and J. Freudenberger, **MG-X** (2005), pp. 333–337.
[26] E. S. Borovik and A. G. Limar, *J. Tech. Phys.* **32**(4) (1962), 441–444. (in Russian).
[27] V. A. Ignatchenko and M. M. Karpenko, *J. Tech. Phys.* **38**(1) (1968), 200–204 (in Russian).
[28] H.-J. Shneider–Muntau and P. Rub, *Coll. Int. Physique sons champs magnetiques intenses*, Grenoble, 1974, p. 161.
[29] M. Date, *J. Phys. Soc. Japan* **39**(4) (1975), 892–897.
[30] G. A. Shneerson, *J. Tech. Phys.* **56** (1986), 36–43.
[31] G. Aubert, *Physica Scripta* **T35** (1991), 168–171.
[32] S. Askenazy, L. Van Bockstal, F. Herlach, and H.-J. Schneider-Muntau, *Meas. Sci. Technol.* **4** (1993), 1058–1064.
[33] A. P. Doroshenko, Y. K. Katrukhin, N. G. Kocharov, and V. P. Chrustaljov, *J. Tech. Phys.* **46**(1) (1976), 209–211 (in Russian).

[34] A. P. Doroshenko, *J. Tech. Phys.* **48**(8) (1978), 1566–1568 (in Russian).

[35] L. Van Bockstal, S. Askenazy, F. Herlach, and H.-J. Shneider-Muntau, *IEEE Trans. On Magnetics* **MAG-30**(4) (1994), 2172–2175.

[36] L. Van Bockstal, G. Heremans, and F. Herlach, *Measur. Science and Technol.* **2** (1991), 1159–1164.

[37] G. Heremans, F. Herlach, L. Van Bockstal, J. Witters, and I. Lefever, *IEEE Trans. On Magnetics* **MAG-28** (1992), 790–794.

[38] L. Van Bockstal, S. Askenazy, and F. Herlach, *IEEE Trans. On Magnetics* **MAG-30**(4) (1994), 2176–2179.

[39] S. Askenazy, *Physica B* **211** (1995), 56–64.

[40] L. Van Bockstal, G. Heremans, L. Liang, and F. Herlach, *IEEE Trans. On Magnetics* **MAG-30**(4) (1994), 1657–1662.

[41] K. Han, R. Walsh, V. Toplosky, and J. Lu, *Supplement Proceedings Volume 2: Materials, Characterisation and Modeling TMS (The Minerals, Metals & Materials Society)*, 2012, pp. 521–528.

[42] Y. M. Eyssa, W. D. Markiewicz, and P. Pernambuco-Wise, *IEEE Transactions om Magnetics* **32** (1996), 2526–2529.

[43] D. Melville and P. G. Mattocs, *J. Phys. D* **5** (1972), 1745–1759

[44] L. Li and H. Herlach, *Meas. Sci. Technol.* **6** (1995), 1035–1042.

[45] S. S. Manson, *Thermal Stress and Low-Cycle Featigue*, New York San Francisco Toronto London Sydney, McGraw-Hill, 1966.

6 Generation of strong magnetic fields in multiturn magnets

In this chapter we consider the magnetic systems acquiring the sufficient strength and thermal stability that provide the conditions for their repeated use.

The development of nondestructive multiturn magnets for the generation of strong magnetic fields has been going on for almost a century. In the 1920s Kapitsa and Wall [1–6] were the first to obtain the pulse fields of tens Tesla in solenoids with noncooling windings operating in conditions of adiabatic heating. This approach is used in the construction of all fields produced by nondestructive magnets in the mentioned range and even higher. In windings with a weakly pronounced skin effect one can choose a conductor cross section in a way that at the given current pulse duration the heating does not exceed the acceptable limit determined by possible damage of an insulation or by reduction of the material yield point. Windings of such magnets manage to cool down during the time between pulses, so that their performance regime can be conditionally named as a regime of rare pulses. Here we will only consider magnets operating under these conditions.

After the first works by Kapitsa and Wall, the induction amplitude of the field produced in nondestructive magnets has been increased approximately from 30 T up to 100 T, which corresponds to almost a tenfold increase of magnetic pressure (up to 4 GPa). The provision of the adequate strength of magnets operating under such high loads is the main problem hampering the obtaining of fields in the megagauss range.

In this chapter we consider examples of nondestructive magnetic systems, with special emphasis on those designed for the generation of near-homogeneous fields. In other words, we are dealing with general-purpose solenoids with a cylinder-shaped winding. Here we will not consider the cooling magnetic systems designed for long-term operation.

It is possible to direct the main ways and stages of technology of the strong fields generation in nondestructive magnets. Most studies are devoted to the construction of multiple-turn magnets. Four types of windings of multiple-turn magnets have been developed: spiral, strip, screw, and disk types. They are schematically shown in Figure 6.1.

In the limits of this paper we limit ourselves to a description of typical magnet constructions designed for the generation of pulse magnetic fields and consider some results of studies devoted to processes leading to the destruction of magnets. The systems for power supply were roughly discussed in Chapter 4. For a more detailed consideration, see the reviews and works mentioned below.

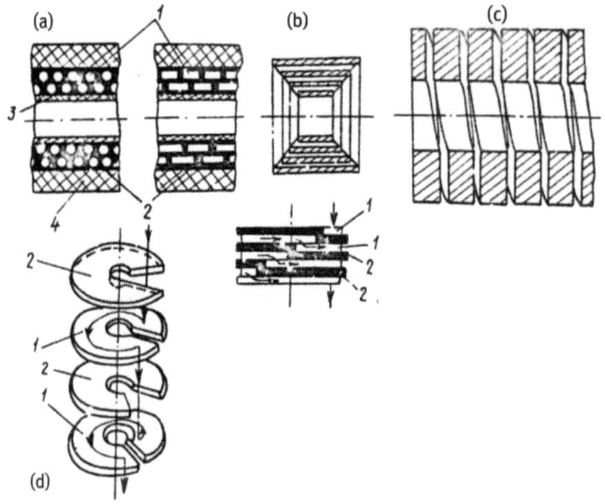

Fig. 6.1: Multiple-turn magnets. **(a)** Solenoids with a spiral winding. To the left, a winding with turns of circular cross section; to the right, with turns of rectangular cross section: (1) conductors; (2) insulation; (3) supporting cylinder; (4) outer reinforcement from stainless steel or strong dielectric). **(b)** Solenoid with a strip winding (insulation is not shown). **(c)** Solenoid with a screw winding (insulation is not shown). **(d)** Bitter disk solenoid: to the left, elements of a winding; to the right, side-view on an assembled winding: (1) conductors; (2) insulation gaskets. The current direction is shown with arrows.

Constructions of solenoids with spiral windings are used and hve been improving to the present day. The advances made in this work is due to achievements in technology, application of new materials, and development of new types of windings. Along with windings, which become monoliths after being impregnated with a binder, magnets consisting of mechanically separated layers are also used. The development of equally-loaded windings opened the possibility of obtaining fields with magnetic pressure essentially exceeding the magnetic ultimate strength of the material. This resulted in achieving fields with induction close to 100 T in large magnets with energy of the order of 10^7 J or more.

The strength of solenoids with a screw winding (Bitter solenoids) is determined by the corresponding strength of the applied material. This was the advantage of Bitter coils compared to polylayer magnets, until modern superstrong materials began to be used in the latter, together with the mentioned construction improvements. At present there are grounds to think that solenoids with a polylayer spiral winding (polyhelix) offer promise for the generation of megagauss fields in a nondestructive magnet. However, the Bitter and single-turn magnets still have their own fields of application. The latter are irreplaceable for the cases where fast-rising pulses of a field are required. These magnets are discussed in Chapter 8.

6.1 Traditional constructions of solenoids with spiral multilayer windings

In the strongest solenoids a spiral winding is made from an insulating wire of circular or rectangular cross section. Magnets of this type were used even back in Kapitsa's experiments [2]. The multilayer winding (Figure 6.2, (1)) is made from cadmium bronze and impregnated with a phenolic resin plastic (make lite). The conductivity of the material is 9 % less than the conductivity of copper, but the yield point is four times higher. The winding is compressed along the radius by a massive reinforcement (8). The fasteners (2) and (7) hold the winding in the axial direction. The ends of the winding are connected with the rings (3) and (6) with the aid of sliding contacts. The current is delivered to the ring (6) and removed from the ring (3). For this purpose the cylinder (4) is used, which connects the ring (3) with the power source and serves, in fact, as an outlet insulated from the ring (6) by the dielectric cone (5). The cylinder has a spiral slit and, in fact, forms one more layer of the winding. Many of the elements of the Kapitsa solenoid have been retained in succeeding constructions. As a rule, the researchers use the external strong reinforcement along with the alloys and composite materials which, on the other hand, should retain the conductivity close to the conductivity of copper and, on the other hand, have a high yield point.

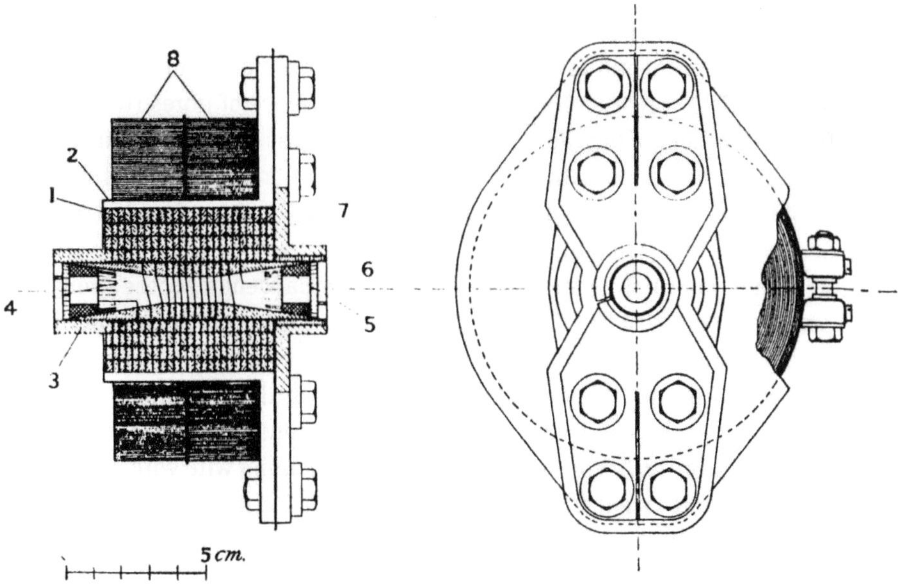

Fig. 6.2: The construction of the Kapitsa solenoid [2].

In all such magnets described in the literature, except the magnets with equally-loaded multilayer windings, mechanical stresses are commensurate with magnetic pressure of the generated field[1]. This pressure is 10^9 Pa in a field of 50 T. Even strong materials such as beryllium bronze have the yield point that is essentially less than the given value. Only new composite conductors and dielectrics (carbon-filled plastics, Zylon and others), developed in the end of the 1980s, are characterized by strength limits of the order of 10^9 Pa. In the past decades much profound experience in the design and technology of manufacturing of multiturn pulse magnets has accumulaed. The list alone of works describing the different constructions and test results, published beore the mid 1990s, contains a few tens of items [4, 5]. Constructions of multiple-turn magnets are thoroughly considered in a book by Lagutin and Ozhogin book [6], om a revoew by Herlach [7] and in the booksbx Herlach and Miura [8, 9].

Magnets with a multilayer spiral winding made from a wire of constant cross section are currently quite often used. When being placed, the winding is impregnated with a binder (mostly an epoxy resin is used) and, when solidified, it turns into a monolithic magnet.

In order to achieve more and more strong fields it is natural to use conductors and insulation fabricated from high-strength materials as well as to improve the technology of placement of turns and of impregnation of windings. Before the new superstrong materials appeared (at the end of the 1980s), a series of technical aspects permitting the enhancement of the winding strength had been developed and used in a wealth of constructions. This purpose was served by application of cylindrical (external and internal) and axial reinforcements, and by the development of optimal constructions of outlets. The presence of pores (weak spots) leads to displacement of turns or to rotation of their cross section, resulting in the possible destruction of insulation. The dense packing of reinforcements and contraction of flanges in the axial direction lead to internal stresses in windings and thus giving favor to the closing of pores.

One of the constructive methods providing the enhancement of the strength of a winding was the use of the preliminary compressing stress, which is produced by packing a rigid external reinforcement. A typical construction is the monolithic magnet with a multilayer winding developed at the University of Oxford [10]. A fragment of the winding in Figure 6.3 shows constructive decisions used in magnets of this type. A winding is fabricated from copper wire with a rectangular cross section and a shell from stainless steel enhancing the strength of wires. A glass-epoxy compound is used here as insulation. The magnet contains the char-

1 Calculations of stresses in windings are considered in Chapter 5.

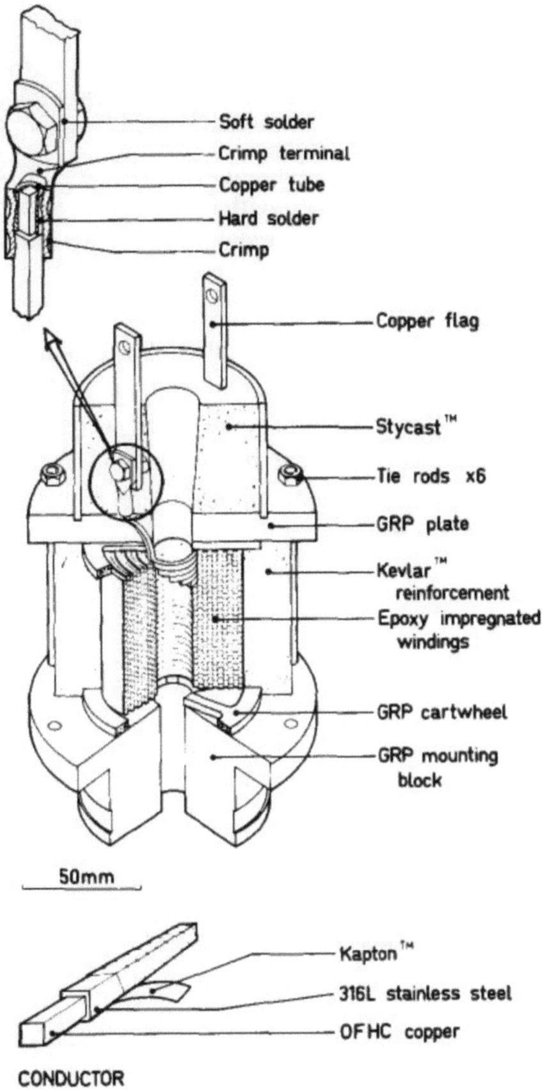

Fig. 6.3: The externally reinforced coil developed in the University of Oxford [10].

acteristic units: the external cylindrical reinforcements, made from stainless steel, and face flanges contracted by bolts.

Other example are the magnets developed in the Kirenskii Institute of Physics (Krasnoyarsk, Russia). The magnet described in Katruhin's paper [11] had a multilayer winding. Each layer was coiled around a separate holder with a wire with rectangular cross section. Wires were insulated with glass cloth and bonded with glue. The silicon-organic polymer was added to the composition of an epoxy-phenol binder. This permitted the electric strength and admissible temperature of an insulation to be increased. Coils were turned over insulation, closely packed, and bonded with glue. The current in an individual layer of the monolithic multilayer winding could be different, depending on the posed problem. An assembled winding was pressed in a steel reinforcement made from steel acquiring the high elastic modulus. In such magnets the fields with induction up to 46 T were reliably generated. In the author's opinion, a maximal strength features a magnet in the case when tangential stress on an inner layer of a winding is equal to a radial stress on an external one.

Additional improvements were made in a magnet described in [12] (Figure 6.4). A winding was fabricated from a copper wire subjected to cooling: it was pulled through dies and then rolled. A wire subjected to the hard head had the ultimate tensile strength (UTS) 320 MPa. A winding of the solenoid consisted of two 25 cm-long coils connected in a series. A glass-epoxy compound was used as insulation. The magnet was operating at cryogenic nitrogen temperature. It was equipped with face and external reinforcements.

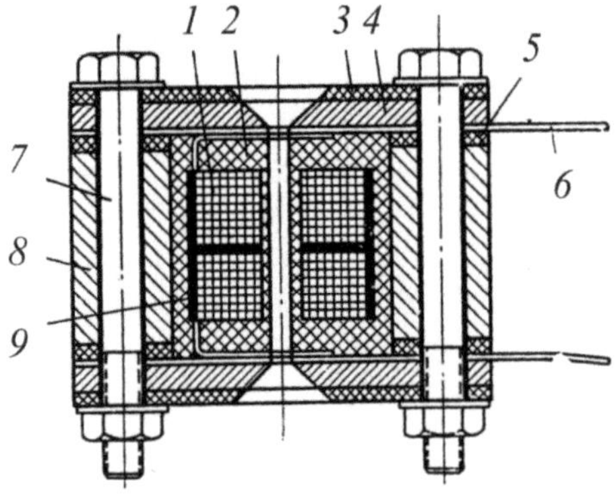

Fig. 6.4: The magnet with the refined construction of the outlets [12].

In a study by Gersford, Müller, and Roeland [13] it was shown that the strength of a magnet can be increased by installing the additional steel cylinder at the definite radius inside of the winding. This cylinder serves as an internal reinforcement.

The magnetic system can be made as mechanically separated layers with current (polyhelix magnet [14]). The winding fragment of such a magnet is shown in Figure 6.5. One can see by Schneider-Muntau the layout of the winding layers. An edge of one winding layer with contact is also shown in the figure.

Outlets and inner connections are important details of a magnet. In the magnet with a monolithic winding developed at the Oxford University [10] outlets have massive contacts, shown in Figure 6.3. A similar approach takes place in the construction of the monolith magnet with the 60 T-field, recently developed in China [15] (Figure 6.6). In both cases one of the outlets was brought out of the limits of

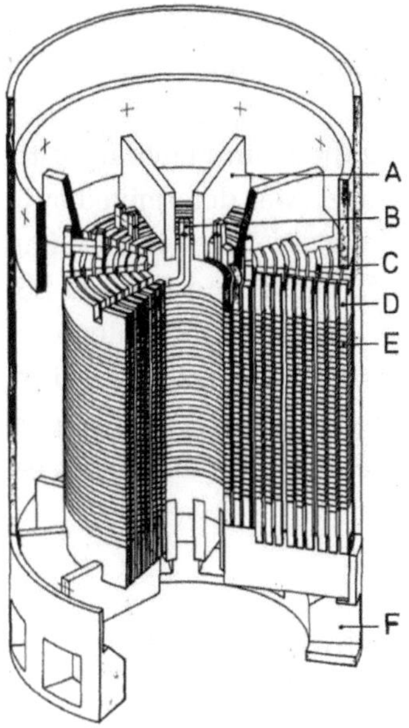

Fig. 6.5: Polyhelix magnet [14]

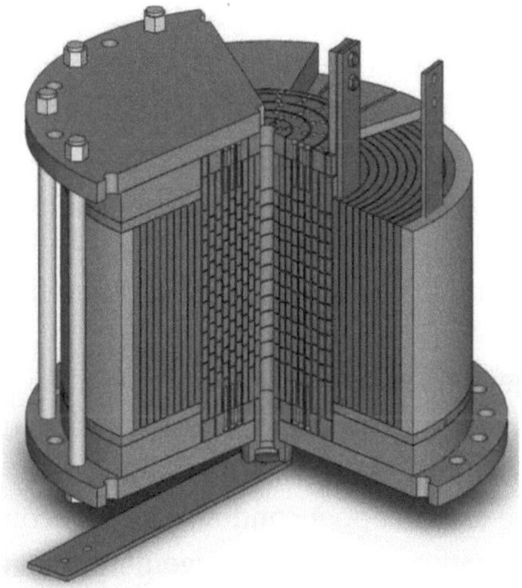

Fig. 6.6: The monolithic magnet with induction 60 T [15].

the winding to the region of weak field, and another is adjacent to the inner edge of the winding, where the strongest field takes place. In the monolith magnets described in [12] and [16], another construction was used: both outlets are on the external edge of the winding in the region of weak field. Such a decision decreases the probability that wires carrying a current to a winding might break. In the first of these magnets (Figure 6.4) two sections of the winding are coiled with one wire, going from the central plane, and finished by outlets placed in the region of a weak field under the facial reinforcements on an external radius of the winding. In the solenoid of Lagutin and Ozhogin [16] a current is delivered to an inner layer of the winding from outside in the face part of the magnet using a conductor packed in a form of a plane spiral.

Up to the beginning of 1980s copper alloys were the main material for the conductors of a winding. The characteristic pulse duration was determined by an admissible increment of temperature and was 10^{-4}–10^{-2} s. Borovik and Limar [19], Katruhin, Zelikman, and Vorohov [12], Gersdorf, Müller, and Roeland [13], as well as others, applied lower initial temperatures (20 or 77 K) in order to increase the pulse duration[2]. A field produced in nondestructive magnets acquiring the monolithic multilayer spiral winding with homogeneous current density, the copper alloy wires and insulation impregnated with epoxy resin did not exceed 50–60 T. This fact is confirmed by given examples.

6.2 Present-day materials used to make windings

Beginning in the 1980s, new materials were used for fabricating windings acquiring high mechanical properties. A substantial advance in this direction was made when in the magnets of Lagutin and Ozhogin wire was used which had previously been used in the production of superconductive magnets [16]. It represents a bundle of filaments from the niobium-titanium alloy filling a copper matrix by 70 %. This conductor acquires a high yield point and, in addition, at normal temperature it has sufficiently low resistance, determined by the matrix material. Application of such conductors permitted the authors of [16] to construct magnets generating fields of 50 T

Several other composite conductors have been developed and are used in present-day magnets. These conductors have mechanical properties, in particular, the ultimate tensile strength (UTS) and elastic modulus which are much higher than in copper, and the conductivity is close to the copper conductivity. The conductors operate in a range of temperatures between close to criogenic nitrogen and room temperature and above. Towards the end of pulse duration their resistance increases but remains permissible.

The most commonly used are conductors of the following four types. The first of them is Cu–Nb microcomposite. Using in magnets the windings from this material, Foner obtained a field with induction of 68.4 T [20]. The Cu–Nb microcomposite is a copper matrix in which filaments of Nb are distributed. In the course

2 The utility of the application of low initial temperatures is considered in Chapter 3, where the admissible values of the action integral for copper windings are also given.

of fabrication of wires the stocks are pulled through dies with with a hole of decreasing size. For example, in the process described in [21], the niobium wires with diameter 1.5 mm were placed in a copper matrix with diameter 76 mm. The material is then extruded, swaged, and drawn to final size (2 mm × 3 mm). In the course of cold casing the material is subjected to so much high pressure that diffusion of niobium into copper occurs. Figure 6.7 shows a fragment of the cross section of a conductor with Nb fibers, taken from [22]. One can see filaments of niobium in the copper material. The authors of [22, 36] suggest that the increase of the UTS happens due to the insertion of Nb-nanowhiskers in copper. Finally, a quasi-epitaxy is performed between the interfilamentary copper and the niobium nanowhiskers.

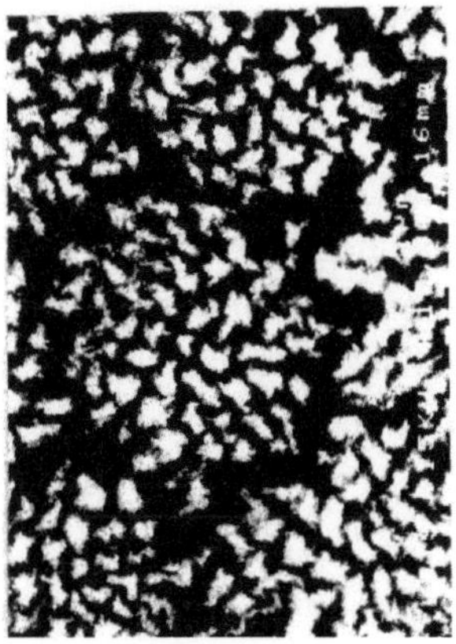

Fig. 6.7: The strucrure of the Cu/Nb conductor [22].

The microcomposite Cu–Ag [23] also exhibits high strength. The Cu–Ag alloy wire is obtained by cold working combined with intermediate heat treatment. According to the authors of [23], both components of alloy were deformed into very fine filaments by cold drawing. The technology of fabrication of Cu–Nb and Cu–Ag conductors is described in [24, 27].

The third microcomposite exhibiting high strength properties together with acceptable conductivity is the Glid Cop material [25, 26]. In the process of its fabrication, the internal oxidization of Cu–Al powder takes place, leading to oxide dispersions in the copper. High strength is achieved due to the presence of 10 nm particles of Al_2O_3 in copper [26].

The characteristic strength parameters are the ultimate tensile strength (UTS) and Young's modulus. For this, the parameters characterizing the conductivity are also of importance. This is the ratio of specific conductivity of the International Annealed Copper Standard (IACS $= 1.724 \cdot 10^8 \Omega \cdot$ m.), and parameter R_{293}/R_{77}, which is the ratio of the material resistance at room temperature to the resistance at initial temperature 77° K. The given parameters are important, since they characterize the energy release and the temperature increment for adiabatic heating of a winding.

The properties of composite conductors depend on their composition and the production technique. This is proved by the studies of Sims and Hill [21], in which microcomposites of various content are compared relative to their characteristic parameters. Table 6.1 presents the data related to the conductors with the highest strength characteristics from those studied in this work. They are in close agreement with the data given in other publications.

One more kind of high strength conductors is macrocomposites. They are Cu-SS conductors from a copper wire with stainless steel jackets [17, 10, 28–32, 49]. (Figure 6.8). High strength is achieved as a result of the elaborate technology, including heating and cold deformation. Examples of such conductors are given in Table 6.1.

The composite conductors feature the reduction of conductivity with increase of the strength. An example is the dependence of conductivity on UTS, obtained in [10, 22, 25, 49]. In Figure 6.9 [23] examples of such dependences for different composition Cu-Ag alloys are presented. Curves ***a***, ***b***, ***c*** correspond to various modes of intermediate heat treatment and cold drawing. The increase of resistance restricts to a degree the acceptable values of the UTS, which should be taken into account in the fabrication of composite conductors.

Table 6.1: Mechanical properties of the conductors and materials used for internal and external reinforcements.

Material	**Ref.**	**77 °K**		**293 °K**		**IACS %**	R_{273}/R_{77}
		UTS, GPa	**E, GPa**	**UTS, GPa**	**E, GPa**		
Conductors							
Cu+18%AG	[21]	1,25	129	1,02	126	71,4	3,06
Cu+18%Nb	[21]	1,27	116	1,05	111	69,5	4,54
Glid Cop AL60	[21]	0,86	143	0,62	118		
Cu/12X18H10T	[31]	>1,25		0,91		51,9	8,38
Materials for reinforcements							
Zylon/epoxy composite filling factor 77,5 %	[37]	4,3	222	3,3	205		
Carbon fibre	[39]			6	170		
S2-Glass fibre	[8]			4,5			
Steel MP35N Co+Ni+Cr+Mo	[46]	2,5	238	2	227		

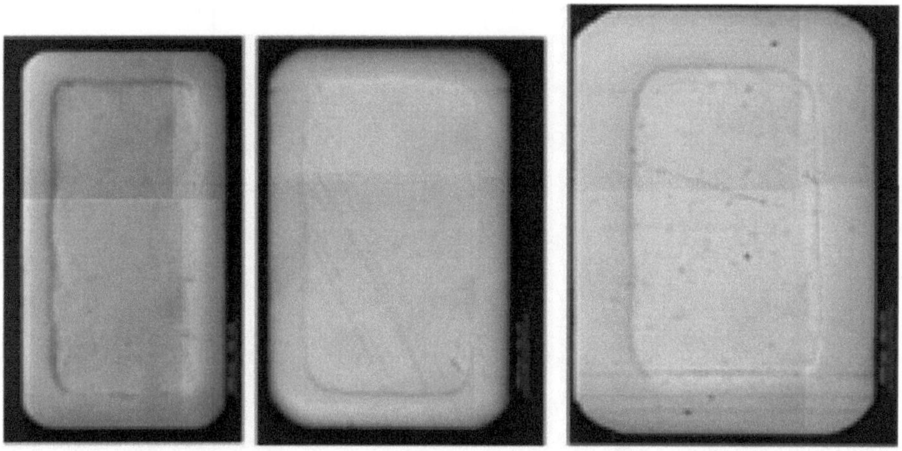

$Cu_{58}SS_{42}$ 2.36 · 4.50 | $Cu_{54}SS_{46}$ 2.65 · 4.50 | $Cu_{42}SS_{58}$ 3.00 · 5.00

Fig. 6.8: Cross section of Cu/SS macrocomposite conductors for pulsed magnets [49].

Besides the above-mentioned materials, the highly strong varieties of stainless steel are also used in present-day magnets. In several magnets a stainless steel cylinder, covered on two sides with an insulating layer, is used for fabrication of internal and external reinforcements (see Table 6.1). In part icular, the high strength characteristics are inherent in the cold-deformed alloy MP35N (35 % Co, 35 % Ni, 20 % Cr, 10 % Mo) [46]. It has the yield point 2125 MPa at room temperature and 2500 MPa at 77 K.

The manufacturing of nondestructive magnets has advanced greatly, due to application of superstrength dielectric materials. Plastics such as Zylon [33] have very high UTS (above 3 GPa). Huang, Frings, and Hennes [37] have shown that this parameter for Zylon/epoxy composite attains 4.3 GPa. In [18] it was indicated that the Cu–Nb conductor, operating at the load of 95 % UTS, and the reinforcement made with Zylon, best realize the potential of both materials, since at the given load they have equal elongations. Table 6.1 contains estimated values of strength parameters of dielectrics, applied in constructions of magnets. One can judge mechanical properties of contemporary materials by the graphs given in [32], and by ones presented in Figure 6.10. They show the dependence of the tension on the relative deformation. It is worthy to note that the modulus of elasticity and UTS of strongest dielectrics are higher than that of the strongest composite conductors.

The particularity of zylon and carbon fiber is the anisotropy of mechanical properties [37]. Deformations occurring in the thin-wall cylinder undergoing ra-

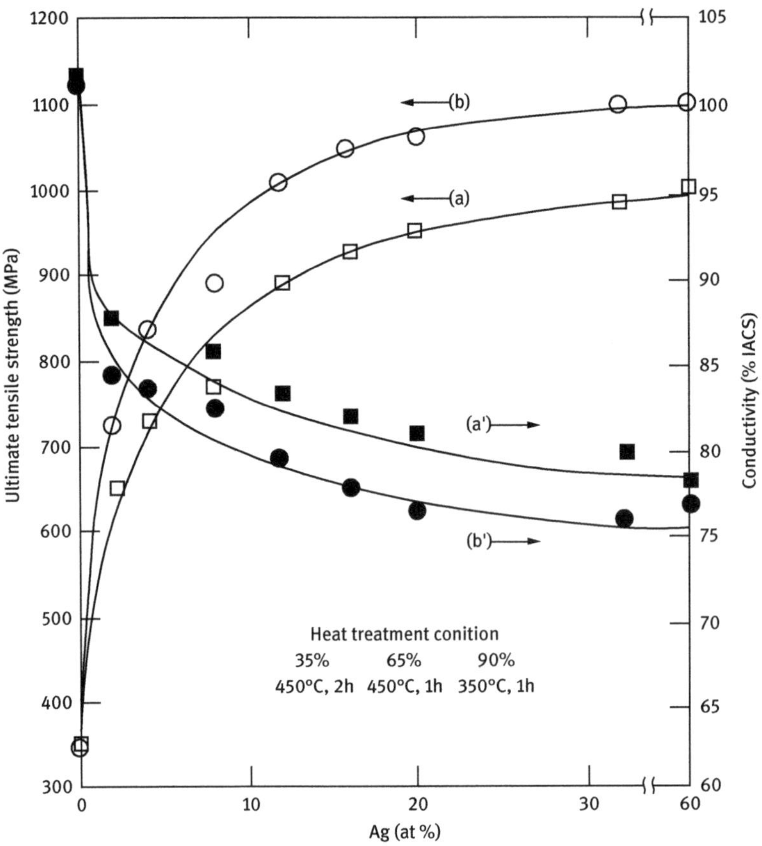

Fig. 6.9: UTS versus conductivity for various Ag contents of intermediate heat treatment and cold drawing[23].

dial forces are characterized by the modulus of elasticity $E_{\text{longitudinal}}$, which is tenfold in comparison with the modulus of elasticity $E_{\text{tangental}}$ in the case of deformations caused by the axial force. According to data in the report of Witte, Jones et al. [39], the ratio $E_{\text{tangental}}/E_{\text{longitudinal}}$ is 2.7G Pa/230 GPa for zylon and 14 GPa/170GPa for carbon fiber.

As mentioned, Lagutin and Ozhogin [16] and Foner [20] were the first to apply composite conductors. Since then, these materials and extremely strong dielectrics have found wide application in the construction of nondestructive magnets. As an example of application of present-day materials we give the parameters of the ones used in construction of a multilayer magnet with a field of 78 T, developed in the National High Magnetic Field Laboratory (NHMFL, USA) [18]. The material of the conductors is the Cu–Nb microcomposite with modulus 120 GPa, UTS ≈ 1120 GPa, and zylon fiber with modulus 580 GPa, UTS 3.3–3.5 GPa. In [49]

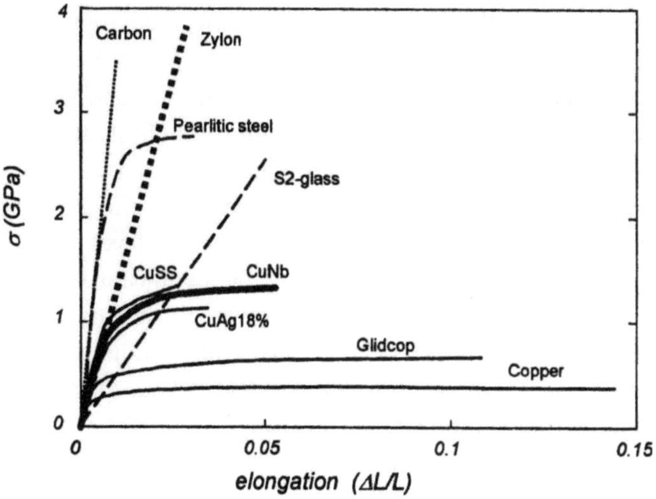

Fig. 6.10: Stress vs. strain curves of materials used for building pulsed magnets [32].

several kinds of magnets with copper/stainless steel composites are described. In the magnet with a winding containing 40 % stainless steel, a field of 78 T was produced. In this magnet the optimized system of reinforcements made with zylon was used.

6.3 Special features of constructions of present-day multiturn monolithic magnets with field of 60–80 T

The examples above and the data presented in [8, 9] show that over the past 10–15 years fields with induction up to 70–80 T have been produced in nondestructive monolithic magnets. This resulted not only from the application of new materials but also because of the effective technical solutions. As in previous works, the initial compression of the winding was used along with strong external and internal reinforcements. The application of equally-loaded windings turned out to be an effective means to increase the strength of magnets.

Present-day magnets are rather complicated multilayer systems with distributed electromagnetic forces. The analytical relations given in Chapter 5 remain valid for preliminary estimations. However, the final analysis and optimization of magnetic systems is conducted with a help of computer simulation using the finite-element method. Witte, Jones, Gaganov, and Freudenberger [39] have calculated stresses in the winding with internal reinforcements, using different computer programs. Numerical calculations have elucidated, in particular, the influence of material anisotropy on the stress distribution. The anisotropy leads to the increase of the von Mises stress in an external reinforcement. This effect is more pronounced in the case of zylon fiber compared to carbon fiber. The experience of using multiturn magnets shows that the application of initial radial compression leads to an increase in magnet strength. A field of radial forces opposite in direction to electromagnetic forces decreases stresses in the winding induced by a discharge. A calculation of stresses arising in the polylayer magnet with a field of 65T, developed at the NHMFL, confirms the significance of this effect [35]. In the given example, the conductors were subjected to an initial stress of 500 MPa and internal reinforcements of 800 MPa. Due to this fact, in the operating regime the loads did not exceed acceptable values: 950 MPa in conductors (Cu–Nb) and 2500 MPa in internal reinforcements (zylon).

The initial longitudinal compression with a help of flanges also plays a positive role. It was used in all known constructions of polylayer multiturn magnets and in Bitter solenoids. Experiments have confirmed the efficiency of this method to increase strength. However, according to the authors of [39], this cannot be explained by considering the calculated results obtained in the cited study. Note that in an initial state in the winding there are always "weak spots", for instance, pores between wires. In the process of initial loading, or in the course of training discharges, these pores close due to plastic deformation. These factors are difficult to take into account in calculations.

In the general case the winding structure of the multilayer multiturn magnet consists of alternating layers of conductor and insulation (Figure 6.11). As shown in Chapter 5, in two-component medium the stress in each layer depends on the relation between the modulus of elasticity of applied materials. In windings where the modulus of elasticity of conductor essentially exceeds that of insulation, the mechanical load is taken by the conductor. This took place in previous constructions, where dielectric layers performed mainly the role of insulation. However, as was pointed out in [13], the magnet strength increased when the additional metallic cylinder, playing the role of internal reinforcements, was inserted between layers. New possibilities of the significant increase of the magnet strength appeared when insulating materials with the yield point of order of a few GPa became available. Van Bockstal, Heremans, and Herlach [40] demonstrated the efficiency of composite internal reinforcements in magnets with commercial soft copper wire. In coils with reinforcements made with S2-glass with the yield point 2.4 GPa a field with induction of about 70 T was attained.

The stress in the multilayer magnet can be significantly reduced if the load is equal in all layers of the winding. Borovik and Limar [41] demonstrated that, using the system of equally mechanically separated conductive layers instead of a winding with constant current density, it is possible to sharply reduce stresses.

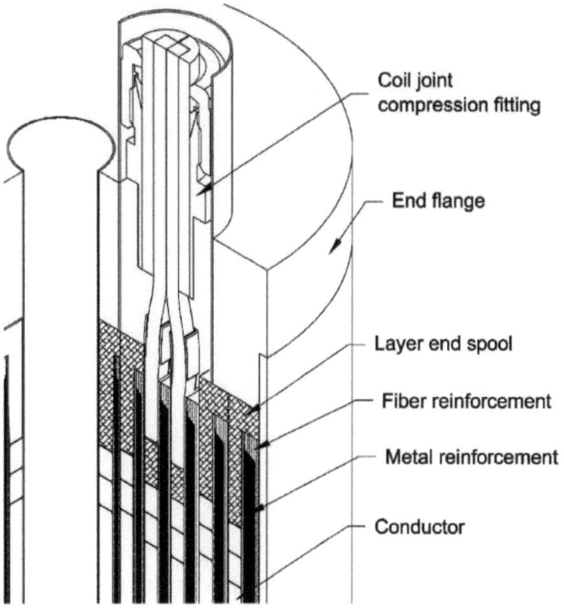

Fig. 6.11: Example of a six-layer coil with “polylayer winding [35].

They calculated the current distribution approximating the discrete distribution by a continuous one, and derived the formula for the aspect ratio given in Chapter 5. The same idea was considered in [42–45] and later in several other publications. The first magnet with an equally-loaded winding was developed by Date [44]. In this magnet the current distribution in layers was determined by solving the system of n equations for currents in mechanically separated layers (n is an amount of layers). Each equation relates the azimuth stress in the given layer to currents in other layers. The ideas of Date were used as a base in the development of Osaka University magnets. In the magnet of a nine-layer winding from Cu–Ag composite, having above 18 mm and an external reinforcement from maraging steel (Figure 6.12) a field of 70 T without failure during the pulse was obtained. Placing in the series one more winding layer of more than 10 mm and a reinforcement in the form of a steel tube with an external diameter of 18 mm, the field was obtained of 80.3 T [48].

In the majority of contemporary monolithic magnets, equally-loaded elements are used in a whole magnetic system or in its part. Each of them is the layer of winding and the dielectric cylinder placed outside. The latter plays the role of not only the insulating insert but perceives the greater part of the load in the layer. In Chapter 5 there are examples of calculations based on the model of continuous current distribution which illustrate the features of systems with internal reinforcements. It was also pointed out there that the equivalent stress of two-component layer is determined by formula (5.24a), $[\sigma'_\varphi] = \lambda\sigma_C + (1-\lambda)\,\sigma_R$,

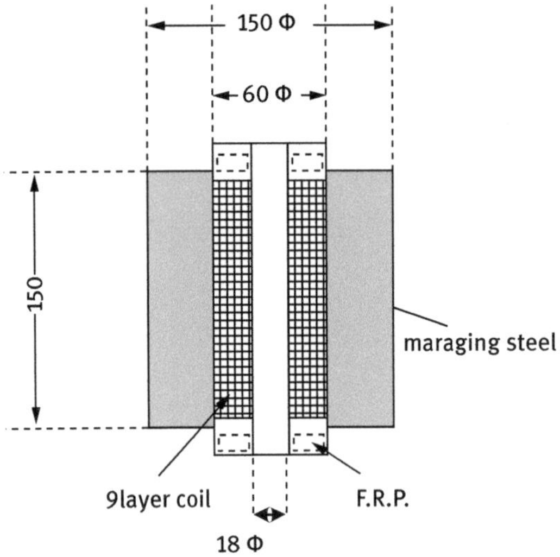

Fig. 6.12: Magnet of Osaka University [48].

where λ is the filling factor and σ_C, σ_R is the hoop stress in conductors and reinforcements, correspondingly. In present-day magnets the cylinder made with zylon fiber or other material with modulus of elasticity above 200 GPa and UTS of the order of 2–5 GPa are applied as internal reinforcements. Thus, the condition $\sigma_R > \sigma_C$ is fulfilled. In those cases when filling factor takes values 0.4–0.5, the reinforcements perceive most of the load. Along this, the above-mentioned composite materials with much higher strength parameters compared to copper (used in the pioneering work of Van Bockstal, Heremans, and Herlach [40]), are also applied in these magnets. They also undergo the definite part of the load, in spite of the use of internal reinforcements.

We noted above that contemporary composite conductors have a specific resistance higher than copper. It increases when the copper percentage is reduced and the composite strength is increased. Therefore, the rational choice of composite conductors used in combination with high strong reinforcements should be made accounting for their strength behavior and heating. The copper percentage in composite conductors is chosen in a way that at sufficiently high strength they had an acceptable specific resistance. It is necessary, because with a properly chosen thickness the heating should not exceed the permissible level. The calculation of the heating of a multilayer winding is made by considering the field diffusion in the transitional regime. The characteristic feature is nonuniformity of heating calculated by Lie and Herlach [48]. The analytical solution of the model problem is given in Chapter 3.

In the development of magnets the important factor that should be taken into account is the electrical voltage of insulation between the layers of the winding.

Not all highly strong dielectrics acquire sufficient electric strength. For instance, in the magnet described by Lagutin, Rossel, Herlach, and van Bruynseraede [38], in the internal part of a Cu–Nb winding zylon was used as a material for internal reinforcements. This material did not provide the needed electrical strength in the external part of the winding where the voltage between layers is higher. Therefore for this part of the magnet the cylindrical reinforcements were fabricated from from an S2-glass composite. They have higher electrical strength. In the external zone of the magnet soft copper with reinforcements is used. The load acting in the external zone is partially accepted by the external steel reinforcement. It comprises the whole unit with the axial reinforcement and forms a dewar filled with liquid nitrogen.

From the preceding it is seen that a reasonable choice of constructive parameters of the magnet can be only made when taking into account the rather complicated combination of the different factors. We showed above how the results of calculations can be used for the estimation of the effect of the individual factors. Present-day developments are based on the multifactor computer optimization of the magnetic system. Further on we give some results of such calculations.

One of the first magnets in which a zylon fibre was used for the containment of the winding was developed in [18]. The winding has eight layers of Cu–Nb wire and consists of two parts. In the inner part, four layers are mechanically separated. The thickness of the conductors and reinforcements are chosen in such a way that all the conductors in the inner layers are equally loaded. The Von Mises stress in these conductors is close to 0.9 GPa. Conductors in an external part are less loaded. The most load in the internal part falls on four cylindrical reinforcements. Three of them have equal stress at about 3.5GPa, and the first one has the stress of 1.8 GPa. These cylinders are much more loaded than the conductors. In the most loaded inner part of the winding, the condition of equality of stresses both in conductors and dielectric cylinders is somewhat fulfilled. The external part of the winding is mechanically separated from the internal part and is subjected to the action of a weaker field. Calculations have shown that the strength of the system of conductors with external reinforcements is sufficient for the containment of this part of the winding. A thickness of insulating layers in this part is small and the layers experience an insignificant part of load.

The authors of [18] pointed out some technological peculiarities of a zylon plastic. Thus its packing factor is 80 %. This makes its vacuum impregnation difficult, and therefore the reinforcements were applied using a wet winding. In the described magnet with a 10 mm bore a field of 77.8 T was obtained.

In other magnets [32, 35, 38, 39, 46, 51] the considered constructive peculiarities are roughly conserved. The detailed description of the magnet with a field of 75 T one can find in [38]. The magnet contains the similar constructive elements

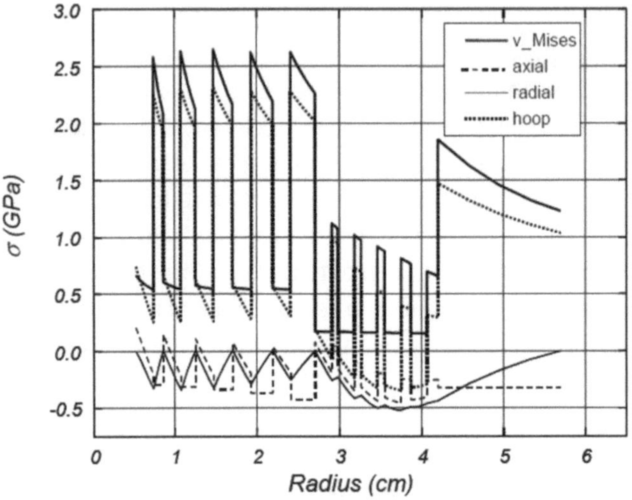

Fig. 6.13: Stress distribution in the winding of the two-section magnet with the magnetic field 75 T [38].

which were considered above. The stress distribution in the winding layers of this magnet is shown in Figure 6.13. Magnet strength is additionally provided by application of zylon as internal reinforcements and conductors with Cu–Nb composite. In addition, in reinforcements of the internal mostly loaded part of the winding the constancy of the Von Mises stress is fulfilled. In this part of the magnet there are five mechanically separated layers. Conductors both in the internal and external part are equally loaded. The Von Mises stresses in the conductors are about 0.5 GPa, and in the reinforcements of the inner part of the winding about 2.5 GPa.

A somewhat different construction of the monolithic magnet designed to generate a field with induction 80 T is described in [50]. In this magnet there is no external reinforcement. The winding consists of 15 conducting Cu–SS layers with internal reinforcements of zylon. As is seen from Figure 6.14, the von Mises stress in 13 internal reinforcements, carrying the most load, is close to 2.5 GPa, and the stress in conductors in the most part of the winding is close to 500 GPa.

6.4 The results of tests of multiturn magnets and investigation of their destruction

The elaboration of the magnets designed for multiple-use includes their tests and studies of their destruction. Strictly speaking, a magnet designed for multiple use must operate in such a regime where it resumes its original shape and size once

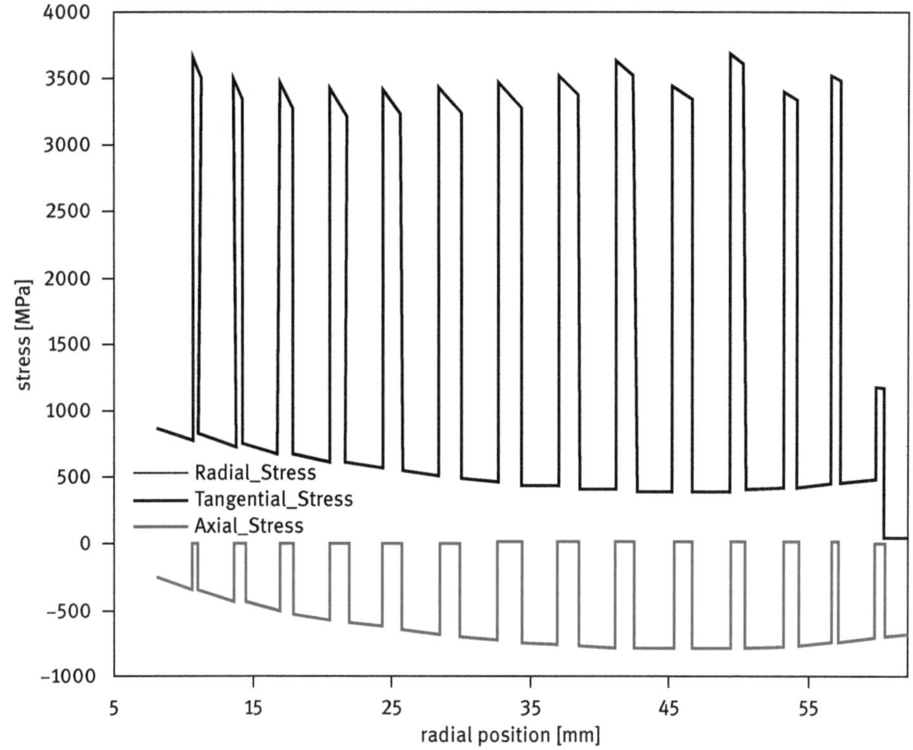

Fig. 6.14: Calculated stresses in the equally-loaded winding of a single-section magnet with the field 80 T [50].

the pulse is finished and the magnet is cooled. However this state may be established not immediately, but after several discharges, in the course of which weak spots in the winding are removed.

The experimental studies of Melville and Mattocs [60] have shown that the calculated and measured values of the elastic deformation in the middle cross section of the magnet within the limits of the small error are in accordance with the calculations of Gersdorf, Müller, and Roeland [13]. The experiments were carried out in a relatively weak field: the induction was about 10 T. However, in these experiments regimes with plastic deformations were also realized.

In a strong field the winding undergoes plastic deformation. As was pointed out in Chapter 5, after the first pulses, a hardening of the winding exerted to the sign alternating load occurs. In such a manner the effect of training manifests itself which is conducted before putting the magnet into operation. With a growing of number of nondestructive pulses, the system arrives at a stationary state. After the stabilization process the stress due to deformation acquires the form of a hysteresis loop. In this case deformations are not "accumulated", and the service life corresponds to the process of few-cycle fatigue. The service life is determined by the development of deformations followed by a failure of weak points.

The statistical data does not exist which would be needed to present the characteristic of the resource of such complicated device as multiturn magnet, in terms of the theory of few-cycle fatigue. The data on the service life of different multiturn magnets are given only in a few of the papers which contain the description of their constructions. The solenoid in [12] failed in the field of induction 44–46 T; however, it may withstand up to 100 pulses in a 40 T field of induction. According to the published data, apparently the magnet described in the paper by Kindo [48] operates in a few-cycle regime. The magnet with a nine-layer winding sustained hundred pulses in a field of 60 T and 500 pulses in a field of 55 T. When using one more layer with an internal reinforcement, such a magnet tolerated five shocks in a field of 75 T and more than 20 shocks in a field of 70 T.

Another pattern was exhibited in experiments by Portugall, Mainson, Billete, et al. [63]. In this work, the "accumulation" of deformation is a possible mechanism of magnet aging. This process can also extend at a constant pulse amplitude, similar to that described in the simplified model considered in Chapter 5. This suggestion is supported by the experiments presented in [63], where the variations of the resistance of the winding were studied. After a series of discharges with increasing induction amplitude, the resistance of the winding increased in consequent pulses of constant amplitude (Figure 6.15). The authors attributed this fact to the aging of the magnet, since the change in resistance is a consequence of residual deformation.

A similar regime is possible in the numerous experiments in which the growth of the inductance in the discharges with growing amplitude of the induction was observed.

The state of the magnet is characterized by an inductance increment due to the change in the arrangement of turns, their rotation (for example [10, 32, 40, 47]). The experiments of Jones, Herlach, Lee, et al. [10] give an example of what was mentioned above. The dependences in Figure 6.16 plotted according to data from that study show the inductance increment vs the induction amplitude. Insulation in the coils has been made with a kapton material, and copper conductors enclosed in a stainless steel shell (Figure 6.3) were used. A sharp growth of ΔL, associated with plastic deformation permits us to determine the induction threshold. A characteristic feature is the significant difference of the threshold inductions of coils 1 and 2. Both coils had turns with rectangular cross sections, the winding was a vacuum impregnated with epoxy resin, and they had bores close in size: 13 and 12 mm, but coil 1 had no external reinforcement. This coil was destructed in one shot at B_{max} = 48 T, whereas the second coil was repeatedly

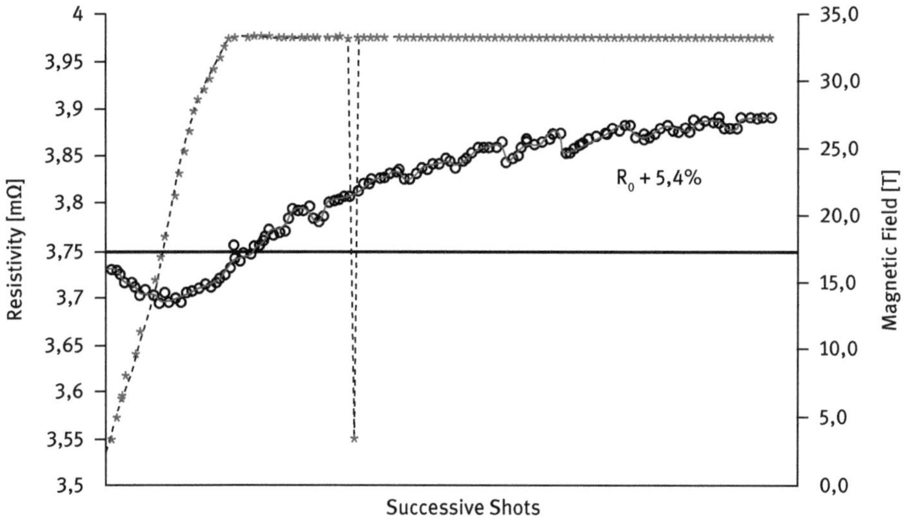

Fig. 6.15: Increase in resistivity of the magnet in the course of its testing [63].

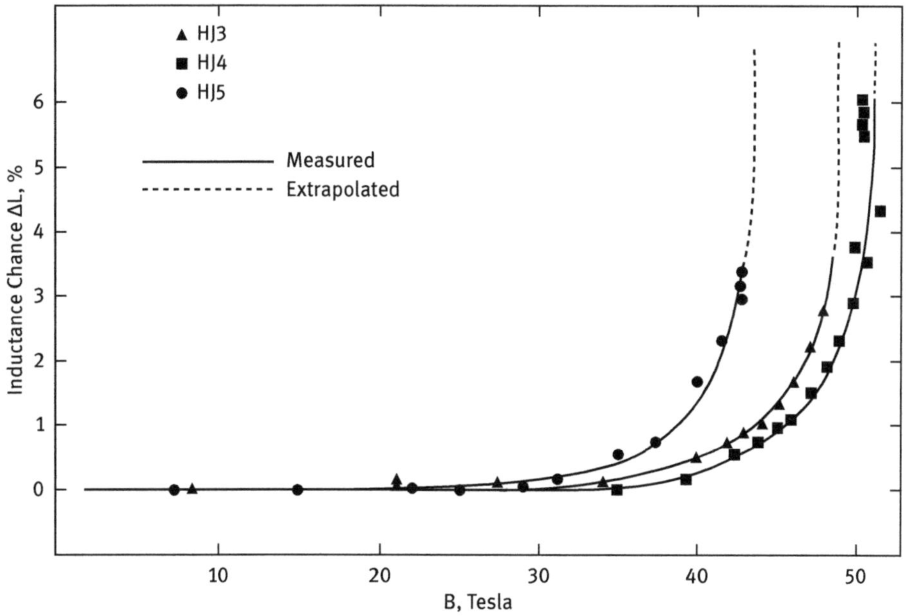

Fig. 6.16: The increase in inductivity of the magnet in the course of its testing [10].

used for generation of a field with induction 51.7 T. Traces of deformation are seen in the photos in Figure 6.16, where sectional views of the coils after failure are

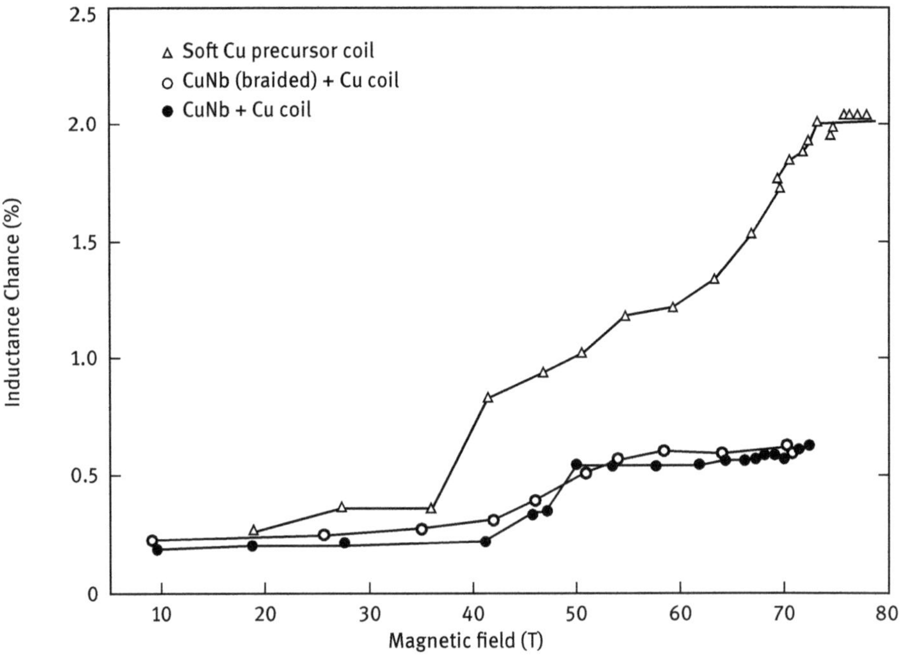

Fig. 6.17: A change in the inductance of the magnet in the course of testing in the field with an increasing amplitude. The upper dependence is obtained for the magnet with copper conductors, the lower dependence for the magnets with five layers of Cu–Nb in the internal part and five layers of copper in the external part of the winding [38].

shown. In upper picture (coil 2) the point of failure is marked. A characteristic feature is the rotation of the coil conductors in the mostly loaded layers of the winding.

The measurement of inductance as a means of monitoring the winding state is used in tests of magnets constructed with high strength materials. In [32] the dependences $\Delta L = f(B_m)$ are built for two coils which produced the field of about 70 T (Figure 6.17) The tests of these magnets with conductors of CuNb composite and reinforcements of zylon have shown that while the field was growing the inductance increment achieved 0.5 % in the field of about 40 T, then with further field increase the stabilization of ΔL took place. This possibly points to the fact that when the induction amplitude goes through a mentioned value, the internal structure of the winding undergoes some changes, after which its properties stabilized. A series of discharges with an increasing field plays the part of trainer. After that, 12 pulses with induction 70 T were obtained in the coil without failure.

The authors of [38] point out that in the course of training the axial compression bolts could be retightened after each increase of the peak field. In the described experiments the coil before discharge had the temperature 77 °K and corresponding resistance, whereas after discharge the resistance was increased by 2–3 times. This fact points to the heating of the winding leading to thermal stresses and to the change of mechanical properties of material. The residual deformation immediately after discharge is the consequence of the action of not only electromagnetic forces but of the mentioned factors as well.

The simulation confirmed the reasonability of coil training using a series of pulses with gradually increasing induction peak. The inductance increment by 1–2 % in this case is an indicator of the expected catastrophic failure. Calculations confirmed that immediately after the pulse the residual thermal stresses could be present in the winding.

The sensitivity of the magnetic system to the training regime is a point in favor of a significant influence of plastic deformations on the stress level in the winding. These deformations can occur during the process of magnet fabrication, in particular, in systems with initial mechanical stresses. Along with this, the change in the mechanical properties of materials at initial cooling and subsequent heating, technological defects, and nonuniformities in material properties are also important factors. These factors can be taken into account in a computer simulation; however, all these factors cannot be taken into account to give a precise picture of the stresses in the magnetic system.

Therefore the magnet generating a superhigh field remains the object where the proficiency and intuition of developers do not play less of a role than rigorous computer simulation.

In a time while the winding is cooling, the magnetic system is filled with liquid nitrogen. The technology for fast refueling of liquid nitrogen was the subject of special studies. In a paper Marshall, Swenson, Gavrilin, and Schneider-Muntau described the cooling system for a bimodular magnet generating a field of 65 T [66]. The cooling time of this system is 30 min.

6.5 Magnets with record fields

In all the monolithic magnets considered above, the achieved field level did not exceed 80 T. Optimization of magnetic systems is possible, as was shown using computer calculations in the study [52]. However, it is difficult to expect stronger fields without increasing the dimensions and energy of the magnet with an equally-loaded winding. In Chapter 5. we presented the dependences for aspect ratios derived using the model of continuous current distribution. They show a sharp growth of the aspect ratio in going from a field of 70–80 T to one at megagauss level. This is illustrated by the example of the system with equally-loaded conductors, where for the aspect ratio the evaluation $\alpha \approx \exp(B_i^2/2\mu_0[\sigma])$ is valid. Here B_i is the induction on the axis. Knowing the aspect ratio α_1 for the magnet with induction $B_{i,1}$, it is possible to evaluate the ratio α_2 for a magnet of similar construction with stronger field $B_{i.,2}$. With a constant strength parameter $[\sigma]$ the model of continuous current distribution admits the following estimation:

$$\alpha_2 = \alpha_1^{\eta}, \tag{6.1}$$

where $\eta = (B_{i,2}/B_{i,1})^2$. For example, for $B_{i,2} = 100$ T, $B_{i,1} = 75$ T and $\alpha_1 = 5 \div 10$ we have $\alpha_2 = 17 \div 60$. Here with the increase of induction by 33 % the aspect ratio increases by many times, leading both to a growth of dimensions and of energy of the magnetic system.

It was shown in Chapter 5 that the length of the magnet with equally-loaded winding should not be less than its external diameter $2R_2$. Therefore, in going to megagauss field the volume of the magnetic system

$$V \geq 2\pi R_1^3 \alpha^3, \tag{6.2}$$

and the field energy sharply increase.

Calculations for the system with internal reinforcements give a similar result. The conditions of equal heating in all winding layers should be provided, i.e., the action integral $S = a_s \delta_m^2 \tau$, should be constant (δ_m is the amplitude of current density, a_s is the dimensionless factor determined by a law of current density changing). In a multilayer winding with increase of the layer radius, the magnetic flux, which is embraced by turns of the layer, increases. The induced voltage also increases. Theoretically it is possible to provide the conditions that the induced stress and heating of conductors be equal in all winding layers (see Chapter 5). For this purpose, at constant current density in all layers an amount of turns should decrease inversely to flux Φ_m. This is principally achievable if in each layer the width is chosen properly. In such hypothetical magnet with layers of equal length the layer currents should increase if the number of ampere turns in each layer is kept constant.

Another way of sustaining an acceptable voltage level in the winding is the splitting of the magnet into several coaxial modules with independent supply, and reduction of the rate of the flux change, produced by external modules. This can be done by increasing the current pulse duration in these parts of the magnet. In its turn the increase of the pulse duration requires the current density in external modules to be decreased in order to conserve the admissible value of the action integral. From what has been said, it is evident that the optimization of magnet performance assumes the splitting of the magnet into several coaxial modules, the rational choice of the supply system and planning of the current turning on in separate parts of the magnet. An example is the magnet developed at the NHMFL and Los Alamos National Laboratory (LANL, USA) [53], designed for generation of long duration pulses with peak induction of 60 T. Its winding consists of 8 coils, connected in three modules and energized independently. A characteristic feature is the different voltage of the supplying sources: 1.2 kV for the first module (1–3 coils), 8 kV for the second (2–7 coils) and 12 kV for the third one (8th coil). The pulse duration in the third module is approximately tenfold compared to the first module.

The three-module system was proposed by Surma [57]. In this work a system of independent supply of all modules was realized in the model experiment. Modern magnets with a field "megagauss" consists of two mechanically unconnected windings, one of them being placed inside the other. The windings are energized from independent sources. In their study, Gersdorf, Roeland, and Mattens [54] showed the expediency of application of the system where the internal and external coil generate fields with equal induction, but the pulse duration of the first coil is much less than of the latter. Ashkenazy [55] developed the analytical theory of the system "coil-in, coil-out" with equally loaded windings, and grounded the application of such a system for the generation of a megagauss magnetic field.

From the calculations it is evident that the most rational system is the system consisting of two modules, each of which produce approximately the same field. The inductance and the energy of the internal magnet are relatively small. The current density is rather high in the conductors of this part of the magnet, but their heating would remain admissible, owing to the small duration of the current pulse. The pulse produced by the external module is much longer. Respectively, the cross section of the conductors, the volume, and the energy of the magnetic field of the external coil is much greater than that of the internal one. The optimal power source for the internal magnet is the high voltage capacity bank which delivers the energy of the order of several megajoules, and for the external magnet whether the capacity bank of much higher energy or synchronous generator operating in the slowing-down regime.

The development of the two-module magnet with a field of 100 T began in the USA in 1990s [52]. The project, developed at NHMFL and LANL, presents the basic constructive features of such a magnet. At the intermediate stage a magnet with a field of 88 T was constructed; this is described in the report by Swenson, Sims, and Rickel [58]. Figure 6.18a shows the general view of the magnet and the construction of its internal part. The magnet consists of two modules. The inner module produces a field of induction at 50.9 T in the cavity 15 mm indiameter, and the external module produces a field at 39.1 T. The inner module of the magnet has eight equally-loaded layers. Each of them consists of the winding made with Cu–Nb wire with a cross section of 3×5.8 mm, insulated with

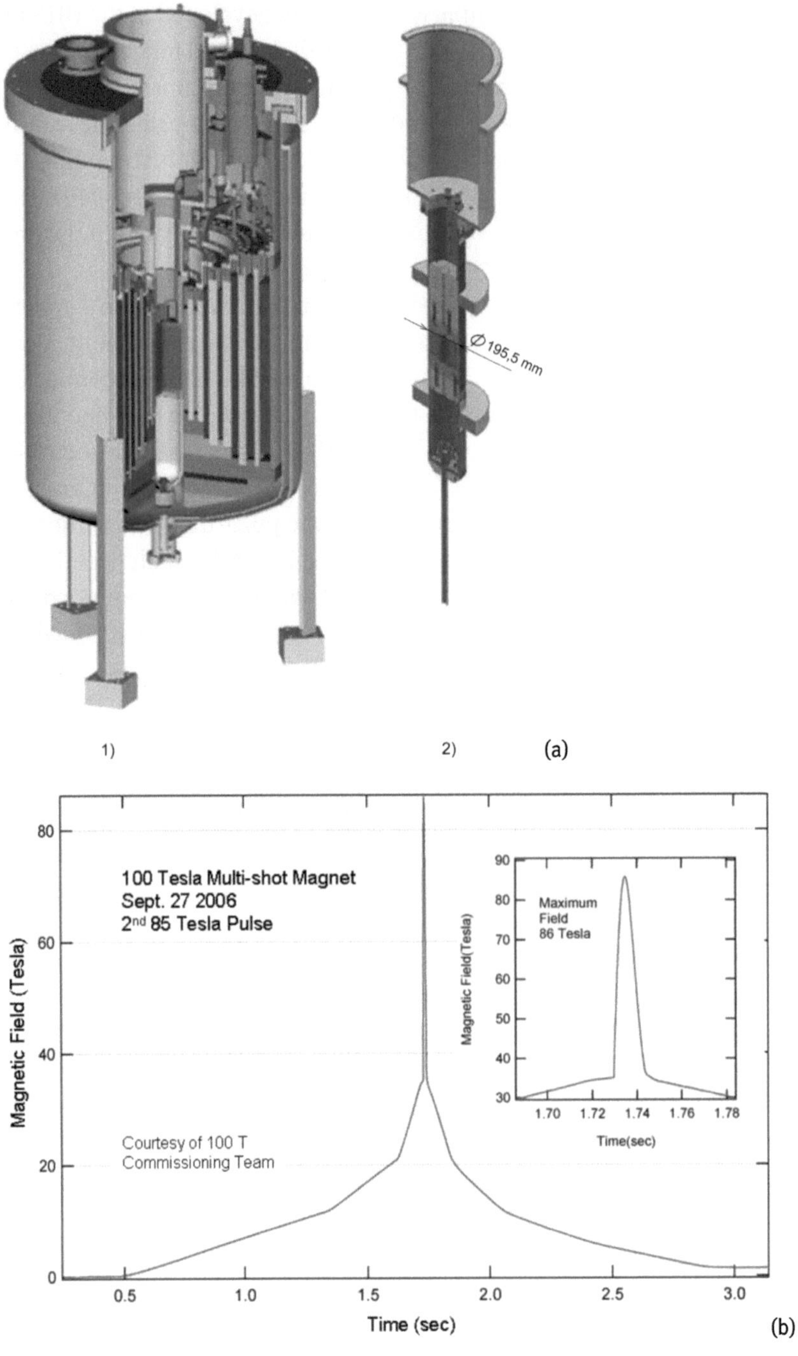

Fig. 6.18: Magnet of the Los Alamos National Laboratory with the field 90 T. [68]

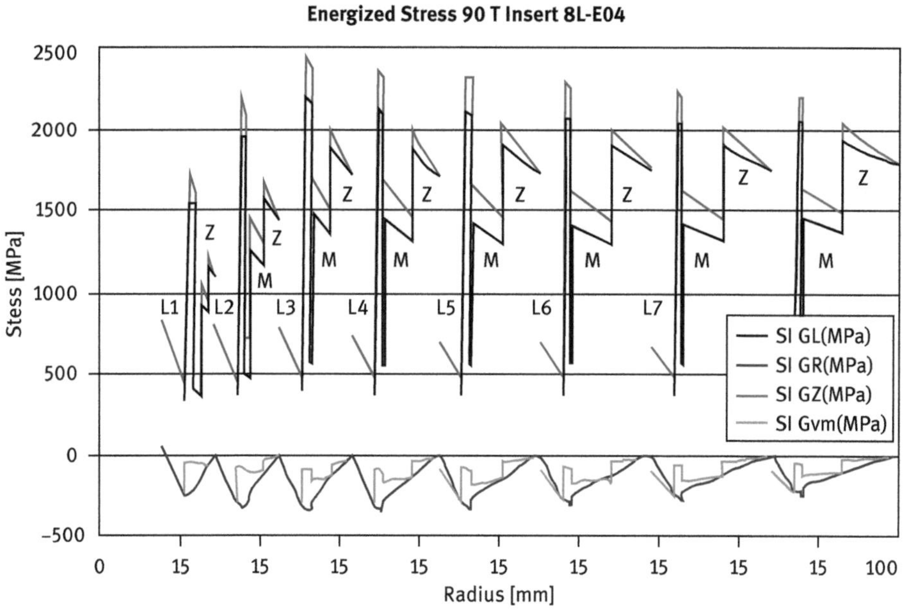

Fig. 6.19: Stress distribution in the inner module magnet with a field of 90 T.

kapton/zylon material. The internal reinforcements are made of the steel layer MP35N and of zylon. The steel reinforcements are insulated from the wires by the layer of kapton/zylon. The magnet operates under conditions when the elements of the winding are equally loaded. This can be seen in the graphs of calculated stresses (Figure 6.19). In Cu–Nb conductors the von Mises stresses are 700–800 MPa, of a zylon/epoxy composite – about 2000 MPa and 1500 MPa in the MP35N metal reinforcements. The characteristic oscillogram of the pulse of two-module magnets is shown in Figure 6.18b: the pulse duration of the internal field is 120 ms, and of the external field is 3 s. Respectively, the power sources are essentially different: to supply the internal module the capacity bank of energy 1.3 MJ is used, and the external module is powered by the motor-generator.

Further developments were completed by the construction of a magnet in the LANL where the record field with the induction of 104T was obtained [59]. To power the internal module, the capacity bank of the energy 2.6 MJ was used in that magnet. The power supply of the external module is the synchronous generator providing the power about 1430 MW.

Somewhat different is the construction of the compact two-module magnet with a field of 91.4 T, developed at the Dresden Magnetic Field Laboratory (DMFL) (Figure 6.20). In this magnet, with the insert radius of 8 mm and external of 160 mm, the gap between the internal and external modules is reduced and the

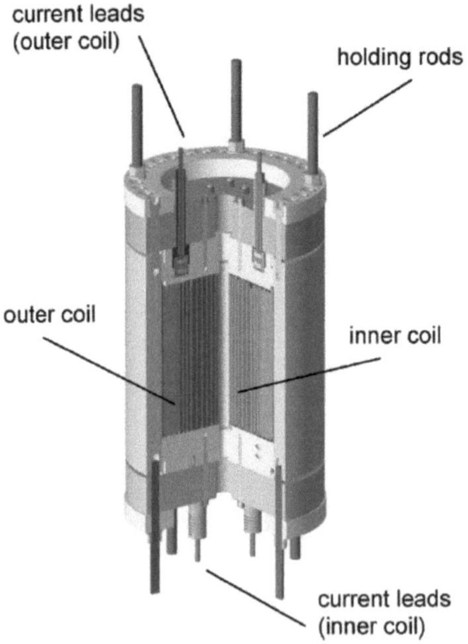

Fig. 6.20: Magnet with a field of 91.4 T, developed at the Dresden Magnetic Field Laboratory [61].

level of stresses in the internal zylon reinforcement achieves 3.4 GPa. The authors of this construction in [61, 62] underline the importance of tightening of the face part of the internal magnet: there in the vicinity of the winding ends an "instability of the thick layers of internal zylon reinforcement against the shear and axial stresses" can occur [62]. The forces causing these stresses are absent in the mid-plane of the magnet, but are essential in the vicinity of the winding edges. In the magnet developed at DMFL the lengths of internal and external modules are equal. It allowed the development of the single system of tightening of the face ends of internal and external modules with a help of massive flanges made from stainless steel. In order to energize the external and internal module, capacity banks of 8.6 MJ and 0.92 MJ, respectively, are used.

The two-module magnets in close proximity of the energy scales are being devloped as a joint effort of several European laboratories [63] and in China [64].

Although the concept of a two-module "megagauss" magnet is the basis of contemporary developments, the discussion of the most rational design of such a magnet is not finished. For instance, Schneider-Muntau and Simson [65] justified the opportunity of constructing a monocoil magnet with the winding of ten mechanically separated layers and inner diameter of 10 mm and a single power source having the energy 1.73 kJ and voltage 24 kV. In such a magnet, energized by a capacity bank, the calculated value of induction is 100 T at a pulse duration of 15 ms.

The price for the doubling induction obtained in going from a one-module to a two-module magnet is the multiple increasing of the dimensions and energy of the magnetic system. Therein lies the above-mentioned feature of equally-loaded magnets with azimuth current: the sharp growth of the winding volume and stored energy with an increase in the calculated induction.

In spite of the use of the strongest materials, the generation of fields with induction of the order of 100 T in a nondestructive magnet remains as the challenge for contemporary technical capacities. A further advance in the technology of megagauss fields requires new approaches. The definite impact can be provided by the application of an additional magnet with the balanced weakly-loaded winding, placed in the solenoid cavity. It is clear from the estimations given in Chapter 5 and in [5.17] that in such a system one can obtain stronger fields, in comparison with the basic magnet, without increasing mechanical stresses. Interesting prospects are opened due to the resumption of studies on the development of quasiforceless magnets. The theoretical aspects of this direction will be considered in next chapter.

6.6 Flat helical solenoids

As mentioned above, the equally-loaded winding has become the basis of the construction of contemporary nondestructive magnets intended for generation of fields of the megagauss range. Yet, as the technology of producing strong fields advance, a number of other constructions have been developed. They are not only of historical interest, but exemplify the effective solution of the problem of generating fields of a few tens of Tesla.

Multiturn solenoids with a winding in the form of a plane spiral are of considerable current use in the technique of stationary and pulsed magnetic fields. Such a magnet made in the form of disks with cuts, separated by insulating layers, and connected in a series (Figure 6.1), was proposed by Bitter [67, 68]. Once insulating layers have been installed, the stack of disks is contracted with flanges and studs when fabricated. Bitters solenoids gained wide application due to their simple construction and high strength due to the ultimate tensile strength of their material. The strongest fields were obtained in magnets made of beryllium bronze. For a long time the strength characteristics of Bitters solenoids were highly competitive with that of multilayer spiral magnets. In recent years, more strong fields are obtained in multilayer magnets due to the application of superhigh strong materials and the constructional improvements described above.

Detailed results of mechanical and thermal calculations of the Bitter coils were considered by Montgomery's [69]. A corresponding formula can be found in Chaper 5. Beginning from the first papers by Bitter in the 1930s, flat helical solenoids have found wide application in the generation of stationary fields. These are, primarily, flat helical solenoids with bores, through which the cooling liquid is pumped. Contemporary magnets of this type are described in the work of Shneider-Muntau et al. [71]. This construction is also used in pulsed magnets, beginning with the study by Furth and Waniek [72]. In their experiments they used a coil with approximately 20 turns per 1 cm. The width of the winding in a radial direction varied from 5 to 125 mm. The magnet can operate at cryogenic nitrogen temperature. Other examples are the pulsed disk magnets described in [72–77]. In several works the contact between the disks was achieved throught longitudinal compressing. The flanges of the magnets were contracted with a help of studs, and in the work by Bird, Bole, at al. [77] with a help of a hydraulic press. Soldering was also used for this purpose in some studies [75].

In pulse magnets a winding in the form of a screw turned from a cylindrical blank is widely used [78–84] (Figure 6.21).

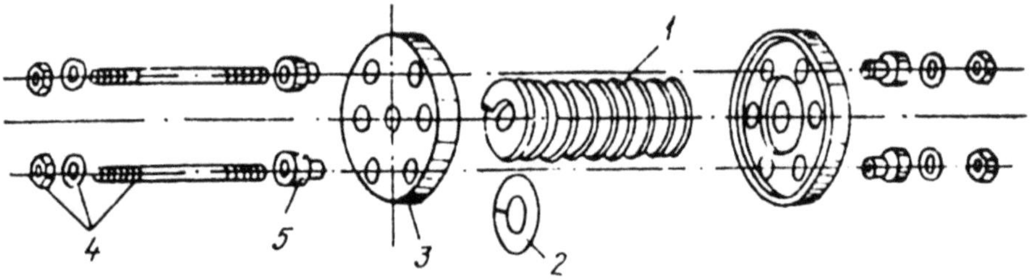

Fig. 6.21: Disassembled disk solenoid with flanges [78]: (1) winding; (2) insulation; (3) flanges; (4) connections and mounting components; (5) insulating gaskets.

To fasten the winding an external cylindrical reinforcement was applied in many works. If its rigidity is not sufficiently high, plastic deformation can occur, producing a gap between the coil and the supporter. This leads to the probable radial expansion of the winding in subsequent discharges and to winding failure. To fasten the winding, Foner and Kolm [78] used a ceramic reinforcement. It has a high rigidity and nonresidual deformation. In the coil with an inner diameter of 4.75 mm and outer of 25.4 mm, they managed to obtain a field of 65 T. The external reinforcement in some discharges could be cracked, but there was the opportunity of its replacing.

The drawing and photos of the magnet developed by Kondorskiy and Susov [81] demonstrate the main constructional elements of the flat helical solenoid, as shown in Figure 6.22. The spiral turned from beryllium bronze is confined by the

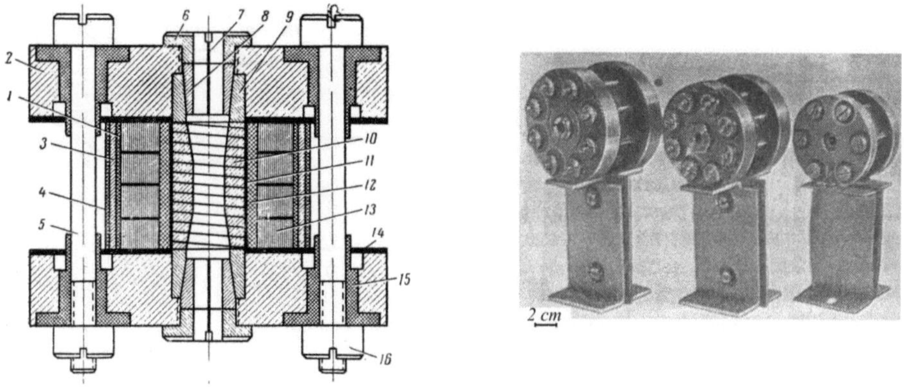

Fig. 6.22: A magnet with a winding in the form of a turned spiral of beryllium bronze, described by Kondorskiy and Susov [81]; (10) winding; (13) cylindrical reinforcement from steel tape; (2) the face reinforcement from stainless steel; (5,6,8,9,16) metal details of fixing an contacts; (11, 12, 14), basic insulation (fluoroplastic, fabric-based laminate, paper-based laminate); (1, 3, 4, 7, 15) dielectric gaskets and bushings.

cylindrical reinforcement and face flanges, contracted with bolts. In this magnet a field of 63 T was generated without failure.

Many authors studied experimentally the behavior of Bitter solenoids in strong fields. In [72] the disks, compressed in an axial and radial direction, resettled the axial deformation; however, in strong fields the inner edges of the disks were bent because of radial forces. The coil of pure copper was noticeably deformed, even in a field of 35 T, and failed in a field of 50 T. With coils of beryllium bronze several sequent pulses of 65 T were obtained. In a field of 85 T the coils failed after the single discharge. The authors point out that at short pulses inertia effects appeared.

The behavior of th einsulation of disk coils was studied in [72], using a great number of samples. The best material permitting operatipn at a voltage up to 8 kV, according to authors, is mica with teflon in oil, and the epoxy resin as a binder. This iw widely used is mica insulation [81]. The fabrication technology of disk magnets is described by Karasik [82]

Beside the mechanical damages, the deposition of copper vapors on the insulation surface due to the heating of disks by pulsed current may be responsible for the overlapping of disk coils. These vapors occur, evidently, as a result of th eoverheating of the disk ribs, where current density is increased.

In the Bitter disks the current is bending the bore and decreasing with distance from the magnet axis. A similar current distribution takes place in the magnetic system shown in Figure 6.23. It has strips in the form of a bellows with bores and cuts [85] (Figure 6.23a).

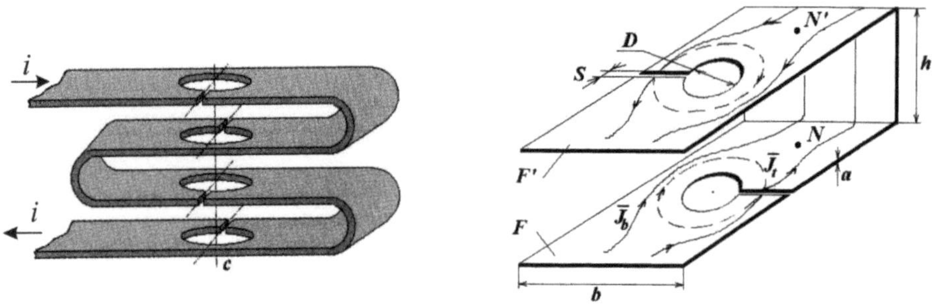

Fig. 6.23: The magnet in the form of a bellows, with bores and cuts [85].

This system can be presented as sequentially connected bifilar elements. The current in each pair of strips is the sum of the two components. One of them (solid lines in Figure 6.23b) is counter directed currents in each sheet from the pair. They bend around the bore and induce the field in the gap, directed parallel to the sheet plane. In the bore the intensity of this field is rather small. The other component of the current embraces the bore, much like the current in Bitter coils does. This current has the same direction in both sheets of the given pair and induces the axial field. This configuration is simple in construction and can be used for the generation of fields with a short rising time.

6.7 References

[1] P. L. Kapitza, A Method of producing strong magnetic fields, *Proc. Roy. Soc. Ser. A* **105** (1924), 691–710.
[2] P. L. Kapitza, Further Developments of The Method of Obtaining Strong Magnetic Fields, *Proc. Roy. Soc. Ser. A* **115** (1927), 658–683.
[3] T. F. Wall, *J. Inst. Elecrt. Engrs.* **64** (1926), 745–757.
[4] S. A. Smirnov, A. V. Georgievskiy, and V. M. Yushnina, *Fisika I tehnika silnih magnitnih poley. Sbornik referatov*, M. Atomizdat, 1970 (in Russian).
[5] G. A. Shvetsov and L. D. Vakulenko (compilers), *Ultrahigh Magnetic Fields. Years 1924–1985 Bibliographic index*, Novosibirsk, Hydrodynamics Institute of SB RAN, 1986.
[6] A. S. Lagutin and V. I. Ozhogin, *Silniye impulsniye magnitniye polya v fizicheskom experimente*, M. Energoatomizdat, 1988 (in Russian).
[7] F. Herlach (ed,), *Strong and Ultrasong magnetic Fields and Their Applications*, Berlin Heidelberg New York Tokyo, Springer-Verlag, 1985.
[8] F. Herlach, Pulsed Magnets, *Rep. Prog. Phys.* **65** (1999), 859–920.
[9] F Herlach and N. Miura (eds.), *High Magnetic Fields. Science and Technology*, New Jersey London Singapore Hong Kong Taipei, World Scientific, 2003.
[10] H. Jones, F. Herlach, J. Lee, et al., *IEEE Trans. on Magnetics* **24**(2) (1988), 1055–1058.
[11] Y. K. Katrukhin, *Pribory I technika experimenta* **5** (1987), 170–171 (in Russian).

[12] Y. K. Katrukhin, L. I. Zelikman, and A. I. Vorochov, *Pribory I technika experimenta* **5** (1974), 177–179 (in Russian).
[13] R. Gersdorf, F. A. Muller, and L. W. Roeland, *Rev. Sci. Instr.* **36**(8) (1965), 1100–1109.
[14] H. J. Schneider-Muntau, *IEEE Trans. on Magnetics* **Mag.17**(5) (1981), 1775–1778.
[15] L. Li, T.Peng, H. X. Xiao, Y. L. Lu , Y. Pan, and F. Herlach, *IEEE Trans. on Applied Supercond.* **22**(3) (2012), 4300304.
[16] A. S. Lagutin and V. I. Ozhogin, *Pribory I technika experimenta* **3** (1981), 195–198 (in Russian).
[17] G. Melville, D. J. Reiner, W. I. Khan, and P. G. Mattocks, Magnetic Fields, *J. Appl. Phys.* **50**(11) (1979), 7771.
[18] L. Li, B. Lesch, V. G. Cochran, Y. Eyssa, S. Tozer, C. H. Mielke, D. Rickel, S. W. Van Sciver, and H. J. Schneider-Muntau, **MG-VIII** (1998), pp. 128–131.
[19] E. S. Borovik and A. G. Limar, *J. Tech. Phys.* **31**(8) (1961), 939–943 (in Russian).
[20] S. Foner, *Appl. Phys. Letters* **49** (1986), 982–983.
[21] J. R. Sims and M. A. Hill, **MG-VI** (1994), pp. 179–186.
[22] S. Askenazy, F. Lecouturier, L. Thilly, and G. Coffe, **MG-VIII** (1998), pp. 132–140.
[23] Y. Sakai, K. Inoue, T.Asano, H. Wada, and H. Maeda, *J. Appl. Phys. Lett.* **59** (1991), 2965–2967.
[24] J. D. Embary, K. Han, J. R. Sims, J. Y. Coulter, V. I. Pantsyrnyi, Shikov A., and A. A. Bochvar, **MG-VIII** (1998), pp. 158–160.
[25] A. V. Nadkami, T. Klar, and W. M. Shafer, A New Dispersion-Strengthened Copper, *Metals Engeneering Quarterly*, (Aug. 1976), 10–15.
[26] A. V. Nadkami, F. C. Laabs, H. L. Dowuing, and C. V. Renand, *Materials and Manufacturing Processes* **7**(1) (1992), 1–13.
[27] J. D. Embury and R. Han, **MG-VIII** (1998) pp. 147–153.
[28] H. Jones, R. G. Jenkins, M. Van Cleemput, R. J. Nicholas, W. J. Siertsema, and W. J. Siertsema, *Physica B* **201** (1994), 546–550.
[29] F. Dupony, S. Askenazy, J. P. Peyrade, and D. Legat, *Physica B* **211** (1995), 43–45.
[30] M. Van Cleemput, H. Jones, M. Van der Burgt, J.-R. Barrau, J. A. Lee, Y. Eyssa, and H. J. Schneider-Muntau, *Physica B* **216** (1996), 226–229.
[31] V. Pantsyrnyi, A.Shikov, A. Nikulin, G. Vedernikov, I. Gubkin, Salunin N., **MG-VIII** (1998), pp. 141–144.
[32] A. S. Lagutin, K. Rossel, F. Herlach, and Y. Bruynseraede, **MG-IX** (2004), pp. 86–89.
[33] K. Yabuki, Poly(p-phenylenebenzobisoxazole) fiber, in: *12th Annual Meeting, Polymer Processing Society*, Sorrento, Italy, May 1996, p. 279.
[34] H. J. Schneider-Muntau and C. A. Swenson 100 T Monocoil Magnets, MG-10(2005), pp.63–64.
[35] C. A. Swenson, W. S. Marchall, A. V. Gavrilin, K. Han, J. Schilling, J. R. Schus, and H. J. Schneider-Muntau, *Physica B* **346–347** (2004), 561–565.
[36] W. A. Spitzig, F. C. Laabs, H. L. Downing, and C. V. Renaud, *Material & Manufacturing Processes***7**(1) (1992), 1–13.
[37] Y. K. Huang, P. H. Frings, and E. Hennes, *Composites: Part B* **33** (2002). pp. 109–115.
[38] A. S. Lagutin, K. Rossel, F. Herlach, and Y. Bruynseraede, *Physica B* **346–347** (2004), 599–603.
[39] H. Witte, H. Jones, V. Galanov, and J. Freudenberger, **MG-X** (2005), pp. 333–337.
[40] L. Van Bockstal, G. Heremans, and F. Herlach, *Mts. Sci. Technol.* **2** (1991), 1159–1164.
[41] E. S. Borovik and A. G. Limar, *J. Tech. Phys.* **32**(4) (1962), 441–444 (in Russian).

[42] V. A. Ignatchenko and M. M. Karpenko, *J. Tech. Phys.* **38**(1) (1968), 201–204 (in Russian).
[43] H.-Y. Shneider-Muntau and P. Rub, Conf. Grenoble (1974), No. 242, Editions du CNRS, 1975, p. 161.
[44] M. Date, *J. Phys. Soc. of Japan* **39**(4) (1975), 892–897.
[45] G. A. Shneerson, *J. Tech. Phys.* **56** (1986), 36–43 (in Russian).
[46] K. Han, A. Ishmaku, H. Garmesyani, V. J. Toplosky, at al., *IEEE Trans. on Appl. Supercond.* **12** (2002), 1244–1247.
[47] K. Rossel, F. Herlach, J. Vanacken, A. S. Lagutin, Y. Brynseraede, Van Humback, *J. Phys. B* **346–347** (2004), 571–575.
[48] K. Kindo, *Physica B* **294–295** (2001), 585–590.
[49] L. Lecouturier, J. Billette, J. Bèard, F. Debray, N. Ferreira, J. M. Tudela, G. Rikken, and P. Frings, *IEEE Trans. on Appl. Supercond.* **22**(3) (2012), 4300404.
[50] P. Frings, J. Billette, J. J. Beard, O. Portugall, F. Lecouturier, and G. Rikken, *IEEE Trans. on Appl. Supercond.* **18**(2) (2008), 592–595.
[51] T. Peng, L. Lie, J. Vanacken, and F. Herlach, *IEEE Trans. on Appl. Supercond.* **18**(2) (2008), 1509–1512.
[52] L. J. Campbell, D. Embury, K. Han, D. M. Parkin, A. G. Baca, K. H. Kihara, et al., **MG-VIII** (1998), pp. 85–98.
[53] H. J. Boeing, L. J. Campbell, M. L. Hongdon, E. A. Lopez, D. G. Rickel, J. D. Rogers, et al., **MG-VI** (1994), pp. 67–74.
[54] R. Gersdorf, F. A. Müller, and L. W. Roeland, *Rev. Sci. Inst.* **36**(8) (1965), 1100–1109.
[55] S. Askenazy, *Physica B* **216** (1996), 221–225.
[56] L. J. Campbell and J. Schilling, in: F. Herlach and N. Miura (eds.), Controlled Maveform Magnens, in: *High Magnetic Fields. Science and Technology*, pp. 153–203, New Jersey London Singapore Hong Kong Taipei, World Scientific, 2003.
[57] M. Surma, *J. de Physique. Coll. C1, Suppl. 1* **65** (1984), 45–48.
[58] A. S. Swenson, J. R. Sims, and D. G. Rickel, **MG-SF** (2006), pp. 207–215.
[59] C. H. Mielke, Report on the conference Megagauss XIV (2012) (to be published).
[60] D. Melville and P. G. Mattocs, *J. Phys. D* **5** (1972), 1745–1759.
[61] S. Zherlitsyn, B. Wustmann, T. Herrmannsdorfer, and J. Wosnitza, *IEEE Trans. on appl. Supercond.* **22**(3) (2012), 4300603.
[62] S. Zherlitsyn, T. Herrmannsdorfer, B. Wustmann, and J. Wosnitza, *J. Low. Temp. Phys.* **170** (2013), 447–451.
[63] O. Portugall, M. Mainson, J. Billete, F. Lecouturier, P. Frings, and G. Rikken, **MG-XI** (2006), pp. 53–57.
[64] T. Peng, Q. Q. Sun, X. Zhang, Q. Xu, H. X., F. Herlach, Y. Pan, and L. Li, *J. Low. Temp. Phys.* **170** (2013), 463–468.
[65] H.-Y. Schneider-Muntau and C. A. Swenson, **MG-X** (2006), 63–67.
[66] W. S. Marshall, C. A. Swenson, A. Gavrilin, and H. J. Schneider-Muntau, *Physica B* **346–347** (2004), 594–598.
[67] F. Bitter, *Rev. Sci. Instr.* **7** (1936), 479.
[68] F. Bitter, *Rev. Sci. Instr.* **7** (1936), 482.
[69] D. B. Montgomery, *Solenoid Magnet Design*, New York, Wiley Interscience, 1969.
[70] V. V. Ahmanov, L. M. Barkov, R. S. Bobovikov, Y. P. Dobretsov, et al., *Pribory I technika experimenta* **4** (1965), 182–187 (in Russian).

[71] H. J. Schneider-Muntau and Y. Nakagawa, Steady State Resisbeve and Hybrid Magnets, in: F. Herlach and N. Miura (eds.), *High Magnetic Fields. Science and Technology*, New Jersey, London, Singapore, Shanghai, Hong Kong, Taipei Bengalore, World Scientific, 2003.
[72] H. P. Furth and R. W. Waniek, *Rev. Sci. Instr.* **27** (1956), 195–203.
[73] A. G. Bontch-Osmolovskiy and K. I. Krilov, *Izvestija Vuzov, Radotechnika* **2** (1959), 159–164 (in Russian).
[74] R. I. Kuskovsky, T. B. Novey, and S. D. Warshaw, *Rev. Sci. Instr.* **32** (1961), 674–682.
[75] R. Evangelist, G. Pasotti, and G. Saierdoti, *Nucl. Instr. and Methods* **16** (1962), 189–194.
[76] T. H. Fields, *Nucl. Instr. and Methods* **20** (1963), 465–476.
[77] M. D. Bird, S. Bole, Y. M. Eyssa, B. J. Gao, and H. J. Schneider-Muntau, *Physica B* **216** (1996), 193–195.
[78] S. Foner and H. Kolm, *Rev. Sci. Instrum.* **27**(7) (1956), 547–548.
[79] I. T. Fakidov and E. L. Zavadskiy, *Fizika metallov I metallovedeniye* **8**(4) (1959), 562–568 (in Russian).
[80] R. Stevenson, *Canad. J. Phys.* **39**(2) (1961), 367–369.
[81] E. I. Kondorskiy and E. V. Susov, *Pribory I tehnika experimenta* **1** (1963), 125–130 (in Russian).
[82] V. R. Karasik, *Fisika i Tehnika silnich magnitnih poley*, Moskow, Nauka, 1964 (in Russian).
[83] L. Hoffman and M. Morpurco, *Nuclear Instr. and Methods* **20** (1963), 489–493.
[84] I. P. Efimov, S. I. Krivosheev, and G. A. Shneerson, **MG-VIII**, 1998, pp. 108–111.

7 Solenoids with quasi-force-free windings

Electromagnetic forces acting on a conductive media are reduced to zero if the following condition is satisfied:

$$\boldsymbol{\delta} = \lambda \mathbf{H}, \tag{7.1}$$

where δ is the current density and λ is the scalar coordinate function. In a region where this condition is valid, the vector of the current density is parallel with that of induction, therefore the equality

$$\mathbf{f} = [\boldsymbol{\delta}, \mathbf{B}] = 0 \tag{7.2}$$

holds, where f is the volume electromagnetic force. Magnetic fields in which electromagnetic forces are absent are referred to as force-free.

Force-free magnetic systems have been the object of much research. Firstly there are the theoretical works of a general character, among which are the initial publications and succeeding papers related to problems of astrophysics, superconductivity, and plasma physics [1–14]. then there are the works concerned with toroidal magnetic systems, among which are inductive energy storages and stellarators. In the windings of these magnetic systems, electromagnetic forces can be drastically decreased, because the current distribution corresponds approximately to the force-free field (see, i.e., [15–22]).

Force-free systems are of great scientific interest with regard to the problem of achieving ultrahigh magnetic fields. A field satisfying equation (7.1) can only exist if the current distribution is continuous in space. A similar situation in possible in plasma, but is excluded in an actual magnet, the winding of which is made of solid conductors separated by insulation gaps. In a system with discrete conductors it is possible to produce a field close to force-free if the vectors of the current density and induction are approximately parallel. Such windings can be said to be quasi-force-free. An actual magnet designed for functioning in such fields, should contain a winding consisting of several equilibrium current layers, which is the discrete analog to a force-free winding. The simplest system of this kind was considered in [23, 24].

In [24] a multilayer quasi-force-free winding with currents varying its direction in each layer was considered. The authors of [25–29] have suggested the method of forming the system of equilibrium current layers and shown that mechanical stresses can be reduced to values of the order $B_i^2/(2\mu_0 N^2)$, where B_i is the induction on the magnet axis and N is the number of layers. One feature of an actual magnet is the existence of an outer zone, where the poloidal current should be closed, and faces, which must be balanced in the same way as the main part of a winding. The author of [29] suggested a method to produce such current distribution outside a winding that the residual stresses are equal in all internal reinforcements, and the magnetic system is of acceptable radial sizes even in the field with induction of about 100 T.

The content of this chapter in certain respects reflects the results of these studies, which are mostly theoretical. In spite of technological difficulties related to realization of windings for quasi-force-free magnets, one can reasonably expect that with their help it would be possible to achieve fields in the megagauss range with no winding destruction.

7.1 Quasi-force-free configurations, an analog of which is a winding of a quasi-force-free magnet

7.1.1 One-dimensional quasi-force-free magnetic systems: the flat layer and cylinder

The simplest example of a force-free configuration is a flat layer (Figure 7.1) with current distribution such that condition (7.2) is valid.

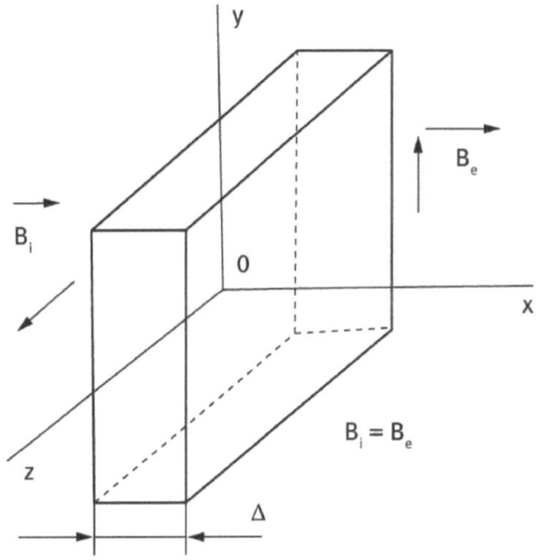

Fig. 7.1: Flat conductive layer with magnetic field.

From this condition two equations result, as projected onto axes *x* and *y*:

$$\mu_0 \delta_y = \lambda B_y = -\frac{dB_z}{dx},$$

$$\mu_0 \delta_z = \lambda B_z = -\frac{dB_y}{dx}. \tag{7.3a}$$

Then comes the equality

$$B_y^2 + B_z^2 = B^2 = \text{const.} \tag{7.3b}$$

This is a conventional condition of plasma equilibrium with zero pressure. In a flat layer, the induction vector of a force-free field changes its orientation, but the module is kept constant. In the particular case when the conditions at the layer boundaries are given by $B_z = B, B_y = 0$ at $x = 0$; $B_z = 0, B_y = B$ at $x = \Delta$, the linear current density $\mathbf{j} = \int_0^\Delta \boldsymbol{\delta}\; dx$ is numerically equal to $\sqrt{2}B/\mu_0$, $\mu_0 j_y = \mu_0 j_z = B$, and vector $\mathbf{j}$ is directed at an angle $\pi/4$ to the vectors of induction outside the layer. One possible induction distribution is one of the form $B_y = B \sin\beta(x)$; $B_z = B\cos\beta(x)$, where the function $\beta(x) = \text{arctg}(B_y/B_z)$ varies arbitrarily within the limits $0 \le x \le \Delta$ of a layer from 0 to $\pi/2$. Then for current components we have

$$\delta_y = \frac{B}{\mu_0}\sin\beta(x)\frac{d\beta}{dx}, \quad \delta_z = \frac{B}{\mu_0}\cos\beta(x)\frac{d\beta}{dx}, \tag{7.4a}$$

and $\lambda = (1/\mu_0)(d\beta/dx)$. In the particular case when $\lambda = \pi/(2\mu_0\Delta) = \text{const}$, the induction components and current density in a layer change as $\sin\lambda\, x$ and $\cos\lambda\, x$, but the current density module is kept constant. In the general case this does not occur. For instance, if conditions $\beta = (\pi/2)(x/\Delta)^2$ and $\lambda = \pi\, x/\mu_0\Delta$ are satisfied, the following equalities take place:

$$B_y = B\sin\left[\frac{\pi}{2}\left(\frac{x}{\Delta}\right)^2\right], \qquad B_z = B\cos\left[\frac{\pi}{2}\left(\frac{x}{\Delta}\right)^2\right], \tag{7.4b}$$

$$\delta_y = \frac{B\pi\, x}{\mu_0\Delta^2}\sin\left[\frac{\pi}{2}\left(\frac{x}{\Delta}\right)^2\right], \quad \delta_z = \frac{B\pi\, x}{\mu_0\Delta^2}\cos\left[\frac{\pi}{2}\left(\frac{x}{\Delta}\right)^2\right]. \tag{7.4c}$$

As an other example let us consider the field in an infinitely long cylinder, the induction vector of which in the cylindrical coordinates has merely components B_z and B_φ, depending, as does the function λ, only upon the r-coordinate. In this case the condition $\text{div}\,\mathbf{B} = 0$ is fulfilled at any choice of $B_\varphi(r)$, $B_z(r)$. Equation (7.2) falls into two equations:

$$\mu_0\delta_\varphi = \lambda B_\varphi = -\frac{dB_z}{dr}, \tag{7.5a}$$

$$\mu_0\delta_z = \lambda B_z = \frac{1}{r}\frac{d}{dr}\left(rB_\varphi\right). \tag{7.5b}$$

With condition $\lambda = a = \text{const}$, the solution of this system is known as

$$B_z = NJ_0(ar), \quad B_\varphi = NJ_1(ar), \tag{7.6a}$$

where $J_0(ar)$, $J_1(ar)$ are the Bessel functions. Such a field is produced by the current with components $\delta_\varphi = (1/\mu_0)a\,NJ_1(a\,r)$ and $\delta_z = (1/\mu_0)a\,NJ_0(a\,z)$ [1]. in addition, $\delta_\varphi/\delta_z = B_\varphi/B_z$, which means that the condition for vectors of current density and induction be parallel is satisfied. In this case the current occupies an unlimited region from $r = 0$ to $r = \infty$.

For the goal of construction of the general purpose magnet, the force-free system is of interest where the region with an axially symmetric current distribution occupies a bounded volume of space in both the axial and the radial direction. In the simplest case one can assume that the magnetic system length far exceeds its radius, and the field in the middle zone (far from the faces) is close to uniform. In order to have the uniform field near to an axis region ($0 < r < R_1$), it is possible, with the help of a long force-free solenoid, to produce the current distribution of the considered type in the range $R_1 < r < R_2$, where aR_1 is the root of function $J_1(x)$, and aR_2 is the root of function $J_0(x)$. It is evidently that the condition $aR_1 < aR_2$ should be valid; therefore if $aR_1 = \lambda_k^{(1)}$, then $aR_2 = \lambda_m^{(0)}$, where $m > k$. Here $\lambda_k^{(1)}$ is the k-root of the function $J_1(x)$ (the zero-root corresponds to $k = 0$), $\lambda_m^{(0)}$ is the m-root of the function $J_0(x)$. Such a choice of boundaries ensures the condition $B_\varphi = 0$ on the inner surface of the winding and $B_z = 0$ on the external one. For instance, for a long solenoid with a force-free winding of an arbitrary inner radius R_1 one can choose an external radius to be equal to $R_2 = (\lambda_2^{(0)}/\lambda_1^{(1)})R_1 = 1.44R_1$. In this case $a = \lambda_1^{(1)}/R_1 = 3.83/R_1$, the thickness of the winding is $0.44R_1$. Let the induction on an axis of such solenoid is B_i. The field in the winding is described by the following equations:

$$B_z(r) = B_i J_0\left(\lambda_1^{(1)} r/R_1\right)/J_0\left(\lambda_1^{(1)}\right), \tag{7.7a}$$

$$B_\varphi(r) = B_i J_1\left(\lambda_1^{(1)} r/R_1\right)/J_0\left(\lambda_1^{(1)}\right), \tag{7.7b}$$

$$B_\varphi(R_2) = 0.85B_i, \quad \boldsymbol{\delta} = \lambda_1^{(1)}\mathbf{B}/(\mu_0 R_1), \quad J_0\left(\lambda_1^{(1)}\right) = -0.40.$$

For the infinitely-long solenoid this solution is not unique. Eliminating the function $\lambda(r)$ from equations (7.5a), (7.5b), we come to the equation

$$B_z\frac{dB_z}{dr} + B_\varphi\frac{1}{r}\frac{d}{dr}\left(rB_\varphi\right) = 0, \tag{7.8}$$

which can also be derived directly from the condition (7.2). If one of the functions B_z or B_φ is prescribed, it is possible to find the other one using equation (7.8). As an example let us consider two other force-free fields produced by a current existing in the region $R_1 < r < R_2$. As in the previous case, in the region $0 \le r \le R_1$ the condition $B_z = B_i$, $B_\varphi = 0$ is fulfilled, and in the region $r > R_2$ we have $B_z = 0$. First we express the induction azimuth component in the region $R_1 < r < R_2$ as

$$B_\varphi = \frac{g(r - R_1)}{r}. \tag{7.9a}$$

From equation (7.8) and condition $B_z(R_2) = 0$ we find

$$B_z = g\sqrt{2\left(\ln\frac{R_2}{r} + \frac{R_1}{R_2} - \frac{R_1}{r}\right)}, \tag{7.9b}$$

and from condition $B_z(R_i) = B_i$ we obtain

$$g = \frac{B_i}{\sqrt{2\left(\ln\frac{R_2}{R_1} + \frac{R_1}{R_2} - 1\right)}}. \tag{7.9c}$$

Here

$$\lambda = \frac{\mu_0\delta_\varphi}{B_\varphi} = \frac{\mu_0\delta_z}{B_z} = \frac{B_i}{r\sqrt{2\left(\ln\frac{R_2}{r} + \frac{R_1}{R_2} - \frac{R_1}{r}\right)}}. \tag{7.9d}$$

In the second case we set the linear law for changing B_z into a wall of the solenoid: $B_z = B_i(R_2 - r)/(R_2 - R_1)$. Equation (7.8) takes the form

$$\left(rB_\varphi\right)\frac{d}{dr}\left(rB_\varphi\right) = -r^2 B_z \frac{dB_z}{dr} = \frac{r^2 B_i\,(R_2 - r)}{(R_2 - R_1)^2}. \tag{7.10a}$$

Upon integrating using condition $B_\varphi(R_1) = 0$, we obtain

$$B_\varphi = \frac{1}{\sqrt{6}r}\cdot\frac{B_i}{(R_2 - R_1)}\sqrt{4R_2\left(r^3 - R_1^3\right) - 3\left(r^4 - R_1^4\right)}. \tag{7.10b}$$

In the limiting case when $R_2 - R_1 = \Delta \ll R_1$, we come to the configuration above, i.e., the current layer can be treated as being flat (Figure 7.1). The same result can be derived from formula (7.6a) under the condition $a\,r \gg 1$, replacing the Bessel functions by their asymptotic expressions.

7.1.2 Two-dimensional force-free configurations satisfying the characteristic boundary conditions

Axially-symmetric force-free magnetic systems, which are an analog of finite length magnets, were observed in [30]. They include the regions $T_{0,1,2}$ and T' (Figure 7.2). The former consists of a near-to-axis working zone of the magnet T_0 and adjacent outer region of the field. The force-free winding is located in region T_1. Here the magnetic field induction has both azimuth and poloidal components.

According to the virial theorem [12, 13], the force-free field in the bounded region T_1 can only exist in the presence of outer conductors carrying current, which are subjected to action of electromagnetic forces. In the system shown in Figure 7.2, outside the force-free winding there is also the outer region T_2, where the

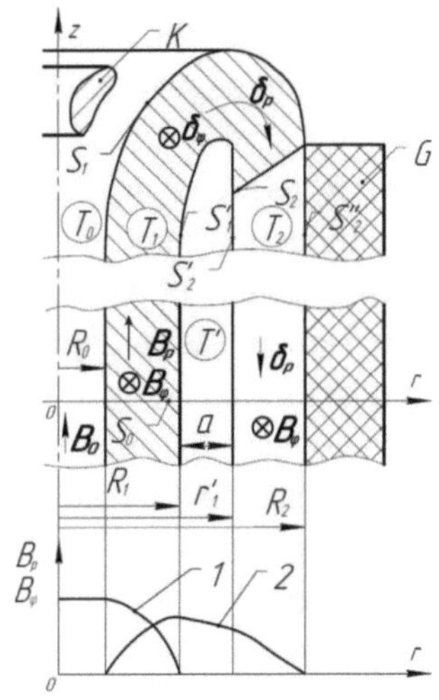

Fig. 7.2: Force-free magnetic system with an outer region T_2, in which the azimuth current is lacking, and bandages, perceiving the load, are located: G is the outer bandage, *K* the contour with an extraneous current. (1) $B_p(r)$; (2) $B_\varphi(r)$.

poloidal current is distributed. The region of the force-free field T_1 borders on the inside ($r = R_0$) with the poloidal field, and it borders on the outside ($r = R_1$) with the azimuth field. At these boundaries, the magnetic pressure has no step. In the region T_2 the electromagnetic force affect the conductors with poloidal current, and the load is perceived by a set of internal reinforcements, located in the region T_2, as well as by outer reinforecement G. In the general case the induction in the region T_2 can have both azimuth and poloidal components. The latter is comparatively small if the azimuth current in the region T_2 is absent. Between the inner and outer regions the intermediate region, T' can exist, containing no conductors with current. In the regions T_2 and T', as in the region T_0, the contours with an extraneous or inducted azimuth current can be positioned. Of some interest is also the configurations without region T' ($a = 0$, Figure 7.2).

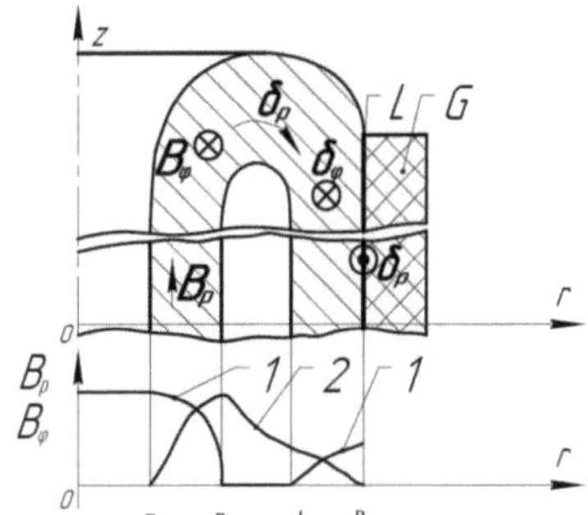

Fig. 7.3: The system in which the magnetic pressure is formed in the layer L at the boundary of the force-free zone and perceived by outer reinforcement G; (1) $B_p(r)$; (2) $B_\varphi(r)$.

The other situation takes place in the system presented in Figure 7.3. It has no region T_2 with poloidal current, in which the radial forces would be perceived by reinforcements.

The poloidal field induction grows in the range $r_1' < r < R_2$ as the outer boundary is approached. At transition over the boundary, the induction and magnet pressure drop sharply to zero. This means that at the outer boundary of the force-free region T_1 a surface current is present. In the actual magnetic system of this kind, the cylinder S_1'' carryng the azimuth current should be located on the outer part of the boundary S_1. The surface of this cylinder experiences the magnetic pressure $B^2(R_2)/(2\mu_0)$, perceived by outer reinforcement. The step of the poloidal field induction at the outer boundary can be avoided if the coaxial diamagnetic screen is located outside (Figure 7.4). In this case electromagnetic forces act on the screen, since the induction step occurs on the screen surface instead of the edge of the winding. Comparing the discussed configurations, it can be seen that the first one has the advantage from the standpoint of ensuring the strength of the magnetic system, since in the region T_2 the load can be perceived by a set of equally-loaded dielectric cylinders (intermediate reinforcements). In addition, the load can be also perceived by an outer reinforecement G. By contrast, in the the configurations shown in Figures 7.3 and 7.4 the load is transmitted only to the outer bandage located outside the current layer, or outer screen.

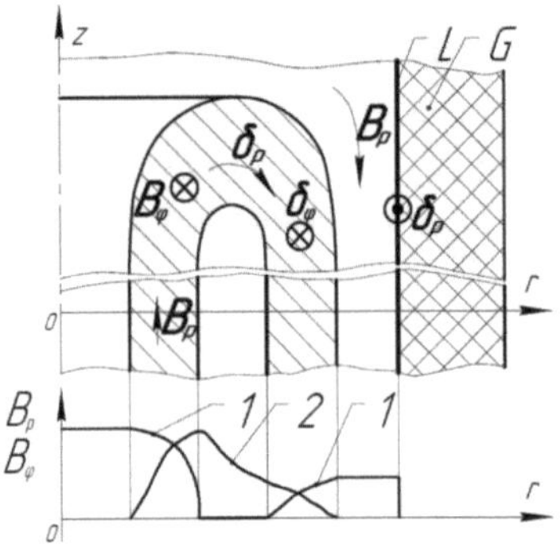

Fig. 7.4: System in which at the boundary of a force-free winding the surface current is absent due to application of closed diamagnetic screen L, perceiving the load]. (1) $B_p(r)$; (2) $B_\varphi(r)$.

In the region T_1 (Figure 7.2) the magnetic field satisfies the Maxwell equations and the additional condition (7.1). The scalar function γ in this equation takes a constant value on each force line of magnetic field. In the axial-symmetric field the components of the current density and magnetic field intensity are expressed through the current function $\psi_i = i/(2\pi)$ and the function of the poloidal field flux

$\psi_B = rA_\varphi$, where A_φ is an azimuth component of the vector potential:

$$\left.\begin{aligned} H_\varphi &= \frac{1}{\lambda}\delta_\varphi = \frac{\psi_i}{r} \\ H_z &= \frac{1}{\lambda}\delta_z = \frac{1}{\mu_0 r}\frac{\partial\psi_B}{\partial r} = \frac{1}{\lambda r}\frac{\partial\psi_i}{\partial r} \\ H_r &= \frac{1}{\lambda}\delta_r = -\frac{1}{\mu_0 r}\frac{\partial\psi_B}{\partial z} = -\frac{1}{\lambda r}\frac{\partial\psi_i}{\partial z} \end{aligned}\right\} . \tag{7.11a}$$

In the region of the force-free field T_1 both functions take the constant values on the same surfaces and are linked by the relation

$$\frac{\mu_0}{\lambda}\nabla\psi_i = \nabla\psi_B. \tag{7.11b}$$

From the Maxwell equation $\mathrm{rot}_\varphi\mathbf{H} = \delta_\varphi$ we derive the following equation which holds in the region T_1:

$$\frac{\partial^2\psi_B}{\partial z^2} + r\frac{\partial}{\partial r}\left(\frac{1}{r}\frac{\partial\psi_B}{\partial r}\right) + \mu_0 r\delta_\varphi = 0. \tag{7.12a}$$

This equation has the counterpart for the function ψ_i

$$\frac{\partial}{\partial z}\left(\frac{1}{\lambda}\frac{\partial\psi_i}{\partial z}\right) + r\frac{\partial}{\partial r}\left(\frac{1}{\lambda r}\frac{\partial\psi_i}{\partial r}\right) + \lambda\psi_i = 0, \tag{7.12b}$$

where $\lambda = f(\psi_i)$. Equation (7.12b) was derived in [6] and used in studies of the toroidal plasma configurations.

In regions T_0, T_2, and T' outside of the located in them contours with induced or extraneous azimuth currents, the function ψ_B satisfies equation (7.12a) in the absence of the term $\mu_0 r\delta_\varphi$:

$$\frac{\partial^2\psi_B}{\partial z^2} + r\frac{\partial}{\partial r}\left(\frac{1}{r}\frac{\partial\psi_B}{\partial r}\right) = 0. \tag{7.13}$$

The functions ψ_i and ψ_B should satisfy the specific conditions at the boundaries S_0, S_1, S_1', S_2, S_1', and S_2''. The boundary S_0 is a plane of symmetry. At this boundary the condition $B_r = 0$ or $\partial\psi_B/\partial z = 0$ is fulfilled. At the boundaries S_1, S_2, S_1' the functions ψ_i and ψ_B should be continuous. The normal derivatives of these functions also should not have a step at these boundaries if the surface current is absent.

The line S_1 is a generatrix of the inner solenoid surface. The normal component of the poloidal current density is absent at this boundary. Hence, the normal component of the poloidal field intensity is equal to zero, and the function of the poloidal field flux takes the constant value $\psi_B(S_1) = \psi_0$. In addition, the current function $\psi_i(S_1) = 0$.

Thus, the solution of equation (7.12b), describing the force-free field in the region T_1, should satisfy two boundary conditions on the S_1:

$$\psi_i(S_1) = 0;$$
$$(\partial\psi_i/\partial n)|_{S_1} = r\lambda B_\tau(S_1)/\mu_0, \tag{7.14}$$

where $B_\tau(S_1)$ is the tangent component of the intensity of the "vacuum" magnetic field at the boundary .

The solution of equation (7.12b) is defined by the chosen function $\lambda(\psi_i)$. Let us assume that the discussed configurations are the face parts of the long magnet ($l \gg 2R_0$). Near the medium plane of the magnet, where the radial component of the induction and current density is absent, the induction in the region near to the axis, T_0, practically does not depend on z: $B_z = B_i$, if $0 < r < R_0$. In this region, as in the magnet of infinite length, these conditions are fulfilled. In order to construct the force-free system with no surface currents, it is necessary that the functions ψ_i and ψ_B and their normal derivatives be continuous, not only far from the magnet faces, but also on all the boundaries.

We shall restrict our consideration to the case when λ = const (linear problem). Then equation (7.11) takes the form

$$\frac{\partial^2\psi_B}{\partial z^2} + r\frac{\partial}{\partial r}\left(\frac{1}{r}\frac{\partial\psi_B}{\partial r}\right) + \lambda^2\psi_B = 0. \tag{7.15}$$

In this equation the parameter λ differs from zero only in the region T_1. In this region the current function ψ_i with condition λ = const can be represented as $\psi_i = (\lambda/\mu_0)\psi_B'$, where $\psi_B' = \psi_B - \psi_0$. At the boundary S_1' the function ψ_i takes the constant value $\psi_i = \psi_{i,0} = i/(2\pi)$, where i is the total poloidal current in the winding. In region T' this value of the current function is kept constant. In region T_2 the azimuth field can be arranged through the choice of the poloidal current distribution in such a way, that the requirement for the continuity of ψ_i at the boundary S_1 be fulfilled. In the general case, in region T_2 both the poloidal component of the volume electromagnetic force $\mathbf{f}_P = [\boldsymbol{\delta}_\varphi, \mathbf{B}_P]$ and the azimuth force $\mathbf{f}_\varphi = [\boldsymbol{\delta}_P, \mathbf{B}_P]$ are present. The latter can be made equal zero if the current distribution is selected so that the condition $\boldsymbol{\delta}_P \| \mathbf{B}_P$. is met.

In constructing a force-free magnetic system, its configuration should be set up so that in all the the considered regions $T_0 + T_1 + T' + T_2$ the condition of the continuity of the function ψ_B' and its normal derivative be fulfilled. At the boundary S_1 we have $\psi_B' = 0$, and at the boundary S_1' this function takes the constant value $\psi_B' = \mu_0 i/(2\pi\lambda)$. Moreover, the solution of equation (7.15) should satisfy the

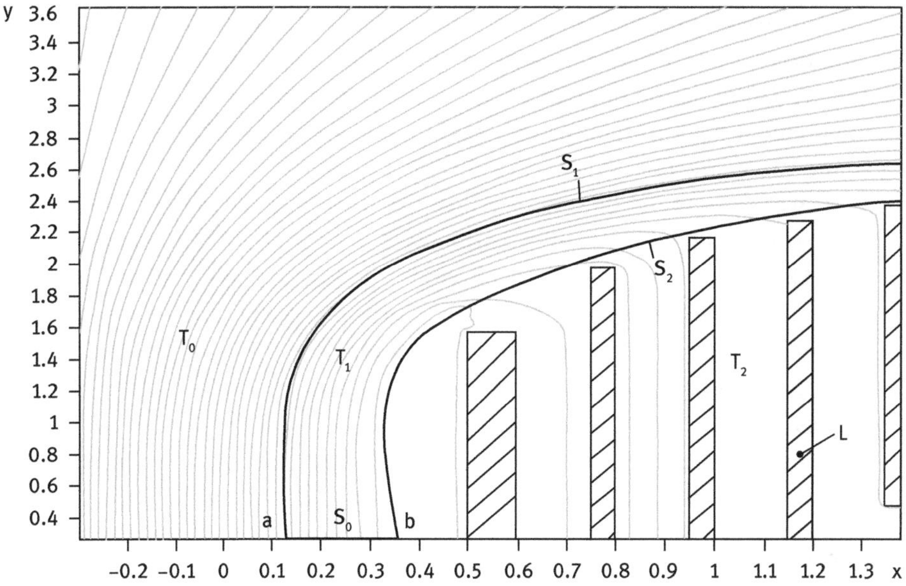

Fig. 7.5: Fragment of a flat force-free field near the boundary of the magnetic system. Lines of the poloidal field are shown. The diamagnetic bodies L are located in the region T_2.

second boundary condition (7.14). The given situation is typical for incorrect problems of mathematical physics: it is impossible to fulfill at once two conditions at the boundary of an arbitrary form, just as in the problems of electrostatics it is impermissible to immediately set the potential and its normal derivative [31].

Two-boundary conditions can be fulfilled at S_1 and S_1' by the following methods, which can be applied in conjunction or separately: by variation of the form of boundaries S_1, S_1', S_2 and location of diamagnetic bodies or contours with current in the regions T_0 and T_2 [6, 17, 28, 30, 32, 33].

An example of a similar configuration is shown in Figure 7.5, where a fragment of a flat force-free field is represented. In this example the boundaries S_1 and S_2 are fixed, and the section S_1' is lacking. At the section of the boundary S_0 the induction distribution is given as the following dependence of coordinate x: $B_y = B_i \cos[\lambda(x - x(a)]$, here $\lambda = (\pi/2d)$, where $d = x(b) - x(a)$. In the region T_2 located are the diamagnetic bodies. Setting on them the values of ψ_B', it is possible, using the test calculations, to attain the approximate coincidence of the normal derivatives $\partial\psi_B'/\partial n$ in fixed points on either side of the boundary S_1. In this example the lines of the poloidal field in the region T_1 are at the same time the ones of the poloidal current. This current is removed from the region of the force-free field, i.,e., from the region of winding, into the outer region T_2.

It is also of interest to set up the problem somewhat alternatively, which opens up additional opportunities for constructing a force free field: we can consider S_2 (Figure 7.2) to be the boundary with a given distribution of the flux function of the poloidal field. In this case the region to be calculated for the function ψ_B' falls into two categories: T_1 and $T_0 + T' + T_2$. Then the continuity of the normal derivative $\partial\psi_B'/\partial n$ at the boundaries S_1 and S_2 can be achieved by not only the above-mentioned methods, but by variation of the distribution ψ_B' at the boundary S_2. Along this, the current distribution function is also subjected to variation, i.e., the poloidal current distribution is formed at the boundary by purposely removing the current from the region T_1 into the region T_2 through the boundary S_2. The current removal is convenient when constructing windings of small thickness.

The choice of configurating force-free windings to satisfy the conditions of continuity both of the poloidal field flux function and its normal derivative has not receive proper attention in the literature. For instance, in [15, 17] toroidal systems were calculated at boundaries for which the condition of the continuity of the function $\partial\psi_B'/\partial n$ was violated. In such systems the step of the tangent component of the poloidal field intensity occurs, i.e., the surface azimuth current takes place.

7.1.3 Features of current distribution in the face zone of a force-free magnet

Let us consider the magnetic system whose boundary S_1 turns into a plane at the periphery of the face part. The simplest example is a system with the periphery part in the form of a layer of thickness d, enclosed between the flat parallel boundaries S_1 and S_1' (Figure 7.6). Let us also assume that the diamagnetic screen is located parallel to S_1 and separated from S_1 by a gap of constant thickness h [28, 32, 33].

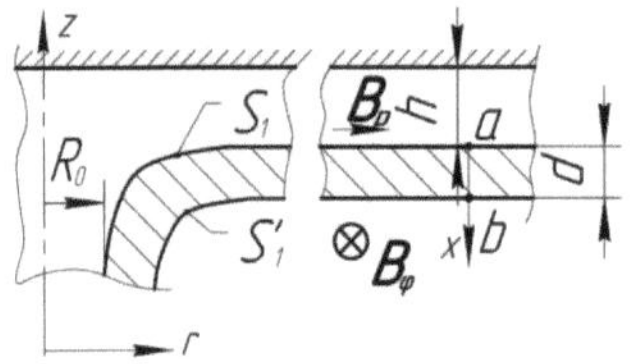

Fig. 7.6: Magnetic system with a flat face screen.

Induction in this gap changes by the law

$$B_r(a) = \frac{\psi_0}{rh}. \tag{7.16a}$$

In the actual magnetic system with a flat faced diamagnetic screen this distribution of the poloidal field takes place if $r - R_0 \gg h$, where R_0 is the inner radius of the magnet.

When there is no current removal from the boundary S_1', the azimuth field induction depends on the radius as

$$B_\varphi(b) = \mu_0 \frac{\psi_i}{r}. \tag{7.16b}$$

Both functions ψ_0 and ψ_i take constant values on the indicated boundaries. Note that $B_r(b) = 0$, $B_\varphi(a) = 0$.

In the case λ = const we have [30]

$$\begin{aligned} \psi_i &= (\psi_0/(\mu_0 h))\sin(\pi x/(2d)), \\ B_r &= B_r(a)\cos(\pi x/(2d)), \\ B_\varphi &= B_r(a)\sin(\pi x/(2d)). \end{aligned} \tag{7.17}$$

Here the condition $B_r^2 + B_\varphi^2$ = const is fulfilled.

The same condition inherent in the force-free field can also take place when there is no screen, but then the current should be removed into the region T_2. Calculations show that ,in a flat layer, the lines of the current and magnetic field intensity, passing through the points with coordinates x_1, r_1, are described by the equation

$$x(r) = \frac{1}{\lambda}\arcsin\left(\frac{r}{r_1}\sin(\lambda x_1)\right). \tag{7.18}$$

These lines are first almost parallel to the upper boundary, then steeply bend and cross the lower boundary at a rectangular angle at the point with coordinate $r = r_1/\sin(\lambda x_1)$ (Figure 7.7). Further, the current is closed beyond the force-free field in region T_2.

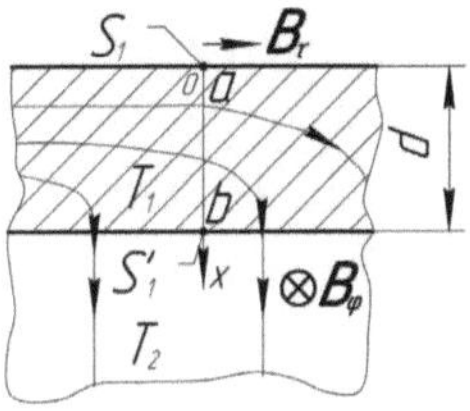

Fig. 7.7: Lines of the poloidal current in a flat layer with current removal.

It was shown in [30] that the induction distribution at the lower boundary of the layer obeys the law

$$B_x = -\frac{1}{\lambda r}\frac{\partial}{\partial r}(rB_r(a)). \tag{7.19}$$

The current density distribution is described by

$$\delta_x = \frac{\lambda B_x}{\mu_0} = -\frac{1}{r}\frac{\partial}{\partial r}\left(rB_r(a)\right). \tag{7.20}$$

With current removal according to this law, in a flat layer a force-free configuration of the field takes place. Along this, the equality $B_r(a) = B_\varphi(b)$ is valid, where points a and b lie on the same normal to the boundary.

7.2 The methods of realization of a quasi-force-free winding. The estimates of residual mechanic stresses in a thin-wall quasi-force-free winding

Let us consider whether or not it is possible to substitute a force-free distribution by a discrete system of conductors in a region far from the faces, where the radial component of the current density and induction is lacking. Here we shall focus on the problem of the approximation of the force-free distribution in systems with conductors of finite, while small, thickness.

7.2.1 Quasi-force-free winding with pairs of the equilibrium current layers (number of pairs $N \gg 1$)

Let us consider the winding which consists of N pairs alternating layers with orthogonal currents in each of them (Figure 7.8). The layers are separated by insulation inserts. The distribution of the axial and azimuth currents, approximating the force-free current distribution, is produced in such a way that in every pair the stretching stress acting on the inner ringed turn ***1*** with azimuth current would be brought to equilibrium by the compressing force acting on the outer layer **2** with axial current. Then each pair is equilibriated while the insulation insert ***3*** is under compression.

Further, to estimate the process we suggest that $h \ll \Delta$ and, in addition, the thickness of each layer is small compared to its radius. Assume also that the thickness of each layer is much less than the penetration depth of the magnetic field. This permits us to consider both components of the current density to be constant in corresponding layers. With this assumption the induction B_z linearly decreases in the section $r_n < r < r_n + \Delta$ and is kept constant in the section $r_n + \Delta < r < r_{n+1}$. The induction B_φ increases in the section $r'_n < r < r'_n + \Delta$ and is kept constant in the section $r'_n + \Delta < r < r'_{n+1}$. In [26] it was shown that the stress module in the

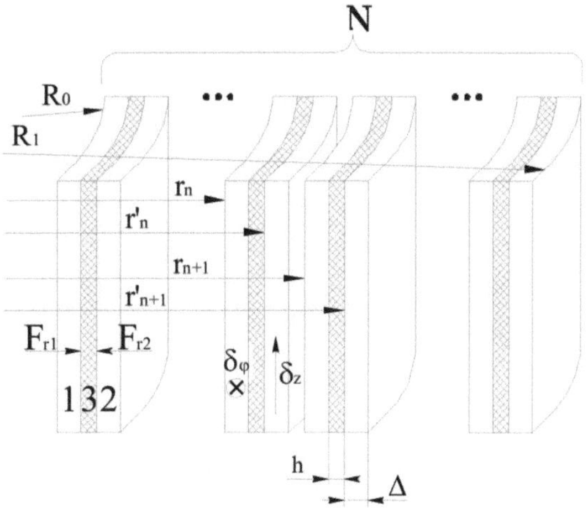

Fig. 7.8: Quasi-force-free winding with pairs of balanced current layers.

insert can be estimated by the formula

$$|\sigma_\varphi|_n = |\sigma_r|_n \approx B_\varphi\left(r'_n\right)\delta_{z,n}\Delta. \tag{7.21}$$

The induction B_φ is a magnitude of the order B_i, where B_i is the field on the magnet axis. The current density δ_φ is the magnitude of the order $2B_I/(\mu_0 d)$, where $d = 2N\Delta$ is the thickness of a magnet winding. From here the estimated value of stress is

$$|\sigma_\varphi| = |\sigma_r| \approx B_i^2/(\mu_0 N). \tag{7.22}$$

Essentially, the stress is the reciprocal of a number of layers N.

More accurate analysis has been carried out for a system of equally-loaded layers. In calculations, the model of the continuous current distribution was used. The magnetic pressure acting on the insulation cylinders is then expressed by the approximate formula

$$\sigma_0 = B_i^2 \frac{\Delta}{\mu_0\left(R_1 - R_0\right)} \approx \frac{B_i^2}{2\mu_0 N}, \tag{7.23}$$

where $N \approx (R_1 - R_0)/(2\Delta)$ is the number of winding layers. Note that the layers are of the same thickness, separated by thin insulation inserts, and carry the axial and azimuth current. The formula (7.23) gives a closer approximation than (7.22). It shows that the stress in the system under consideration is N times smaller than the magnetic pressure calculated for a field on the solenoid axis.

Figure 7.9 represents the numerical results for the system containing five pairs of equally-loaded layers at field induction $B_i = 100$ T.

A comparison with the model of the continuous current distribution shows that this model gives a workable approximation for estimating of the actual current distribution in a winding, in spite of the relatively small number of layers. The

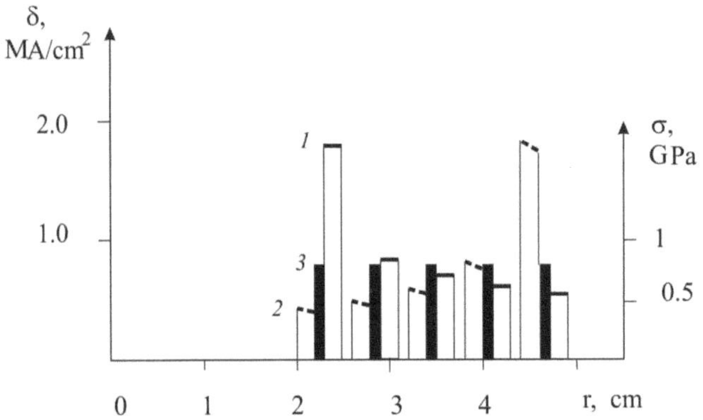

Fig. 7.9: Distribution of mechanical stresses and current densities in the quasi-force-free winding consisting of five pair of layers (computer simulation): (—) = δ_z; (••) = δ_ϕ; (■) = $/\sigma_r/ = /\sigma_\phi/$.

discrepancy takes place near the boundary, where analytical calculation gives an infinite growth of the current density. As for stresses, the result of the numerical calculations (which is $\sigma_r = \sigma_\varphi \approx 8 \cdot 10^8$ Pa) practically coincides with that obtained using (7.23).

The model of continuous current distribution can also be used when the alternative initial propositions are chosen for determining the winding construction. By way of example, we refer to the magnetic systems in which the density of the axial or azimuth current are kept constant in the limits of winding. The results of calculations for such systems are given in [26].

7.2.2 Multilayer magnetic systems with variable direction of current in each layer

An approximation of the current distribution corresponding to the force-free field can be formed as a system of layers with a current discretely varying its direction [26] (Figure 7.10a). In Section 5.4 it was shown that stresses in thin-wall cylinders with a currentcan be drastically decreased if the cylinder is inserted in the outer field. This situation can be realized in multilayer windings, where each layer is in a field produced by its own current as well as by the other layers. In the particular case of a single layer, a sharp drop of stresses can occur if the current is directed at an angle of about $\pi/4$ to an axis [23, 24].

Figure 7.10b shows the vector diagram of currents in N layers of the winding, the thickness of which is much less than the radius. The diagram is plotted for the case when the vectors of the linear current density **J** in the n-layer are inclined at an angle $\pi(n - 1/2)/(2N)$. Here the layers are numbered from the inner layer

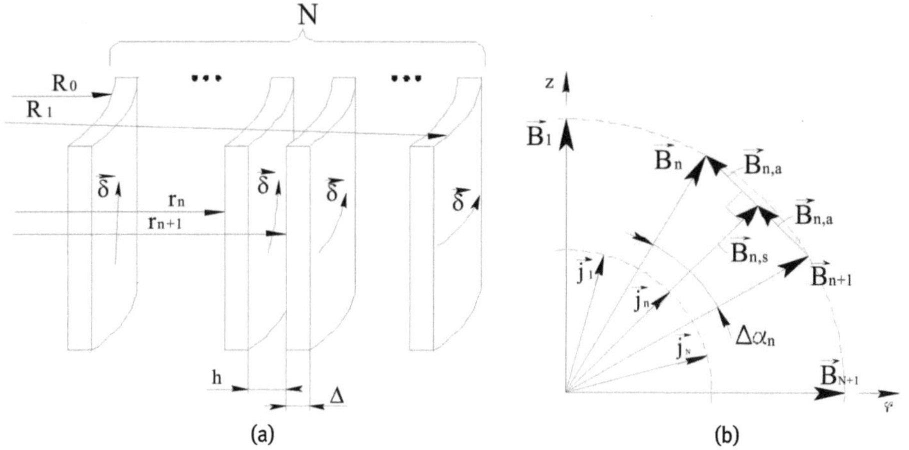

Fig. 7.10: Quasi-force-free winding with currents of variable direction; (a) currents in the winding layers; (b) vector diagram of linear currents and induction.

(closest to the solenoid axis). In the considered case, in the adjacent layers the angles between vectors **J** are similar (they equal $\pi/(2N)$), as are equal modules. With such a distribution of currents, the induction vectors are similar in module as well and change the inclination in respect to the axis z from zero to $\pi/2$.

Here the condition is fulfilled, which links the vector **J** and induction step at the transition through the n-layer:

$$\mathbf{J}_{\mathrm{n}} = \frac{[(\mathbf{B}_{n+1} - \mathbf{B}_n), \mathbf{e}_r]}{\mu_0}, \tag{7.24a}$$

$$|\mathbf{J}_n| = B_i 2 \sin(\pi/(4N)), \tag{7.24b}$$

where the unit vector $\mathbf{e}_r$ is directed along the radius, i.e., is normal to the winding layers. To simplify further consideration, we can disregard the curvature of the thin layers and assume them to be flat. In the middle plane of the layer the vectors of current density $\boldsymbol{\delta}_n = \mathbf{J}_{\mathrm{n}}/\Delta$ and induction are parallel. Here the radial force $f_r = \delta_\varphi B_z - \delta_z B_\varphi$ is equal to zero. To the left and right of this plane, f_r changes its sign which results in the partial compensation of the volume force components $\delta_\varphi B_z$ and $(-\delta_z B_\varphi)$.

At the edges of the layer an angle between vectors $\boldsymbol{\delta}_n$ and **B** is $\pi(4N)$, and the volume force assumes the value

$$f_r(1) = -f_r(2) = \frac{|j_n| B_i}{\Delta\mu_0} \sin\left(\frac{\pi}{4N}\right). \tag{7.25}$$

The azimuth stress in the middle plane of the layer is estimated to be $\sigma' \approx f(1) \cdot \frac{\Delta}{2}$.

In a system with a large number of layers, $\sin(n/(4N)) \approx \pi/(4N)$; therefore, according to (7.24) and (7.25), the following approximation for the azimuth stress is valid:

$$\sigma_\varphi \approx \frac{B_i^2}{\mu_0} \cdot \frac{\pi^2}{16N^2}. \tag{7.26}$$

Essentially the stress is the reciprocal of the layer number squared. More rigorous calculation of stresses can be made similarly to the calculations in Section 5.4 for a single layer. Let us introduce a coordinate x, which is counted from the inner boundary of the n-layer: $x = r - r_n$. The axial component of the current density in this layer does not depend on x and equals $\delta_{z,n}$, while the azimuth one is determined by the relation $\delta_\varphi = \delta_{\varphi,n} \cdot r_n/(r_n + x)$. The further calculations differ from those made in Section 5.4 only by changing from radius R_1 to r_n.

The volume force changes its sign in a layer, and its integral $F_{r,n}(x)$ attains a maximum in a point with the coordinate $x = \Delta/2$. The current density in a layer is of the order of $B_i/(\mu_0 \Delta N)$, where B_0 is the induction on an axis. Hence $F_{\max,n}$ is of the order of $B_i^2/(\mu_0 N^2)$, and the values of stresses are given by

$$\sigma_r = -4\left(F_{\max,n}/\Delta\right)\left(x - \frac{x^2}{\Delta}\right) \tag{7.27a}$$

$$\sigma_\varphi = -4\left(F_{\max,n}/\Delta\right)\left[\theta\left(x - \frac{x^2}{\Delta}\right) - (1-\theta)\, a_n \frac{\Delta}{6}\right]. \tag{7.24b}$$

The modules of both expressions have a maximum value in the middle of the layer ($x = \Delta/2$):

$$|\sigma_r|_{\max} = F_{\max,n}, \tag{7.28a}$$

$$|\sigma_\varphi|_{\max} = F_{\max,n}\left[\frac{\theta}{3} + \frac{2}{3}\right]. \tag{7.28b}$$

Both components of the stress tensor differ from the maximal value of the resultant of volume forces $(F_n)_{\max} = a_n \Delta/4$ only by numerical factors close to unity.

As an example we shall consider a winding with small curvature, so that the terms Δ/r_n can be ignored. Actually, this is the system of N flat conducive layers of thickness Δ. For simplicity we assume that the thickness of isolations inserts between the layers is negligibly small. In the limit $N \to \infty$ there is the force-free field, in which $|B| = \sqrt{B_z^2 + B_\varphi^2} = \text{const}$. Let us express currents $\delta_{z,n}$ and $\delta_{\varphi,n}$ in terms of inductions on each side of the n-layer: $\delta_{\varphi,n} = -(B_{z,n+1} - B_{z,n})/(\mu_0 \Delta)$; $\delta_{z,n} = (B_{\varphi,n+1} - B_{\varphi,n})/(\mu_0 \Delta)$. Then the relation for maximal values of resultants forces in the balanced layers takes the form

$$(F_n)_{\max} = \frac{1}{4\mu_0}\left[B_{z,n}^2 + B_{\varphi,n}^2 - \left(B_{z,n} \cdot B_{z,n+1} + B_{\varphi,n} \cdot B_{\varphi,n+1}\right)\right]. \tag{7.29}$$

At the inner boundary we have $B_{z,1} = B_0$, $B_{\varphi,1} = 0$, and at the outer boundary $B_{z,N+1} = 0$; $B_{\varphi,N+1} = B_0$, where N is a number of layers. In the above example the induction components have been described by a formula following from (7.24):

$$B_{z,n} = B_i \cos \frac{(n-1)\Delta\pi}{2d}, \quad B_{\varphi,n} = B_i \sin \frac{(n-1)\Delta\pi}{2d}, \tag{7.30a}$$

where $d \approx N\Delta$ is athewinding thickness. In this example

$$(F_n)_{\max} = \frac{B_i^2}{4\mu_0}\left(1 - \cos\frac{\Delta\pi}{2d}\right). \tag{7.30b}$$

In another way, the resultants are equal in all the layers, and the winding is equally loaded. The maximal value of volume force in each layer takes place at its inner boundary ($x = 0$) and is

$$f_{\max} = \frac{B_i^2}{\mu_0\Delta}\left(1 - \cos\frac{\Delta\pi}{2d}\right). \tag{7.30c}$$

In the particular case of a single layer $\Delta = d$, $F_{\max} = 0.5B_i^2/(2\mu_0)$. For two and three layers we have, correspondingly, $F_{\max} \approx 0.15B_i^2/(2\mu_0)$ and $F_{\max} \approx 0.065B_i^2/(2\mu_0)$. These estimates show that even the comparatively simple single-layer as well as the two- and three-layer system with currents of variable direction can provide a drastic decrease in layer stresses in comparison with the magnetic pressure of afield with induction B_i. In the first example, the current density vector in a layer is directed at the angle $\pi/4$ to an axis z, and in the second and third ones the direction of the vector changes by similar steps at the transition from one layer to the next. The same pattern takes place in the general case of N numbers. At $N \gg 1$ we have that $\Delta\pi/(2d) = \pi/(2N) \ll 1$, hence

$$F_{\max} \approx \frac{B_i^2}{2\mu_0} \cdot \frac{\pi^2}{16N^2}, \quad f_{\max} \approx \frac{B_i^2\pi^2}{8\mu_0\Delta N^2}. \tag{7.31}$$

The order of magnitude $F_{\max}$ corresponds to the above estimate: $F_{\max} = \text{const}/N^2$. This result is confirmed by numerical calculations for operating systems with no assumption of a small conductor curvature. Figure 7.11 shows the results of computer simulation for an equally-loaded winding with five cylindrical layers.

Comparison with Figure 7.9 shows that with comparable design parameters, stresses in layers with currents of variable direction are approximately seven times less than in a system with pairs of current-carrying layers. At $B_\text{i} = 100$ T the maximal value of the resultant $F_{\max}$ is 1.110^8 Pa. Note that the calculations using (7.31), which disregard the curvature of the layers, give the close result ($F_{\max} = 0{,}9710^8$ Pa).

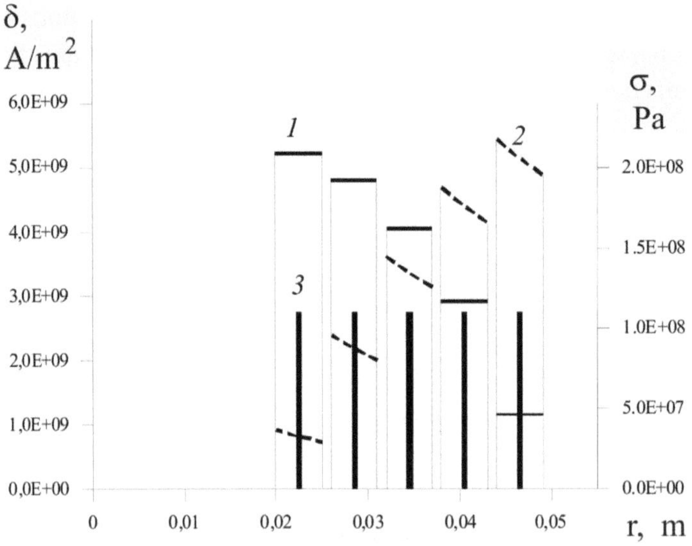

Fig. 7.11: Distribution of the resulting volume force and current densities in the quasi-force-free winding consisting of five layers with inclined currents (computer simulation).

Below we represent the results of calculations of stresses in the layers of windings with a thickness small enough that the above formuls relating the magnitudes $F_{\max}$ and B_i could be used.

Azimuth and radial stresses in the equilibrium layer is determined by (7.31). For a winding with fixed ends, the following relations are valid (see Chapter 5):

$$\theta = \mu/(\mu - 1), \quad \sigma_z = \mu(\sigma_r + \sigma_\varphi).$$

The von Mises formula for such a winding can be written in the form

$$\sigma_M = \frac{F_{\max}}{3\sqrt{2}} \left[(\theta - 1)^2 + ((\theta + 2) - \mu(\theta + 5))^2 + (3 - \mu(\theta + 5))^2\right]^{1/2}. \tag{7.32}$$

It is convenient to introduce the strength parameter of a quasi-force-free winding:

$$\eta_1 = \frac{2\mu_0 \sigma_M}{B_i^2}. \tag{7.33a}$$

The dimensionless number η_1 characterizes a level of stresses that affect a winding material. The acceptable conductor stress $\sigma_1 = B_{M1}^2/(2\mu_0)$ has to satisfy the condition $\sigma_1 \geq \sigma_M$. The induction on a magnet axis is restricted by the condition

$$B_i \leq B_{M1}/\sqrt{\eta_1}, \tag{7.33b}$$

where B_{M1} is “the strength magnetic limit” of a winding material.

The obtained formuls allow us to formule this condition for a thin quasi-force-free winding with N layers as

$$B_i \leq \lambda_0 N B_{M1}. \tag{7.33c}$$

Here λ_0 is a numerical factor. Calculations based on the von Mises strength criteria make it possible to find the value of the parameter λ_0. In a single-layer system $\lambda_0 = \sqrt{2}G$, where for a number G we have $G = (F_{\max}/\sigma_M)^{1/2}$. We have $G \approx 1.7$, if $\mu = 0.33$, and therefore in this system $B_i = 2.4B_{M1}$. If in the multilayer system the current distribution over the layers is approximately given by sin law, we obtain $\lambda_0 = (4/\pi)G \approx 2.16$.

7.3 Configurations of magnetic systems with equilibrium windings with zero thickness

The formulas in Section 7.1 describing the field in a force-free flat layer can be approximately used, as well, in the case when a layer is not flat, but its thickness is small compared to the characteristic dimensions of the magnetic system. The configuration of thin winding can be approximately calculated, assuming its thickness z to be equal to zero. With this assumption, at the boundary S_1, which merges with S_1' (Figure 7.7), the following conditions have to be satisfied:

$$\psi_B = \psi_0 = \text{const}, \quad B_\tau(a) = B_\varphi(b). \tag{7.34}$$

The second of these is the equilibrium condition of each boundary element, which follows from (7.4) [30]. Without current removal, the current function at the boundary of the equilibrium layer takes the value

$$\psi_i(b) = r(b) B_\tau(a)/\mu_0 = R_0 B_i/\mu_0 = \text{const}, \tag{7.35}$$

where R_0 is the inner radius of magnetic system, and B_i the induction on an axis.

A number of equilibrium force-free systems of zero thickness have been studied in [34–36]. Without current removal, the problem of construction of an equilibrium force-free layer of zero thickness is the counterpart in mathematical terms to the problem of streams in hydrodynamics [38]. In the considered incorrect problem of mathematical physics, it is necessary to build the boundary under two conditions: the function of the poloidal field flux should be constant, and the given law describing the change of the tangent-to-boundary induction component of this field $B_\tau(a) = \text{const}/r$ should be satisfied [31].

7.3.1 One-modular configurations

The simplest example of a one-modular configuration is the coaxial magnetic system containing one module with a free boundary, as shown in Figure 7.12. The force-free region of small thickness $R_1 - R_0 \ll R_0$ is located on the section up to point b with coordinate $r(b) = R_2$. in the shown section, the function of the poloidal field flux ψ_B is constant, and the induction should obey the condition $|B| = \mu_0 i/r$, where i is the poloidal current. Further away, the poloidal current is closed at the wall of the cylinder of radius R_2. In the cylindrical section only the former condition is fulfilled. Magnetic pressures on each side of the winding (in points ***M*** and ***N***) are made equal because of the choice of a winding configuration.

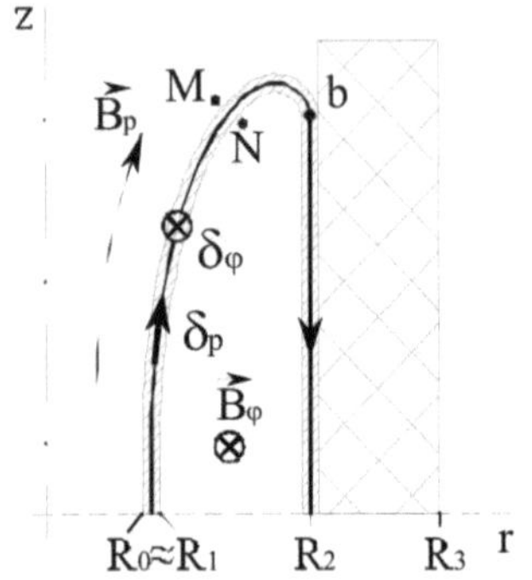

Fig. 7.12: Coaxial magnetic system with a free boundary and outer bandage: (δ_p and δ_ϕ are the poloidal and azimuth current components).

In a similar plane problem the induction absolute value is constant at the face. In this case the shape of the boundary can be analytically calculated, using the conformal mapping. In the plane field the R_1 is a half distance between parallel conductors with the opposite currents. The maximal dimension of the face section with a constant induction module corresponds to the condition $R_2 = 2R_1$ [39, 40].

Examples of analytical solutions for a free boundary of the coaxially symmetric system are missing in the literature[1].

Numerical methods allow us to find the shape of the face using the special-purpose procedure of iterations. With $R_2/R_1 \gg 1$ the equilibrium figure cannot be constructed, since the asymptotical dependence for the poloidal field induction is $B_p = \text{const} \cdot r^{-2}$, while $B_\varphi = \text{const} \cdot r^{-1}$. The calculation method developed by Shishigin is described in [37], where the equation of regression is also given, which allows determining the coordinates of points of the profile of the thin equilib-

1 The equilibrium layer for the coaxially symmetric system is analytically calculated in [32] however, the corresponding field can be formed only in the presence of exterior diamagnetic forces.

rium winding. A number of calculations made using the method of iterations have shown that the iterative scheme is converged if the aspect ratio $\alpha = R_2/R_1 \leq 1.64$ (Figure 7.12). At condition $\alpha > 1.64$ it is not possible to construct the equilibrium configuration. In this case the magnetic pressure at the outer boundary is given by

$$P_M(R_2) = \frac{B_\varphi^2(R_1)}{2\mu_0} \cdot \left(\frac{R_1}{R_2}\right)^2 \approx \frac{B_i^2}{2\mu_0} \cdot \left(\frac{R_1}{R_2}\right)^2, \tag{7.36a}$$

where B_i is the induction at an axis in the region of a uniform field. At the possible largest value $R_2/R_1 \approx 1.64$ the magnet pressure is $0.37B_i^2/(2\mu_0)$. The counterpart stress in the reinforcement can be calculated by the formulas which were used for a single-turn magnet. The largest in the reinforcement are the azimuth stresses. Therefore in comparison of different magnets it is convenient to introduce the conditional strength parameter of the magnet outer zone $\eta_2 = 2\mu_0\sigma_\varphi(R_2)/B_i^2$. This parameter characterizes the sophistication of the magnet system configuration: the less the conditional strength parameter the more high field can be attained not going out of the limits of acceptable load for a winding material. In the case under consideration the strength parameter takes the value

$$\eta_2 = 0.37\frac{R_3^2 + R_2^2}{R_3^2 - R_2^2}. \tag{7.36b}$$

It is reasonable to suggest that the value of an aspect ratio $\alpha \approx 1.64$ restricts the region of existence of equilibrium configurations in the considered coaxial system, which is a prototype of the single-module magnet with thin quasi-force-free winding. In the pulse magnetic field the placement of conductive bodies (diamagnetic screens) near the winding, which results in forcing out the field due to a sharply pronounced skin effect, opens up additional opportunities for constructing equilibrium configurations [34].

An example is the magnetic system with a flat screen and outer reinforcement (Figure 7.13). Previously we have shown that a face diamagnetic screen allows obtaining an equilibrium configuration if the poloidal current is closed over the surface of a cylinder of radius $R_2 \gg R_0$. The magnetic pressure of azimuth field acts at the cylinder of radius R_2 and is perceived by outer dielectric rainforcement.

At condition $r \gg R_1$ the cylinder surface is a plane parallel to the screen and spaced at distance $h = R_0/2$, where $R_0 \approx R_1$ is the inner radius of the thin-wall magnet. The method of iterations permits to construct the configuration in which the condition $B_{p,r}r = \text{const}$ is fulfilled not only far from the axis, but also on the transition section up to point b. The limit value of $r(b)$ is approximately $1.54R_1$.

In the considered system an induction maximum on the screen surface is about $0.4B_i$, and the total force experienced by the screen can be calculated using

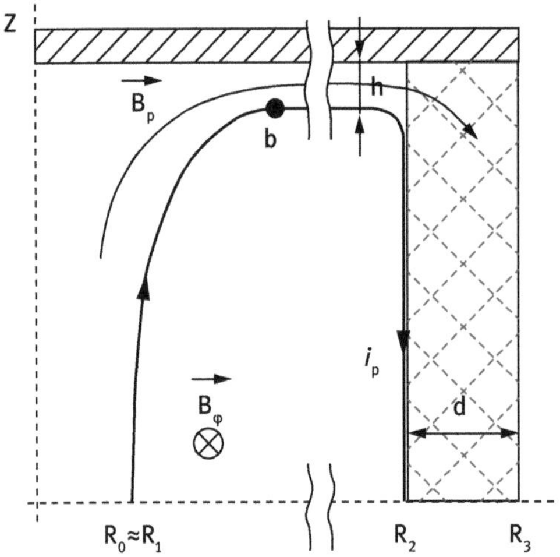

Fig. 7.13: Magnet with thin quasi-force-free winding and flat diamagnetic screen.

the following approximated formula, which is valid at $R_2 \geq 2R_1$:

$$F \approx \frac{B_i^2 \pi R_0^2}{\mu_0}\left(\ln\frac{R_2}{R_1} - 0,86 + \frac{1,6R_1}{R_2}\right). \tag{7.27}$$

In the field of induction 100 T this force at $R_0 = 2 \cdot 10^{-2}$ m is of the order 10^7 H and weakly depends on the ratio R_2/R_1. Therefore, for holding the screens in position a special-purpose construction should be used which can handle the large axial force: for instance, a set of studs. The stresses in constructions elements could be decreased if the effective pulse duration is smaller than the period of elastic vibrations of the system holding the screens in position (see Chapter 8). The period of vibrations can be increased by positioning additional weights to increase the screen mass.

Obviously, thid construction with flat screens is not unique. Application of faces and screens of other shapes can result in lower values of an axial force. However, it is unlikely that the optimization of a face screen would provide a significant decrease of force acting on it.

As noted in Section 7.1, it is possible to construct a totally equilibrium force-free configuration with a coaxial screen (Figure 7.4). A thin-wall winding with a coaxial screen is shown in Figure 7.14.

An outer cylindrical part of a winding located in such screen will be equilibrium if the condition $|B_\varphi(R_2)| = |B_p(R_2)|$ is fulfilled. Accounting for $|B_\varphi(R_2)| = B_i R_1/R_2$, a $|B_p(R_2)| = B_i R_1^2/(R_S^2 - R_2^2)$, we obtain $R_S = \sqrt{R_2^2 + R_1 R_2}$.

For the magnet positioned in a coaxial screen, the limiting value of an aspect ratio $R_2/R_1 \approx 2.12$. This corresponds to a totally equilibrium module, when both boundary conditions (7.34) and (7.35) are valid, not only at the outer boundary but

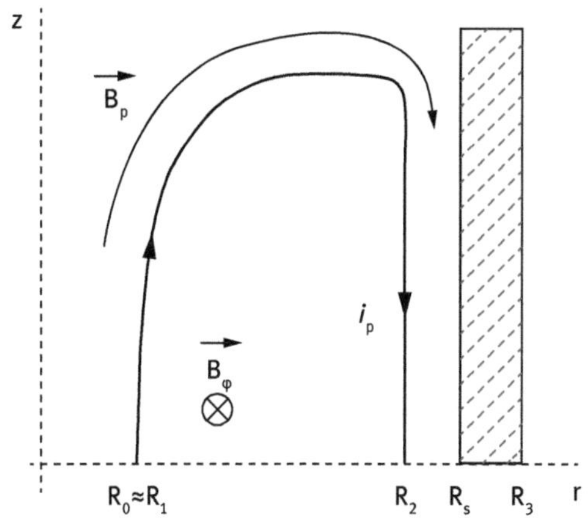

Fig. 7.14: Thin-wall winding in a coaxial screen.

also at the face part of winding. In this case we have the ratio $R_S/R_1 \approx 2.57$. The magnetic pressure onto the cylindrical screen is $0.23 \cdot B_i^2/(2\mu_0)$, and the strength parameter takes the value

$$\eta_2 = \frac{R_S^2 - R_3^2}{R_S^2 + R_3^2}. \tag{7.38a}$$

At lower values of the aspect ratio the outer boundary of the module is not in equilibrium; at larger values the configuration with a face part in total equilibrium does not exist.

The above considered magnetic system is a particular case of a more general configuration in which a winding with a totally or partially equilibrium force-free layer of small thickness is located in a region between two coaxial cylinders on the surface of which the condition $\psi_B = 0$ is prescribed (Figure 7.15).

The flux of a poloidal field formed in a region $R_0 < r < R_1$ bends around the winding and passes through the region $R_2 < r < R_3$. The configurations of the equilibrium section ab have been calculated at different values of ratios R_0/R_1 and R_1/R_3. The results are represented in Figure 7.16, which shows the dependences.

The curve plotted for the case $R_0 = 0$ (the inner cylindrical screen is lacking) in an extreme case $R_1/R_3 = 0$ (the point m in Figure 7.16) describes the considered magnet configuration without an inner cylindrical screen. At other values of this parameter the dependence corresponding to the condition $R_0 = 0$ characterizes the system in which the magnet with the equilibrium face part is located in the aligned cylindrical diamagnetic screen (Figure 7.14).

The curve E in Figure 7.16 is the geometric locus of numbers R_0/R_1, corresponding to the totally equilibrium modules. The parameters R_0/R_1 with values close to unity, correspond to configurations in which the field is not much different from the flat one. In this case the aspect ratio for the totally equilibrium module

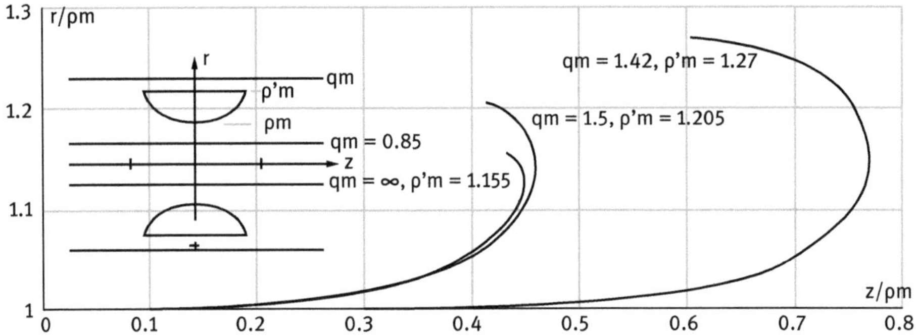

Fig. 7.15: Thin-wall magnet (or m-module of a multimodular magnet) located between two coaxial diamagnetic cylinders. Notations are: $q_{m-1} = R_0/R_1$; $\rho_m = 1$; $\rho'_m = R_2/R_1$; $q_m = R_3/R_1$, where R_0 is the radius of inner cylinder, R_1 the inner radius of the magnet, R_2 its outer radius, R_3 the radius of the outer cylinder. The curves are plotted for $R_0/R_1 = 0.85$.

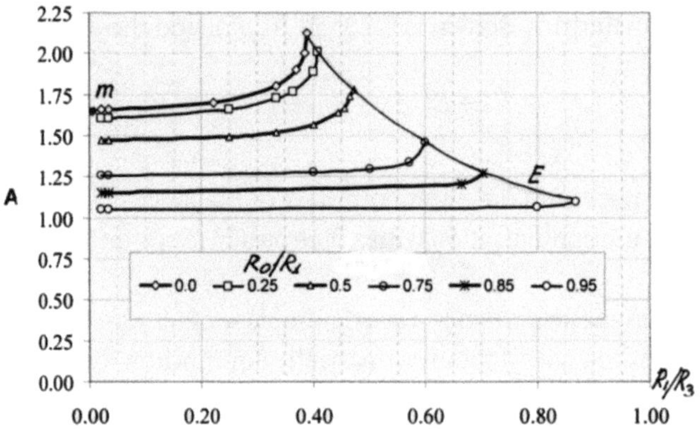

Fig. 7.16: Dependence of the aspect ration R_2/R_1 on ratios R_0/R_1 and R_1/R_3, characterizing the system in which the module with a quasi-force-free winding is located in a gap between the coaxial diamagnetic cylinders.

is close to unity and can be calculated by the formula used in electrostatics [40]: $R_2/R_1 \approx 3 - 2R_0/R_1$.

7.3.2 Multimodular systems

Application of multiply connected configurations consisting of the counter-switched equilibrium coaxial modules extends significantly the capabilities for decreasing the loads acting on the outer reinforcement. A case in point is the system of two thin-wall coaxial modules with opposing currents, separated by radial

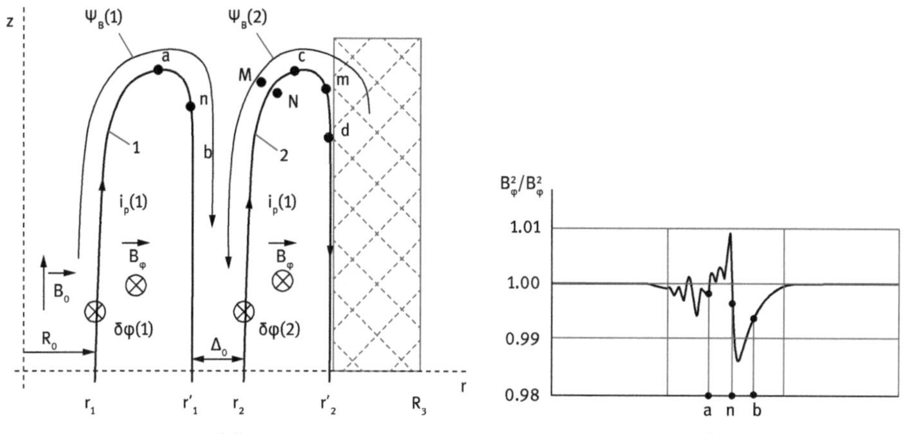

Fig. 7.17: (a) Two-module thin-wall coaxial magnet with the totally equilibrium first module and partially equilibrium second one; (b) changing of the ratio of magnetic pressures of the azimuth and poloidal fields in the vicinity of a point *n*.

gap Δ_0. In this context, the paper [35,] studies the case when at the edges of the gap the function of the poloidal field flux takes the same module and the opposite in sign values: $\psi_B(1) = -\psi_B(2)$ (Figure 7.17a).

The section of the second module below the point *m* is not equilibrium; however, the rest of its boundary and the boundary of the first module is equilibrium throughout. The equilibrium configuration of the first module is much like the system using cylindrical screen described previously. The calculations based on the method of iterations have shown that a two-modular system exists if the ratios of boundaries radii are $r_1/r_1'/r_2/r_2' = 1/2.03/2,86/3.03$. Also, the length of the second module differs little from that of the first. The curve plotted in Figure 7.17b shows the change of magnetic pressure at displacement over the edge of equilibrium section. This curve, to some extent, characterizes the accuracy which we have in the construction of the equilibrium figure using the method of iterations. The azimuth field induction on a nonequilibrium section is $B_\varphi(r_2') = B_i(r_1/r_1') \cdot (r_2/r_2')$, where B_i is the induction on the axis of the magnet-working region (far from the edge). Here $B_\varphi(r_2') = 0,427B_i$, and the corresponding magnet pressure at the outer boundary is $B_\varphi^2(r_2')/(2\mu_0) = 0,18B_i^2/(2\mu_0)$.

Thus, in this case the strength parameter takes the value

$$\eta_2 = 0.18\frac{R_3^2 + R_S^2}{R_3^2 - R_S^2}\,. \tag{7.38b}$$

The numerical coefficient in this formula can be decreased to 0.12 with the change of the flux function value $\psi_B(2)$ by $\psi_B(2) = -8\,\psi_B(1)$.

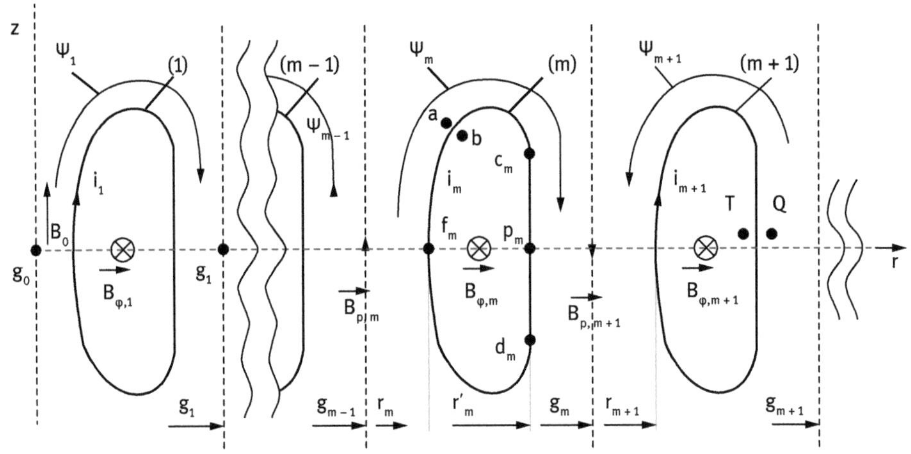

Fig. 7.18: Multimodular magnet with a quasi-force-free winding.

A system consisting of a large number of counter-switched coaxial modules, is presented in Figure 7.18. Here we see the surfaces where the poloidal field flux equals zero.

At distances from the face exceeding the gap between the two modules, these surfaces practically do not differ from the coaxial cylinders. Therefore, for approximated calculations of the modular system of large length one can change these surfaces by diamagnetic cylindrical screens of radius g_m. Thus, it is possible to approximate the multimodular system by the set of elements shown in Figure 7.18. Each of them contains the totally or partially equilibrium module positioned between two coaxial screens.

In the particular case when all modules except the last are totally equilibrium, the induction in the last gap (in a point ***P***) is

$$B(P) \approx B_i/(\alpha_1\alpha_2,\ldots,\alpha_M), \tag{7.39}$$

where M is the number of modules, and $\alpha_k = r_k'/r_k$ is the aspect ratio of the corresponding module [36].

For two- and three-modular magnets, the stresses in the outer bandage have been calculated in the context of the model with intermediate diamagnetic cylinders. The strength parameters are determined by following formulas:

$$\text{For } M = 2 \qquad \eta_2 = \frac{R_3^2 + \alpha_2^2 R_2^2}{R_3^2 - \alpha_2^2 R_2^2} \cdot \frac{1}{\alpha_1^2\alpha_2^2}, \tag{7.40a}$$

$$\text{For } M = 3 \qquad \eta_2 = \frac{R_4^2 + \alpha_3^2 R_3^2}{R_4^2 - \alpha_3^2 R_3^2} \cdot \frac{1}{\alpha_1^2\alpha_2^2\alpha_3^2}. \tag{7.40b}$$

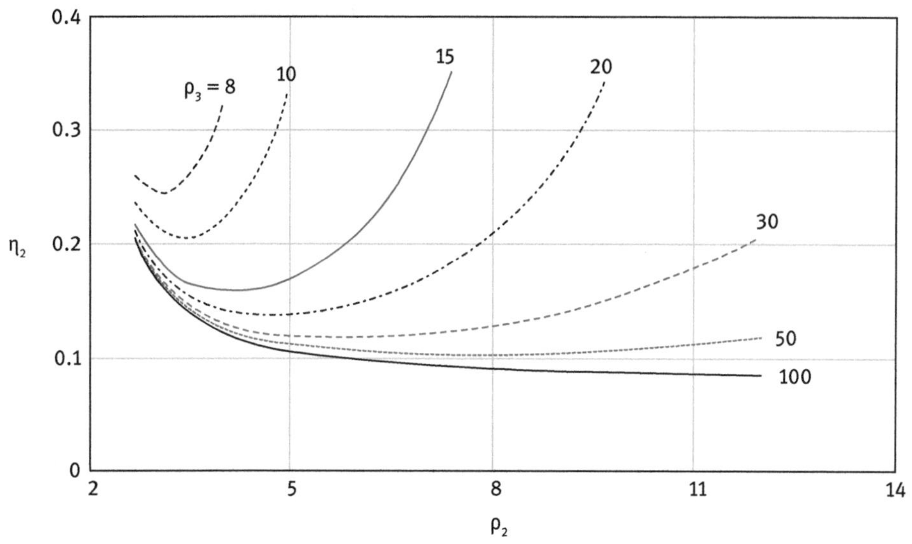

Fig. 7.19: Dependence of the strength parameter of the two-module quasi-force-free magnet on the ratio $\rho_2 = R_2/R_1$, plotted at different ratios of the parameter $\rho_3 = R_3/R_1$.

For each value of ratio R_3/R_0 a value R_2/R_0 can be found at which the strength parameter takes the minimum value $\eta_{2,\min}$. For instance, at $R_3/R_0 = 20$ the minimum value of η_2 is 0.14 (Figure 7.19).

Figure 7.19 shows the values of the minimal strength parameter calculated for different ratios of the outer to the inner radius of the magnetic system. It is seen that the asymptotic value of the strength parameter η_2, corresponding to the large values of the outer radius, is close to 0.08.

Figure 7.20 also demonstrates the results of optimization for a three-modular magnet [36]. Analysis shows that the strength parameter decreases noticeably compared to a two-modular magnet, when the ratio of the outer to the inner radius becomes more than 15.

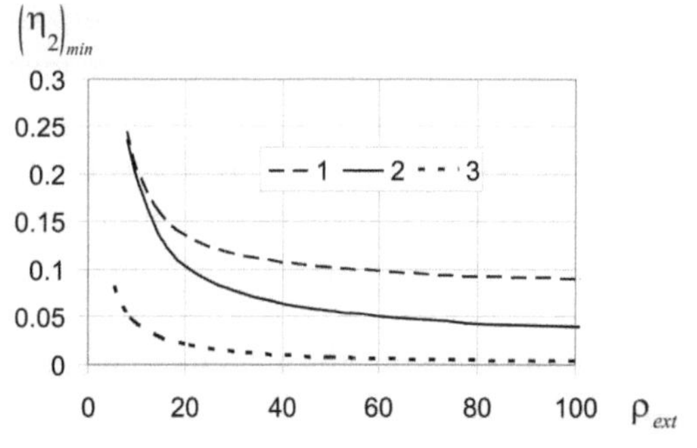

Fig. 7.20: Dependence of the minimal strength parameter of the multimodular system with outer bandage on the ratio of its outer to its inner radius $\rho_{ext} = R_3/R_1$; (1) two-modular system; (2) three-modular system; (3) number of modules $M \gg 1$.

The authors of [42] estimated the asymptotical dependence of the aspect ratio of a multimodular magnet on the strength parameter. It is given by

$$\alpha \approx 1/\eta_2. \tag{7.41}$$

7.4 Thin-wall quasi-force-free magnets with current removals

For configurations in a form of a thin layer, at the boundary of which the condition

$$\frac{\partial}{\partial \tau}\left(rB_\tau\left(a\right)\right) < 0 \tag{7.42a}$$

is fulfilled, the equilibrium can be achieved by the removal of current from the winding (region T_1 in Figure 7.2). Then there is no need to construct an equilibrium configuration satisfying two boundary conditions. The normal component of the current density at the boundary of the equilibrium layer is determined by

$$\delta_n\left(b\right) = -\frac{1}{r\left(b\right)}\frac{\partial \psi_i\left(b\right)}{\partial \tau} = \lambda B_n\left(b\right)/\mu_0. \tag{7.42b}$$

We note that this expression for δ_n is the extension of equation (7.20) to the case when the winding is not a flat layer, but its thickness is small compared with other magnet sizes. The equality (7.42b) is the boundary condition for estimating the poloidal field in the region T_2 (Figure 7.2). In this region the cylindrical conductors with axial current can also be located as the internal reinforcements holding these conductors. Note that in the limiting case of a small-thickness winding the poloidal field induction in the region T_2 determined by boundary $B_n(b) = (d/2\pi)\mu_0\delta_n$, condition is much less than the induction of the azimuth field.

7.4.1 Systems with equally-loaded internal reinforcements

The least radial sizes are intrinsic to the magnetic system, in the outer zone of which the equally loaded cylindrical conductors with reinforcements are positioned [29]. For instance, for the cylinder with quasi force-free winding the magnetic field with induction B_i is produced in the region T_0 $(0 < r < R_0)$, quasi force-free winding by itself is stacked in the region 1 $(R_0 < r < R_1)$, and the inverse poloidal current is distributed in the region 3 $(r_1 < r < R_2)$ (Figure 7.21). In the region of inverse current the magnetic field induction has only azimutal components.

In this system, except for the internal reinforcements, the outer reinforcement can be used for an additional fastening of the winding.

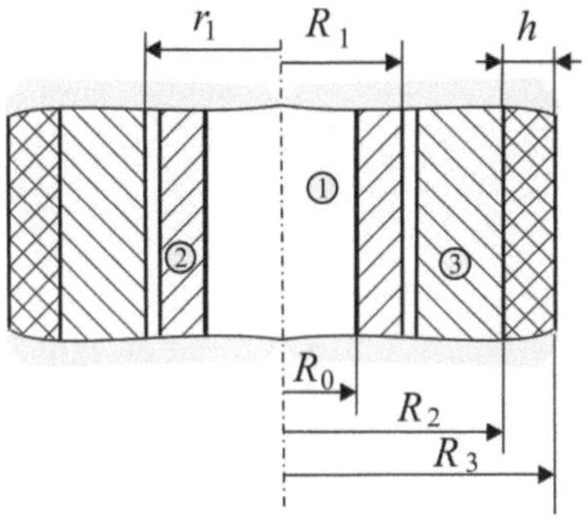

Fig. 7.21: Solenoid with a quasi-force-free winding and axial inverse current passing through region 3.

The simplest model of this system was based on the assumption that the number of bandages is large, their thickness is small ,and they do not interact mechanically [29]. In the context of this model the poloidal current distribution on the radius is assumed to be continuous. In an equally-loaded system the stresses in all layers are equal to $\sigma_\varphi = \sigma_2 = B_{M2}^2/(2\mu_0)$, where B_{M2} is the "magnetic strength limit" of the reinforcements material. The current function satisfies the equation

$$\sigma_2 \approx rB_\varphi\delta_z = -\mu_0\psi_i\frac{d\psi_i}{dr}\cdot\frac{1}{r}. \tag{7.43a}$$

The solution takes the form of

$$\psi_i(r) = \left(C - r^2\sigma_2/\mu_0\right)^{1/2}, \tag{7.43b}$$

where the integration constant C can be found from the condition for the inner boundary of the equally-loaded region of the magnetic system: $\psi_i(r_1) = B_\varphi(r_1)r_1/\mu_0 \approx B_\varphi(R_1)R_1/\mu_0 = \psi_{i,0}$, where r_1 is the inner boundary of the region in which the poloidal current is closed (Figure 7.21). Thus, we have the expressions for C and $\psi_i(r)$:

$$C = (\psi_{i,0})^2(1 + \eta_2'/2), \quad \psi_i(r) = \psi_{i,0}[1 + (\eta_2'/2)(1 - r^2/r_1^2]^{1/2}, \tag{7.43c}$$

where $r \geq r_1$. Consider further a winding of small thickness for which the following conditions are satisfied: $R_1 - R_0 \ll R_0$, $B_\varphi(R_1) \approx B_i$.

In this case,

$$\eta_2' = 2\mu_0\sigma_2/B_\varphi^2(r_1) = \eta_2(r_1/R_1)^2(B_i^2/B_\varphi^2(R_1)), \tag{7.44}$$

where $\eta_2 = 2\mu_0\sigma_2/B_i^2$ is the strength parameter of the outer zone of the magnet.

At full closing of the current, in the depth of the winding outer zone $B_\varphi(R_2) = 0$, and we have the following expressions for functions of the flux, induction, and the aspect ratio $\alpha = R_2/R_1$:

$$\frac{\psi_i(r)}{\psi_{i,0}} = \sqrt{\frac{R_2^2 - r^2}{R_2^2 - r_1^2}} = \sqrt{1 + \frac{\eta_2'}{2}(1 - \rho_1^2)} = \sqrt{1 + \frac{\eta_2}{2}\left[\left(\frac{r_1}{R_1}\right)^2 - \rho^2\right]}, \tag{7.45a}$$

$$B_\varphi(r) = B_\varphi(r_1)\frac{r_1}{r}\sqrt{\frac{R_2^2 - r^2}{R_2^2 - r_1^2}}, \tag{7.45b}$$

$$\alpha = \frac{R_2}{R_1} = \sqrt{\left(\frac{r_1}{R_1}\right)^2 + \frac{2}{\eta_2}}, \tag{7.45c}$$

where $\rho_1 = r/r_1$, $\rho = r/R_1$. The maximal value of ρ is $\rho_{\max} = R_2/R_1$. In a magnetic system with no gap between the quasi-force-free winding and the outer zone, the equalities $r_1 = R_1$, $\eta_2' = \eta_2$ are valid. In the most interesting case, when $\eta_2 \ll 1$, the aspect ratio practically does not depend on r_1.

The strength parameter of the equally-loaded winding can be calculated using the formula

$$\eta_2 = \frac{2R_1^2}{R_2^2 - R_1^2}. \tag{7.46}$$

For instance, in a system with aspect ratio $\alpha = 5$, we have $\eta_2 = 1/12$; whence it follows that application of the equally-loaded thick-wall frame with an inverse current distributed over its depth results in a drastic decrease of the stress in the region 2, in reference to the magnetic pressure of the generated field.

In case considered above, the axial current density increases infinitely at the edge of the calculated region (at $r = R_2$):

$$\delta_z = -\frac{B_{M2}}{\mu_0\sqrt{2(R_2^2 - r^2)}}. \tag{7.47}$$

This effect is lacking in systems with an outer conductive layer, over which a part of the current is closed. The outer layer should be secured by an additional reinforcement. The paper [34] examines such magnets. They are more compact in comparison with magnets with intermediate reinforcements alone, and yet in this case the following estimate remains valid for the aspect ratio:

$$\alpha \approx \sqrt{\frac{2}{\eta_2}} \tag{7.48}$$

7.5 Comparative estimates of the residual stresses and sizes of magnets with a quasi-force-free winding and loaded outer zone

By setting the equilibrium winding a a layer of zero thickness we can calculate similar systems of discrete equilibrium layers with poloidal and azimuth currents. When constructing magnets designed for producing high fields, we are confronted with a complex problem: the configuration and current distribution should be chosen in such a way that the desired strength be present in all magnet parts. In an equilibrium winding, as we have shown, residual stresses are determined by a chosen number of layers and can be reduced to values not exceeding the strength limit σ_1 of the material. The elements of the outer zone of a magnet system where the azimuth current is lacking are not equilibrium; however the mechanical stresses in them should not be larger than the admissible value of σ_2. Certain possibilities of forming the outer zone of an finite-length magnet are considered in [36].

The system with an equilibrium outer zone considered in Section 7.4 applied to the magnet of infinite length or region remote from the face parts of the magnet. This system has the least radial size, and this advantage can be intrinsic as well for a magnet system of finite length if the field in the outer zone is identical to that for the magnet of infinite length, and the dependence of the azimuth field induction is described by the formulas presented in Section 7.4. The profile of such a magnet should be constructed so that along with the boundary ψ_B = const the equilibrium condition $|B_p(M)| = |B_\varphi(N)|$ is fulfilled. Calculations show that in the absence of the face screen, using the current removal, one can construct the one-modular thin winding so as to provide both the equilibrium condition and constant azimuth stresses in the outer zone with cylindrical reinforcements. Such configuration exists only at small aspect ratios (about two and lower).

The equilibrium of the face part of the magnet with an equally-loaded outer zone can be achieved in the system with face diamagnetic screen. For this purpose the gap in the section $R_1 < r < R_2$ should be calculated so that the induction module of the poloidal field in the gap and azimuth field in the outer zone would be the same. For this purpose the gap thickness can be calculated using the approximative formula ignoring the edge effect at $r \approx R_2$:

$$h/R_0 = (R_2/2)(R_2^2 + R_1^2 - r^2)^{-1/2}. \tag{7.49}$$

The winding equilibrium can also be provided through the transformation of the gap $h(r)$ in the case with discrete current removal in a system with few reinforcements.

It is interesting to compare the asymptotic values of the aspect ratios for three systems: the quasi-force-free system with an equally loaded outer zone, the multimodular system, and the equally-loaded magnetic system of traditional construction with azimuth currents considered in Chapter 5. The aspect ratio α_I of the first system is determined by (7.48). In a system with a large number of equally-loaded modules a relatively small strength parameter could be used; however, the aspect ratio in a magnetic systems of this type is higher compared to a magnet with an equally-loaded outer zone. For a multimodular magnet the aspect ratio has been estimated in [36] and [41]. In the calculations it was suggested that the number of equally-loaded modules with inner reinforcements is much more than unity, and the distances between them are small relative to the inner radii of the modules. With small values of strength parameter η_2 the aspect ratio of the multimodule magnet was estimated to be

$$\alpha_{II} \approx 1/\eta_2. \tag{7.50}$$

The aspect ratio of a magnet with azimuth currents α_{III} is given by (5.31). Thus, in the case when the condition $\eta_2 \ll 1$ is satisfied, the following proportion of the aspect ratios of magnetic systems takes place:

$$\alpha_I/\alpha_{II}/\alpha_{III} \approx \left(\frac{1}{\eta_2}\right)^{1/2} \bigg/ \left(\frac{1}{\eta_2}\right) \bigg/ \exp\left(\frac{1}{\eta_2}\right). \tag{7.51}$$

In the case where $\eta_2 = 0.1$, the aspect ratios are in the relation $3.12/10/(2.2\cdot 10^4)$.

The given comparison of the aspect ratios shows that multimodular magnets have an aspect ratio a few times higher than that of magnets with face screens. With this, both modifications of magnets with quasi-force-free windings acquire significantly lower aspect ratios compared to traditionally constructed magnets, provided that the strength parameter is small relative to unity. Thus, the application of magnets with a quasi-force-free winding opens the possibility of attaining ultrahigh fields in the case when the magnetic strength limit of a material is much less than the magnetic pressure of the generated field.

7.6 Design methods of quasi-force-free magnets

The given data do not exhaust the problems of the theory of quasi-force-free magnets. The conditions of the equilibrium of windings need refinement to take into account the real characteristics of a winding, namely, its final thickness, discreteness, heating, and the skin effect on the distribution of forces. These questions are to some extent considered in the publications mentioned above. In particular, in the [28, 35] it is demonstrated that in the case of a sinusoidal pulse the condition

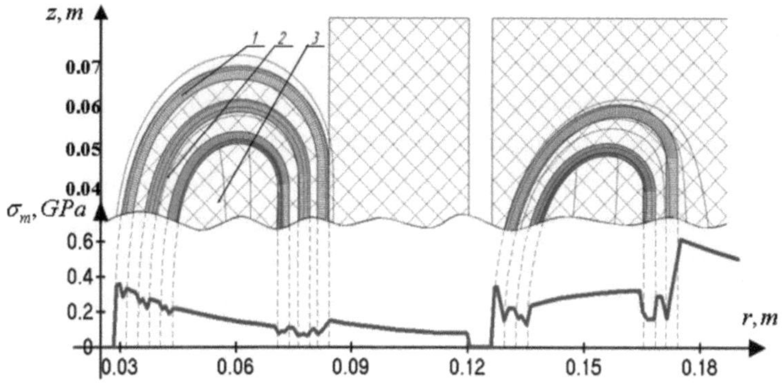

Fig. 7.22: Distribution of the magnetic field and mechanical von Misses stress in two-modular five-layer systems. Induction on an axis of the coil B_0 = 100 T.

of the least heating is satisfied, provided the classical thickness of the skin layer is close to the thickness of the conductor. In this regard the system under consideration is similar to the conventional multilayer winding discussed in Chapter 3. Then, mechanical stresses will differ little from the stresses calculated without taking into account the skin effect.

Figure 7.22 shows the configuration of a two-module magnet with a three-layer winding in the first module and two-layer winding in the second. The calculation was carried out using the 2-D finite-element model. In this example the finite thickness of the conductors was taken into account.

In an actual system the lines of the poloidal field are not rigorously parallel to the current layers. This small departure from the "ideal" configuration does not manifest itself noticeably in the distribution of electromagnetic forces. The calculated stresses due to a field induction of 100 T do not exceed 530 MPa.

More accurate calculations should take into account the discreteness of the winding in the azimuth direction and the presence of outlets. This requires calculation using the 3-D model. An example of such a calculation is presented in [42].

The sensitivity to small departures from equilibrium conditions is the characteristic feature of equalized windings. However, this system is sensitive to small actions compensating these departures. By filling the gaps between conductors with an elastic insulating material, it is possible to avoid the stresses caused by an insignificant violation of the equilibrium conditions of the conductors, so that the conditions become closer to the calculated ones [27, 42]. Note that the correction of the magnetic system may be produced by a variation of the form of the conductors and also of the screens, as well as by displacement of the conductor with small azimuth current on the magnet axis.

The experimental checking of the theoretical results requires constructing models with real magnets which exactly realize the calculated configurations. Figures 7.23 and 7.24 show the models of one-layer magnets with an external reinforcement and cylindrical screen. In order to produce the windings of the magnets shown in Figures 7.23 and 7.24, a casting on burned patterns was used. The windings were produced with a high degree of accuracy by the prototype method. In a weak field, the small deformations were studied with the help of a laser interferometer. A comparison with the calculation has shown that for a unipolar pulse there is a cor-

Fig. 7.23: A single-layer magnet with three turns and an external dielectric reinforcement made from aluminum alloy by casting for burnt-out models.

Fig. 7.24: Casing of the winding for the single-layer eight-turn magnet designed for operation with a coaxial screen.

respondence between the calculated and measured displacements of the winding surface due to the elastic deformation [43]. The experimental studies are in their initial stage. The problems nearest to solution are the ultimate correction of the magnetic systems, which is proved by measurements of the induction distribution, and the conduction of experiments in a strong field.

7.7 References

[1] S. Lundquist, *Ark. Fys.* **2** (1950), 361–365.
[2] R. Lust and A. Schluter, *Z. Astrophys.* **34** (1954), 263–282.
[3] S. Chandrasekhar, *Proc. Natl. Acad. Sci. USA* **42** (1956), 1–5.
[4] A. Schluter, *Z. Naturforsch.* **A12** (1957), 855–859.
[5] E. N. Parker, *Phys. Rev.* **V.109** (1958), p. 1440.
[6] G. K. Morikava, in: *Proc. of the Controlled Thermonuclear Conference*, Washington, 1958, pp. 428–432.
[7] Y. P. Emetz and Y. P. Kovbasenko, *Technicheskaja Electrodynamika* **6** (1981), 3–6 (in Russian).
[8] H. P. Furth, M. A. Levine, and R. W. Waniek, *Rev. Sci. Instr.* **28**(11) (1957), 949–958.
[9] A. N. Lebedev, *J. Tech. Phys.* **34** (1964), 812–817 (in Russian).
[10] Y.P. Emetz and Y. P. Kovbasenko, *J. Tech. Phys.* **53** (1983), 1425–1429 (in Russian).
[11] A. M. Campbell and J. E. Evetts, *Critical currents in superconductors*, London, Taylor and Francis LTD, 1972.
[12] C. L. Longmire, *Elementary Plasma Physics*, Interscience Publ., New York London, J. Wiley & Sons, 1963.
[13] V. D. Shafranov, *Voprosi teorii plazmi* **2** (1963), 92–131 (in Russian).
[14] B. B. Kadomcev, *Kollektivniye yavleniya v plasme*, Moscow, Nauka, 1988 (in Russian).
[15] G. J. Buck, *J. Appl. Phys.* **36** (1965), 2231–2235.
[16] Y. M. Eyssa and R. W. Boom, *IEEE Trans. on Magn.* **MAG17**(1) (1981), 460–462.
[17] Y. Bi and L. Yan, *IEEE Trans. on Magn.* **MAG19**(3) (1983), 324–326 .
[18] Y. Miura, M. Sakota, and R. Shimada, *IEEE Trans. on Magn.* **30**(4) (1994), 2573–2576.
[19] S. Nomura, T. Osaki, J. Kondoh, et al., *IEEE Trans. Appl. Supercond.* **9**(2) (1999), 354–356.
[20] S. Nomura, D. Ajiki, C. Suzuki, and N. Watanabe, *IEEE Trans. on Appl. Supercond.* **11**(1) (2001), 20–24.
[21] D. R. Wells and R. G. Mills, in: *High Magnetic Fields*, Chap. 2, pp. 44–47, Cambridge, Mass., Technol. Press, 1962.
[22] V. E. Bikov, A. V. Georgievskiy, V. I. Koryavko, and Y.A. Litvinenko, Bessiloviye toroidalniye magnitniye sistemi, preprint of Kharkov Phys.-Tech. Institut, 1976, pp. 76–39 (in Russian).
[23] A. A. Kusnetsov, *J. Tech. Phys.* **31** (1961), 650–655 (in Russian).
[24] M. Z. Claude and A. Mailfert, *Bull. Soc. frans.electriciens* **4** (1963), 33–37.
[25] E. L. Amromin, V. Y. Khosikov, and G. A. Shneerson, *Plasma Devices and Operations* **4** (1998), 321–326.
[26] G. A. Shneerson, O. S. Koltunov, and V. Y. Khosikov, *J. Tech. Phys.* **7**(1) (2002), 110–116 (in Russian).

[27] G. A. Shneerson, A. I. Borovkov, O. S. Koltunov, D. S. Mikhalyuk, and V. V. Titkov, **MG IX** (2004), pp. 602–615.

[28] G. A. Shneerson, O. S. Koltunov, et al., *Physica B* **346–347** (2004), 566–570.

[29] G. A. Shneerson, *J. Tech. Phys.* **56**(1) (1986), 36–43 (in Russian).

[30] G. A. Shneerson, *J. Tech. Phys.* **78** (2008), (in Russian).

[31] S. K. Godunov, *Uravneniya Matematicheskoy Fisiki*, Moscow, Nauka, 1971 (in Russian).

[32] G. L. Hand and M. A. Levine, *Phys. Rev.* **127**(6) (1962), 1856–1857.

[33] M. A. Levine, in: *High Magnetic-Fields*, p. 277–280 Cambridge, Mass., Technol. Press, 1962.

[34] G. A. Shneerson, I. A. Vecherov, D. A. Dyegtev, O. S. Koltunov, S. I. Krivosheerv, and S. L. Shishigin, *J. Tech. Phys.* **78**(10) (2008), 1278–1288 (in Russian).

[35] G. A. Shneerson, O. S. Koltonov, D. A. Dyegtev, V. V. Titkov, S. L. Shishigin, and I. A. Vecherov, *IEEE TRANS. on Plasma Science* **38**(8) (2010), 1726–1730.

[36] G. A. Shneerson, O. S. Koltunov, and D. A. Dyegtev, **MG-XIII** (2010), pp. 200–211.

[37] S. L. Shishigin, *Electrichestvo* **7** (2007), 22–27 (in Russian).

[38] M. I. Gurevitch, *Teoriya struy idealnoy dzidkosti*, Moscow, Nauka, 1979 (in Russian).

[39] A. B. Novgorodcev and A. F. Fatchiev, *Izvestiya Vuzov, Energatica* **2** (!982), 17–21 (in Russian).

[40] G. A. Shneerson, *Fields and Transients in Superhigh Pulse Current Device*, // Nova Science Publishers, Inc., New York, 1997.

[41] G. A. Shneerson and D. A. Dyegtev, *Tech. Phys. Letters* **36**(12) (2010), 23–29 (in Russian).

[42] G. A. Shneerson, O. S. Koltunov, S. I. Krivosheev, Y.E. Adamyan, A. N. Berezkin, A. P. Nenashev, and A. A. Parfentyev, *IEEE TRANS. on Plasma Science* **38**(8) (2010), 1731–1739.

[43] G. A. Shneerson, O. S. Koltunov, A. N. Berezkin, A. P. Nenashev, I. A. Vecherov, and A. A. Parfentyev, **MG-XIV** (2012), to be published.

8 Generation of strong pulsed magnetic fields in single-turn magnets. Magnetic systems for the formation of pulsed loads

Single-turn magnets are the most simple in construction (Figure 8.1). For them one can use the conductors with high strength characteristics, such as beryllium bronze, tantalum, or alloy steel.

The magnetic ultimate strength of these metals can reach 80 T. The definite advantage of single-turn magnets is the absence of forces compressing the insulation.

In order to obtain a field with induction of the order of 10 T and above in magnets with characteristic sizes of the order of several cm, it is necessary to produce a current with an amplitude of hundreds of kA. The generation of such currents became possible in the late 1950s. At that time, as part of an exploratory program on controlled thermonuclear fusion, technologies have been mastered which provide the production of strong currents with the aid of high-voltage, low-inductance capacitor banks. Furth, Levine, and Waniek [1, 2] were the first to use these technologies. Later, studies and developments were done in two directions. The first is the production of megagauss fields in magnets which are destroyed in the course of discharge. We will consider this subject in the next chapter. In the present chapter we will dwell on the problems concerning the construction of magnets designed for repeated use. An example of this is the work of Kolb [3],

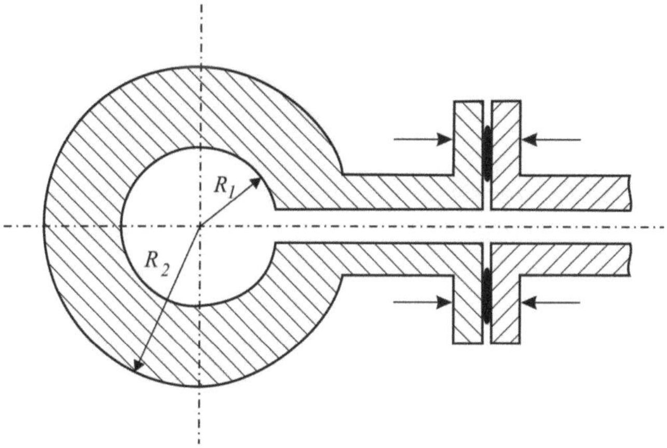

Fig. 8.1: The single-turn solenoid. The forces holding the ends of the slot are shown with arrows.

where in the beryllium-bronze single-turn magnets with diameter of 1 cm and length of 1.6 cm a field with induction up to 70 T was repeatedly generated. Extensive studies with single-turn magnets have been also conducted by Andrianov, Demichev, et al. in 1970 at the Kurchatov Institute of Atomic Energy [4]. Their studies have shown that solenoids from tantalum have the highest parameters: the researchers managed to obtain a field with induction of up to 100 T. The single-turn magnets found application in experiments where the pulse field was used for the production of large pulsed fields [5]. Japanese physicists [6] and others who studied magnetic cumulation used the field of single-turn magnets for the compression of conducting shells (see Chapter 10). Magnets of this type are also used extensively in the plastic processing of metals, and in devices for the acceleration of conducting bodies.

8.1 Mechanical stresses in a single-turn magnet operating under the condition of a sharply pronounced skin effect

The pulse period in single-turn magnets does not as a rule exceed several tens of microseconds. Under such conditions the penetration depth of a field into a conductor is small compared with the thickness of the magenet wall. This suggests that the inner surface of the cylinder is subjected to pressure which can be calculated by integrating the volume force over the normal to the boundary:

$$P = \int_{R_1}^{r(M)} B \cdot \left(-\frac{1}{\mu_0}\frac{\partial B}{\partial r}\right) dr = \frac{B_i^2}{2\mu_{0\,i}} - \frac{B(M)^2}{2\mu_0}. \tag{8.1}$$

In this formula point M is at a distance from the boundary, and we can assume the induction at this point to be equal to zero. The magnetic field acts on the cylinder in the same manner as gas under the pressure $P_M = B_i^2/(2\mu_0)$.

When calculating the stresses we assume that the edges of the insulating slit are fixed. This is necessary in order to avoid failure of the magnet caused by concentration of strengths at its inner side opposite the slot. An example of such a failure is shown in Figure 8.2.

Then we can use formulas (5.10) derived for the solid cylinder. In these formulas one should take the approximate condition $f_r = 0$. It is valid in a wall except for the skin layer, the thickness of which is assumed to be zero. On the inner boundary ($r = R_1$) the condition $\sigma_r' = P_M$ is fulfilled, while on the external one ($r = R_2$) we have $\sigma_r' = 0$. Then we can find the constants C_1 and C_2, and come to the following

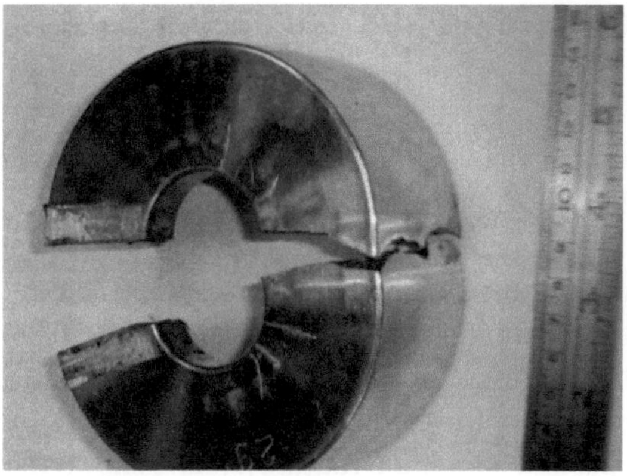

Fig. 8.2: Failure of a thick single-turn magnet with nonfastened slot edges [16].

expressions for stresses:

$$\sigma'_{r,\varphi} = \mp\frac{B_i^2}{2\mu_0} \cdot \frac{R_1^2}{R_2^2 - R_1^2}\left(\frac{R_2^2}{r^2} \mp 1\right). \tag{8.2}$$

The stress peaks on the inner surface. In the limit case, when the condition $h = R_2 - R_1 \ll R_1$ is valid, we have

$$\sigma'_r(R_1) = -\frac{B_i^2}{2\mu_0}, \quad \sigma'_\varphi(R_1) = \frac{R_1}{h} \cdot \frac{B_i^2}{2\mu_0}. \tag{8.3}$$

In the opposite limiting case, when the thickness of a turn is large ($R_2 \gg R_1$), the stresses take the values

$$\sigma'_{r,\varphi} = \mp\frac{B_i^2}{2\mu_0} \cdot \frac{R_1^2}{r^2}, \tag{8.4}$$

$$|\sigma'_r(R_1)| = |\sigma'_\varphi(R_1)| = \frac{B_i^2}{2\mu_0}. \tag{8.5}$$

Neglecting the axial stress, one can evaluate von Mises stress on the surface. We give its value for a turn of a large thickness to be

$$\sigma'_M = \sqrt{3}\,\frac{B_i^2}{2\mu_0}. \tag{8.6}$$

According to (8.3) the stresses in thin-wall single turn magnets with large ratio R_1/h become much higher than the magnetic pressure. Application of the thick-wall ($R_2 \gg R_1$) one-turn solenoid permits the generation of a field closely approaching the strength magnetic limit of the given material without application of the additional precautions for strengthening its construction, as well as the above-mentioned fixation of the slit edges.

8.2 Assessing the strength of single-turn magnets at short pulses

In the case of pulses of longer duration compared to the period of natural elastic oscillations, we can apply the relations obtained for stresses in a static regime. Moreover, we can introduce the instantaneous values of induction $B_i(t)$ in the formulas for the instantaneous stress values. It should be, however, taken into account that the strength characteristics of the materials (the yield point, the allowed stresses), determined by the evolution of ruptures in a solid state, also depend on the duration of the force action: they grow as the pulse shortens. The experiments investigating the copper and aluminum rupture [7] have shown that the time from the instant of the load application until the onset of the rupture obeys to the dependence

$$\tau = \tau_0 \exp\left[\left(U - \gamma'\sigma\right)/kT\right], \tag{8.7a}$$

where T is the temperature, σ is the stress in the specimen, and U, τ_0, and γ' are constants. From here we can find the dependence of the destroying stress σ_C on the duration of the load action (the less τ the higher σ_C):

$$\sigma_C = \frac{1}{\gamma'}\left(U - kT\ln\frac{\tau}{\tau_0}\right). \tag{8.7b}$$

For instance, in experiments with single crystals of aluminum at room temperature the rupture stress was 5.2 kg/mm^2 for an action duration of 10^2 s and 6.5 kg/mm^2 for a duration of 10^{-1} s [7].

In [8] the strength limit of copper was determined at pulses of 800 and 2000 μs. The force was produced when the current pulse proceeds through the turn of a circular wire placed in a uniform magnetic field. The measurements of the rupture stress were carried out at different temperatures under conditions of gradually increasing load. The specimen was ruptured by a few pulses, the corresponding stress limit being denoted as σ_i (Table 8.1).

The regime of the cyclic load was also studied, where the specimen was subjected to the recurring action of pulses of constant amplitude. In this case the fatigue effect of the material revealed that the strength limit was reduced with increasing number of pulses tending to the some value σ_∞. In Table 5.1 are listed the values σ_i, σ_∞ and the static strength limit σ_{st}. The stress σ_{st} is essentially lower than the strength limit σ_i and higher than σ_∞. The stress σ_i increases with shortening pulse durations, while the strength limit at the cycling load does not depend on pulse duration. The cooling of specimens increases their strength at all loading regimes.

Table 8.1: Strength limits of the materials at different temperatures and loading conditions.

Method of loading	Copper, MPa at T (°K)			Aluminum, MPa at T (°K)			Pulse duration s
	243	77	20.4	243	77	20.4	
Static							
σ_{st}	270	400	470	135	230	390	–
σ_{∞}	150	240	250	75	200	220	$8\,10^{-5}$
σ_{∞}	150	240	250	–	–	–	$2\,10^{-3}$
σ_{i}	320	410	470	–	–	–	$2\,10^{-3}$
σ_{i}	550	870	990	220	530	600	$8\,10^{-5}$

One of the purposes of the first experiments on the generation of megagauss fields in single-turn magnets [9] was to determine the induction value B_S, at which the residual deformations appear, and to compare it with the expected threshold value. The simplest method to determine the threshold induction B_S was the extrapolation of the dependence of residual deformation on the induction amplitude, "rearward" to zero. It was found that for the pulse duration of the order of several microseconds the measured threshold induction is higher than the induction B_M calculated using the tabulated values of the static ultimate strength. The "dynamic" values of the ultimate strength $\sigma_S = B_S^2/(2\mu_0)$ obtained in such a way are given in the Table 8.2.

Table 8.2: Tensile strength at microsecond pulses.

Material	B_ST	σ_S MPa
Beryllium bronze	74	2200
Steel 40	74	2200
Kirit*	60	1400
Steel 3	58	1100
Brass	50	1000
Copper	46	840
Wood's alloy	27	

* Kirit is the composit of copper and tungsten, used for production of contacts of high voltage switches; steel-3, a low-carbon steel; steel-40, a tool steel of enhanced strength.

In [9] the effective pulse duration was much larger than the period of the dominant of the eigen elastic oscillations for a thick-wall magnet of an inner diameter 3.4 mm. In this case the rise of the threshold compared with σ_C can be explained by the decrease in dynamic ultimate strength.

Table 8.2 shows that of the tested materials beryllium bronze and steel 40 acquire the highest magnetic limit of strength. Comparison of the data in Tables 8.2 and 8.3 shows that copper dynamic yield σ_S, corresponding to the induction B_S, significantly exceeds both the static yield point σ_0 and the dynamic yield point σ_i obtained at a slower load change.

Recent results verify the decrease in the ultimate strength in the case of pulses of μs length [9, 10]. The fact that the ultimate strength and the corresponding threshold induction depend on the pulse duration was proved in experiments carried out by Gordienko and Shneerson [11]. For a pulse with the rise time 31 μs the threshold of destruction of steel magnets was determined to be 46 T. This is considerably less than a threshold of 58 T in the case of a pulse with a rise time of 3.9 μs.

The data of Table 8.1 shows that the cooling of examples increase their strength in all regimes of loading, whereas Joule heating may cause a significant decrease in ultimate strength. Churaev [12] gives the temperature dependences of this parameter for some metals and alloys. For copper, aluminum, and titanium the ultimate strength decreases linearly with temperature: $\sigma = \sigma_C(1 - \Delta T/T_k)$, where σ_C is the ultimate strength at initial temperature, ΔT is the increment of temperature relative to the initial temperature (20 °C), and $T_k = 2/\alpha_\rho$, α_ρ is the temperature coefficient appearing in the linear dependence of the specific resistance on temperature: $\rho = \rho_0(1 + \alpha_\rho \Delta T)$. Such a dependence for aluminum is shown in Figure 8.3a. For such metals as copper, tungsten, molybdenum, titanium, and rhenium this dependence is valid at least up to the temperature $T_k/2$ (Figure 8.3b). The temperature dependence for such metals as iron, tantalum, and niobium is of a more complicated character. The initial strength of these metals retains up to the temperature close to T_k.

Along with the increase of the strength limit, with actions of short duration the strength of the magnetic assembly also increases, which is related to the inertia effects [9,13]. At the action of the force the effective time τ, which is much less than the period of the natural oscillations of the system, elastic deformation begins after force action is terminated. Because of that, the amplitude of the elastic oscillations and, hence, the amplitude of stresses, is determined by the impulse of the force, but not its maximal value. As an example, let us consider a thin-wall cylinder on the unit surface of which the force $B_i^2/2\mu_0$ affects from the inside. The equation of the cylinder elastic oscillations is written as:

$$m' \frac{d^2 \Delta R}{dt^2} + K\Delta R = 2\pi R \cdot \frac{B_i^2}{2\mu_0}, \qquad (8.8a)$$

where R is radius of the cylinder, ΔR is the change of the radius, $m'=2\pi Rh\gamma$ is the mass per unit length of the cylinder (γ is the density of the matter, h is the thickness of the cylinder's wall), K is the rigidity of the system. Turning to the

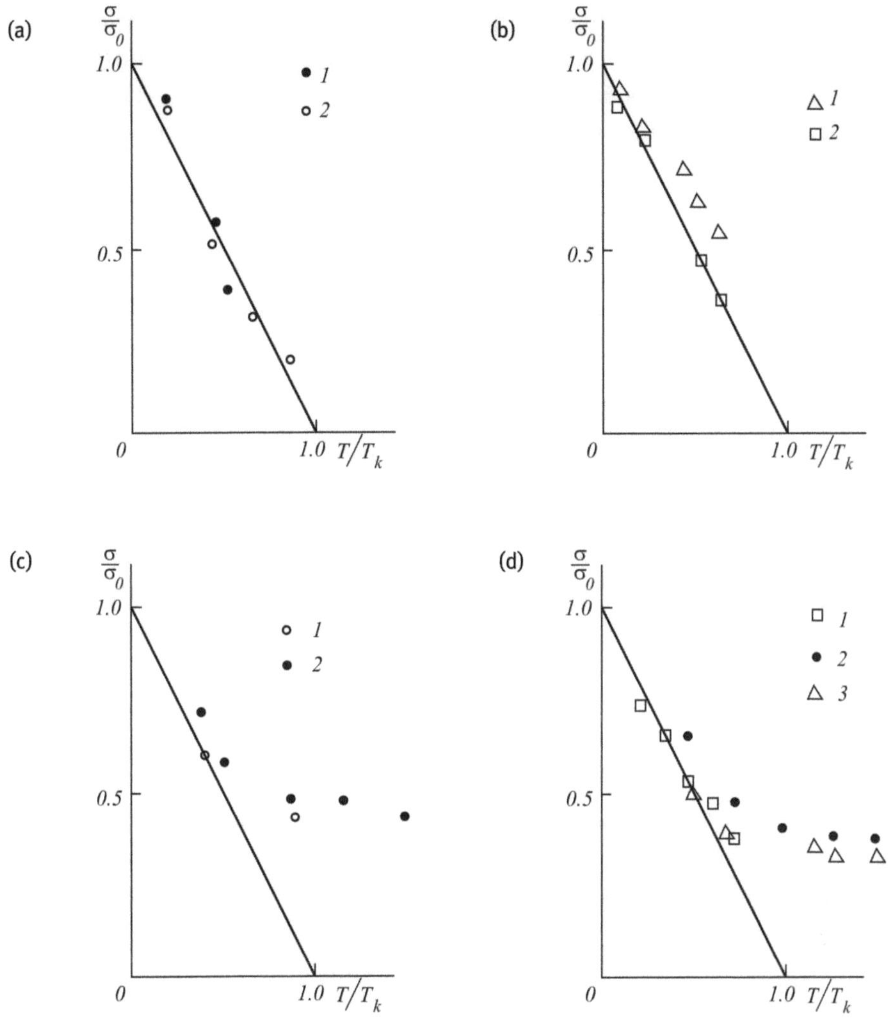

Fig. 8.3: The temperature dependences of the ultimate strength of metals [11]. Curves are plotted for modifications of metals different tensile strength at normal temperature: (a) Al, $T_k = 470\,°\mathrm{C}$, $\sigma_0 = 146$ MPa(1), 90MPa(2); (b) Cu, $T_k = 462\,°\mathrm{C}$, $\sigma_0 = 240$ MPa(1), 470MPa(2); (c) Mo, $T_k = 462\,°\mathrm{C}$, $\sigma_0 = 1510$ MPa(1), 650 MPa(2); (d) W, $T_k = 400\,°\mathrm{C}$, $\sigma_0 = 1500$ MPa(1), 1470MPa(2), 740MPa(3).

relative deformation $\varepsilon = \Delta R/R$, we come to the equation

$$m'\frac{d^2\varepsilon}{dt^2} + K\varepsilon = \frac{\pi B_i^2}{\mu_0}. \tag{8.8b}$$

The rigidity K could be found while considering the static regime when, on the one hand, $\varepsilon = B_i^2 R/(2\,\mu_0 hE)$ (see (5.8)) and, on the other hand, $\varepsilon = \pi B_i^2/(\mu_0 K)$.

From here $K = 2\pi hE/R$. During the action of the force deformation has no time to continue further; therefore ε is small, and the second term in equation (8.8b) can be ignorfed. During this time interval the motion of the cylinder walls is purely inertial, and towards the end of the force action the walls attain the velocity

$$\frac{dR}{dt} = \int_0^{\tau} \frac{B_i^2 R dt}{\mu_0 m'}. \tag{8.9}$$

If the field is decaying for a time much less than the period of the free elastic oscillations, one can suppose that the walls obtained the velocity

$$\frac{dR}{dt} = \frac{\pi R}{\mu_0 m'} \int_0^{\infty} B_i^2 dt = \frac{B_m^2 \pi R}{\mu_0 m'} \tau_{eff}. \tag{8.9b}$$

Here τ_{eff} is the effective pulse duration that for sin weakly-decaying oscillations equals 1/4 of the exp time constant in the case when the current amplitude decays exponentially. Then the cylinder performs free-elastic oscillations, the initial value of velocity being determined by the formula (8.9b). Thus, the initial value of the relative deformation is

$$\left.\frac{d\varepsilon}{dt}\right|_{t=0} = \frac{\pi B_m^2 \tau_{eff}}{\mu_0 m'} = \varepsilon_m \omega_0, \tag{8.10a}$$

where $\varepsilon_m = \sigma_m/E$ is the amplitude of the relative deformation, $\omega_0 = \sqrt{K/m'} = (1/R)\sqrt{E/\gamma}$ is the circular frequency of the fundamental mode of free oscillations. From here we find the maximal stress

$$\sigma_m = E\varepsilon_m = \frac{B_m^2 R}{2\mu_0 h} \omega_0 \tau_{eff}. \tag{8.10b}$$

From this formula follows that azimuth stress differs by the factor $\omega_0 \tau_{eff} \ll 1$ from the value that it could have at the static regime of deformation in a field with induction B_m. In this manner in the regime of short pulses the effect of strengthening of the construction appears in itself, which is determined by inertia of the material and not concerned with increasing of the allowed stress. For instance, in the thin-wall copper turn with characteristic dimension $R = 10$ cm, $h = 1$ cm, the frequency of the natural axial-symmetrical elastic oscillations $\omega_0 = 4.5 \cdot 10^4$ s. Hence at the generation of pulses with effective duration of $2.2 \cdot 10^{-6}$ s ($\omega_0 \tau_{eff} = 0.1$) it is possible to increase by ratio $\sqrt{10}$ the allowed the induction amplitude compared to the regime of slow loading. It is interesting to note that the stress amplitude σ_m in (8.10b) does not depend on the magnet radius, since ω_0 is proportional to $1/R$.

In thick-wall coils ($R_2 \gg R_1$) the frequency of elastic oscillations is $\omega_0 \approx (1/R_1)(E/\gamma)^{1/2}$. Therefore for σ_m the assessment is valid that

$$\sigma_m \approx \sigma_{st}\omega_0\tau_{eff} \approx \frac{B_m^2\tau_{eff}}{2\mu_0 R_1}\sqrt{\frac{E}{\gamma}}. \tag{8.10c}$$

Thus, the σ_m is proportional to $1/R_1$. From here it can be seen that at a given allowed stress of the material the amplitude of the induction is proportional to $(R_1)^{1/2}$ and $(\gamma)^{1/4}$.

8.3 Thermoelastic stresses in single-turn magnets

A single-turn magnet is prone to mechanical stresses produced by both electromagnetic forces and by the heating of the medium due to a pulsed current. In the quasi-static regime, the first of these stresses are described by formulas (8.2)–(8.6). Although these formulas have been derived in an approximation of the ideal conductivity, they adequately describe the stress state beyond the skin layer, i.e., in the main area of a thick-wall magnet. In contrast, the thermoelastic stresses exist in the skin layer, i.e., in the zone whose thickness is as a rule much less than the thickness of the turn. Unlike the stresses due to the electromagnetic forces, thermoelastic stresses reach a maximum at the end of the pulse, when a current in the skin layer is damped out.

In a cylindrical magnet with a large length, with free ends the thermoelastic stresses are the largest on the internal boundary of the turn. They are described by the following formula [14]:

$$\sigma_\varphi(R_1) = \frac{\alpha_0 E}{1-\nu}\left[\frac{2}{R_2^2 - R_1^2}\int_{R_1}^{R_2} T(r)\,r\,dr - T_e\right], \tag{8.11a}$$

$$\sigma_z(R_1) = \frac{\alpha_0 E}{1-\nu}\left[\frac{2\mu}{R_2^2 - R_1^2}\int_{R_1}^{R_2} T(r)\,r\,dr - T_e\right], \tag{8.11b}$$

where T_e is the temperature increment on the surface. The first term is small compared to the second, provided that the thickness of the skin layer Δ satisfies the condition $\Delta \ll R_1$. On this assumption we have

$$\sigma_\varphi(R_1) = \sigma_z(R_1) = \frac{-\alpha_0 E q'(R_1)}{(1-\nu)\cdot C_V}, \tag{8.12}$$

where E is the elastic modulus, C_V is the heat capacity per unit volume, and ν is the Poisson ratio.

The formulas describing the energy $q'(R_1)$ which is released due to the Joule heating are given in Chapter 3, where it is shown that the volume energy density released on the boundary of the conductor in the process of the discharge can be expressed in the form

$$q'(R_1) = \vartheta \frac{B_m^2}{2\mu_0}, \tag{8.13}$$

where B_m is the induction amplitude.

As was shown in Chapter 3, in the case of a "standard" pulse in the form of one half-wave of a sinusoid $\vartheta = 2.42$. In the case of a unipolar pulse corresponding to the critical regime of the discharge when a change in induction of the external field is described by the formula $B_e = B_m(t/t_m)\cdot\exp(1 - t/t_m)$, the parameter ϑ has the close value $\vartheta = 2.84$.

The parameter ϑ sharply increases when a pulse has the form of a weakly-damping sinusoid. When the damping is small, the heating is calculated by formula (3.49) at condition $\beta/\omega \ll 1$. Under this condition the parameter ϑ sharply increases according to $\vartheta \approx (\omega/2\beta)$. For example, in the case when the ratio of two neighboring amplitudes is equal to 0.8 [15] we have $\vartheta = 7.04$.

Along with the magnetic ultimate yield point, determined by formula (5.2), it is possible to introduce the threshold induction B_T. We will find its value from the condition that the azimuth thermoelastic stress is equal to the yield point of the material σ_S. In order to calculate B_T we use formulas (8.12) and (8.13) [14]:

$$B_T = \left[\frac{2\mu_0\sigma_S(1-\mu)C_V}{\alpha_0 E\vartheta}\right]^{1/2} = B_S\Lambda, \tag{8.14}$$

where

$$\Lambda = \left[\frac{(1-\mu)C_V}{\alpha_0 E\vartheta}\right]^{1/2}. \tag{8.15}$$

Parameter Λ is the ratios of the threshold induction B_T to the magnetic yield point B_S. Table 8.3 contains this parameter for some metals. The table contains data for the "standard" pulse and a weakly damped sinusoid ($B_{m,n+1}/B_{m,n} = \exp(-\pi\beta/\omega) = 0.8$). Note that the threshold induction B_T in the second case is much less than in the first one. Table 8.3 shows that the threshold due to the thermoelastic stresses may be lower than the magnetic yield point. Thus, in single-turn magnets the residual deformation due to the heating appears in weaker fields compared with the deformation due to electromagnetic forces.

These relations provide a basis for evaluating the resource of single-turn magnets in the framework of the accepted models for the development of plastic deformation. The simplified scheme of this process corresponding to the regime of uniaxial loading is shown in Figure 8.4. [14]. The skin layer under heating tends

Table 8.3: Ratio $\Lambda = B_T/B_S$ for different metals.

Material	Λ	
	"Standard" pulse	Pulse with $\pi\beta/\omega = 0.8$
Copper	0.62	0.36
Brass	0.64	0.37
Beryllium brass	0.64	0.37
Stainless steel	0.74	0.28
Aluminum	0.55	0.21
Tantalum	0.72	0.27

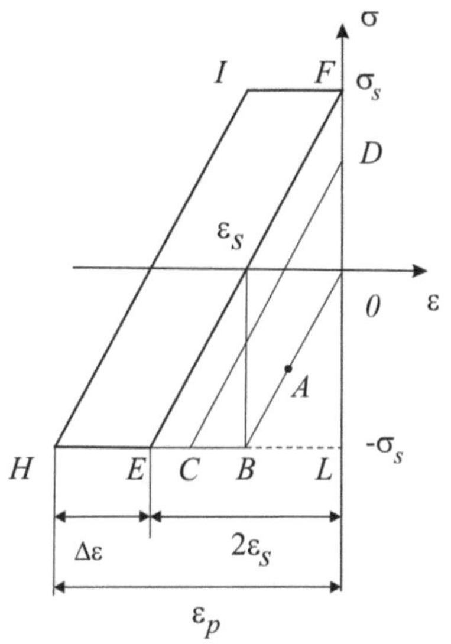

Fig. 8.4: Loading diagram at plastic deformation caused by thermoelastic stresses in the skin-layer.

to increase its length, but the connection with the remaining part of the medium, where the heating is absent, prevents elongation. As a result, the thermal expansion is compensated by the oppositely directed (negative) relative deformation.

In the case of elastic deformation the state of a medium at the end of heating corresponds to the point ***A*** on the stress-strain diagram. Upon cooling, the system returns to its initial state (the trajectory ***OAO***). However, if the stress exceeds the given limit, then the plastic deformation develops with further increasing stress (the trajectory ***OABC***). In the simplest models it is assumed that in the process of plastic deformation, exceeding the threshold value $\varepsilon_S = \sigma_S/E$, the stress remains equal to σ_S. Upon cooling, the transition to another state, along the straight line ***CD***, takes place. In calculations described in [14] it is assumed that the line ***CD*** is parallel to ***OA***. On this assumption, once loading is finished and the system returns in the state $\varepsilon = 0$, the residual stress $\sigma_1 = \varepsilon \cdot E$ appears. Every cycle of heating causes the increment of the temperature T_e and the corresponding deformation $\varepsilon = -\alpha\, T_e$. When the condition $|\varepsilon| < 2\varepsilon_s = 2\sigma_S/E$ is satisfied, then, with the repeated cycles of loading, the state of the system will be described by the points on the straight line ***CD***. However, when the condition $|\varepsilon| > 2\sigma_T/E$ is satisfied, then, in the process of heating the point on the stress-strain diagram shifts along the line ***OBH***, where point ***H*** lies to the left of point ***E***. Further, at the unloading during the process of cooling, the point on the diagram shifts along the line ***HIF***. The further process, when the heating-cooling cycle is repeated, is described by a figure ***FEHIF*** on the stress-strain diagram. In the given idealized scheme the upper and lower boundaries of this figure lie on the lines $|\sigma| = \sigma_S$.

The real diagram differs from the idealized one. Nevertheless, under the condition $2\sigma_S > \alpha E\, T_e/(\mu - 1)$ or

$$T_e > T_0 = \frac{2\sigma_S\,(\mu - 1)}{\alpha\, E}, \tag{8.16}$$

the real stress-strain diagram, like the idealized one, forms the closed figure shifted to the region of negative deformations.

In the regime of plastic deformation a medium absorbs the energy, which is characterized by the area of the parallelogram ***HIFE***. The irreversible changes are accumulated in a medium and cause its destruction after a certain number of cycles.

For each component of the deformation tensor one can introduce the increment of the relative deformation $\Delta\varepsilon = \varepsilon_P - \varepsilon_e$, where ε_P is the plastic deformation (line ***LH***) and $\varepsilon_e = 2\sigma_S/E$ is the elastic part of the deformation (line ***LE***). The quantity $\Delta\varepsilon$ is a characteristic parameter useful for the evaluation of the resource of the system, i.e., of the number N of cycles preceding the destruction. The empirical dependence for this number was obtained by Manson and Koffin [15]. It can be expressed by formula (5.47). A certain complexity to the calculation is the fact that, in a general case, one should consider $\Delta\varepsilon$ as an invariant of the tensor of the plastic deformation

$$\Delta\varepsilon = \sqrt{\frac{2}{3}}\sqrt{\left(\varepsilon_{P,\varphi} - \varepsilon_{e,\varphi}\right)^2 + \left(\varepsilon_{P,z} - \varepsilon_{e,z}\right)^2 + \left(\varepsilon_{P,r} - \varepsilon_{e,r}\right)^2}. \tag{8.17}$$

Such a calculation was carried out by Karpova and Titkov [14]. They calculated the stress field in the zone of the plastic flow and in the external zone, where the elastic deformation takes place. The temperature distribution in the radial direction is approximated by the dependence $T(r) = T_e \exp(x/\Delta)$, where $x = r - R_1$, and Δ is the parameter which has the meaning of a thickness of a skin-layer. The rearrangements of formulas (see [14]) for the case $\Delta \ll R_1$ give the following expression for the plastic components of the deformation at the boundary of the turn:

$$\varepsilon_{P,r} \approx \alpha \left(3T_e - \frac{1+\mu}{1-\mu} T_0 \right), \quad \varepsilon_{P,\varphi} \ll \varepsilon_{P,r} \,. \tag{8.18}$$

In the region of plastic deformation the Poisson coefficient $\mu = 2$. Therefore in the limiting case when heating is absent and $T_e = T_0$, we have $\varepsilon_{P,r} = 0$.

The data in Table 8.3 does not take into account the temperature dependence of the yield point. Some metals exhibit a decrease of the yield point when heated. Hence, the given estimates require significant correction. As a result, one can obtain the reduction of the calculated resource for single-turn magnets.

The dependences of the threshold induction on the number of pulses was calculated in [16]. The calculations, confirmed by experiments, relate to the case where the induction amplitude was 26 T at the frequency 20.5 kHz, and the ratio of the second current amplitude to the first one was 0.57. Table 8.4 gives the resource values for some materials.

Table 8.4: Resources of single-turn magnets according to data from [16].

Material	Resurs
Steel	2 000
Cu+Zr	30 000
Cu+Be	1 000 000

8.4 The destruction of single-turn magnets. The problem of erosion

Many of the experimental and technological installations operate in fields of induction which are significantly less than the damage threshold B_M. If the inertial effects are negligible, a failure of a thin-wall magnet with fastened layers, due to electromagnetic forces, may occur when the mechanical stress (see formula (8.3)) exceeds its ultimate strength. The authors of [10] describe the test results of magnets from low-carbon steel with a wall thickness of 5 mm and a diameter of 30 mm (the length is also 30 mm). The experiments were carried out at a frequency of current oscillations of 65 kHz and an effective pulse length of about 10 μs, and prove the role of slot fastening. Without fastening, the magnet holds up under tens of discharges in a field of 16 T and fails after five discharges in a field of 21 T, because of the stresses concentrated at the inner side of the magnet opposite the slot. Fastening the slot edges makes possible the generation of a field of 21 T in tens of discharges. Such a magnet fails in one discharge, when the induction is 28 T. Note that in these experiments the corresponding ultimate strength for a low-carbon steel-3 was about 1000 MPa, which exceeds the static limit and differs little from the value given in Table 8.2.

The experiments with copper thin-wall magnets described in [17] demonstrate the edge effect. Due to field enhancement, these magnets are deformed to a larger extent near the edges, compared with their middle parts (Figure 8.5). The computer calculation thoroughly describes the magnet configuration after a discharge. It is of interest that the edge parts of the magnet are similar in form to the figures with edges where the magnetic pressure is constant. These are the so-called figures of equilibrium which were considered in [18].

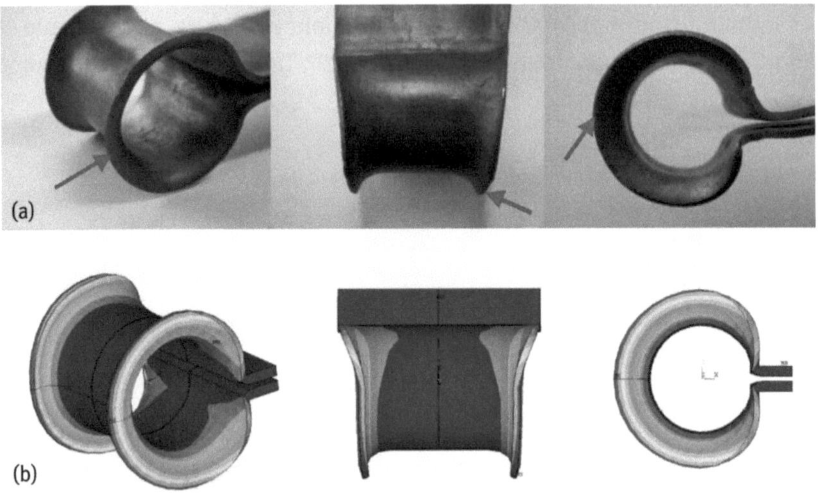

Fig. 8.5: Edge effect as appeared in plastic deformation of the thin-wall copper magnet [17]; (a) experiment; (b) the result of a three-dimensional computer simulation.

Photos of thick-wall single-turn magnets show the regions of the enhanced heat near the edges of the slot and magnet (Figure 8.6). In these regions the edge effect manifests itself. Its action can be decreased by rounding off the angles (see Chapter 1). Along with this, a decrease in heat can be seen near the angles. The current density is lower in these spots. It is evident that there is a resemblance between this effect and the process considered in the model problem on the diffusion of the poloidal field into the region with a rectangular edge (see Chapter 3, formula (3.34)).

An estimation, using formula (8.6), of the ultimate strength of a thick-wall magnet for the case of single short pulses leads to high values of threshold induction, even without taking into account the effects of inertia. This is clear from

Fig. 8.6: Traces of the heating of a thick-wall steel magnet after a pulse.

Table 8.2. However, the permissible value significantly decreases for a magnet intended to operate for a long time. The Koffin–Martin formula (5.47) can be used to evaluate a life expectancy under conditions when after some cycles of plastic deformation failure occurs, which manifests itself in the form of a body crack crossing the magnet wall. A failure just of this kind takes place in some cases. However, this can be preceded by the specific process of the erosion of a single-turn magnet characterized by the appearance and development of regularly spaced radial cracks. As seen from Figure 8.7, longitudinal scratches, specially made on the solenoid surface, become seeds of growing cracks.

The appearance of fine initial cracks during the magnet operation still remains the subject of different hypotheses. Their appearance may be attributed, for example, to thermal stresses in the skin-layer zone described above.

The nuclei of the regularly disposed cracks may be fractures or indentations on the surface due to fragile destruction or plastic deformation, as a result of the

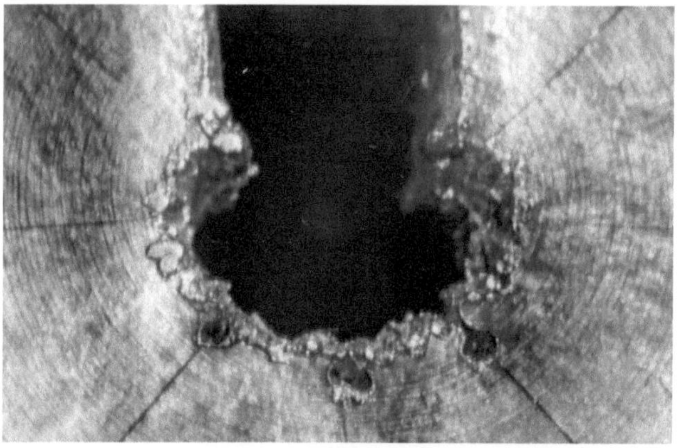

Fig. 8.7: Local disturbances in the form of cracks resulting in the erosion of a single-turn magnet.

fast cooling of the material after discharge, when the heat is removed into the depth of the wall of the solenoid. There are also other possible reasons for the appearance of initial cracks. Analogous phenomena occur during the welding of metals; these are examined in the literature on the theory of these processes. It could be due to a slip of the crystals with the destruction of the boundaries between them, resulting in the formation of so-called chemical inhomogeneities caused by impurities at the boundaries. In the literature on the theory of welding it is noted that "these phenomena may play a decisive role in such processes as the formation of cracks in weld joints when repeatedly heated (reheat cracking)" [22]. Such cracks are visible in the photograph in Figure 8.8.[1] Thermal oxidation may play an essential role in the formation of cracks [16].

Fig. 8.8: Cracks on the surface of a beryllium bronze magnet.

When the magnet surface is heated up to a temperature near the melting point, disturbances in the form of waves can appear in the skin layer, pointing to the development of MHD instability in melted or softened metal (Figure 8.9). These disturbances give rise to cracks.

All the mentioned and other phenomena, developing during fast heating and succeeding cooling, cause the appearance of small radial cracks resulting in the erosion of the surface of the magnet. A current thickens at the bottom of a crack.

1 The photos in Figures 8.6, 8.8, and 8.10 were presented by researchers of the Laboratory of the Applied Electrodynamics headed by V. V. Ivanov (Institute of Electrophysics RAS, Ekaterinburg).

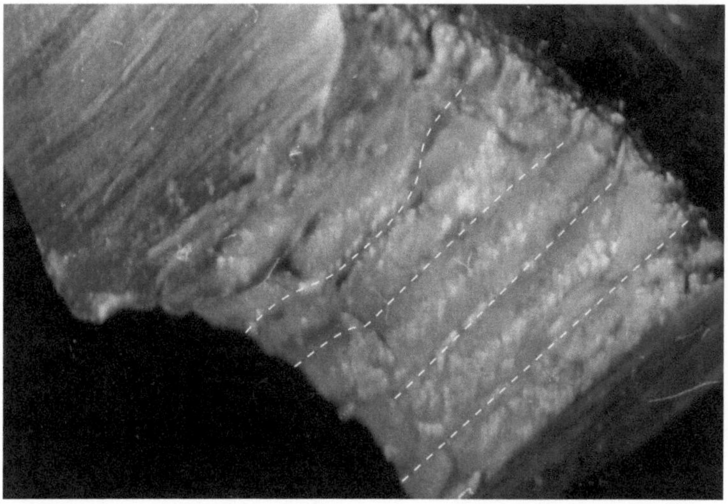

Fig. 8.9: Traces of the development of MHD-instability on the surface of a single-turn magnet.

This causes the higher heating of the region adjacent to the bottom, where zones of melted metal are formed [10, 42]. This is evident in the photographs in Figure 8.10. The melted metal is thrown outward under magnetic pressure. The result is the destruction of the surface. The result is the growth of several regularly-spaced cracks.

Fig. 8.10: The traces of heating of a thick-wall single-turn magnet near the edges of cracks.

Fig. 8.11: Radial slots in a single-turn magnet after 15 discharges.

Studies have shown that the depth of cracks in steel and brass solenoids attained 3–4 mm after 7–15 discharges at a relatively weak field of 56 T (the diameter of solenoid was 6.8 mm, the current rise time 3.9 µs) This process is sharply pronounced in the case of alloys such as brass, low-carbon steel, and beryllium bronze. A steel magnet after 15 discharges is shown in Figure 8.11. One can clearly see the failure of the boundary. We mentioned above that slot growth occurs due to the increase of current density near their ends. Along with this, the current density drops in the regions between them. As a result, a system in which slots alternate with less destroyed regions of the boundary appears: the "saw effect" [2] takes place.

Erosion noticeably reduces the field level at which single-turn solenoids can be repeatedly used. The record values of induction are reached in tantalum solenoids, where the field achieved 100 T at a rise time of about 2 µs. Here, as in the study by Andrianov, Demitchev, Eliseev, and Levit [4], the strongest field with a high rate of rise time was used for the compression of zirconium cylinders in order to hermetically seal the heat-releasing elements in atomic reactors.

Photographs of the samples with traces of erosion demonstrate a stronger destruction in areas where there is a strengthening of the field due to the edge effect.

We indicated above (Chapter 3) that the surface temperature of single-turn magnets can be reduced by application of a weakly conducting coating. This can result in the retardation of erosion and in the increase of the threshold field of the magnet designed for long operation. The role of this coating was experimentally confirmed by Farinski, Karpinski ,and Novak [19] and justified in calculations in [18, 20, 21].

In steel samples, as was mentioned, the erosion begins in fields comparatively weaker than the yield point. Thus combined systems are promising, in which a strong metal, although prone to erosion, is used as a nondestructive element of the construction (external reinforcement). Its surface can be protected with a replaceable gasket. Such a device can be used for the generation of a field with induction close to the magnetic ultimate strength of the material and significantly exceeding that of the gasket material. The latter is partially destroyed in the discharge, but may be replaced after one or several pulses.

8.5 Special construction features of single-turn magnets and their power supplies

The weak spots in the construction of single-turn magnets are the insulation of the radial slit and the current leads. Because of the high heating of the conductor surface, it is necessary to protect the insulation from thermal destruction. The decrease in temperature of the surface in the time of the pulse can be evaluated using the formulas of Chapter 3. In the case of the "standard" pulse, which has the form of a sinusoidal half-wave, the surface temperature can be calculated with the aid of formula (3.48e). For example, for the amplitude of the current linear density 100 kA/cm, the temperature increment at the boundary of a skin layer is several tens of degrees. It can be significantly greater if the pulse has the form of a slightly damping sinusoid. In this case one can use formula (3.49) to calculate the volume density of the thermal energy. In [23] the example of when the first amplitude of the current pulse is twice as much the second amplitude was examined. Under these conditions, the temperature increment of the conductor surface is 200 K in the time of the pulse, provided that the amplitude of the linear current density is 163 kA/m for copper and steel and 140 kA/m for aluminum.

A positive factor is the faster transfer of heat to the wall than to the insulation, after the discharge is ended.

If the slot width of the turn (the dimension taken along the force line) is equal to the length of the magnet, then the induction in the slot and in the magnet will be approximately equal. Hence, the forces acting on the edges of the slot and the forces in the magnet are about the same (as calculated per unit of surface). As already mentioned, with edges of the slot unfastened destruction is possible, due to the concentration of stresses near the turn of the internal surface opposite the slot. In order to prevent failure of the turn, a special fastening of the magnet is required, e. g., with beams installed from outside of power buses and contracted with bolts through insulating pads; see Figure 8.12.

In this very simple system the net cross section of the bolts S should be chosen according to the condition $S > F_m/[\sigma]$, where $[\sigma]$ is the admissible stress of the

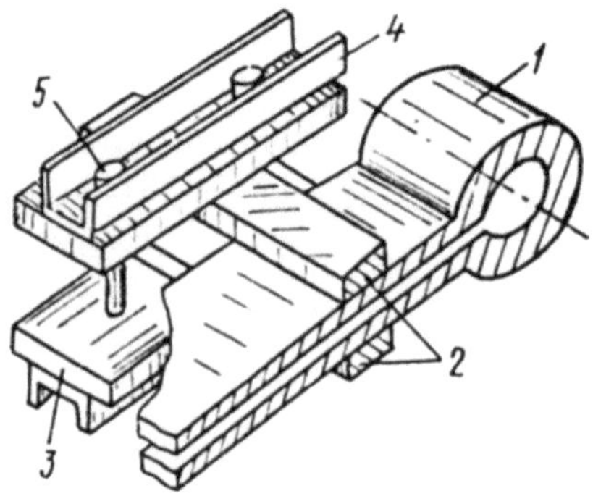

Fig. 8.12: Single-turn magnet with ties; (1) magnet; (2) beam; (3) insulating pad; (4) channel; (5) contracted bolts.

material of the bolts, and $F_m = \mu_0 i_m^2 l/b$ is the amplitude of the force acting on the buses of width b and length l at the current of amplitude i_m. When the pulses are short, to increase the strength of the current lead system it is possible to use the inertial confinement, similar to that which was described above as applied to the thin-wall magnets. A decrease of stresses can take place under condition $\omega\tau \ll 1$, where τ is the effective pulse duration, and ω is the circular frequency of the elastic vibrations of the buses tightened by bolts. For n_0 bolts of length l_0 and of cross section S_0, the maximum tensile stress is [10, 13]

$$\sigma_m = 2F_m\tau\sqrt{\frac{E}{2n_0S_0l_0M}}, \qquad (8.19)$$

where E is the elasticity modulus of the materials of bolts, and M is the mass of the buses and additional load which can be used to increase the mass of the system

A more convenient system is the hydraulic press, similar to one used in experiments by Goto, Miura, Takeyama, and Sakakibara ([24], Figure 8.13) and in other works.

When connecting the buses with the magnet, it is necessary to avoid sparking during contact and thus its rupture. When the flat gasket is compressed between electrodes, contact spots appear, resulting in the disruption of the electrodes due to the thickening of the current in the vicinity of the spots. Experience shows that the most reliable contacts are linear ones (Figure 8.1). These could be made from a high plasticity material, e.g., annealed copper wire, which is placed in a groove 1–2 mm wide along the line of contact. In this case it is necessary to produce high pressure leading to deformation of the wire. This results in linear contact along the whole groove without the appearance of contact spots. According to the data [23] the admissible linear current density in such a contact is about 50 kA/cm with a pulse duration of the order of 10^{-4}s. With such a linear current density in a "standard" half-sine pulse, formula (3.48c) gives the increase in temperature at the skin-layer boundary of a few tens. It can be significantly higher for a slowly

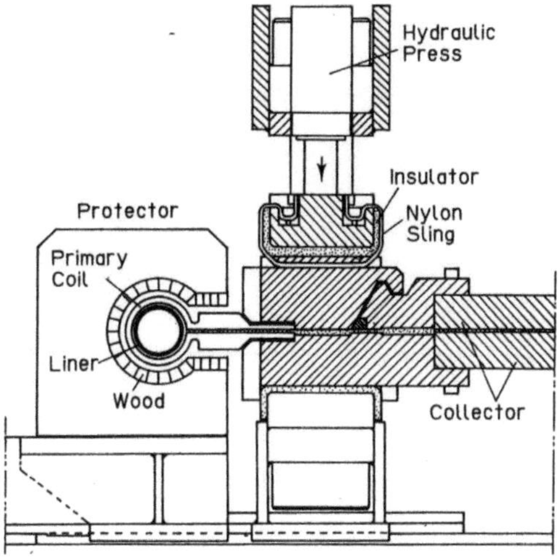

Fig. 8.13: Supporting system of a single-turn magnet with a hydraulic press [24].

decaying sine pulse. The corresponding formulas for the evaluation of the contact temperature are given in Chapter 3 and in [18].

In order to provide high efficiency for the energy transfer to the magnetic field, the inductance of the magnet should far exceed that of the energy storage (see Chapter 4). In the 1960s the high voltage capacitive storages with energy in the order of megajoules were developed for the supply of long single-turn magnets and other devices. They were used in studies of high temperature plasma. In such batteries the storage consists of a large number of capacitors and switches. They are connected in parallel, and their combined contribution to the total inductance can be relatively small. The problem of low-inductance current leading to a far-removed load has been solved in large-scale construction with a help of many cables and/or bifilar buses. An example of this are the batteries described in [25–27, 47].

The necessity of supplying magnets with the contact and bore in the order of several cm and with inductance in the order of several nH complicates the problem of developing a low-inductance generator. Usually, its energy is of the order of 10^4 J, and the number of capacitors is not large [3, 5, 28–30, 40]. Therefore special-purpose low-inductance capacitors and switches must be used. The generator, used by Olson, Bandas, and Kolb [3], consists of 50 kV low-inductance capacitors. The use of rail spark-gap switches and bifilar buses makes obtaining an inductance of 9 nH possible. This results in a rather effective transfer of energy to the 16 nH load. Generators with the specially developed low inductance 50 kV capacitors were used by Andrianov, Demichev, et al. [28] and Botcharov, Zayentz, Popov, et al. [29]. The authors of the first of these works used vacuum switchers

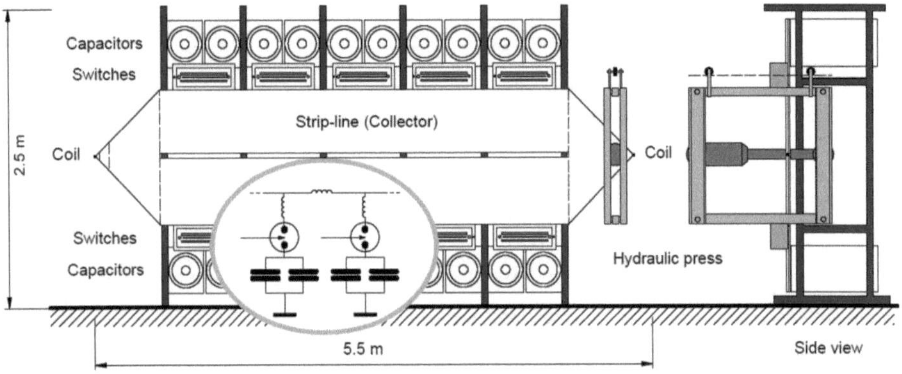

Fig. 8.14: Low-inductance generator used in [5] and a fragment of its equivalent circuit.

developed purposely for this task, and in the second work solid-state switchers were used.

The battery discharge in a single-turn magnet was studied by Portugall, Putham, et al. [5]. They used a generator in which the capacitors were connected to the assembled sufficiently long buses (Figure 8.14). Analysis of the discharge process proceeding in this generator should be made with a help of the equivalent circuit with five elements. This includes inductances of bus regions between the switching spots of capacitors. This feature of the circuit shows itself as high-frequency oscillations on the current curve. The high-frequency vibrations are damped quickly in the circuit with resistance; however they can lead to an overvoltage on capacitors.

Application of spatial buses enhances the capabilities of the low-inductance lead of the current to a small-sized load. An example of this is the generator developed at the Saint Petersburg Polytechnical University (Figure 8.15) [40]. The low inductance of the storage is attained by means of specially-built capacitors and solid commutators. In addition, this generator is provided with a spatial bus arrangement which permits the current to be free to spread with a dense storage layout. The calculation methods of bifilar buses, including their spatial modifications, are considered in [18].

These capacitor storages are examples of devices which not only provide effective transfer of the energy into the load, but also a small rise time of the current. The latter requirement is important for the generation of ultrahigh magnetic fields (see Chapter 8). For this, in definite experiments step-down transformers can be used. In this case the requirements for the generator inductance can be reduced, leading to a decrease in voltage on the load and, correspondingly, to an increase in the current rise time. This factor is of somewhat lesser significance in many

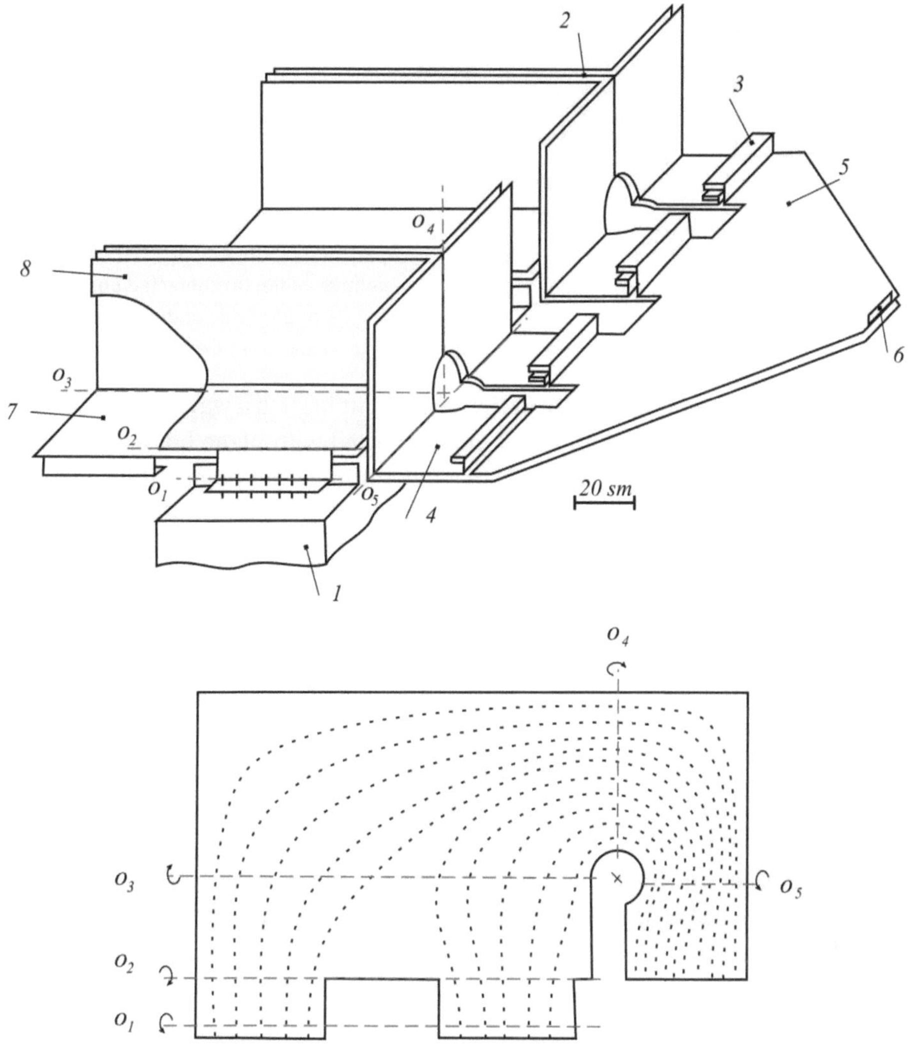

Fig. 8.15: Low-inductance generator with specially-configured buses [40]. The position of the bending of the buses is shown with broken lines.

technical installations where the field amplitude is 20–30 T and the pulse length is tens of microseconds and more. In such installations step-down transformers may be applied. A transformer itself may introduce an additional inductance in the discharge circuit which reduces its efficiency

There are some advantages intrinsic to the cable transformer. This is described in [31] and [10]. The main element of such transformers is the high-voltage coaxial cable wound as a spiral (Figure 8.16). On each turn of the spiral the con-

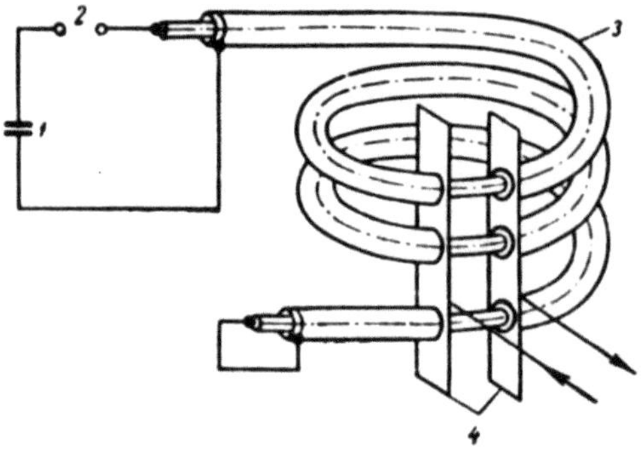

Fig. 8.16: Cable transformer in the discharge circuit of the capacitor and the equivalent circuit: (1) the battery of capacitors; (2) the switch; (3) the high voltage cable; (4) contacted buses on the secondary side of the transformer.

ducting shell on the small length of the cable is cut in such a way that the cuts are one above the other and their edges are connected with plane busses or cables. The cable is switched on to the capacity bank through the commutating gap.

On discharge of the capacity bank, in the shell of each turn there an EMF is enduced, producing a secondary current. When using the autotransformer circuit (Figure 8.16), the primary current flows along the cable conductor and through the end regions of the shell, collecting buses and the load. As a result, the sum current of all turns and the primary current flow through the load.

A cable transformer has some advantages compared to transformers of other types.

1. The leakage inductance of the transformer is significantly reduced, since it is determined only by the cable inductance. The latter may be decreased by winding parallel branches (entries) of the cable.
2. The problem of insulation between the windings is automatically solved, since in this case its role is played by the cable insulation itself.
3. The cable conductor (the primary winding of the transformer) is practically unloaded from the action of electromagnetic forces.

This device, combining the transformer and single-turn magnet, is the flux concentrator [32–37], three versions of which are shown in Figure 8.17. The secondary winding is a single-turn magnet, the cross section of which can be chosen in a way so as to provide amplification of the field (flux concentration) in a working region.

In the flux the concentrator turns of the primary winding may be placed in slots, as is shown in Figure 8.18. This enables the unloading of the primary winding from the axial and radial electromagnetic forces. By properly choosing the gaps between the turn and slot walls, it will be possible to achieve of magnetic field intensity equality from all sides of the turn and to make the resulting elec-

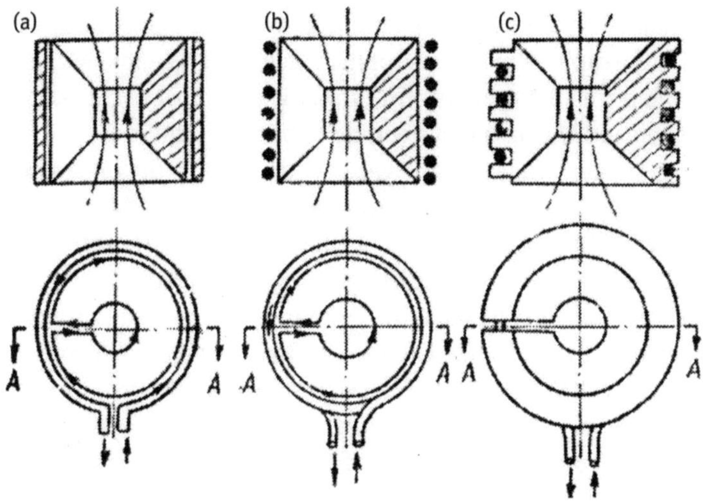

Fig. 8.17: Three versions of flux concentrators, distinguished by the construction of the primary winding: (a) the primary winding is a single-turn; (b) a multiple-turn spiral winding; (c) a multiple-turn spiral winding, placed in slots.

tromagnetic forces acting on the turn equal to zero. In such a device the forces act only on the solid secondary turn of the transformer.

Figure 8.18 shows the field pattern in a magnet with a flux concentrator.

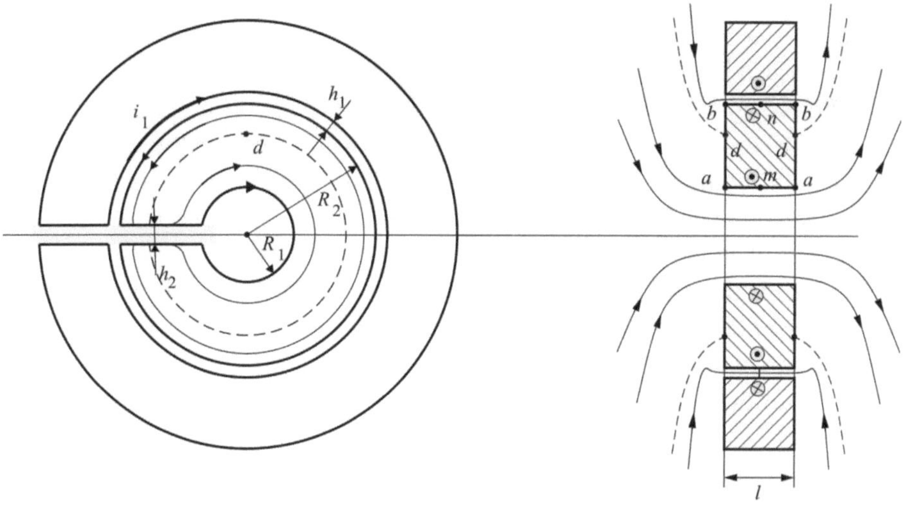

Fig. 8.18: For estimation of influence of a slit upon the field induction at a hole of the concentrator.

In the case of a gap with infinitesimal width, the overall flux closes through the orifice. The field is the same as in the case of the single-turn solenoid with the current $i_1 w_1 + i_2$, where i_1 is the primary current, w_1 is a number of turns, and i_2 the secondary current. In the actual system, as opposed to the conventional single-turn magnet, there are two insulating gaps, h_1 and h_2. On the orifice surface ***am*** and on the part of concentrator surface ***ad*** the induced secondary current coincides in direction with the primary current. On the surface of gap h_1 ***bn*** and on the adjacent part of the concentrator surface ***db*** the current is directed oppositely. The closure of a portion of the flux through the gaps and the flow of a portion of the current from the gaps to the end surface reduce the field, compared to an ideal system with slots of zero width. The effect of the slots can be shown by an example of the simplest system in which the rectangular secondary turn has a length equal to that of an external single-turn magnet of large width. In the absence of slots, neglecting the end effects, the induction in the center can be assumed to be $B_1 = \mu_0 iw/l$. The role of the correction caused by the radial slot h_2 was considered in Chapter 2. The corrections, taking into account the effect of both slots, have been approximately calculated in [18]. When $R_1 \ll R_2$ the formula for the induction, calculated with regard to the effect of the slots, has the form

$$B_2 = \frac{\mu_0 iw}{l}\left(1 - \frac{h_2}{\pi l}\ln\frac{4R_2}{h_1} - \frac{2h_1}{l}\right). \tag{8.20}$$

Here the first term is responsible for the radial, and the second for the annular slot.

In the particular case, when $h_1 = h_2 = 0.05\,l$, $R_2 = 20h_2$, flux losses lead to an decrease in induction of around of 17 % compared to B_1.

The parameters of the magnetic systems with flux concentrators were calculated by several authors, mainly for systems with coaxial cylinders. Ditz [36] arrived at a formula for the calculation of the inductors with concentrators. An extensive reference for the calculation of inductors is contained in the book by Belyy, Fertik, and Khimenko [37]. The formulas given in these publications were derived using the simplified field pattern, and the edge effects and corrections responsible for the mentioned influence of slots were ignored.

The calculation, taking into account the edge effects, gives the following expression for the inductance of the magnet with the concentrator, shown in Fig-

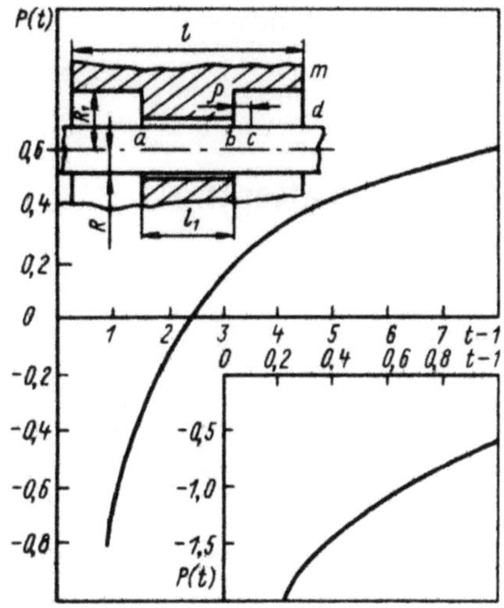

Fig. 8.19: For the calculation of the inductance of the concentrator.

ure 8.19a [18,38][2]:

$$L = \frac{2\pi\mu_0 R h}{l_1}\left[1 - \frac{h}{2R} + \frac{2\pi\, h(l - l_1)R}{l_1(R_1^2 - R^2)} + \frac{4h}{\pi\, l_1}\left(P(t) - C_0 + \ln\frac{\pi\, R}{4\, h}\right) + \frac{4h\pi\varepsilon' R}{l_1 R_1}\right]^{-1} \tag{8.21}$$

where $t = R_1/R$, $C_0 = 0.577$ is the Euler constant,

$$P(t) = \int_0^\infty \left[\frac{K_0(x)\, I_1(tx) + I_0(x) K_1(tx)}{K_1(x)\, I_1(tx)\; - I_1(x)\, K_1(tx)} - \frac{2}{x(t^2 - 1)} - \frac{x}{1 + x}\right]\frac{dx}{x},$$

$K_0(x)$, $K_1(x)$, $I_0(x)$, $I_1(x)$ are the Bessel functions of an imaginary argument. The graph of the auxiliary function $P(t)$ is shown in Figure 18.19b. In this formula the terms of the order of h/R, h/l_1 are taken into account. The first three terms are taken based on the simplified model of the field. The rest take into account the edge effects similar to those described in the examples of Chapter 3 and in [18]. The last term is responsible for the edge effect in the vicinity of the point m (Figure 6.38b). It is similar to that used for taking into account the edge effect in a long thin-wall cylinder (see Chapter 3). When $l_1 \gg R_1$ we have $\varepsilon' \approx 0.62$.

The well-known drawback of the single-turn solenoids, operating under conditions of the pronounced skin effect, is stronger heating as compared to multiturn solenoids, since the current is not of necessity distributed over the cross section of a winding (as is the case of the multilayer coils), but concentrated in the surface

2 This formula does not take into account the effect of the slot.

layer. Heating limits the multiple use of single-turn magnets no less than mechanical strength does.

8.6 Deformed single-turn magnets restored after the discharge

Along with the development of nondestructive magnets, some researchers studied the possibility of constructing "restorable" systems. They change somewhat in size under magnetic pressure, but may be restored in time between discharges. Note that a similar effect occurred in the Kapitsa magnet [2 of Chapter 6], in which a sliding contact was used.

The belt coil may be referred to such systems. Its winding is similar to a spiral spring, locked at one end. The other end is not rigidly locked. Therefore the winding is free to uncoil as a result of electromagnetic forces. This can result in a slight decrease of stresses in turns [39]. Once the discharge is over, the magnet returns to its initial state.

One version of a deformed magnet is the massive single-turn solenoid consisting of two parts [27]. They are separated mechanically, but connected by a conductor, not precluding deformation (Figure 8.20). Under discharge, the parts of the magnet come apart, but because of their large mass, displacement occurs slowly. Contact between the parts of the magnet is not impaired. With a sufficiently short pulse, the induction amplitude is close to the calculated one. In order to protect the surface from erosion, the replaceable bronze pad was placed in this magnet. In experiments with the 30 mm-bore s field of 70 T and duration of about 50 μs was generated.

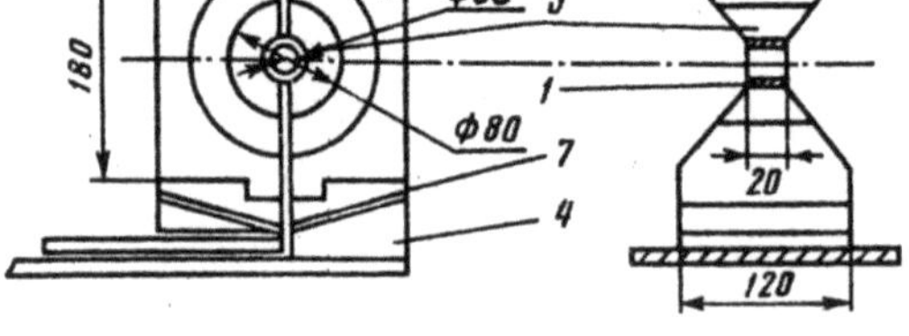

Fig. 8.20: The single-turn magnet with massive movable parts [27].

Mercury turns can be assigned to the magnets of multiple application. These were studied in [41]. Before the discharge, the liquid fills the cavity of the chamber with dielectric walls. During the discharge the liquid moves under magnetic pressure. However, with such a turn the field of 34 T with a rise time of 25 μs was obtained [41]. The magnet may be restored in the time between the discharges. This possibility also exists for single-turn magnets made from fusible material. The authors of [43,] studied magnets of wood alloy. They obtained a field of 50 T in a rise time of nearly 4 μs in magnets with an initial diameter of 3.4 mm and length of 5.5 mm. The failure of the magnet occurred due to the ejection of melting metal from the skin layer. This process ("slow" explosion of a conductor) will be discussed in the next chapter.

8.7 Magnetic systems used for deformation of solids and the study of their properties

As was noted, the electromagnetic forces acting upon the conductors in pulsed magnetic fields can be used for the deformation and acceleration of bodies. Omitting the technical aspects of this processes, we shall consider the typical constructions of the specifically used magnetic systems.

In the 1960s the first publications appeared describing devices for the deformation of conductors with the help of a pulsed magnetic field. The main advantages of electromagnetic metal forming (EMF) are the high adaptability of the process, the absence of mobile components, the possibility of processing in a vacuum and in an atmosphere of protecting gases, insignificant exploitation costs, and the simplicity of automation. Industrial installations commonly consist of a storage capacity, with a set of solenoids and changeable concentrators permitting a number of different operations to be fulfilled: the setting and reduction of pipes, flaring of pipes, cold welding, forming of components from flat conducting sheets, and others. The main types of installations, the calculated methods and features of technology, are described in [46, 38] and also in a number of surveys, such as [48–50]. Recent publications, in particular the *Proceedings of the International Conferences on High Speed Forming* (ICHSF, 2004–2012) and the *Proceedings of the Conference on EMF* in Samara (Russia, 2007), describe different types of the magnetic systems used in practi ce and the achievements both in industry and in computer simulation of forming processes.

Aside from the technical questions, we restrict ourselves to a description of the characteristic magnetic systems used in the EMF. Magnetic systems, used in experiments on plasma compression and heating which are of Z- and θ-pinches

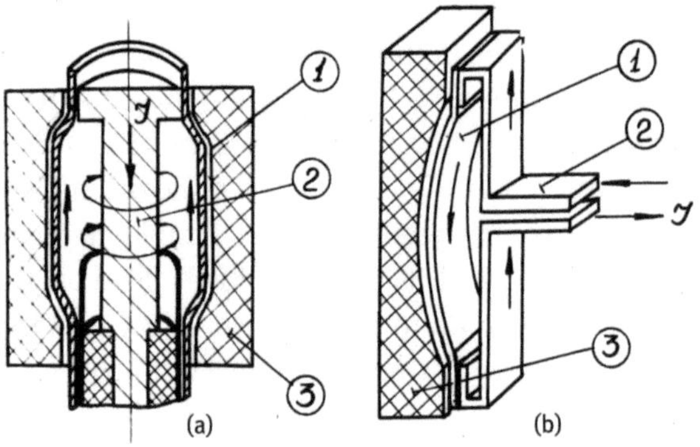

Fig. 8.21: Devices for plastic processing of metals by a strong pulsed magnetic field with a direct delivery of the current to the processed article [44, 45]: (a) the coaxial system for flaring of pipes; (b) the magnetic hammer for processing of flat sheets; (1) the processed article; (2) the device for a current delivery; (3) the matrix.

(Figure 4.10), preceded the technological EMF installations. In Chapter 10 we shall consider these systems as applied to the experiments on magnetic dynamic cumulation. The configurations of the magnets are diverse, but many of them present to some extent the further development of the systems described in an earlier publication by Furth and Waniek [45].

In conductive devices (Figure 8.21) the current is delivered directly to the processed article, as is done in Z-pinches (Figure 4.10). In these devices it is necessary to provide a high-current contact of the external circuit and processed body, but there is no necessity for the high conductivity of the material. In systems of the induction type, like iin a θ-pinch, the current is induced in the article (Figs. 8.22 and 8.23). The problem of contact does not exist in this case. Therefore multiturn magnets may be used, which reduces the requirements for the inductance of the capacity bank. The modifications of the magnetic systems shown in Figure 8.22 and their analogs are used for the deformation of flat sheets. The devices in Figure 8.23 apply for the compression or expansion of cylinders by the pressure of the magnetic field.

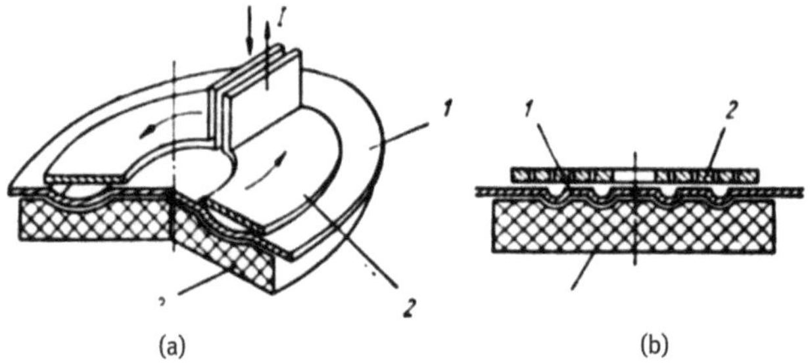

Fig. 8.22: Devices applied in the plastic processing of flat sheets by a strong pulsed magnetic field: (a) the single-turn solenoid over a flat sheet; (b) the multiturn solenoid over a flat sheet; (1) the processed article; (2) the solenoid; (3) the matrix.

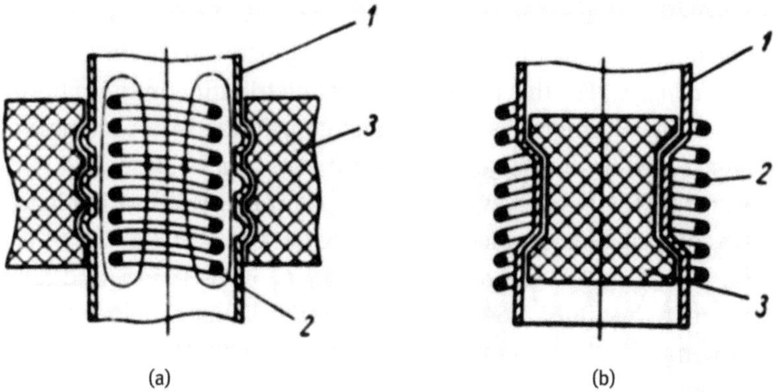

Fig. 8.23: Devices applied in the plastic processing of cylinders by a strong pulsed magnetic field: (a) the system for flaring of pipes; (b) the system for the compression of pipes; (1), the processed article; (2) the solenoid; (3) the matrix.

Magnets for EMF are commonly designed for the field generation with induction under 20–30 T, so that a relatively simple means may be used to provide long-term industrial exploitation. In the majority of the cases multiturn solenoids with a strong dielectric reinforcement of the winding are used. Single-turn magnets are also applied [51], as are devices with transformers and magnetic flux concentrators (see [36, 37, 46]).

Evidently, devices in which the field does not penetrate through the processed detail but is concentrated in the gap between the article and inductor are the most effective. In this case the induced current has the same surface density as the ideal conductor. This is true for the common conductors, if the time constant of the secondary contour (detail) sufficiently exceeds the oscillation period of the current in the circuit. In the case of materials with poor conductivity this condition is not fulfilled. Therefore it is necessary to apply additional coatings from well-conducting materials (for, example, to set-up the copper insert when processing articles from stainless steel [37]).

Another option for excluding the penetration of a field in the cavity of the cylindrical article is the application of a single-turn inductor together with a low-inductance capacity bank. The latter should have a relatively low capacity and, consequently, a high voltage of up to 50–100 kV. A high frequency of the discharge in the circuit may provide the necessary conditions for effective action.

As an example, we discuss the choice of processing frequency in the device for the compression of thin-wall metallic pipes in the field of a cylindrical inductor (Figure 8.23b). The penetration process of the external field of induction B_e into the closed cylinder with a radius R and wall thickness d is described in detail in the [13, 18, 46]. When $d \ll R$ this process could be separated into two stages. The first stage is characterized by the time of the field diffusion by the length d. This is a “skin time” $\tau_1 \approx d^2\mu_0/\rho_0$. For this time the current distribution in the cylinder wall is close to uniform. During the second stage the field tends to become stable in the cylinder cavity. In this stage the field diffusion can be described with reasonable accuracy by the plane-wave model [52, 53]. If the wall heating is relatively low, the second stage is characterized by the time $\tau_2 = \mu_0 Rd/(2\rho_0)$[3]. For a copper cylinder with $R = 10^{-2}$ m, $d = 5 \cdot 10^{-4}$ m, $\tau_2 = 4 \cdot 10^{-4}$ s, and for a stainless steel cylinder, this time is 20–50 times less. In the latter case for effective processing a frequency on the order of 100 kHz is necessary. The requency requirements f could be even higher if the induction exceeds the value $(2h/R)^{1/2}B_0$. In these regimes the conductivity falls, due to heating by eddy currents, and the penetration time of the external field into the cylinder cavity decreases [54, 18]. In the above-mentioned papers of Andrianov, Demichev, et al. [28], the application of a low-inductance high-voltage capacity bank enables the compression of pipes from poorly conducting zirconium.

Specific peculiarities feature the inductors used in the cold welding of metals. The main feature is that the process remains the same as in explosion welding. The magnet configuration is chosen in such a way as to provide the impact of the conductor, accelerated by magnetic pressure, with the immobile conductor at a small angle. It leads to welding at the place of impact and to the displacement of the place along the path, on which the touching and welding of two articles occur. Cold welding in a pulsed magnetic field has been described in many studies (for example in [51, 55]) and has been discussed at several conferences. In particular, the process and technology of cold welding were analyzed comprehensively in a series of papers presented at the Fourth Conference in Ohio. [56–59]. Here we give a scheme of this process according to [59] (Figure 8.24). In the course of cold welding the characteristic wave-shape boundary is formed. The results of a metallographic examination of the welded interface, as presented in [56] (Figure 8.25), give us a rough idea of this process.

The important task in the construction of inductors for full-scale production is to make them long lived. For this purpose it is necessary to decrease the heating and provide the cooling. An example of this is a magnetic system with a single-turn inductor, as described in [60]. In order to reduce the heating, a copper layer is applied to the surface of a steel single-turn magnet. Due to intensive air cooling (velocity of air flow 20m/s) the temperature is kept at 60 °C with a magnet operating frequency of 0.1 1/s. Water cooling is used in single-turn magnets of the firm

3 Note, that the time τ_2 exceeds the skin-time τ_1 approximately by ratio R/d.

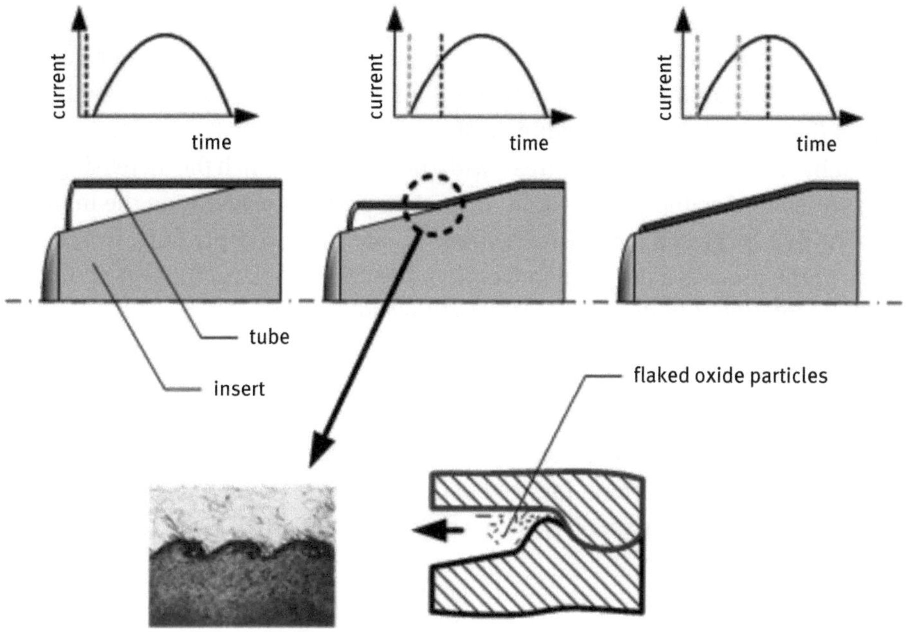

Fig. 8.24: The process scheme of cold welding [59].

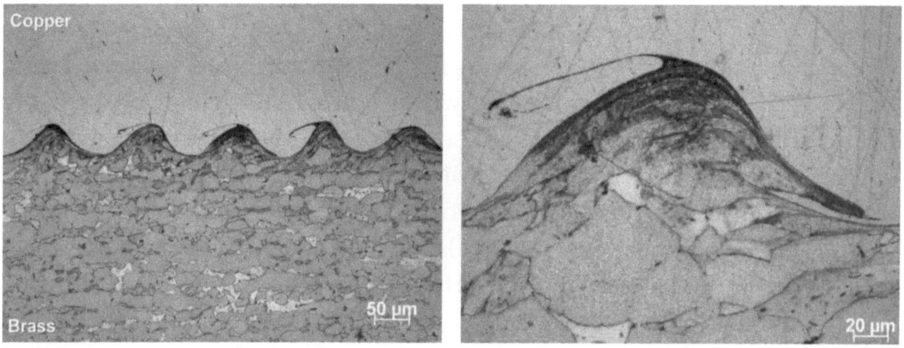

Fig. 8.25: *Left*: the configuration of a weld seam (copper–bronze); *right*: a section of the boundary [56].

"Pulsar" (Israel) [61]. These magnets have a lifetime of at least 50 000 pulses when operating in a field with the induction of 20 T at a pulse frequency of 200/hour.

Among the devices for the inductive deformation we can set aside those designed for the "attraction" of metallic bodies to the inductor, just as a magnet attracts a ferromagnetic body. The interaction between the field and eddy currents, induced by it in the conductor always causes the displacement of the conductor in the region of weaker field. Therefore the pulsed field should be formed such that the field is weaker far from the inductor than in the vicinity of it. An example of this is the system shown in Figure 8.23b, in which a thin-wall cylinder is placed coaxially with a magnet. By contrast, with the device, in which the increasing field compresses the cylindrical shell, in the system under consideration the inductor current first increases for time exceeding τ_2, and then sharply falls. In the first stage of the process a duration longer then the "skin time"τ_1, the penetration of the field into the cylinder is described by the equation

$$B_e = B_i + \tau_2 \frac{dB_i}{dt}, \tag{8.22a}$$

where B_i is the induction of the internal field. If this process lasts much longer than the time τ_2, the current, induced in the cylinder wall, with the linear density $j_\varphi = (1/\mu_0)(B_e - B_i)$, is small. So is the magnetic pressure acting on the cylinder wall:

$$P_M = \frac{(B_e^2 - B_i^2)}{2\mu_0} = \frac{\tau_2}{2\mu_0}\frac{dB_i}{dt}(2B_i + \mu_0 j_\varphi). \tag{8.22b}$$

If the current in the inductor switches off for a time less than the time τ_2, the current is induced in the cylinder wall. As a result, in the shell cavity, for a time of order τ_2, the field penetrated in it is maintained. The shell wall is subjected to an outwardly-directed force. The results for the field first linearly increased up to the value $B_{e,m}$ and then was instantly switched off, as shown in Figure 8.26 [18]. In the particular case of a field with a rise time of $40\tau_2$ the compressing magnetic pressure attains 0.3 $B_{e,m}^2/(2\mu_0)$. After switching off the current in the inductor, the pressure inside the cylinder increases practically up to $B_{e,m}^2/(2\mu_0)$.

The problem of field diffusion is also urgent in using in machines of thin metallic sheets. In this case the process is also characterized by the times τ_1 and τ_2. The first time is the determining time of the equalization of the current distribution in thickness and is the same as for thin-wall cylinders. The second time can be estimated by the formula $\tau_2 = \mu_0 gd/(2\rho_0)$, where g is the parameter having dimensions of length. It is determined by the magnet configuration and the distance from the magnet to the sheet. In [18] we are presented with the solutions of a number of problems on field diffusion through thin flat sheets. One example of this is the field of a wire placed at distance h from the sheet.

The simplest system may be supplemented with estimates taking into account the temporal characteristics of the field diffusion. If the time of switching on and off the current far exceeds the skin time τ_1, the model of the thin conducting film may be admissible [62, 18]. It is assumed that the current is uniformly distributed over the thickness of the sheet. The electric field intensity in the sheet and the

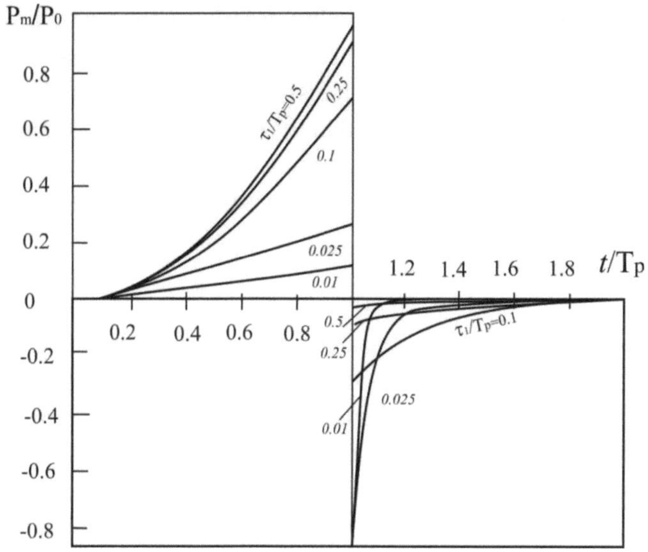

Fig. 8.26: The pressure acting on the wall of the cylinder as the external field first slowly increases and then terminates (T_u is the rise time of the induction up to the value $B_{e,m}$, $R/2d = 5$, $P_0 = (B_{e,m})^2/2\mu_0$).

current linear density satisfy the equality

$$E_x\big|_{z=0} = -\left.\frac{\partial A_x}{\partial t}\right|_{z=0} = \frac{j_x\rho_0}{d}. \tag{8.23a}$$

Along with it we have

$$j_x = \frac{1}{\mu_0}\left(B_{y1} - B_{y2}\right) = \left(-\frac{\partial A_{x1}}{\partial z} + \frac{\partial A_{x2}}{\partial z}\right)_{z=0}, \tag{8.23b}$$

where A_{x1} is the vector potential of the field over the sheet, A_{x2} – under the sheet.

The vector potential in regions (1) and (2) satisfy the Maxwell equation, and the equalities (8.23a) and (8.23b) give the boundary equation on the plane $z = 0$:

$$-\left.\left(\frac{\partial A_x}{\partial t}\right)\right|_{z=0} = \frac{\rho}{\mu_0 d}\left.\left(-\frac{\partial A_{x1}}{\partial z} + \frac{\partial A_{x2}}{\partial z}\right)\right|_{z=0}. \tag{8.23c}$$

The solution of the problem of field penetration through the sheet was also obtained by Maxwell. It is given in the book by Smythe [62]. It was shown that after switching of the current in the inductor the vector potential in regions (1) and (2) is described by the formulas

$$\begin{aligned} \vec{A}_1 &= f(y, h - z) - f(y, z + h + u_0 t), \quad z > 0, \\ \vec{A}_2 &= f(y, h - z) - f(y, h - z + u_0 t), \quad z < 0, \end{aligned} \tag{8.23d}$$

where $u_0 = 2\rho_0/(\mu_0 d)$. Here $f(y, (h - z))$ is the function describing the vector potential distribution in the region $x < h$ in the absence of currents in the sheet.

The solution of this problem shows that after a fast current rise up to a given value i_0, the current of the opposite sign is induced in the sheet. Further, the total current in the sheet is maintained, but it is driven out in a region remote from the wire. In this case the current density under the wire decreases in accordance with the dependence

$$j_x = -\frac{i_0}{\pi h(1 + t/\tau_2)}. \tag{8.23e}$$

Here $\tau_2 = h/u_0$,where $u_0 = 2\rho_0/(\mu_0 d)$ is the characteristic velocity. After the current is switched on, the field can be presented as the superposition of the fields of two conductors. One of them is a wire with a given current, and the second one is a "virtual" wire with an opposed current, which is moved downward with the velocity u_0 from the initial position at the coordinate $y(0) = -h$. In this problem the characteristic length is given by the equality $g = h$. In the other problem dealing with a field of the ring-turn of radius R placed at a distance h from the plane of the sheet, the total current induced in the sheet is given by the formula

$$i = -i_0\left[1 - \frac{h + u_0 t}{\sqrt{(h + u_0 t)^2 + R^2}}\right]. \tag{8.23f}$$

Here the current in the sheet is equal to $i_0/2$ during a time $t_{1/2} = (3^{-1/2}R - h)/u_0$, which can be considered to be an estimate of the time τ_2.

Attraction to the inductor is also possible in the case of nonferromagnetic flat sheets. This technology is used for the external straightening of articles made from thin sheets and is described by Furth and Waniek [45] and Turenko, Batygin, and Gnatov [63]. The formation of the force which attracts the thin nonferromagnetic sheet to the inductor can be considered by the simplest example, shown in Figure 8.27. In this figure the two conductors with the equally-directed currents i_0 and i_1 are placed at distances h and g over the thin conducting sheet. When the current i_0 is switched off in a time much less than the characteristic value τ_2, the current i_2 is induced in the sheet, which is of the same sign as the current i_1. Interaction of the two currents i_1 and $i_2$of the same direction produces the attracting force.

In this example, switching off the current i_0 is equivalent to switching on the current equal $(-i_0)$. Further on the process is described by the mentioned equations. The superposition of the fields generated by the current $(-i_0)$ and by the initial current i_0 is equivalent to a field of two "virtual" wires with currents i_0. One of them, from the initial position with coordinate $x = h$, displaces upwards with the velocity u_0, and the other, from the position $x(0) = -h$, moves with the same velocity downwards. The first of these conductors is responsible for the field

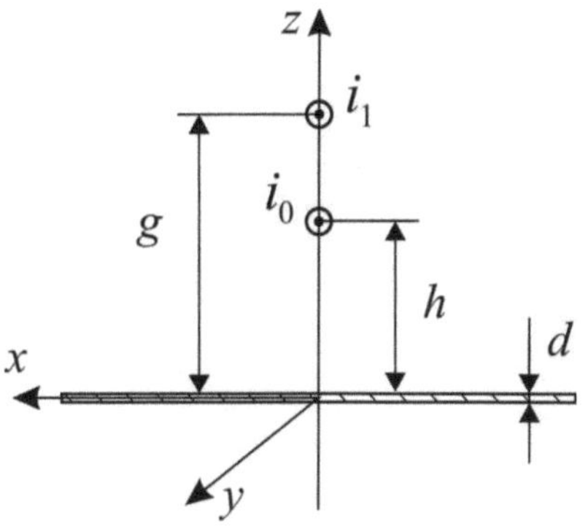

Fig. 8.27: Formation of the force attracting a thin sheet to the inductor.

under the sheet, the second over the sheet. The calculations lead to the following expression for the linear density of the induced current and for the "magnetic pressure":

$$j_x = \frac{i_0 (h + u_0 t)}{\pi \left[(h + u_0 t)^2 + y^2\right]}, \tag{8.24a}$$

$$P_M = j_x B_y = \frac{\mu_0 \left|i_0 i_1\right| g (h + u_0 t)}{2\pi^2 \left[(h + u_0 t)^2 + y^2\right] (g^2 + y^2)} \tag{8.24b}$$

The inductors used in the processing technology of flat sheets are described in [63].

In the particular case when the equalities $g = h$, $i_0 = \Delta i$, $i_1 = i_0 - \Delta i$, take place, the system under consideration presents a single conductor, in which the current endures a jump from the initial value i_0 to $i_0 - \Delta i$. The induced current can be determined by formula (8.24a), which includes a change in the current Δi:

$$j_x = \frac{\Delta i (h + u_0 t)}{\pi \left[(h + u_0 t)^2 + y^2\right]}. \tag{8.25a}$$

When calculating the magnetic pressure, one should use the condition $g = h$ and replace the current i_1 by its value after a jump $i_0 - \Delta i$:

$$P_M = j_x B_z = \frac{\mu_0 \left|\Delta i (i_0 - \Delta i)\right| h (h + u_0 t)}{2\pi^2 \left[(h + u_0 t)^2 + y^2\right] (h^2 + y^2)} \tag{8.25b}$$

The pressure now becomes proportional to the product $(i_0 - \Delta i)\Delta i$. It is maximal at the condition $\Delta i = i_0/2$,, i.e., as the current decreases to half the initial value.

The compression of metal tubes by strong magnetic fields results in high pressures when the tubes collapse or undergo impact with an immobile body. The experiments with liners being accelerated up to high velocity give us the possibility of realizing under the laboratory conditions the technological processes

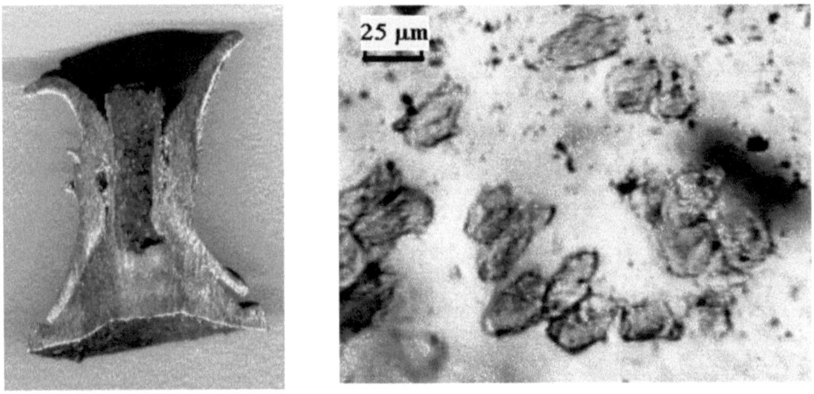

(a) (b)

Fig. 8.28: The experiments on the synthesis of diamonds: (a) liner after compression; (b) Diamond crystals [64].

which require sufficiently high pressures. Experiments by Fridman, Makarevich, Rakhel, and Rumyantsev on the synthesis of diamonds are an example of this [64]. In these experiments a tube filled with graphite was the subject of compression due to a metal cylinder accelerated in the configuration of a Z-pinch. The velocity of the internal boundary reached 430 m/s, and the calculated value of the pressure at the instant of impact was 29 GPa. The results of experiments presented in Figure 8.28 are indicative of an appearance of diamond crystals after the compression of graphite. In the experiments described in [65], the accelerated metal liner was used for compaction of thin-wall tubes from nano-sized ceramic powders.

The pressure produced by the pulse magnetic field can be used for the formation of acoustic waves and for studies of the mechanical properties of materials exerted to short-time loads. The specific property of loads produced by electromagnetic forces is the possibility of controlling the form and duration of the pressure pulse. This allows us to investige deformation and destruction under controlled conditions.

In Section 8.2 we touched on the relationship between the yield point and the time of action of the pulse pressure. The cited experiments confirm that the threshold of strength increases when the duration of the action decreases, which is in accordance with the theory developed by Regel, Slutsker, and Tomashevsky [7]. Studies of the plastic deformation produced by the compression of thin-wall cylinders of the theta-pincy geometry have shown that the yield point of aluminum at pulses of $10^{-5} - 2 \cdot 10^{-4}$ s is 2.5 times greater than the static yield point and in this interval of times changes but slightly [66]. The compression of cylinders in the longitudinal field is accompanied by a loss of stability, resulting in the formation of ripples. Studies of this process gave the data on the development of instabilities in the nonlinear stage characterizing by a change of the mode composition of ripples [66, 67].

The authors of [68–70] studied the development of plastic deformation in copper cylinders subject to compression in the θ- and Z-pinch configurations under conditions of the unchanged axial symmetry. They obtained the data needed for a numerical simulation of plastic deformation under the conditions when the velocity of the relative deformation attained the values of the order of 10^5–10^6 1/s.

The expansion cylinder created in the configuration shown in Figures 8.22a and 8.24a is not accompanied by the development of instabilities. This gives us the possibility of studying plastic deformation under more simple conditions with conservation of the symmetry of the sample. In these experiments the simultaneous measurement of the induction and radius makes it possible to construct a model describing the behavior of a material at short-term intensive loads.

The authors of [71] have shown that the rheological Maxwell model can be used to describe deformation. According to this, in the thin-wall cylinder the azimuth stress is related to the relative deformation in the following way:

$$\sigma + \tau \frac{d\sigma}{dt} = E\tau \frac{d\varepsilon}{dt}, \tag{8.26a}$$

where τ is the relaxation time. The motion of the body is given by

$$h\gamma_0 \frac{d^2R}{dt^2} = \frac{B_i^2}{2\mu_0} - \sigma \frac{h}{R}, \tag{8.26b}$$

where h is the thickness of the wall, γ_0 is the density of the medium, R is the radius, and B is the experimentally measured induction of the field which was generated by a magnet placed inside a cylinder. Optical measurements made it possible to register the change in the radius, which is related to the relative deformation through formula $\varepsilon = \ln(R/R_0)$, where R_0 is the initial radius. The measurements of the induction give the starting data for the calculations of the motion with the particularly chosen dependence of the relaxation time τ.

Studies of expansion of a thick-wall cylinder under the controlled conditions of a Z-pinch configuration were carried out in [72]. The results of these studies enhanced the possibility of constructing the justified models of intensive plastic deformation in addition to the data obtained in experiments on compression [69, 70].

Special-purpose magnetic systems were developed to investigate the formation and behavior of cracks in the fragile destruction of solids. In these experiments the possibilities presented by controlled action were used. The layout of the experiment is shown in Figure 8.29. The discharge of the capacity bank produces a pulse current in the flat conductors placed inside the slit. The magnetic pressure acting on the edges of the slit is $P_m = \alpha\, B_1^2/(2\mu_0)$, where B_1 is the induction in the gap h between the conductors. Under the condition $h \ll b$ (b is the width of the conductors) the field is practically uniform in the gap: $B_1 = \mu_0\, i/b$. To take into account the additional pressure due to the heating of the conductor by the pulse current, the coefficient $\alpha \approx 1.1$ was introduced by authors of [30].

In the experiments described in [73] some examples of the use of the described methods for the investigation of the destruction processes for solids are given. The process of the formation and growth of cracks in the samples of various dielectric materials was registered experimentally.

Figure 8.29 shows the length of the crack as a function of the amplitude of the pressure for the different pulse times. The application of the magnetic-pulse method for the formation of controlled pulses of pressure elucidated a number of general regularities in the process of destruction under a short-time action. The presented data confirm that this process has a threshold character. The results of the experiments in Figure 8.28 show that the amplitude of the pressure of the destructing pulse increases with a decrease of its duration according to the dependence of the form

$$\sigma_d = \sigma_c \cdot \left({}^{T_p}\!/_{\tau_C}\right)^{\alpha}, \tag{8.27}$$

where σ_d is the threshold destructing load, σ_c is the durability of a material in static, T_p is the pulse time, $\alpha \approx \sqrt{2}$, and τ_c is the characteristic parameter, the so-called “structural time of destructing” [75]. Owing to the described method, other parameters featuring the theory of crack development were established: the characteristic dimension of the destruction, the surface, and specific energies.

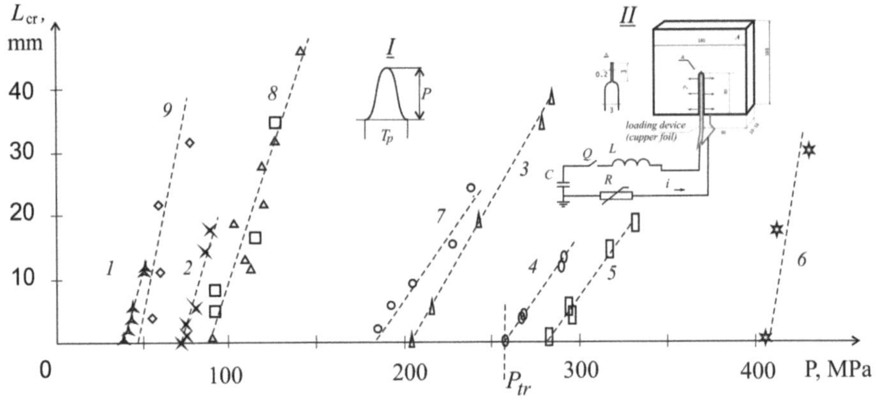

Fig. 8.29: The length of cracks as a function of the magnetic pressure amplitude.
(1) limestone, $T_p = 4.4$ µs; (2) marble, $T_p = 3.6$ µs; (3) gabbro, $T_p = 3.6$ µs; (4) sandstone, $T_p = 3.6$ µs; (5) granite, $T_p = 3.6$ µs; (6) polymer composite $T_p = 1.5$ µs; [74]; PMMA: (7) $T_p = 1$ µs; (8) $T_p = 2$ µs; (9) $T_p = 4.3$ µs; [73]. Pulse shape I and loading scheme II are shown.

The possibility of forming the pressure pulse with given time variations increase when the additional field B_2 parallel to the field B_1, produced by an external source, is used. In this case the pressure on the edges of the slit is described by the formula

$$P_m = \frac{B_1}{\mu_0}\left(\frac{B_1}{2} + B_2\right). \tag{8.28}$$

The details of the experiments are described in [76]. In particular, the instantaneous induction B_1 is proportional to the pressure provided if $|B_1| \ll |B_2|$ and the field B_2 changes slightly.

Configurations of the magnetic system similar to those considered above are used to study the isentropic compression of solids. The magnetic field is produced in the gap between the plates. Due to the impact, a isentropic flow takes place. Optical methods are used to detect the velocity of the external boundary of regions with different thickness. From this data one can find the characteristic parameters of the high-stress equation of the material state. In order to obtain a strong pulsed magnetic field, the authors of [30, 77] (and others as well) produced a current by means of a discharge of a capacitive storage, while Tasker, Goforth, and Oona [78] used the magnetic explosion generator to produce a current.

8.8 Magnetic systems for the acceleration of conductors

Electromagnetic acceleration of conductors is a rapidly advancing field in electrophysics. Along with the systems used in technology, devices for the acceleration of conductors by electromagnetic forces, depending on the type of current delivery, can be of the conduction or induction type. In the first of these, current is delivered to the accelerated body through mobile contacts. The body is sliding along the contact buses, therefore such a device is referred to as a railgun (Figure 4.4). Numerous publications and reports presented at the specialized conferences have been devoted to this topic. According to the content of the reviews, the technique of acceleration of bodies with a mass of several kg up to 2–3 km/s has been worked out. In practice acceleration is complicated by the rupture of the body due to Joule heating. The current density sharply increases near the angular points where the current is concentrated on the narrow region with a characteristic dimension $\rho/(\mu_0 u)$, where u is the velocity of the body. The disruption of contacts is the chief cause preventing the attainment of the velocity calculated according to the ideal model. Due to electric explosion of the contact domain, plasma, shunting the accelerated body, is formed. This effect also takes place in the case of the acceleration of small dielectric bodies when current flows in the layer of the conductor, positioned on the rear side of the sample. The consideration of the arising problems and of the ways of their solution is far beyond the scope of this book.

Bondaletov and coworkers used an accelerator in which a body moves in the strong transverse field of the single-turn magnet with induction 60–70 T [79]. The force accelerating the body is proportional to the product of the field induction and the current in the conductor. Therefore it is possible, enhancing the field, to deteriorate the current in the conductor up to values at which the heating of the body is relatively low. The authors of [79] used a single-turn solenoid in the form of a slot disrupting after each discharge. The similar accelerating system allows the speed-up of the body up to the velocity of several kilometers per second.

In practice the induction acceleration of bodies in a pulsed magnetic field is applied. For this purpose the same magnetic systems can be used as for the deformation of conductors. Configurations like those given in Section 8.6 find applications in devices for the induction acceleration of solids as well, as serving the drivers of the different instrumentation: the switching off equipment, the systems of fast gas puffing, and others. For example, in [80] the magnetic field of the solenoid is used for the induction acceleration of the body, which is a press for the compression of powder materials.

Earlier (in Chapter 4) we considered the conditions of an optimal energy transfer from the storage to the accelerated body. These conditions may be useful in the problems on the conductive and inductive accelerations of conductors, when the equivalent circuit with an alternating inductance can be used.

For several applications induction acceleration of small bodies is used. For example, this method is applied for the separation of scrap metal, and for the recovery of nonferrous metals in the process of ore enrichment. Here the estimation of forces acting on an accelerated body is an urgent problem. Further, we give these estimations under the assumption that the accelerating process occurs in pulsed magnetic fields under the conditions of a sharply pronounced skin effect.

Accelerated metallic pieces can vary in shape. The estimations are appropriate for three characteristic groups of bodies, differing by the relationship of their sizes.

Quasi-spherical bodies. Their three sizes are close (for example, differ no more than twice). Among these are spheres, spheroids, Rankin ovoids [18], parallelepipeds, etc.

Rods. These are bodies, the longitudinal size of which (length) are much larger than two other (cross) sizes.

Plates. Their two sizes (length and width) are much larger than the third size (thickness).

It is known that the electromagnetic force appears at the interaction of induced currents with the external field $\mathbf{B}'$, while the interaction of these currents with their own field $\mathbf{B}''$ gives a resultant force which is equal to zero [81]. With a sharply pronounced skin effect the expression for the total force has the form

$$\mathbf{F} = \int_S \left[\mathbf{j}(s), \mathbf{B}'(s)\right] ds, \tag{8.29}$$

where the vector product of the external field induction and current surface density is presented, and the range of integration is the body surface S. In the points s of this surface the equality $|\mathbf{j}(s)| = |\mathbf{B}_\tau|$ is valid, where $\mathbf{B}_\tau$ is the induction tangent component of the resulting field $\mathbf{B} = \mathbf{B}' + \mathbf{B}''$, which in the approximation of the ideal conductivity satisfies the condition $B_n = 0$ on the surface of the body. A few examples of the calculation of forces for the axial-symmetrical bodies are given in [18, 82]. Further on we will consider examples relating to the case where the characteristic size of a body satisfies the $d \ll |B'|/|\nabla B'|$ (quasi-uniform) field. In all theexamples the force acting on a body can be presented in the form

$$\mathbf{F} = -GV \cdot \nabla\left(\frac{B^2}{2\mu_0}\right), \tag{8.30}$$

where G is the dimensionless number (geometric factor), and V the volume of a body. In [83] one can find the geometric factors of small bodies for two cases: (1) in the vicinity of a body the angle between the vectors $\mathbf{B}'$ and $\nabla B'$ equals zero or π, and (2) these vectors are perpendicular. For an ideally conducting sphere in both cases the geometric factor $G = 3/2$. For other quasi-spherical bodies the values of this factor are of the order of unity. Tables 8.5 and 8.6 contains the geometric factors for plates and rods, differently oriented relative to the induction vector ant its gradient. In the above cases the force is parallel with the axis z. It is determined by integrating the vector product $[\mathbf{j}, \mathbf{B}']_z$ over the body boundary.

In calculations of a plane field it is assumed that the induction has only two components, B'_z andB'_y, independent on the coordinate x, and the symmetry relatively to the plane $y = 0$ takes place. This calculation was performed with the help of the approximate formula for B', which can be obtained using the expressions forB'_y, valid in the vicinity of the point $y = 0$, which coincides with the center of gravity of a body. At conditions $\mathbf{B}' \uparrow\uparrow \nabla|B'|$ and $\mathbf{B}' \uparrow\downarrow \nabla|B'|$ we have the equality $B'_y(0) = 0$. Further, we can use the equation $\mathrm{div}\mathbf{B}' = \partial B'_z/\partial z + \partial B'_y/\partial y = 0$. This leads to

$$B'_y \approx -y\frac{\partial B'_z}{\partial z}. \tag{8.31a}$$

With another mutual orientation of vectors $\mathbf{B}'$ and $\nabla B'$ the expansion for B'_y takes place:

$$B'_y(z) \approx B'_y(0) + \frac{\partial B'_y}{\partial z} z, \tag{8.31b}$$

$$B'_y \approx -y \frac{\partial B'_z}{\partial z}. \tag{8.31c}$$

These formulas enable the calculation of the force acting on a thin flat body, oriented in parallel with a magnetic line (Table 8.5, Nr. 1, 3; the reference point is in the middle plane of plates). Such a body weakly disturbs the magnetic field, and therefore on the body surface the induction of the resultant field $B_\tau \approx B'_\tau$. In this case the current linear density for these configurations is numerically equal to B'_τ/μ_0, and the current lines are orthogonal to the vector B'_τ and aligned parallel to the axis x. In configuration 1 the B'_y takes values $B'_y = \mp(h/2)(\partial B'/\partial z)$ on the upper and lower surfaces, and the force is determined by the formula

$$F_z = -h \frac{\partial B'}{\partial z} \cdot \frac{B'}{\mu_0} S = -\frac{\partial}{\partial z}\left(\frac{B'^2}{2\mu_0}\right) V, \tag{8.32}$$

where S is the square;, and V is the volume of the plate. In this case $G = 1$.

It is easy to show that this coefficient has the same value in configuration 3 of Table 8.5. This geometric factor also remains in the case where a width of the plate (the size b, perpendicular to the axis z) is much less than its length l. Differently, the samples in the form of rods, extended along a line of force, are also characterized by the factor $G \approx 1$, regardless of the mutual orientation of the vectors $\mathbf{B}'$ and $\nabla B'$ (Table 8.6).

In the special case the plane of the plate is oriented perpendicular to the vector $\mathbf{B}'$ (Table 8.5, configuration 2, $[\mathbf{B}, \mathrm{grad}|B|] = 0$). We will restrict ourselves to the example of a round disk with a width h. When it is installed perpendicularly to the lines of a quasi-uniform field with induction $\mathbf{B}'$, the current linear density on both surfaces of the disk has only the azimuth component j_φ, which is distributed according the law [18]

$$j_\varphi = \frac{2B'_r r}{\pi\mu_0\sqrt{R^2 - r^2}}, \tag{8.33a}$$

where R is a radius of the disk. For this system the force F_z is determined by the formula

$$F_z = 2\int_0^R B'_r j_\varphi 2\pi\, r dr, \tag{8.33b}$$

where the value of B'_r can be found from the condition

$$\mathrm{div}\mathbf{B}' = (1/r)(\partial(rB'_r)/\partial r) + \partial B'_z/\partial z = 0,$$

$$B'_r = -(r/2)\,\partial B'_z/\partial z. \tag{8.33c}$$

Table 8.5: The geometric factor of a thin circular plate in the magnetic field.

№	Orientation of vectors **B** and grad|*B*|	Position of a plate	*G*
1	[**B**, grad |*B*|] = **0** **B**∥**z** Plane field	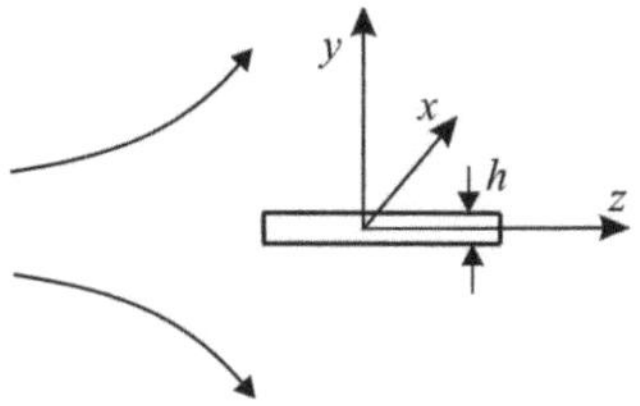	1
2	[**B**, grad |*B*|] = **0** **B**∥**z** Axially symmetric field	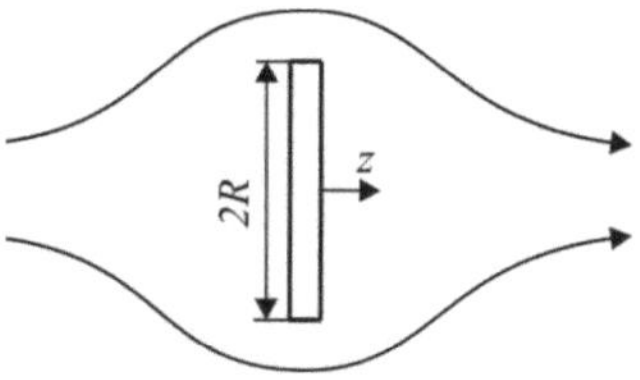	$\frac{8}{3\cdot\pi}\frac{R}{h}$
3	**B**⊥grad |*B*| **B**⊥**z** Plane field	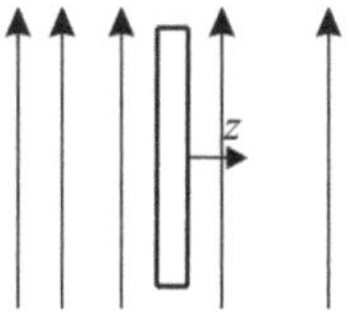	1
4	**B**⊥grad|*B*| **B**⊥**z** Weakly disturbed axially symmetric field	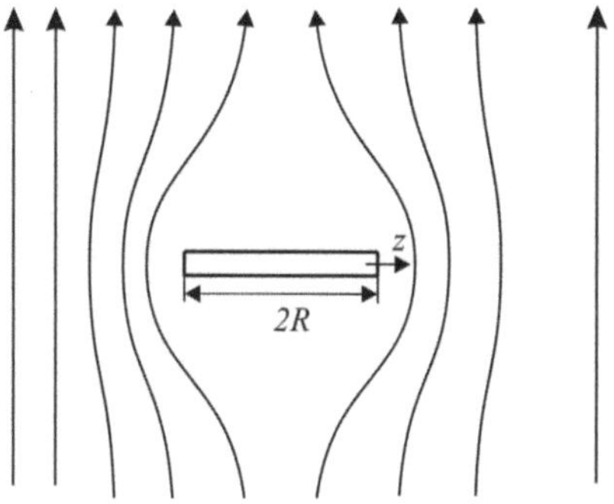	$\frac{8}{3\cdot\pi}\frac{R}{h}$

Using (8.33a–c), we find the value of the factor:

$$G = 8R/(3\,\pi\,h) \tag{8.34}$$

One can show that the factor G also takes the same value for configuration 4 in Table 8.5 when the condition $\mathbf{B}'\perp\nabla|B'|$ is fulfilled. Calculations made in a similar way show that the geometric factor G is also a number of the order of unity a in the

Table 8.6: Geometric factor for the cylindrical conductor in the magnetic field.

№	Orientation of vectors B and grad$\|B\|$	Disposition of the conductor	G
1	$[\mathbf{B}, \text{grad}\,\|B\|] = \mathbf{0}$ $\mathbf{B}\|\mathbf{z}$ Axial-symmetric field	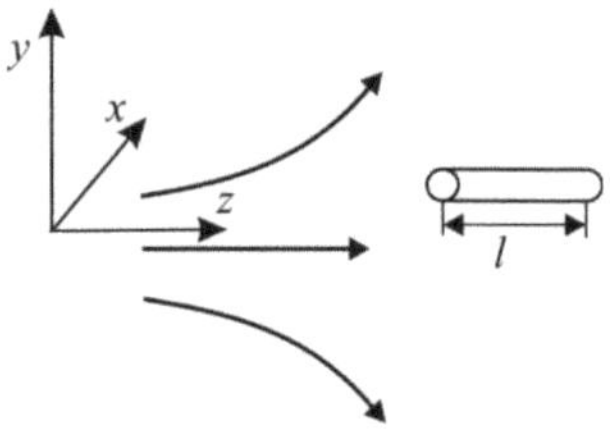	**1**
2	$[\mathbf{B}, \text{grad}\,\|B\|] = \mathbf{0}$ $\mathbf{B}\|\mathbf{z}$ Flat field	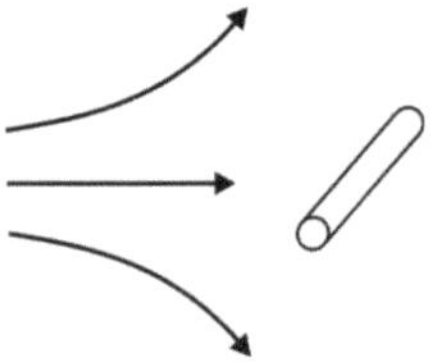	**2**
3	$[\mathbf{B}, \text{grad}\,\|B\|] = \mathbf{0}$ $\mathbf{B}\|\mathbf{z}$ Flat field	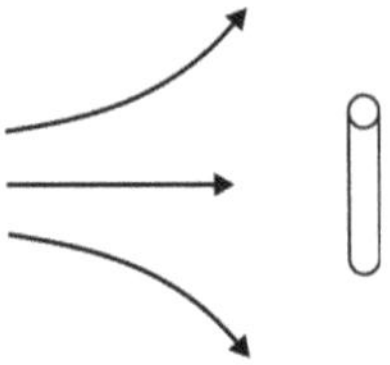	$\frac{16}{3\pi}$
4	$\mathbf{B}\perp\text{grad}\,\|B\|$ $\mathbf{B}\perp\mathbf{z}$ Flat field	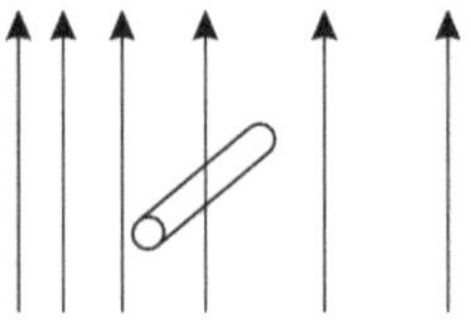	$\frac{16}{3\pi}$
5	$\mathbf{B}\perp\text{grad}\,\|B\|$ $\mathbf{B}\perp\mathbf{z}$ Flat field	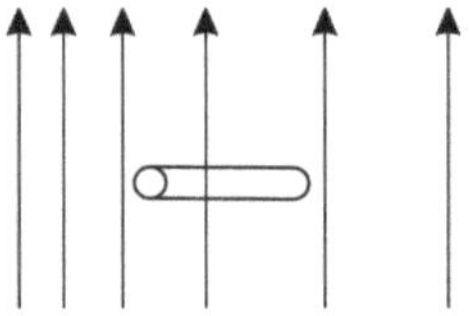	**1**
6	$\mathbf{B}\perp\text{grad}\,\|B\|$ $\mathbf{B}\perp\mathbf{z}$ Flat field	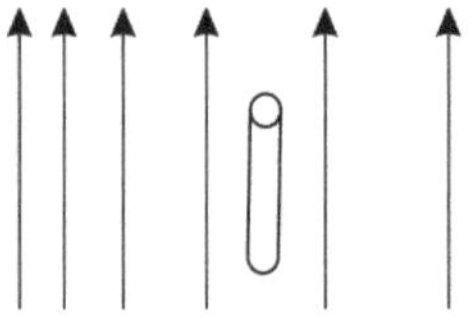	**1**

case where the samples in the form of metallic rods are oriented perpendicular to the magnetic lines of the field $\mathbf{B}'$ (Table 8.6).

If the magnetic system produces the force directed under the angle α to the horizontal plane, then after a short-time impact a body travels along the ballistic trajectory the distance

$$D = \left(G\frac{B_m}{\mu_0} \cdot \frac{\partial B_m}{\partial z} \cdot \frac{\tau_{eff}}{\gamma} \right)^2 \frac{\sin 2\alpha}{g}, \tag{8.35}$$

where $g = 9.8\ \text{m}^2/\text{s}$ is the acceleration of gravity; B_m is the induction amplitude, and $\tau_{eff} \approx (\int_0^\infty B(t)dt/B_m^2)$ is the pulse effective duration. This distance is inversely proportional to the square of the body density. Such a device can be used for the-separation of small conducting bodies by density [83, 84].

Let us examine the requirements for magnetic systems which provide the conditions for attracting nonferromagnetic bodies. We restrict ourselves to the case of small bodies, the size of which is much less than the gradient length $|B|/[\text{grad}B]$. Assume also for simplicity that a body is placed on the axis of the axial symmetric magnetic system. The problem of the attraction of small nonferromagnetic bodies to the inductor can be solved through the formation of a field which is enhanced with a distance from the magnetic system.

One example of this is the system shown in Figure 8.30. Here the field is produced by two flat coaxial contours with radii R_1 and R_2. For estimating we can consider the case where the condition $w_1 i_1 = -w_2 i_2 (R_1/R_2)$ is fulfilled. Then, the induction in the point $z = 0$ turns into zero, and the induction along the axial line varies as

$$B(z) = \frac{\mu_0 i_2 w_2}{2} \left(\frac{R_2^2}{(R_2^2 + z^2)^{3/2}} - \frac{R_1^3}{R_2 (R_1^2 + z^2)^{3/2}} \right). \tag{8.36}$$

The induction maximum is in the point with coordinate $z_m = R_2 (\frac{p^{6/5} - p^2}{1 - p^{6/5}})^{1/2}$, where $p = R_1/R_2$. At the condition $p = 0.5$ we have $z_m = 0.573\, R_2$. Figure 8.30 shows the region in which a force attracting a small body to the conductor can be produced.

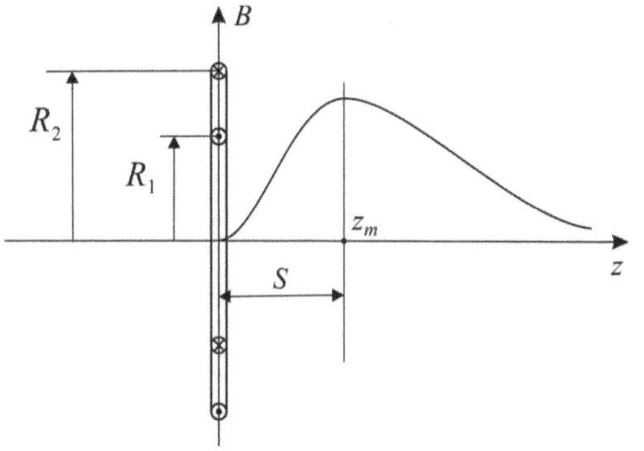

Fig. 8.30: The magnetic system in which an attractive force of a small nonferromagnetic body to the inductor is produced; S is the length of the attraction region.

Another version of a magnetic system with an attraction of a small nonferromagnetic body is shown in Figure 8.31. Here the magnet is placed inside the closed cylinder with high conductivity. In order to generate a sufficiently strong field in the zone behind the accelerated body, the current in the magnet should be rather strong, since the field produced by it is partially compensated by the field of currents, induced on the screen. The authors of [85] considered this magnetic system as a prototype of the device for the recovery of the conducting nonferromagnetic bodies from the eye apple with minor traumas compared to the common surgery.

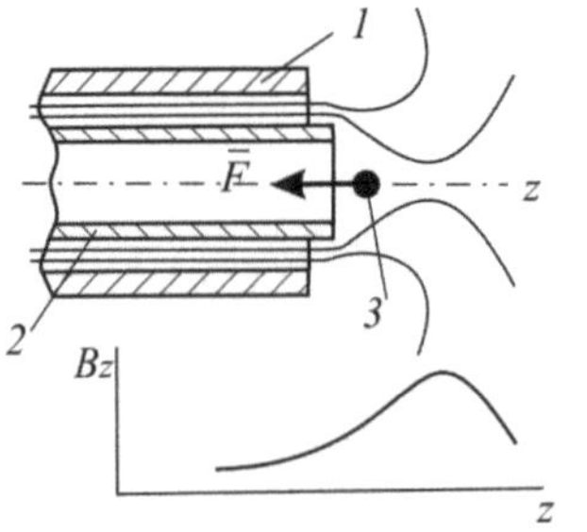

Fig. 8.31: Prototype of the device for attraction of a nonferromagnetic conducting body to the inductor: (1) solenoid; (2) closed screen; (3) accelerated body.

8.9 References

[1] H. P. Furth and Waniek R. W., *Rev. Scient. Instrum.* **27** (1956), 195.
[2] H. P. Furth, M. A. Levine, and R. W. Waniek, *Rev. Sci. Instr.* **28**(11) (1957), 949–958.
[3] N. T. Olson, J. Bandas, and A. C. Kolb, *J. Appl. Phys.* **50**(11) (1978), 7768–7770.
[4] A. M. Andrianov, V. F. Demichev, G. A. Eliseev, and P. A. Levit, Stability of Single-Turn Coils in Pulsed Magnetic Fields up to 1 MGs, preprint of Kurchatov Institute 2025, Moscow, 1970 (in Russian).
[5] O. Portugall, N. Puhlmann, Y. U. Müller, S, I. Barczewski,, and M. von Ortenberg, *J. Phys. D. Appl. Phys.* **32** (1999), 2354–2366.
[6] S. Takeyama, **MG-XIII** (2010), pp. 163–169.
[7] V. R. Regel, A. I. Slutsker, and O. E. Tomashevskiy, *Kinetichyeskaya priroda procynosti tverdih tel*, Moscow, Nauka, 1974 (in Russian).
[8] E. S. Borovik, M. S. Mamedov, and V. G. Volotskaya, *Fisika metallov I metallovedenie* **19**(3) (1965), 451–455 (in Russian).
[9] G. A. Shneerson, *J. Tech. Phys.* **32** (1962), 1153–1156 (in Russian).
[10] P. N. Dashuk, S. L. Zayents, V. S. Komelkov, G. S.,N. N. Kuchinskiy Nikolaevskaya, P. I. Shkuropat, and G. A. Shneerson, in: V. S. Komelkov (ed.), *Tekchnika bolshih impulsnih tokov i magnitnih poley*, Moscow, Atomizdat, 1970.
[11] V. P. Gordienko and G. A. Shneerson, *J. Tech. Phys.* **35**(6) (1965), 1084–1090 (in Russian).
[12] V. A. Churaev, *Fizika Tverdogo Tela (Solid State Physics)* **33**(2) (1991), 474–478 (in Russian).
[13] H. Knoepfel, *Pulsed High Magnetic Fields*, Amsterdam London, North-Holland Publishing Company, 1970.
[14] I. M. Karpova and V. V. Titkov, *J. Tech. Phys.* **65**(6) (1995), 54–63 (in Russian).

[15] S. S. Manson, *Thermal Stress and Low-Cycle Featigue*, New York San Francisco Toronto London Sydney, McGraw-Hill Book Co., 1966.
[16] Y. Livshiz, A. Izhar, and O. Gafri, **MG-XI** (2006), pp. 242–245.
[17] D. F. Rankin, B. M. Novak, and I. R. Smith, *IIEE Proc.-Sci. Meas. Technol.* **3** (153), 130–138.
[18] G. A. Shneerson, *Fields and Transients in Superhigh Pulse Current Device*, New York, Nova Science Publishers, 1997.
[19] A. Farinski, L. Karpinski, and A. Nowak, *J. Techn. Phys.* **20**(2) (1979), 265–280.
[20] Y.E. Adamyan, V. V. Titkov, and G. A. Shneerson, *Izvestiya Academii Nauk SSSR, Energetika i Transport*, **5** (1984), 104–107 (in Russian).
[21] I. M. Karpova, V. V. Titkov, and G. A. Shneerson, *Izvestiya Academii Nauk SSSR, Energetika i Transport* **3** (1988), 122–127 (in Russian).
[22] N. N. Prohorov, *Fizicheskiye Processi v Metallah pri swarke*, Vol. 1, Moscow, Metallurgiya, 1968 (in Russian).
[23] B. E. Fridman and F. G. Rutberg, *Pribory i Technika Experimenta* **2** (2001), 70–78 (in Russian).
[24] T. M. Goto, N. Takeyama, and S. Sakakibara, **MG-IV** (1987), pp. 149–158.
[25] W. L. Baker, J. H. Degnan, and R. E. Reinovsky, **MG-III** (1984), pp. 39–49.
[26] R. E. Reynovsky, I. R. Lindemuth, W. L. Atchison, J. C. B. Cochrane, and R. J. Faehl, **MG-IX** (2004), pp. 399–405.
[27] Y .E. Adamian, A. N. Beryoskin, S. G. Bodrov, Y. N. Bocharov, et al., **MG-VI** (1992), pp. 25–34.
[28] A. M. Andrianov, V. F. Demichev, G. A. Eliseev, and P. A. Levit, *Pribory i Technika Experimenta* **1** (1971), 112–114 (in Russian).
[29] Y. N. Botcharov, S. L. Zayentz, P. G. Popov, E. L. Litvinova, A. I. Kruchinin, G. S. Kuchinskiy, and G. A. Shneerson, *Pribory i Tehnika Experimenta* **1** (1981), 167–169 (in Russian).
[30] h. Mangeant, F. Lassale, J. Petit, and M. Bavay, **MG-IX** (2004), pp. 445–449.
[31] V. B. Gaase and G. A. Shneerson, *Pribory i Technika Experimenta* **6** (1965), 105–110 (in Russian).
[32] G. I. Babat, *Indukcionniy nagrev metallov i ego primeneniye*, Moscow, Gosenegoizdat, 1946 (in Russian).
[33] Y. B. Kim and E. D. Platner, *Rev. Sci. Istr.* **30**(7) (1959), 524–533.
[34] M. N. Wilson and K. D. Srivastava, *Rev. Sci. Istr.* **36**(8) (1956), 1096–1100.
[35] V. R. Karasik, *Fisika i Tehnika silnich magnitnih poley*, Moscow, Nauka, 1964 (in Russian).
[36] H. Ditz, H.-J. Lippman, and H. Schenk, *ETZ-A* **88** (1967), 475–480.
[37] I. V. Belyy, S. M. Fertik, and L. T. Khimenko„ *Electromagnetic Metal Forming Handbook*, translation from *Spravochnik po Magnitnoimpul'snoy Obrabotke Metallov* (in Russian), translated by M. M. Altynova, Material Science and Engineering Dept., Ohio State University, 1996.
[38] V. P. Knyazyev and G. A. Shneerson, *Izvestiya Visshih Uchebnih Zavedeniy, Energetika* **4** (1971), 34–39 (in Russian).
[39] M. M. Karpenko, B. P. Hrustalev, and M. Y. Kanevskiy, *Tech. Phys. Letters* **1**(2) (1975), 78–81 (in Russian).
[40] Y. N Botcharov, S. I. Krivosheev, N. G. Lapin, and G. A. Shneerson, *Pribory i Technika Experimenta* **2** (1993), 92–95.
[41] E. Quercigh and A. J. Hertz, *J. Nucl. Instrum. Methods* **20** (1963), 494
[42] N. T. Olsen, A. C. Kolb, and N. R. Pereira, **MG-III** (1984), pp. 108–114.
[43] V. P. Gordienko and G. A. Shneerson, *J. Tech. Phys.* **34**(2) (1964), 376–378 (in Russian).

[44] D. H. Birdsall, F. C. Ford, H. P. Furth, and R. E. Riley, *American Machinist. Metalworking Production* **105**(20) (1961), 117–121.
[45] H. P. Furth and R. W. Waniek, *American Machinist. Metalworking Production* **106**(18) (1962), 50–53.
[46] R. Winkler„ *Hochgeschwindigkeitsbearbeitung – Grundlagen und technische Anwendung elektrisch erzeugter Schockwellen und Impulsmagnetfelder*, pp. 307–333, Berlin, VEB-Verlag Technik, 1973.
[47] P. Y. Emelin, F. G. Rutberg, and B. E. Fridman, *Pribori i Technika Experimenta* **5** (1993), 109–115 (in Russian).
[48] G. Weimar, *Werkstatt and Betrieb* **96**(12) (1963), 893–900.
[49] E. J. Bruno (ed.), *High Velocity Forming of Metals*, ASTME, 1968.
[50] G. Zittel, in: *Proceedings of 4th International Conference on High Speed Forming (ICHS)*, Ohio, USA, 2010, pp. 2–15, Dearborn MI, ASTIME, 1968.
[51] V. F. Demichev, *Atomnaya Energiya* **73**(4) (1992), 278–284 (in Russian).
[52] M. G. Vitkov, *J. Tech. Phys.* **35**(3) (1965), 410–413 (in Russian).
[53] B. NovgorodcevA and G. A. Shneerson, *Trudi Leningradskogo Polytechnicheskogo institute* **273** (1966), 139–151 (in Russian).
[54] G. A. Shneerson, *J. Tech. Phys.* **35**(12) (1965), 2234–2239 (in Russian).
[55] 8.55. V. Shribman, A. Stern, Y. Livshitz, and O. Gafri, *Welding Journal* **81** (2002), 33–37.
[56] K. Faes, T Baaten, W. De Waele„ and N. Debroux, in: *Proceedings of 4th International Conference on High Speed Forming (ICHS)*, Ohio, USA, 2010, pp. 84–96.
[57] Y. Zhang, S. Babu, and G. Daehn, in: *Proceedings of 4th International Conference on High Speed Forming (ICHS)*, Ohio, USA, 2010, pp. 97–107.
[58] E. Uhlmann and A. Ziefle, in: *Proceedings of 4th International Conference on High Speed Forming (ICHS)*, Ohio, USA, 2010, pp.108–116.
[59] A. Elsen, M. Ludwig, R. Schaefer, and P. Groche, in: *Proceedings of 4th International Conference on High Speed Forming (ICHS)*, Ohio, USA, 2010, pp.117–125
[60] S. Golovashchenko, N. Bessonov, and R. Davies, in: *2nd International Conference on High Speed Forming*, 2006, pp. 141–151.
[61] Y. Y. Livshitz, privat communication.
[62] W. R. Smythe, *Static and Dynamic Electricity*, 2nd edn., New York, McGraw-Hill, 1950.
[63] A. N. Turenko, Y. V. Batygin, and A. V. Gnatov, *Teoriya i 'eksperiment magnitno-impul'snogo prityazheniya tonkostennyh metallov*, Har'kov: Izd. HNADU, 2009 (in Russian).
[64] B. E. Fridman, I. P. Makarevich, A. D. Rakhel, and B. V. Rumyantsev, **MG-IX** (2004), pp. 438–444.
[65] S. Paranin, V. Ivanov, S. Dobrov, A. Nikonov, and V. Khrusov, **MG-IX** (2004), pp. 132–136.
[66] V. G. Belan, S. T. Durmanov, I. A. Ivanov, V. F. Ltvashov, and V. L. Podkovirov, **MG-III** (1984), 218–220.
[67] V. T. Mikchelsoo and G. A. Shneerson, *J. Tech. Phys.* **40**(10) (1970), 2198–2208 (in Russian).
[68] J. H. Degnan, W. L. Baker, M. L. Alme, C. Boyer, J. S. Burf, J. D. Beason, et al., *Fusion Technol.* **27** (1995), 115–123.
[69] J. Petit, Y. A. Alexeev, S. P.Ananyev, and M. N. Kaseev, **MG-VII** (1997), pp. 569–576.
[70] Y. A. Alexeev, M. N. Kazeev, J. Petit, A. Hagland, **MG-IX** (2004), pp. 119–124.
[71] V. P. Knyasyev and G. A. Shneerson, *J. Tech. Phys.* **40**(2) (1970), 360–371 (in Russian).
[72] J. Petit, Y. A. Alexeev, S. P. Ananiev, M. N. Kazeev, V. F. Kozlov, and Y. S. Tolstov, **MG-XI** (2006), pp. 225–241.

[73] S. I. Krivosheev, N. F. Morozov, V. Y. Petrov, G.A Shneerson„ Materials *Science* **32**(3) (1996), 286–295.

[74] J. R. Klepachko, S. A. Atroshenko, P. Chevrier, et al., *Engineering Fracture Mechanics* **75**(1) (2007), 136–152.

[75] N. F. Morozov and Y. V. Petrov, *Problems of distruction of solid bodies*, Sankt Petersburg, Russia, SpbGU Publishers, 1997.

[76] S. I. Krivosheev and G. A. Shneerson, **MG-SF** (2006), pp. 407–412.

[77] M. Bavay, J. Mervini, and R. B., **MG-SF** (2008) Spielman, pp. 425–430.

[78] D. G. Tasker, J. H. Goforth, and H. Oona, *IEEE Trans on Plasma Physics* **38**(8) (2010), 1828–1834.

[79] V. N. Bondaletov, E. N. Ivanov, C. A. Kalihman, and Y. P. Pichugin, **MG-III** (1984), pp. 234–238.

[80] G. A. Taber, B. A. Kabert, A. T. Washburn, T. N. Windholtz, C. E. Slone, K. N. Boos, and G. S. Daehn, in: *5th International Conference on High Speed Forming* – 2012, pp. 125–134.

[81] L. D. Landau and E. M. Livshitz, *Electrodynamica sploshnih sred*, Moscow, Nauka, 1991.

[82] G. Y. Sermons, *Dynamics of solids in the electromagnetic field*, Riga, Zinatne, 1974.

[83] S. I. Krivosheev, S. P. Nenashev, and G. A., *Izvestita RAN, Seriya Energetika* **3** (2004 Shneerson), 131–140 (in Russian).

[84] A. P.,S. I. Nemashev Krivoshees, V. N. Tisenko, et al., Method and Apparatus for Sorting Non-ferrous Metals, patent USA, Nr. 5823354, date of filing 15 Jan 1997.

[85] Y. S. Astakhov, O. G. Dyaliashvili, and G. A. Shneerson, *IEEE Transactions on Magnetics* **28**(1) (1992), 659–662.

9 Generation of ultrahigh magnetic fields in destructive single-turn magnets

The magnetic fields with induction so high that partial or total destruction of the magnet occurs in one pulse are naturally referred to as ultrahigh. In practice, however, the fields above 100 T (1 MG) are said to be ultrahigh (or megagauss). The difference between these two definitions is not essential if we are dealing with hard materials (beryllium bronze, tantalum, steel with elevated yield point, etc.) which have a magnetic ultimate strength of close to 100 T.

In 1957 the well-known paper by Furth, Levine, and Waniek [1] initiated a large stream of works concering the generation of a megagauss-valued magnetic field in single-turn electromagnets. Such magnets are widely used in technology and physical experiments.

The problem of the generation of a megagauss-valued magnetic field in destructible small-volume single-turns is considered in the monographs [2–7] and reviews [8–14]. In this chapter we consider the qualitative picture of processes causing the destruction of one-turn magnets in the course of generating ultrahigh magnetic fields, as well as the mathematical models of these processes, in order to analyze the experimental data and requirements for the performance of devices delivering magnetic fields of the desired intensity. Below we will assume that walls of a turn are immovable.

9.1 Physical processes accompanying the generation of megagauss magnetic field in single-turn magnets

The studies in several areas of high-density high-temperature plasma, the solution of certain problems in solid state physics, and finally, the challenges of the novel technology inspires R&D of devices for the generation of ultrahigh magnetic fields generated by a direct discharge of a capacitor bank into a single-turn magnet.

In the majority of experiments presented in Table 9.1, turns of rather small dimensions have been used, the length of which did not exceed a few mm, and the inner diameter was, as a rule, less than the length. Such turns have the highest initial geometric factor B/i, and with a current rise time of the order of a few microseconds, field levels up to 360–390 T have been achieved at currents of order of 1 MA, in spite of the magnet destruction in the course of the discharge. The major factor determining the achieved level is the expansion of the inner radius of a turn

Table 9.1: Magnetic fields obtained in single-turn magnets (a summary of experimental data).

Referens		d mm	l mm	h mm	material	I_m MA	dI/dt MA/us	t_1 us	$i(t_B)$ MA	B_m T	t_B us	B_∞ T
Furth H. P., Levine M. A., Waniek R.W [1]	1957	3.2	6		Cu	1.1	0.4	11		160	6.4	133
Shneerson G. A. [15]	1962	3.4	5.5		st	1.37	0.55	3.9	1.37	155	3.9	156
Forster D. W., Martin J. C. [16]	1967	5	15	1.25	Cu	3.5	5			260	1/0	470
Shearer J.W et al. [17]	1966				Cu					210		
Shearer J.W [18]	1969	3	8		Ta	8.8	4.6			355		450
Andrianov A. M. et al. [19]	1970	2.1	2.5		Al	1.36	0.85	2.66		165	1.65	143
Andrianov A. M. et al. [19]	1970	2.1	2.5		Cu	1.54	1	2.8		200	1.9	210
Andrianov A. M. et al. [19]	1970	2.1	2.5		Ta	1.54	0.95	2.9		310	2.3	240
Knoephel H., Luppi R. [20]	1972	5	10		Ta	2.5		7		180	5	182
Herlach F, McBroom R. [21]	1973	2.5	5	2.1	Cu	1.3	1.43	2.2		204	1.6	251
Shneerson G.A [22]	1974	3.4	2.6		st	2.65	0.9	4.6		200	2.6	200
Bocharov Y. N. et al. [23]	1979	3	5	1.5	Cu	1.17	1.23	1.5	1.17	206	1.5	233
Bocharov Y. N. et al. [23]	1979	3	3	1.5	Cu	1.15	1.2	1.5	1.15	280	1.5	230
Bocharov Y. N. et al. [23]	1979	2	2	1.5	Cu	0.99	1.03	1.5		350	1.5	213
Gennadiev N.N et al. [24]	1979	2.3	5	0.5	Cu	1.15	1.39	1.3		220	1.3	247
Gennadiev N.N et al. [24]	1979	2.3	2.5		Cu	1.15	1.27	2.25		180	1.45	236
Andrianov A. M. et al. [25]	1982	3	3	0.5	Cu	0.66	0.83	1.25	0.66	190	1.25	191
Andrianov A. M. et al. [25]	1982	3	5	0.5	Cu	0.77	0.9	1.3	1	195	1.4	199
Andrianov A. M. et al. [25]	1982	3	7	0.5	Cu	1.1	1.11	1.55		200	1.7	221
Bocharov Y. N. et al. [26]	1982	2	2	1.5	Cu	1.18	1.68	1.1	1.13	390	0.97	270
Andrianov A. M. et al. [27]	1984	2	2	1.5	Cu	0.745	1	1.35		215	1.16	210
Nakao.K. et al. [28]	1985	2	2	1.9	Cu	1.35	2.22	1.38	1.1	263	0.84	312
Nakao.K. et al. [28]	1985	10	10	3	Cu	2.07	1.34	2.75	2.03	152	1.44	243
Goto T. et al. [29]	1987	3	3–4	1,8–3,75	Ta	2,5	1,3	2,6		280	2,6	356
Von Orterberg M. et al. [30]	1994	10	10	3	Cu	1.41				103		
Von Orterberg M. et al. [30]	1994	5	5	2	Cu	1.41				196		
Krivosheev S. I. et al. [31]	1997	2	2.73	1.93	Cu	1.26	2.12	0.93	0.94	310	0.47	305
Krivosheev S. I. et al. [31]	1997	2	1.82	1.9	Cu	1.19	2.29	0.82	1.1	360	0.57	317
Miura N. et al. [32]	1998	10	10	3	Cu	2.8	1.25	3.00	2.50	210	2.50	234
Mackay K. et al. [33]	1998	0.05	0.03	0.025	Cu	0.014	0.239	0.022	0.004	53	0.02	103
Lemke R. W. et al. [34]	2004				Cu	17	85			1000		
Novac B. M. et al. [9,35]	2004	2	2	1.6	Cu		1.25		0.77	240	0.9	238
Boriskin A. S. et al. [36]	2004	5	5	1.5	Cu	2.4	2.8571	0.8	2.4	290	0.7	354

Notations in the Table: d, an inner diameter of the magnet; l, length; h – thickness of the wall; I_m, a current amplitude; $dI/d\,t$, rate of a current rise; t_i, a time of maximum current; B_m, an induction amplitude; t_B, a time of maximum induction; $i(t_B)$, – the value of current at a time of maximum induction; B_∞ – induction calculated from formula (9.14).

under destruction. In coils of small dimensions this effect leads to a decrease of the geometric factor.

The data in Table. 9.1 corresponds to a field level which, in most cases, is lower than the sublimation threshold B_{S3} (for copper it is about 370 T; see Table.3.1). Therefore in the first part of this chapter (Sections 9.1–9.4) we will not consider the processes accompanied by an expansion of a medium and by a loss of conductivity. They are treated in Sections 9.5 and 9.6.

The experiments with turns of small dimensions can be divided into two groups. The first group includes experiments with magnets conditionally named thick-wall [1, 15, 17–20, 22, 24], having an external radius much larger than the inner one (Figure 9.1a). Usually thick-wall turns were fixed to an outer clamp and partially destroyed in a single discharge. The second group [16, 21, 23–33, 35, 36] presents the results of experiments with "thin-wall" turns with a wall thickness 2–3 mm or less (Figure 9.1b). These magnets, as a rule, have been totally destroyed in a discharge. Such a discharge is featured by the absence of the destruction of a measuring coil or another object, placed in an orifice of the magnet, because the substance injection occurs in radial and axial directions. It is also worthwhile to distinguish a third group of experiments, which used single-turn magnets with relatively large initial dimensions (length, inner radius), of the order of 1 cm or more [16–18, 32, 34]. This group includes both "thin-wall" and "thick-wall" magnets. The geometric factor of such magnets is less sensitive to the radius change occurring during a time of the order of less than 1 µs. In the experiments of this group at currents of the order 10^7 A it is possible to obtain field levels of the order 10^3. In some experiments (e. g., [37]) the field was produced in the vicinity of the cylindrical conductor, carrying a current, in the configuration of Z-pinch. The general feature of these systems and of the systems of the third group is the

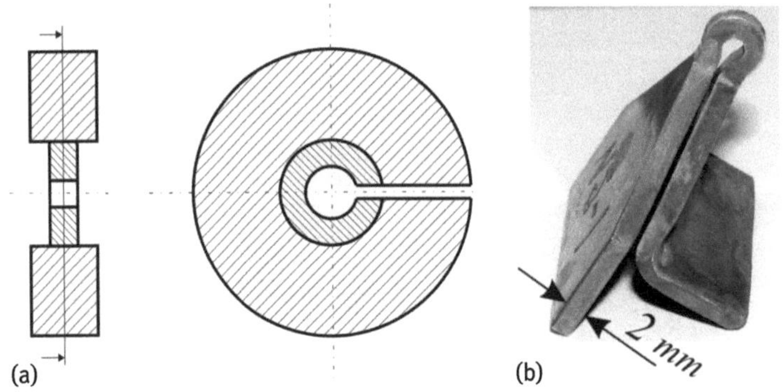

Fig. 9.1: (a) Scheme of thick-wall turn in a massive frame. (b) Photo of a thin-wall turn.

conservation of the geometric factor in spite of the destruction of the conductor surface in the process of discharge.

The data of Table 9.1 shows that in order to generate fields above 150 T, current pulses of the order 1 MA with a rise time in the μs range are required. The generation of megagauss fields in single-turn coils with the fast current rise turned out to be possible due to a low-inductance high-voltage capacitor bank designed in the 1950s. Furth, Levin, and Vaniek [1] were the first to realize this possibility.

The basic qualitative features of processes involved in the generation of ultrahigh magnetic fields of μs duration in thick single-turn magnets of small volume were studied in early 1960s [1, 15, 17–19]. As an example, Figure 9.1 shows the simplest device containing the thick wall disk-shaped magnets with an orifice and insulated slot. In these experiments the magnets were placed in a rigid massive frame and switched on to a capacity bank. In the discharge the partial destruction of the magnet, accompanied by a splash of luminous explosion fragments, occurs.

The photographs of thick-wall magnets after a discharge, obtained from [15], demonstrate their violent disruption (Figure 9.2). The induction amplitude of the field produced in single-turn magnets is determined by the concurrence of two processes. On the one hand there is the current rise in time. On the other hand there is the increase of the effective characteristic dimensions of a turn due to the field diffusion in a conductor, and conductor disruption as a result of the combined action of electromagnetic forces and heating.

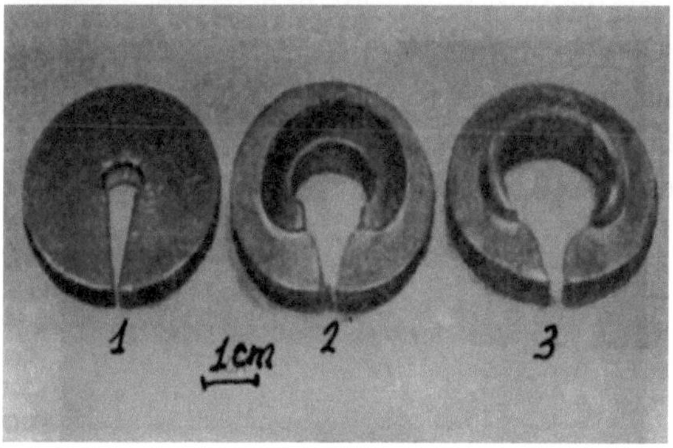

Fig. 9.2: Photographs of copper thick-wall single-turn magnets after a discharge for different values of the first induction amplitude: (1) 63; (2) 93 T; (3) 114 T. Current rise-time is 3.9 μs.

The magnetic pressure $P_M = B^2/2\mu_0$ in the field with induction over 70 T exceeds the static yield limit of most of hard metals, while the surface of a conductor in a single half-period is heated above 1000 K° (see Chapter 3). In the experiments by Furth, Levin, and Vaniek [1] the solenoids of beryllium bronze with initial inner diameter of approximately 6 mm revealed the weak indications of melting and deformation in a field of 70 T. In a field with an induction of 120 T the intense surface fusion was accompanied by the residual increase of the inner diameter by 1.5 times. On the fused surface the traces of liquid metal instabilities could be seen. In a field with induction of 160 T the solenoid with an inner diameter of about 3 mm was intensively deformed. The authors suggest that the deformation of the solenoid is caused by metal flow , while the retardation of the deformation is due to its inertia.

The main purpose of the studies in [15] was to elucidate the relative role of plastic deformation and metal ejection in the process of magnet disruption, as well as to evaluate the speed of the radius increase. The experiments were made with single-turn solenoids of various materials: copper, steel 3, steel 40, aluminum, duralumin, kirit (kirit is a compound made by baking tungsten and copper powders), beryllium bronze, brass, and wood alloy (for the study of disruption caused solely by ejection). The initial dimensions of solenoids are as follows: the inner diameter was 3.4 mm, external 40.5 mm, length 5.5 mm.

In the investigated range of fields (the peak field achieved in steel solenoids was 155 T at the rise time of 3.9 μs) the dependence of the induction amplitude on initial voltage remained linear in the error limits (7–8 %). The expansion of the inner radius up to the moment of the current first maximum was not so essential as to reveal itself in the decrease of the induction amplitude below the calculated value. Comparing the values of the induction and its rise rate in a deformed turn with the corresponding values obtained in weak fields (i.e., with the absence of deformations) one can easily calculate with an error of 20–25 % the instant effective radius of an orifice and evaluate the average velocity $\langle u \rangle$ of its expansion. The velocities for the interval $\pi/2 < \omega t < 3\pi/2$ are given in Table 9.2

For comparison, Table 9.2 presents the values of the Alfven velocity $u_A = B_{m1}/\sqrt{\mu_0 \gamma_0}$ where γ_0 is a material density. In the conditions of these experiments the average velocity of the radius expansion is significantly lower than the Alfven velocity. We will further show that the Alfven velocity characterizes the steady

Table 9.2: Experimental values of the rate of decrease of the inner radius for single turn thick-wall magnets [15].

Material B_{m1}, T	$\langle u \rangle$, m/s	u_A, m/s	
Copper	93, 114	300, 330	760, 900
Brass	90, 127	300, 400	720, 1130
Low-carbon steel	85,118	160, 240	760, 1060
Tool steel 40	143,117, 155	350, 240, 350	1340, 920, 1380
Aluminum	53, 80	270, 400	910, 1370
Beryllium bronze	90,115	120, 32	480, 850

hydrodynamic flow of an incompressible fluid. As judged from the data in Table 9.2, in the experiments under consideration such a flow could not be developed within the time of the discharge.

In solenoids of the length of the order ≤ the diameter the displacement of the material in an axial direction is essential. If the role of the ejection is not significant (as is shown below, this is a case mainly of plastic and well-conducting metals), then, due to the displacement, the axial cross section of the solenoid after the discharge takes a characteristic shape (Figures 9.3 and 9.4) resembling that of the cylindrical sticker after a hit on a solid target.

The increase of the duration of the current pulse front leads to the development of violent deformation prior to the current first maximum. In [38] the deformation of a thick-wall single turn magnet was studied when a strong magnetic field was rising relatively slowly. At a rise time of 31 μs the intense plastic deformation of the steel turns resulted in a significant decrease (down to 0.4) in the ratio of the first amplitude of the induction to its calculated value, B_{m1}/B_{m0}. The peak induction in this case did not exceed 70 T, while the calculated value was about 170 T. Intense residual deformation took place in the case, where in the magnets made from tool steel 40 appeared in the field with induction amplitude $B_S = 46$ T. This is noticeably lower than the corresponding value for short pulses. The velocity of the radius expansion was close to $u_A\sqrt{1-2\mu_0\sigma_S/B_{m1}^{\;2}}$, where $\sigma_S = B_S^2/2\mu_0$.

Although, due to the deformation of the insulating slot the shape of the orifice of the single-turn coil after the discharge differs from the circular one, the size of

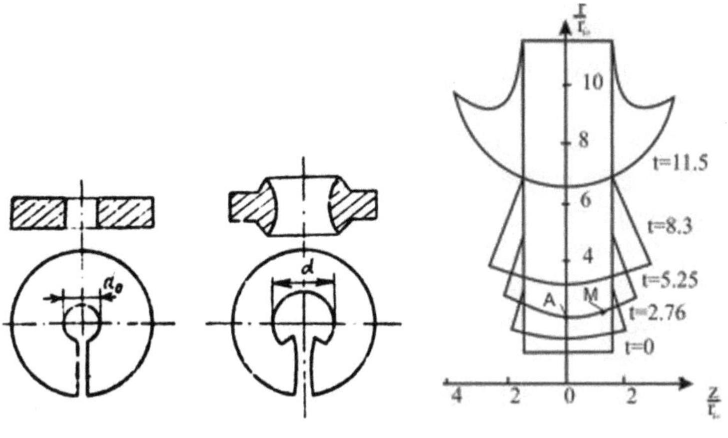

Fig. 9.3: A single-turn magnet before and after the discharge, and the picture of the flow subjected to destruction, resulting from the computer simulation on the approximation of a flow of incompressible liquid.

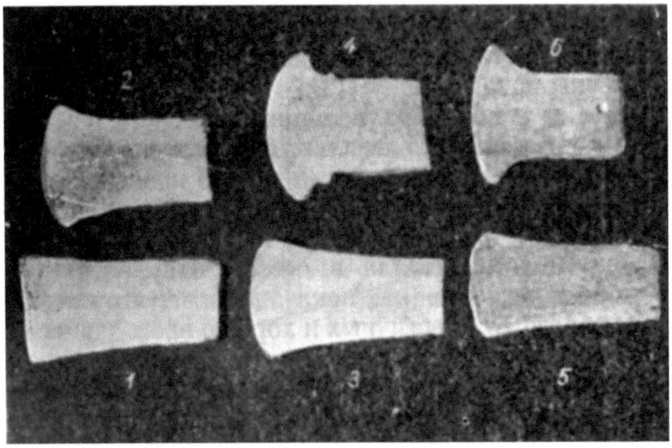

Fig. 9.4: The radial cuts of the samples of the thick-wall single-turn magnets after a discharge. The sample material and the induction amplitude are, relatively: (1) low-carbon steel, 85 T; (2) low-carbon steel, 129 T; (3) copper, 93 T; (4) copper, 114 T; (5) brass, 90 T; (6) brass, 130 T.

the orifice could be characterized by the largest dimension d in the direction perpendicular to the slot (see Figure 9.2). Extrapolating the dependence $d = f(B_{m1})$ until intersection with the straight line $d = d_0$, where d_0 is the initial inner diameter of the solenoid, one can evaluate the "magnetic limit" of the material strength B_S which is the field in which at a given form and pulse duration the disruption of the coil occurs. The values B_S presented in the previous chapter (Table 8.2) for the conditions of experiments in [15],

The majority of experiments with thick wall-coils of small dimensions are featured by the fact that the distance which the shock wave travels during the current rise time exceeds the length or thickness of a turn. For instance, in copper at the sound velocity of about 4.5 km/s and time of 1 μs the elastic wave passes 4.5 mm, whereas the turn length is usually less than that. In this case the increase of the radius and length is determined mainly by the medium flow through free boundaries, and the compressibility has little effect on the deformation process.

Studies of deformed samples show that along with those (copper, aluminum samples) that evidence the obvious traces of plastic deformation, there were also such that were destroyed mainly due to the ejection of metal. In this regard the wood alloy is the most significant: the samples, in spite of the low strength properties, do not practically show plastic deformation. although an increase in diameter caused by discharge was rather significant. Before the experiments the grid was applied on the ends of the magnet. The lack of distortions on the grid and of the changes of a shape of the longitudinal cross section proves that the plastic deformation of the samples of wood alloy is low. The dependence of the induction amplitude on the current amplitude sharply deviated from a straight line, because the expansion of the inner radius began prior to the moment of the current first maximum.

In order to characterize the relative role of matter ejection and plastic deformation one can use the criterion [15]

$$K_d = \frac{\Delta m \cdot S}{m \cdot \Delta S}, \tag{9.1}$$

where m, S are the initial mass and square of the face surface of the cylindrical solenoid of rectangular cross section, Δm, ΔS are the decrease of these values due to the discharge. The coefficient K_d is equal to unity at the absence of mechanical deformation and zero at the absence of ejection. Although the coefficient characterizes the resultant effect, it can be used in order to evaluate the role of ejection in the process of discharge in solenoids produced from various materials.

Experimental data shows that in high-plastic metals (copper, aluminum) the relative amount of the ejection matter is low, e.g., after the experiment in the field of 114 T the coefficient K_d for the copper samples is close to 0.1. For brass and steel samples it was already in the range 0.4–0.7 in fields of about 100 T. It seems reasonable to say that the splashing of molten metal is the basic mechanism of ejection, as demonstrated by traces in the form of solidified jets on the surface of samples after the experiment. This explanation is confirmed by comparing the induction obtained in the experiment with B_{S2}, characterizing the metal melting threshold. By analogy with electrical explosion of conductors [40], the breakup of the current layer at temperatures lower than the vaporization temperature can be called “slow explosion”. as opposed to processes associated with the metal vaporization – “fast explosion”. These processes are featured by the essentially higher threshold B_{S3} (see Table 3.1). In the case of fusible metals, the electric explosion can be violently developed prior to deformation and resulted in an essential decrease of the field level compared with the calculated level, due to the expansion of the inner turn. This happened in experiments with samples from wood alloy, where the coefficient K_d was equal to unity, and the process was conditioned entirely by splashing of molten metal in an axial direction. The traces of the molten metal are clearly seen in Figure 9.5.

One more example of experiments with thick-wall magnets is the experiment with biconic samples (Figure 9.5) where the flow was close to spherical [22]. In these experiments the current amplitude was up to 2.6 MA and the rise time about 4.5 μs. At the largest currents the induction amplitude achieved 200 T and was essentially less than the calculated value. As is seen in Figure 9.6, the magnetic field achieved a maximum before the current did, which indirectly showed a significant increase of the inner radius on the current pulse front. Direct evidence was

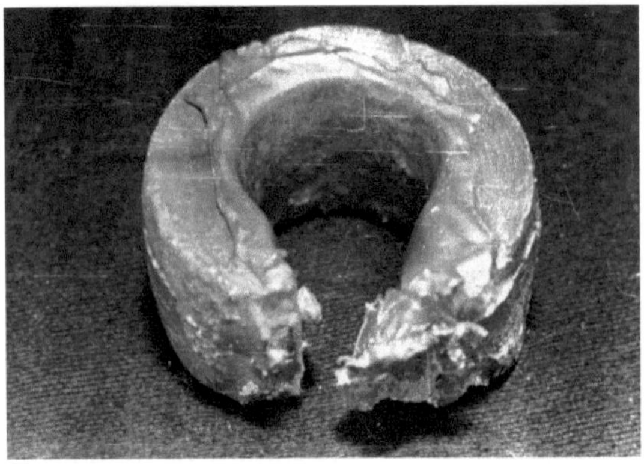

Fig. 9.5: A single-turn magnet from wood alloy after the action of a 50 T field.

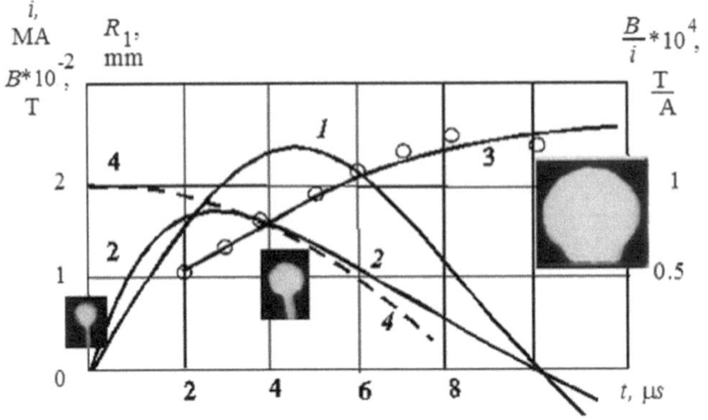

Fig. 9.6: The parameters of the process of destruction of the thick-wall biconical turn; (1) current i; (2) induction B_0 in the center; (3) internal radius R; (4) geometric factor B_0/i. The x-ray photographs are shown for the initial, intermediate, and final stages of coil destruction.

obtained from the x-ray photos of the solenoid orifice. Figure 9.6 shows the temporal dependence of the effective inner radius that can be found with the help of the instant value of the geometrical factor $G = B/i$ determined on the induction oscillogram. This dependence shows that the instant speed of the inner radius expansion achieved 600 m/s. In these experiments the basic mechanism of the deformation was hydrodynamic flow. The role of splashing was insignificant.

Another situation was in the experiments by Shearer and in other studies with the thick-wall relatively long magnets [17, 18]. In an earlier work [17], with the calculated value of the first current amplitude of $I_{m1} = 12$ MA (a rise time $t_m = 4.8$ μs), the field of 210 T was generated. In [18] the field was produced in a 1 cm-width slot ending in a cylindrical orifice in which the measuring probe was placed. A slot was cut in a copper sheet switched on to the generator buses. An even stronger field was obtained in tan experiment with an even larger current and shorter rise time ($B_m = 310\,\mathrm{T}$, $I_{m1} = 25\,\mathrm{MA}$, $t_m = 0.7\,\mu\mathrm{s}$). In these works the flow was approximately one-dimensional, and the shock wave was formed in the medium. Behind the shock front the medium density was greatly increased as Shearer demonstrated by x-ray measurements [18].

In fields with induction $B > B_{S3}$, evaporation of the medium and theformation of metallic plasma begins to play a similar part when it occurs in "fast explosion" of wires [39]. This effect, a "fast explosion" of a skin layer, reveals itself in fields of the order 1000 T; however in a thin-wall turn the evaporation may take place even in a weaker field. Forster and Martin [16] consider the electrical explosion to be the basic process leading to the destruction of solenoids in their experiments with fields achieving 250 T. In later experiments Andrianov, Demichev, et al. observed the evaporation of edge regions of thin turns [24, 25, 27].

The x-ray photos derived from [49] (Figure 9.7) show the evaporation of the turn and its breakup in a few fragments due to the development of MHD instability in the late stage of discharge. A similar picture of the turn breakup was observed in [31], where the explosion of the various-length single-turn coils with an inner diameter of 2 mm and wall thickness of 1.5 mm was investigated (Figure 9.8). This specially developed installation with rather low inductance [41] provided the ini-

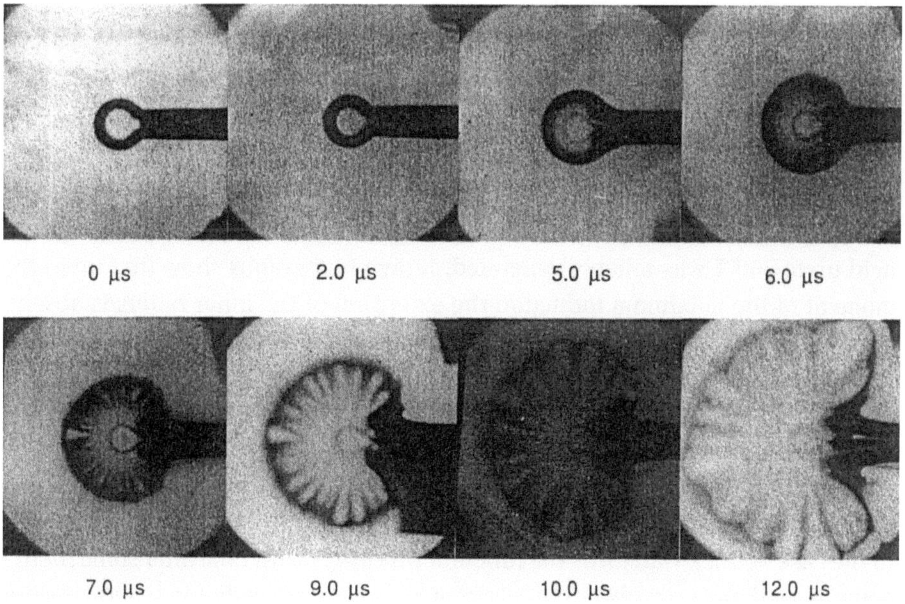

Fig. 9.7: A typical picture of an explosion of a single-turn coil in the megagauss magnetic field. The sizes of the coil are: diameter 8 mm, wall thickness 2.8 mm, length 8mm [49].

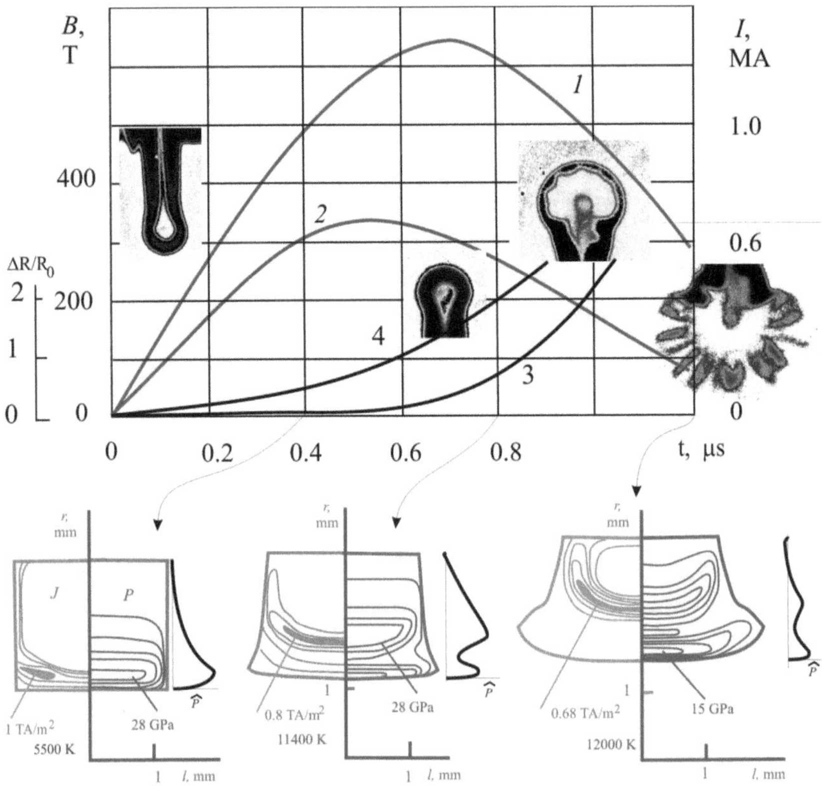

Fig. 9.8: Explosion of the thin-wall turn: (1) current *i*; (2) induction *B*; internal radius *R*; (3) experiment; (4) MHD-calculation. The x-ray photographs and the results of 2D-calculations from [31].

tial current rise up to $3 \cdot 10^{12}$ A/s. At the explosion of thin-wall turns a magnetic field up to 360 T was reliably generated. X-ray photographs show that up to the moment of the maximum induction the extension of the inner radius is absent. This fact, found in other works as well, can be attributed to the expansion of the medium when heated by a current in a skin layer.

The behavior of a coil in an ultrahigh magnetic field is characterized by the temporal changing of the geometric factor $G(t) = B(t)/i(t)$ and its value at the maximum induction $G_m = B_m/i(t_B)$, where t_B is the moment when the induction reaches a maximum. The dependence $G(t)$ was established in many experiments. In the case of thick-wall turns the function $G(t)$ drops with time after some inertia delay. Nonetheless, the examples given in Figure 9.9. show that in several experiments with thin-wall turns (group II) the function $G(t)$ was not decreasing. In the literature at least two effects are reported which lead to retardation of the decrease of $G(t)$, or even to a short-term increase of the geometric factor. The first one is a

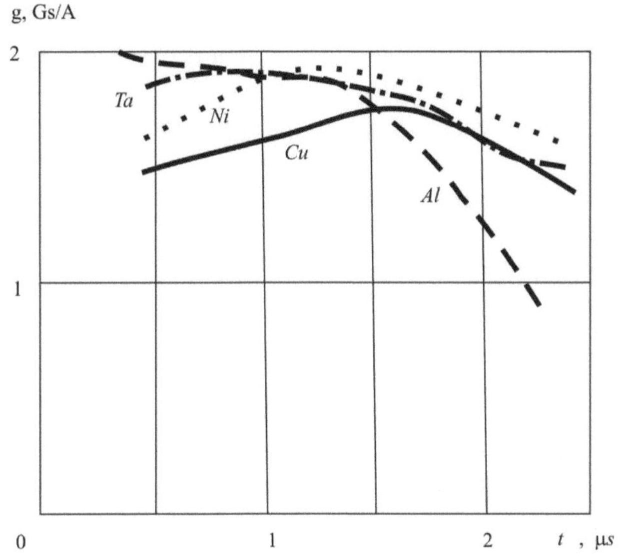

Fig. 9.9: Time functions of the geometric factor $G(t)$ for turns from different materials [24, 27].

decrease in the effective length of the thin-wall magnet because of the loss of conductivity of the edges, due to the Joule heating and expansion [24, 27]. The other is the above-mentioned "swelling" of the magnet walls when heated, resulting in the displacement of the inner edge to the axis and the retardation of the inner radius growth. In addition, one should take into consideration the possibility of a decrease in length as well as an increase in the geometric factor, due to wall deformation under the action of opposite axial forces [42, 43]. In the experiments with solenoids of group III, the geometric factor is less sensitive with respect to the radius change. Therefore the ratio G_m/G_0, where G_0 is the initial geometric factor, may remain close to unity, in spite of the displacement of the magnet wall.

9.2 Modeling problems illustrating the role of different factors leading to the destruction of single-turn magnets

The coil destruction is the result of the two- or even three-dimension MHD-flow of the conductor accompanied by field diffusion, heating, and ejection of material. These processes occur in combination, so that rigorous analysis only is possible using numerical experiments. However, in certain experiments these or other factors could be dominant, and their action can be considered separately in the framewok of the modeling problem representing the given process. The study of simple models is not only useful, but necessary for an understanding of specific features of the each of the phenomena under consideration. It is possible to distinguish the following modeling problems:

1. Expansion of the ideally conducting short turn in the approximation of non-compressible liquid.
2. Expansion of the ideally conducting long magnet under the action of the shock wave induced by the magnetic field.
3. Nonlinear diffusion of the field into the immobile conductor.
4. Ejection of the metal through the end surfaces ("slow explosion")
5. Evaporation of the metal and loss of conductivity in expansion ("fast explosion").

Each of the physical processes can be considered within the framework of the modeling problem as a factor diminishing the induction which can be achieved. Here one can take into account the properties of the electrical circuit which connects the magnet with the energy source. The calculation of the motion of the field-conductor boundary is sufficiently complicated, even in the framework of the simplified model; therefore it is reasonable to sparate out the more simple limiting cases. For example, one can choose to supply a source of unlimited power and infinite capacity, but with the inductance and resistance negligibly small compared to a magnet. Another model is the current source, i.e., the generator, with transitional impedance much larger than the magnet. In the first case at the input of the magnet the voltage is prescribed, in the second case it is the current.

Geometrically, the limiting case is a long- or short-turn one. In the first case the length is so much larger than the diameter, that one can ignore the influence of the edge effects on the induction in the greater part of the magnet working volume as well as on the character of the flow. In the second case the turn length is equal to the inner diameter or less. If the outer radius is much larger than the inner one (thick wall solenoid), the induction and current are coupled by the relation

$$B(0,\ t) = \frac{a\mu_0 i\,(t)}{R_1\,(t)}. \tag{9.2}$$

In a thin disk we have ($l/2R_1 = 0$) $a = 1/\pi$. The same value a, at condition $0 \leq l/2R_1 \leq 1$, can be used for the calculation (with an error not exceeding 16 %) of the field in the center of a turn with a large ratio of the outer radius to the inner one. With constant coefficient a the factor $G = B/i = a\mu_0 R_1$ decreases practically inversely as the turn is expanding. In a long solenoid the induction is $B = \mu_0 i/l$ where l is the length. In this case the geometrical factor $G = \mu_0/l$ remains constant, in spite of the increase of the inner radius.

In a circuit with a given current the correlation between the calculated and the achieved induction significantly differs for the two above-mentioned types of single turn magnets. In the second type, the induction is equal to the calculated one and could be arbitrary high, while in the first type the induction remains finite at unlimited current growth and, as it will be shown below, its asymptotic value is determined by the current rise time.

Therefore, for long magnets the analysis will be confined to a system with a source of unlimited power. For the short systems the case of the "point" turn with a small initial radius (1–2 mm) and low inductance (≈ 1 nHn), could be of interest, since the storage needed for the power supply of such a magnet could acquire a modest amount of energy (of the order of 10^4 J). However, it is difficult to make a magnet with lower inductance than the inductance of the turn. Therefore in order to supply such coils, generators with much higher inductance than the inductance of the solenoid are used, what provides the increased energy storage. The processes in the circuits of these generators can be described with a help of the model of the current source.

9.3 Hydrodynamic flows in single-turn solenoids. Application of the model of a noncompressible liquid with ideal conductivity to the description of the deformation of a thick-wall turn

Solenoids made of good conducting materials usually function under the condition of a sharply pronounced surface effect. Therefore in the first approximation they can be considered as ideally conducting. In this case the deformation can be described by the solution of continuous medium equations in conjunction with the equation of metal state and with the boundary condition $P(A) = B_\tau^2/2\mu_0$, where A is the point on the boundary, and B_τ is the induction component tangential to the boundary in the point A.

The solution is essentially simplified for the modeling problems. The simplest of them is the model of noncompressible liquid, and it could be used for calculating the flow, if the following condition is fulfilled: the propagation time of the sound wave at a distance of the order of a solenoid length is much less than the rise time of the field up to maximum. In this case one can assume the sound velocity to be infinitely high, i.e., the material is considered as a noncompressible medium.

If the outer radius is assumed to be constant, which is the case when a thick wall solenoid is fixed in the rigid rim, then the inner radius can increase only

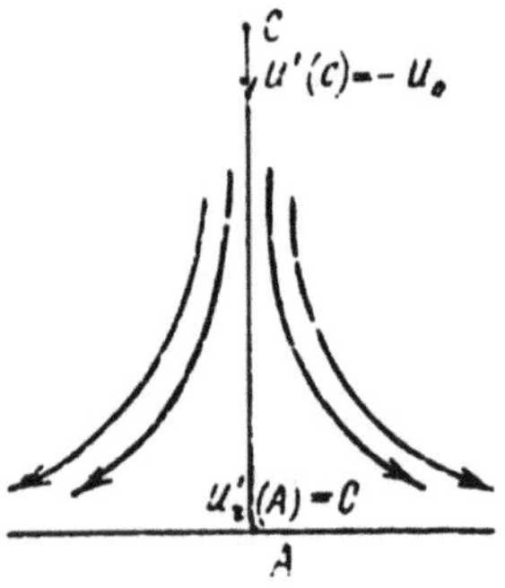

Fig. 9.10: The fluid in a system of coordinates related to the conductor boundary.

due to the displacement of metal through the end faces. The flow is essentially two-dimensional, as is schematically shown in Figure 9.10. In the system of reference related to point *A* on the surface, the elements of the medium flow past the observer and diverge laterally toward the ends. It corresponds qualitatively to a similar picture of the deformation that occurs at the impact of a sticker on a solid target (Figures 9.3, 9.4). In the considered model the target is the field-conductor boundary moving with speed u_0 toward the conductor.

Let us consider the stationary flow in the chosen system of reference. The equation of motion along the radius in the plane of symmetry has the form

$$\gamma_0 u_r' \frac{\partial u_r'}{\partial r_1} = -\frac{\partial}{\partial r_1}(P + P_M), \tag{9.3}$$

where γ_0 = const is the medium density, P is the pressure in the medium, P_M is the magnetic pressure, and $r_1 = r - R_1$, $u_r' = u_r - u_0$. Taking into account the given properties of the medium, let us use the simplest "liquid–solid" model. Aassume that the medium with the constant density does not resist to the shift, i.e., behaves as the ideal noncompressible liquid in region 1, where $|P| > \sigma_S$, being rigid in region 2, where $|P| < \sigma_S$. Here the σ_S can be considered to be the yield point of the material for the given duration of the action. This model was used by Lavrent'ev in the theory of cumulation [44]. The point *C* on the boundary of the regions is immovable in the system of reference connected with point *A*. In the region $r_1 > r_1(C)$ where the medium is immovable in the laboratory system of reference, its elements move with velocity $-u_0$ relative to point *A*. Integration of equation (9.3) over r_1 gives the relation (the analog of Bernoulli law)

$$\frac{1}{2}\gamma_0 \left(u_r'\right)^2 + P + \frac{B^2(r)}{2\mu_0} = \text{const.} \tag{9.4a}$$

At the point $A(r = R_1)$ we have $B = B(R_1)$, $P = 0$, $u'_r = 0$; point C: $B = 0$, $P = \sigma_S$, $u'_r = -u_0$, hence,

$$\gamma_0 \frac{u_0^2}{2} + \sigma_S = \frac{B^2(R_1)}{2\mu_0}, \tag{9.4b}$$

where $B(R_1)$ is the induction near the inner boundary of the solenoid. From here we obtain the relation for the velocity of the boundary shift at the stationary flow:

$$u_0 = \frac{B(R_1)}{\sqrt{\mu_0 \gamma_0}} \sqrt{1 - \frac{2\mu_0 \sigma_S}{B^2(R_1)}}. \tag{9.5}$$

In the limit $B(R_1) \gg 2\mu_0\sigma_S$ the value u_0 becomes equal to the Alfven velocity $u_A = B(R_1)(\mu_0\gamma_0)^{-1/2}$. We can use the derived relation to connect the induction with the current parameters. We should also consider two limiting cases: the solenoid connected to the source of the infinite power with voltage U_0, and the point solenoid switched into the circuit with the given current.

In the first case we first consider the solenoid in the form of a slot (Figure 9.11a). Equating U_0 and $d\Phi/dt = 2u_0 gB$, we obtain the relation connecting the voltage of the source and the induction in the slot:

$$U_0 = 2gB\sqrt{\frac{B^2}{\mu_0\gamma_0} - \frac{2\sigma_S}{\gamma_0}}. \tag{9.6}$$

From here in the limiting case, when $B \gg (2\mu_0\sigma_S)^{1/2}$, we have

$$B = \left(\frac{U_0}{2g}\right)^{1/2} (\mu_0\gamma_0)^{1/4}. \tag{9.7}$$

The dependence $B = \text{const} \cdot U_0^{1/2}$ is rather characteristic for the field in the slot. It will be also encountered in other models describing the explosion. As an example, let us consider the induction produced in an asymptotic regime in a slot

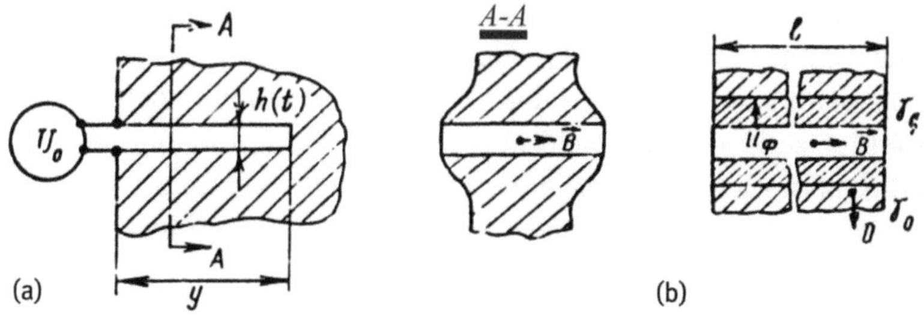

Fig. 9.11: The slot, to the edges of which the source of power U_0 is switched on: (a) the 2D-flow in a short single-turn magnet; (b) the 1-D flow (the model of a shock wave).

with 1 cm length, to the ends of which a voltage 10^4 V is applied. If $\gamma_0 \approx 9.10^3$ (copper), then $B \approx 200$ T. In order to produce the field with induction of 1000 T one should provide the average electric intensity $E_{CP} = U_0/2g$ close to 10^5 V/cm (for copper).

For a long solenoid with a round orifice with the initial radius $R_1(0) = R_0$ the equation determining the radius increase at the stationary flow has the form (at $B \gg (2\mu_0\sigma_S)^{1/2}$)

$$\frac{dR_1}{dt} = \frac{B}{\sqrt{\mu_0\gamma_0}} \tag{9.8a}$$

and the voltage at the input of the solenoid

$$U = \frac{d}{dt}\left(\pi R_1^2 B\right). \tag{9.8b}$$

If the solenoid is connected to a source of infinite power, then $U = U_0$ and the solution of equations (9.8a) and (9.8b) at condition $R_1(0) = R_0$ has the form

$$\frac{R_1(t)}{R_0} = \left[1 + \frac{3}{2}\left(\frac{t}{t_f}\right)^2\right]^{1/3}, \tag{9.9a}$$

$$\frac{B}{B_f} = \frac{t}{t_f}\left[1 + \frac{3}{2}\left(\frac{t}{t_f}\right)^2\right]^{-2/3}, \tag{9.9b}$$

where $t_f = R_0(\frac{\pi R_0}{U_0})^{1/2}(\mu_0\gamma_0)^{1/4}$, and $B_f = (\frac{U_0}{\pi R_0})^{1/2}(\mu_0\gamma_0)^{1/4}$. The induction reaches a maximum at the moment $t_m = \sqrt{2}t_f$:

$$B_m = 2^{-5/6}B_f = 2^{-5/6}\left(\frac{U_0}{\pi R_0}\right)^{1/2}(\mu_0\gamma_0)^{1/4}. \tag{9.9c}$$

At the moment t_m the inner radius is $R_1(t_m) = 4^{1/3}R_0 = 1.59R_0$. The average electric field intensity on the turn circuit, $E_{CP} = U_0/2\pi R_0$, is (as in a case of a slot) the basic parameter determining the induction amplitude.

Table 9.3 presents the calculated results of the average electric intensity needed so that field inductions of 500, 700, and 1000 T can be achieved in a solenoid of copper and tantalum. The table also contains the values of the source voltage needed for generating the field in a solenoid with the initial radius of 5 mm (the radius at the moment of the induction minimum is close to 8 mm) and the values of the magnetic field energy at the moment t_m. The voltage exceeds 200 kV for the induction 10^3 T.

As related to the experiments with thick-wall rectangular turns one should use the approximation of the point turn, assuming the induction and current to be related by (9.2). Let us search for a solution corresponding to the stationary

Table 9.3: Parameters of the generator required for obtaining a field in the turn under destruction, with the given induction amplitude. Nominators of the fractions correspond to the equation $dR_1/dt = V_A$ (the model of the stationary two-dimensional flow of incompressible liquid), denominators correspond to $dR_1/dt = V_A(2\lambda)^{-1/2}$ (model of the strong shock wave).

Parameters	Copper			Tantalum		
B_m, T	500	700	1000	500	700	1000
$\langle E \rangle$, V/cm	37/21	74/42	150/86	32/18	64/36	102/71
U_0, êV	117/67	230/132	470/269	100/57	200/113	320/223
W'_m, kJ/m	197	380	786	197	386	796
t_m, μs	2/2,6	1,5/1,9	1/1,3	2,3/3	1,7/2,2	1,2/1,5

motion of the boundary at the velocity equal to the Alfven one. Then from equations (9.2) and (9.3) we can find $B(0,t) \approx B(R_1,t)$. The qualitative behavior of the curve $B(0,t)$ follows directly from these equations: first, the inner radius is close to its initial value, the induction increases proportionally to the current, and then with increasing R_1 the induction decreases compared to the calculated value, i.e., a decrease of the geometric factor B/i takes place. The limiting regime is of interest when the current amplitude is unlimitedly large. In this regime the induction reaches a maximum at the pulse front when the current still grows linearly as $i = i't$.

The asymptotic induction value can be evaluated taking $B(0,t) = B_\infty = \text{const}$ and assuming $R_1 \gg R_1(0)$. Then $R_1 = tB_\infty(\mu_0\gamma_0)^{-1/2}$, and

$$B(0,\ t) = \frac{a\mu_0 i' t}{R_1} \approx B_\infty \approx \frac{a\mu_0 i' \sqrt{\mu_0\gamma_0}}{B_\infty}. \tag{9.10a}$$

From here,

$$B_\infty = \left(ai'\right)^{1/2} \mu_0^{3/4} \gamma_0^{1/4}. \tag{9.10b}$$

The calculation of the expansion of a thin turn with radius R gives an analogous expression for the induction B_∞, on the assumption that the radius increases with a velocity equal to the Alfven velocity. The equation for the velocity is then of the form

$$\frac{dR}{dt} = \frac{i}{l + 2\varepsilon' R}\sqrt{\frac{\mu_0}{\gamma_0}}. \tag{9.11a}$$

In deriving this equation, we used formula (2.26) for the induction in the center of a thin turn, which is valid for a sharply pronounced skin effect. Under the condition $l/(2R) > 1$ the quantity $2\varepsilon'$ may be assumed to be 1.24, which gives an error less than 6 % in calculation of the induction in an average cross section of the turn. Further calculations result in the following dependence for the induction:

$$B = \frac{\mu_0 i}{\left((l + 2\varepsilon' R_0)^2 + 4\varepsilon' \sqrt{\frac{\mu_0}{\gamma_0}} \int\limits_0^t i(t)\, dt\right)^{1/2}}. \tag{9.11b}$$

When the current increases linearly, in the limiting case $R \gg R_0$ for the induction we obtain expression (9.10b) where $a = 1/(2\varepsilon') \approx 0.8$.

Note that the model of steady flow is applicable in this case, since the velocity of the inner radius expansion is constant. In the general case the flow is nonstationary, and instead of the equation (9.3) one can search for the solution of the Euler equation, which for the turn mid-plane has the form

$$\gamma_0 \left(\frac{\partial u_r}{\partial t} + u_r \frac{\partial u_r}{\partial r}\right) = -\frac{\partial}{\partial r}(P + P_M). \tag{9.12}$$

As applied to the conditions of the experiments with the biconic solenoids, rather precise results can be derived from the model built on the simplified assumption of the spherical character of the flow. It is suggested that the velocity at the point with coordinate r and the boundary velocity $\dot{R}_1$ are related by $u_r = \dot{R}_1 R_1^2/r^2$. The calculated results are presented in Figure 9.12 [22]. The qualitative behavior of the curves, calculated for the sinusoidal current pulse, corresponds to the experiment: the induction amplitude is less than the calculated one and achieved before current maximum. With the linear current rise, when the first term in the left part of equation (9.11) is small in comparison with the second, a steady flow takes place, and a boundary motion occurs with the velocity

$$\dot{R}_1 = \frac{B(R_i)}{\sqrt{3\mu_0\gamma_0}} \approx \frac{B(0,\ t)}{\sqrt{3\mu_0\gamma_0}}, \tag{9.13a}$$

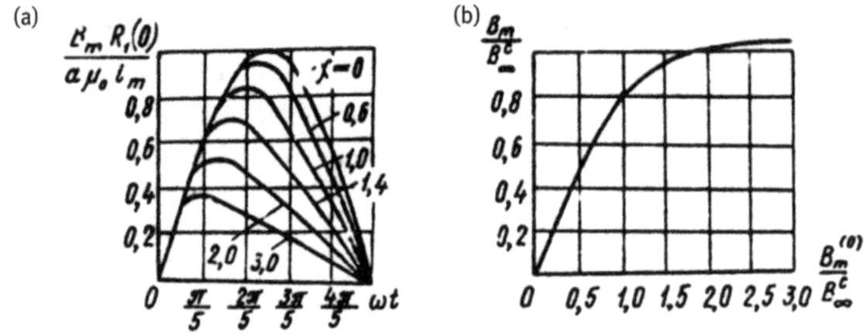

Fig. 9.12: Induction in biconical magnets: (a) dependences for the induction amplitude at different values of the characteristic dimensionless parameter $\chi = (3/2)^{1/4} B_m^{(0)}/B_\infty$, where $B_m^{(0)} = \mu_0 a i_m / R_1(0)$ is the calculated value of the induction amplitude; (b) the dependence $B_m/B_m^{(0)} = f(B_m^{(0)}/B_\infty)$.

differing from the Alfven by a factor $1/\sqrt{3}$. For the copper biconic solenoid, instead of formula (9.12) we have

$$B_\infty \approx (ai)^{1/2} (3\gamma_0)^{1/4} \mu_0^{3/4}. \tag{9.13b}$$

As applied for the copper biconic solenoids with parameter $a = 0.2$, formula (8.13) takes the form

$$B_\infty \approx 2,1 \cdot 10^{-4} \sqrt{i'} [T, A/s]. \tag{9.14}$$

It has been shown that in the experiments from [15] the flow had no time to reach the steady state, and therefore the velocity of the radius increase was less than the Alfven velocity. On the contrary, in experiments with biconical turns a flow with constant velocity was established before the induction reached its amplitude value. Along with estimates based on the steady flow assumption, it is of interest to estimate the time it takes for a steady regime to be established. Further we will show that the Alfven velocity characterizes the steady hydrodynamic flow of an incompressible medium.

For the spherically symmetric flow, when the above-mentioned coordinate dependence for velocity takes place, it is possible to transform equation (9.11) into the form

$$\gamma_0 \left[R \frac{d^2 R}{dt^2} + \frac{3}{2} \left(\frac{dR}{dt} \right)^2 \right] = P_M (R_1) - P_\infty. \tag{9.15}$$

To do this, it is necessary to integrate both parts of equation (9.11) over the radius R_1 up to infinity. It can be assumed that when $r \to \infty$ the pressure is σ_S, which is close to the yield point of the material.

Equation (9.15) is the well-known Rayleigh equation used in the theory of cavitation [44]. Further, we consider the initial stage of the process when the velocity is still low, and the radius is close to its initial value. In this stage the first term in the left-hand side of (9.15) plays the decisive role. The time function for the internal radius of the winding can be written in the form $R_1 = R_0 + bt^\beta$, where the second term is much less than the first term. The magnetic pressure is $P_M(R_1) \approx \mu_0 (ai't)^2/(2R_0^2)$. Further, we obtain the equation

$$R_0 \gamma_0 \left[\beta (\beta - 1) bt^{\beta-2} \right] \approx \frac{\mu_0}{2} \left(\frac{ai't}{R_0} \right)^2 - \sigma_S. \tag{9.16}$$

The calculation problem is the determination of the time moment t_0, when the radius found from equation (9.16) will be equal to the value calculated by the formula

$$R_1 = R_0 + \frac{ai't^2}{R_0} \sqrt{\frac{\mu_0}{3\gamma_0}}, \tag{9.17a}$$

which describes the expansion with constant velocity. The time moment t_0 is the estimation of the duration of the first stage, when the flow with the Alfven velocity is still not stationary. Near the time t_0 in a superstrong field, the condition $P_M \gg \sigma_S$ may be assumed to be satisfied. Then we have $\beta = 4$, and

$$R_1(t) \approx R_0 + \frac{\mu_0 (ai')^2 t^4}{24\gamma_0 R_0^3}. \tag{9.17b}$$

Equating equations (9.17a) and (9.17b), we have

$$t_0 = \left[24 \cdot \sqrt{\frac{\gamma_0}{3\mu_0}} \cdot \frac{1}{ai'}\right]^{1/2} R_0. \tag{9.18}$$

For the estimation it is convenient to introduce the calculated value of the induction amplitude $B_{m,0} = \frac{2ai'\mu_0 t_m}{\pi R_0}$, where t_m is a rise time of the sinusoidal current up to a maximum. The ratio t_0/t_m follows from the derived formula

$$\frac{t_0}{t_m} = \nu \left(\frac{R_0}{t_m u_{AO}}\right)^{1/2}, \tag{9.19}$$

where $u_{AO} = B_{m,0}/\sqrt{\mu_0\gamma_0}$ is the Alfven velocity corresponding to the calculated induction amplitude, and the numerical factor $\nu = (48/\pi)^{1/2}(1/3)^{1/4} = 2.97$. This formula shows that the pause preceding the radius rise with the Alfven velocity increases as the radius increases and as the current rise time decreases, and the pause decreases as the calculated induction amplitude increases.

The calculations performed on the assumption of incompressible liquid with an ideal conductivity can be used for description of the deformation of a winding. The authors of [45] solved the equations of hydrodynamics with regard for the boundary condition followed from the assumption of ideal conductivity. At every step of time the field satisfying the condition of the absence of the induction normal component at the boundary was calculated. Thus it was possible to determine the tangential component, the displacement of the medium elements, and to form a new boundary. The program developed by V. V. Titkov enabled the description of the medium flow which causes the formation of the specific configuration described above (see Figure 9.3). Concurrently, these calculations for the biconical magnets gave the values of velocity close to the measured and evaluated values obtained in the model of spherical flow.

The approximate formulas obtained in the model of stationary flow are convenient because they enable evaluating the requirements for the energy sources needed for the generation of multimegagauss fields in thick-wall turns with a high initial geometric factor. Here, as opposed to the estimates given in Table 9.3, we use the model of current source.

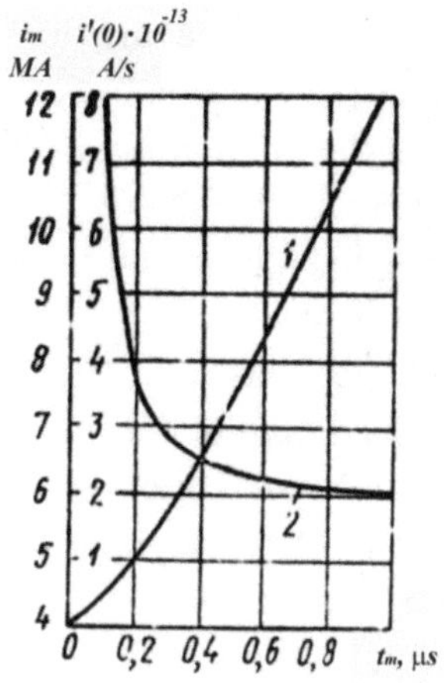

Fig. 9.13: Parameters of the current pulse required for producing a 1000 T-field in a biconical single-turn magnet. (1) i_m; (2) $(di/dt)_0$.

For example, let us calculate the source parameters needed for generating the field with induction 10^3 T in the biconic solenoid with aninitial length of 1.5 mm and initial diameter of 2 mm (parameter $a = 0.2$). Figure 9.13 shows the curves permitting the connection of the current pulse parameters: the amplitude i_m, initial rise velocity i' and rise time t_m in the example under consideration. With a small rise time the induction amplitude is close to the calculated one. With growing t_m the expansion of the turn occurs before current maximum, and tor achievr the desired induction it is necessary to increase the current. Also, with growing t_m the asymptotic regime is set down which can be described by formula (9.12). In the regime when $i_m = 7.4$ MA, $t_m = 0.5\,\mu$s, $i' = 2.3 \cdot 10^{13}$ A/s the value B_m determined by formula (9.14): $B_m \approx B_\infty$. In the circuit with total inductance 5 nHn, the given initial steepness and current can be provided if the capacity bank with energy of 140 kJ and voltage of 115 êV is used. This example shows that in order to generate the field with induction of the order of hundreds of Tesla in the tiny turn, it is necessary to use the source with the relatively small energy storage and fast current rise. The parameters of such a scale can be reasonably achieved owing to up-to-date high-voltage impulse technology.

9.4 Electrical explosion of turns of small thickness. Evaluation of the induction achieved in the destruction of turns with small initial dimensions

Numerical modeling of a two-dimensional magneto-hydrodynamical flow enables us to describe the destructive process of a thin-wall turn and compare the results with the experimental data, without using the assumption on ideal conductivity and the model of incompressible liquid. This is due to the fact that for a turn thickness of the order of 1 mm it is impermissible to assume the skin-layer to be equal to zero, and, moreover, it is necessary to take into account the peculiarities of the fluid in which the phase transition occurs and the conductivity changes.

To describe the processes occurring in a one-turn magnet Mielke and Novac [46] used a simple model: they divided the turn body into elements in the form of rectangular rings. Further on, one should solve the system of equations describing the transitional process in the system of magneto-coupled circuits. This model describes well the two-dimensional nonlinear diffusion of the field into the conducting material and takes into account edge effects. It was of considerable use in calculations of the devices working in conditions of a pronounced skin effect, including the magnetic systems used in magnetic forming [47]. In [46] the radial displacement of elementary rings, when exposed to electromagnetic forces, was taken into consideration. The actual flow in thick-wall turns has a more complicated two-dimensional character, as was shown above. In this case one has to describe with the necessary completeness the pressure distribution and changes of the turn profile in the course of the medium flow in the axial direction. A more rigorous description requires the joint solution of the hydrodynamical equations and Maxwell's equations [31, 43]. In a thin magnet the nonlinear field diffusion leading to the current "tightening" to a mid-plane should be taken into account. A peak in the radial distribution of hydrodynamical pressure can be formed in this plane, in the depth of the skin layer, which diminishes the rise of the inner radius. In this case the assumption of ideal conductivity is inapplicable and the complete description of the process is required. In current studies the method of finite elements is most used to describe two- and three-dimensional MHD flows. Along this, the results obtained by other methods are still meaningful. Here we will consider the flow in a thin-wall turn, as given by Krivosheev, Titkov, and Shneerson [31]. The system of equations was solved in this work describing the medium flow under the action of the electromagnetic forces and pressure gradient. It consists of the equations of continuum mechanics and Maxwell's equations. The first group of equations written in cylindrical coordinates has the form

$$\gamma \frac{D u_r}{Dt} = -\frac{\partial P}{\partial r} + f_r; \tag{9.20a}$$

$$\gamma \frac{D u_z}{Dt} = -\frac{\partial P}{\partial z} + f_z; \tag{9.20b}$$

$$\frac{D\gamma}{Dt} + \gamma\left(\frac{\partial u_r}{\partial r} + \frac{\partial u_z}{\partial z} + \frac{u_r}{r}\right) = 0; \tag{9.20c}$$

$$\gamma \frac{DE}{Dt} = P\frac{D\gamma}{Dt} + \delta^2 \rho; \tag{9.20d}$$

$$P = P(\gamma, T); \tag{9.20e}$$

$$E = E(\gamma, T). \tag{9.20g}$$

Here D is an operator of a substantional derivative, (9.20a) and (9.20b) are equations of motion of the nonviscous liquid, (9.20c) is the equation of continuity, (9.20d) is the energy equation, and (9.20e) and (9.20g) are state equations. In this equations P and E are the pressure and inner energy per unit mass of a substance, γ and T are the mass, temperature u_r and u_z are theradial and axial velocity components, f_r and f_z are radial and axial components of the electromagnetic force, δ is the current density, and ρ is the specific resistivity. The values f_r, f_z, and j should be found in the course of solving the second problem: a determination of the induction in the solenoid volume at any instant of the time span of the experiment. To describe the magnetic field, the flux function was used, which satisfies the equation

$$\frac{\mu_0}{r\rho}\left(\frac{U(t)}{2\pi} - \frac{D\psi}{Dt}\right) + \frac{\partial}{\partial r}\left(\frac{1}{r}\frac{\partial\psi}{\partial r}\right) + \frac{\partial}{\partial z}\left(\frac{1}{r}\frac{\partial\psi}{\partial z}\right) = 0, \tag{9.21a}$$

where $U(t)$ is the voltage produced by an external source at current inputs of the solenoid; μ_0 is the magnetic constant. Beyond the solenoid cross section $\rho = \infty$, and instead of (8.16a) we have the equation

$$\frac{\partial}{\partial r}\left(\frac{1}{r}\frac{\partial\psi}{\partial r}\right) + \frac{\partial}{\partial z}\left(\frac{1}{r}\frac{\partial\psi}{\partial z}\right) = 0. \tag{9.21b}$$

At the solenoid boundary the conditions of continuity of the normal and tangent induction components must be satisfied. Physical property of the medium are given by a specific appearance of the equations of state (9.20e), energy (9.20g), and specific resistance ρ depending on density γ and temperature T. In conditions of high temperature MHD processes in metals the three-term equations of state are usually used [48]:

$$P(\gamma, T) = P_x(\gamma) + P_T(\gamma, T) + P_e(\gamma, T), \tag{9.22a}$$

$$E(\gamma, T) = E_x(\gamma) + E_T(\gamma, T) + E_e(\gamma, T). \tag{9.22b}$$

These contain three terms of e pressure and three terms of compression energy describing the "cold" motion of ions:

$$P_x(\gamma) = c_0^2\gamma_0(\gamma/\gamma_0)^2(\gamma/\gamma_0 - 1), \tag{9.23a}$$

$$E_x(\gamma) = \frac{1}{2}c_0^2(\gamma/\gamma_0)^2(\gamma/\gamma_0 - 1)^2, \tag{9.23b}$$

the thermal motion of ions:

$$P_T(\gamma, T) = \Gamma\gamma C_\rho T, \tag{9.24a}$$

$$E_T(T) = C_\rho T, \tag{9.24b}$$

and motion of free electron gas:

$$P_e = \frac{1}{2} E_e \gamma, \tag{9.25a}$$

$$E_e = \frac{1}{2} \beta_0 \left(\gamma/\gamma_0\right)^{-1/2} T^2, \tag{9.25b}$$

where c_0 and γ_0 are the velocity of sound and mass density in a metal at normal conditions, $\Gamma \cong \Gamma_0 = 2$ is the Gruneisen parameter, C_p is the specific heat of the ion component of a metal, and β_0 is the coefficient of the electron specific heat.

These equations of state are applicable in the range of temperatures of the order of several hundreds degrees up to 50 000 K. In developing of the physical model of a conductor in a ultrahigh magnetic field the most complicated problem is the description of the electric resistivity in a broad range of densities and temperatures. At field levels above approximately 450 T (for copper) the specific heat exceeds the sublimation energy. Therefore in a this section a simplified description of the conductance was used, where with decreasing mass density of a conducting material up to some critical value, for example, $\gamma_k \cong 0,1\gamma_0$ the conductivity disappeared instantly. At densities corresponding to the solid and liquid phase, the conductivity can be described with a help of relation (3.4a). Thus we have

$$\begin{aligned} \rho &= \rho_0(1 + \beta\, q') && \text{at} \quad \gamma > \gamma_k, \\ \rho &= \infty && \text{at} \quad \gamma \le \gamma_k \end{aligned} \tag{9.26}$$

where $q' = E\gamma$.

Taking various values of γ_k in sampling numerical calculations (for example, $\gamma_k/\gamma_0 = 0.1$, 0.5, 0.9), one can ascertain that variations of this parameter appear only slightly in the ultimate results. So, one can conclude that the loss of conductance occurring due to the liquid–vapor phase transition is not a basic factor determining the picture of the MHD flow of a solenoid in the considered regimes. Therefore, in numerical calculations the parameter $\gamma_k = 0.1\gamma_0$ was used. Such an approach is also confirmed by more detailed studies of one-dimensional flow at electrical explosion, which we will consider in the next chapter. It is worth noting that the intensity of electrical field at the boundary of mediums in the fast-rising field can approach 10kV/cm and even exceeds this value. In such a strong field the ionization processes can begin in both metal vapors and air. In the this model this effect is not taken into consideration.

For numerical integrating of the system (9.20)–(9.21) we used the Lagrangian finite-difference mesh, consisting of quadrangle cells with nods moving together with the substance. In this case the artificial pressure is added to the actual equation of metal state, which is proportional to $(D\gamma/Dt)^2$ and designed for spreading of the shock steps on the finite-difference mesh. The solenoid boundary is free from the action of surface forces and therefore the condition $P = 0$ is fulfilled on its surface.

Thus, the mechanical group of equations (9.20a–g) enables us, along with velocities and thermodynamical characteristics, to find the position of the nods of the Lagrangian mesh at any instant of time t if the forces f_r and f_z are known. The last are determined in the course of solution of the equation of the electromagnetic field at given values of coordinates r and z of mesh nods.

The modeling of electromagnetic fields in a solenoid of finite dimensions during the electroexplosion processes required the development of a special technique. This technique was developed by Titkov. It is based on the combination of the mesh method for the solution of equation (9.21a) in the conducting volume and method of boundary elements for the solution of the external problem (in nonconducting volume).

A peculiarity of the numerical modeling of electromagnetic fields generated by single-turn magnets, as applied to the conditions described above, is the fact that from the experimental data not the voltage $U(t)$ on the solenoid is known, but its total current $i(t)$. This complicates the application of equation (9.21a), where $U(t)$ appears. Therefore at every step in time in the course of iterations the voltage should be chosen in a way as to provide the current value known from the experiment.

Figure 9.8 shows the calculated results derived for the experiments with the single-turn magnets of small dimensions (0.3 cm length, 0.2 cm inner radius, 0.2 cm wall thickness). In these experiments the field level of 360 T was achieved at the current 1MA with a rise time up to a maximum 0.8 μs. According to calculations, the turn shape changes in the process. There is an axial flow accompanied by the appearance of zones of low density near the edges. The field of velocities exhibits a two-dimensional flow character: near the angles the radial and axial components of velocity are close in value. In a mid-plane the dependence of the velocity on the radius has a maximum shifted in the depth by fractions of mm from the inner surface. The axial displacement of the medium into the skin-layer zone leads to a diminishing of the link of this layer with the rest of the medium. This manifests itself in the formation of a local minimum of pressure. The current distribution in a short thin-wall turn heated by current differs significantly from the case of one-dimensional diffusion into the immovable wall. The spatial distribution of the current density is affected by the two-dimensional character of the diffusion and faster penetration of the current into the medium near angles. At the moment of time 1,2 μs the temperature in this region achieves 14 000 °K. Due to reduction of the conductivity in the vicinity of angular points, the current concentrates in the middle part of a turn, which, in its turn, leads to higher heating in this region as well. As a result, compared to the case of one-dimensional diffusion, a faster radial displacement of the current density peak occurs. This process continues for the whole current rise time. The distribution of electromagnetic forces in a symmetry plane exhibits a sharp maximum, the position of which is shifted together with a current wave.

The medium density is noticeably reduced near the angles, where the medium expansion is faster. Note that in the process of heating and expansion of a conducting material, the point characterizing the medium state on the plane $P = f(T, 1/\gamma)$ passes above the critical point. This is true for all elements of the medium, including the boundary element.

At some distance from the boundary in the turn body the pressure peak is formed, and in the region between the peak and boundary, where $P = 0$, there is a drop in pressure, which reduces the action of electromagnetic forces. Along with this, the motion of the elements close to the boundary becomes slower. In the region between the boundary and peak a "reverse" flow is formed in reference to the maximum pressure point. The velocity of the medium element on the inner surface (the velocity of coil inner radius rise) is significantly less than its maximum value. In the mid-plane near the inner boundary the density is less than in the turn thickness: the radial dimension of the zone where the density is less than $0.8\gamma_0$ is 0.5 mm up to the moment 1.2 μs. Thus "swelling" of the medium occurs, which results in a slowing of the inner radius rise, and even in its decrease. This effect is clearly seen on the x-ray photographs presented in [49] (Figure 9.7). A reverse motion of the boundary takes place in spite of the fact that the medium is subjected to intense magnetic pressure.

Thus there are grounds to believe that the observed displacement of the boundary and the delay of the radius rise, as opposed to calculation where ideal conductivity is assumed, are caused by the explosion of the skin layer in the medium with actual conductivity. The edge effect is of considerable importance, since it leads to a decrease of the turn effective length and to a increase of the current density in the middle part of a turn.

In the published reports on experiments with thick-wall turns this effect, namely, the delay of the radius rise, as opposed to a model where ideal conductivity is assumed and inertial delay is considered, hast not been pointed out. As was mentioned above, in thick-wall turns the heating of the skin layer occurs less intensively. It is possible that in these experiments the radius rise began after an additional short delay caused by the "swelling" of the skin layer [49], but this effect only weakly affected the inner radius rise. More detailed computational analysis of an abundance of available experimental data, analogous to the above-described, will be the task of future studies. Their findings could provide the supplementary data on the influence of the initial dimensions, turn configuration, and rate of current rise on the hydrodynamic flow and behavior of the geometric factor.

The available data and calculated results show that the explosion of a small thin-wall turn is a combination of several mutually-connected processes which depend in different degrees on the material properties and turn dimensions. Insignificant variations of the initial conditions and of the character of the current rise can noticeably change the course of the process in this nonlinear system. Therefore even at similar initial conditions the experiments can differ significantly in the measured geometrical factor.

It is reasonable to compare the experimental data with the simplest evaluation of the induction amplitude given in equation (9.14). It describes the particular case considered above of the flow in a biconic magnet. Along this, the dependence of the form (9.10b) $B_m = \text{const} \cdot \mu_0^{3/4} \gamma_0^{3/4} (di/dt)^{1/2}$ follows from considerations of the dimension for magnets with other configurations. The dependence is valid under the following conditions: the current increases linearly, there is the hydrodynamic flow of incompressible ideally conducting medium; the geometric similarity is retained with time; the inner current radius is much more than the initial radius. Another words, this dependence describes some kind of asymptotic regime of the flow. The similar picture is realized in experiments with thick-wall turns, in which the condition $G_m/G_0 < 1$ is justified. The induction reaches a peak at the pulse front, when the current still increases linearly as $i = i't$. Also, due to a radius rise the peak induction is achieved before the peak current, and the induction amplitude is less than its calculated value. The calculated results are presented in Table 8.1 and Figure 9.14.

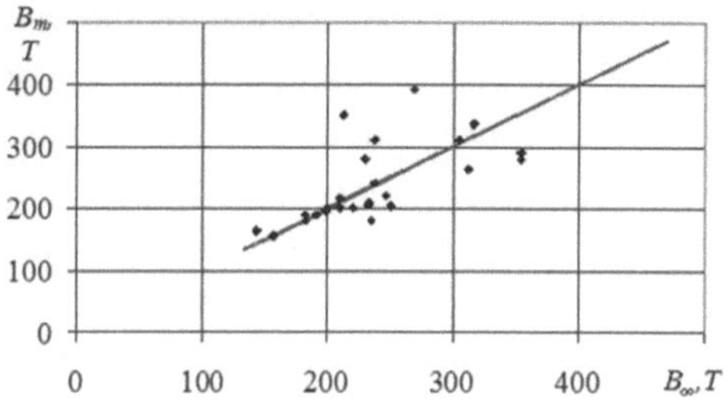

Fig. 9.14: The experimental values of the induction amplitude in copper turns, as given in Table 9.1, in comparison with the values calculated by formula (9.14).

It is worth noting that the given formula is inapplicable if the geometrical factor is retained close to its initial value up to the moment of a current peak. This occurs in a weak field and in the case of coils with a relatively large initial diameter (of the order of 10 mm or more) and at μs-duration pulses. In such turns the radius increase to the moment of a current peak is small compared to its initial value. The given evaluation is inapplicable as well in the case of magnets of large length, the deformation of which does not lead to a reduction of the geometrical factor (for example [34]). For other cases the experimental results are shown in Figure 9.14, where the dependence given by formula (9.14) is also presented. It describes the mentioned particular case and does not take into consideration the diversity of the initial geometric factors, but gives the estimating value of induction in destructive thick-wall turns. These estimates are somewhat reduced, since they ignore the inertial delay of establishing of the flow close to the stationary one. Formula (9.14) does not consider the peculiarities of an explosion of thin-wall turns which cause a delay of the geometric factor decrease. However, it does take into account the basic effect – the presence of hydrodynamic flow. Therefore, our very simple evaluation here is suitable for a significant body of experiments, including the experiments with thin-wall magnets.

9.5 One-dimensional hydrodynamic flow in the wall of a single-turn magnet. Shock wave in conductors initiated by superstrong magnetic fields

Although most of the experiments on the generation of superhigh fields is performed with short coils, it is important to consider the flow in the magnets of length much larger than the distance which the sound wave passes in the metal for the characteristic time of the process. Here we assume the thickness of the wall to be unlimited, and so the reflection of the wave from the external boundary can be disregarded. For the current rise time $t_m = 1$ μs and 10 μs the running time of the sound wave in copper is 4.5 and 45 mm, respectively. Thus, even for $t_m = 1$ μs the model of the shock wave is valid if the length of the turn and thickness of its wall is only a few centimeters. This model will be considered in this chapter. The shock wave causes the displacement of the inner boundary of the magnet. The additional factor affecting the velocity of displacement of the effective field-conductor boundary is the bursting of melted metal and the loss of conductivity at the phase transition (evaporation), resulting in the disruption of the skin layer. These processes, by analogy with the electric explosion of wires, could be defined as the electric explosion of the skin layer. They will also, to some extent, be considered in the following sections of this chapter.

In an ideally conducting medium the magnetic field plays the role of a piston forming the wave. Although the flow initiated by the variable magnetic field in a general case is nonstationary, we will use the relations describing the stationary process. Comparison with the numerical calculations made in that rigorous treatment shows that such an approximation is of acceptable accuracy.

It is known that u_f, the velocity of the flow behind the shock wave front propagating in nonperturbed medium, and the velocity of the shock wave front D are coupled by the linear relation

$$D = C_1 + \lambda u_f, \tag{9.27}$$

where C_1 is close to the sound velocity (see Table 3.1), and λ is the numerical parameter (for copper $C_1 = 4{\cdot}10\,\text{m/s}$, $\lambda = 1.53$ [50]). In a stationary regime, u_f equals the velocity of the boundary field-conductor displacement, and the pressure on the front is equal to the magnetic pressure. Let us use one of the Rankin–Hugonio conditions:

$$P_f = \frac{B_e^2}{2\mu_0} = \gamma_0 D u_f \tag{9.28a}$$

or

$$\frac{B_e^2}{2\mu_0} = \gamma_0 u_f \left(C_1 + \lambda\, u_f\right). \tag{9.28b}$$

Further on we find

$$u_f = \frac{C_1}{2\lambda}\left[\sqrt{1 + \frac{B^2}{\bar{B}^2}} - 1\right], \tag{9.29}$$

where $\bar{B} = C_1(\mu_0\gamma_0/2\lambda)^{1/2}$. In case of copper $\bar{B} \approx 250$ T.

In the limiting case when $B \ll \bar{B}$, the velocity of the boundary caused by the compressibility of a medium is proportional to B^2:

$$u_f \approx \frac{C_1 B^2}{4\lambda\,\bar{B}^2} = \frac{B^2}{2\mu_0\gamma_0 C_1}. \tag{9.30a}$$

In this case the velocity is significantly less than the Alfven velocity: for instance, in a field with induction 100 T the wall of the copper solenoid displaces with a velocity of about 80 ms, while Alfven velocity is $V_A \approx 10^3$ ms. In short solenoids, due to the flow through the faces considered in (9.2), the displacement velocity becomes close to V_A. The above estimate shows that in relatively wea fields the additional contribution to the displacement velocity, connected with the compressibility, is small. In a field with induction $B \gg \bar{B}$ the velocity of the boundary movement becomes proportional to Alfven velocity:

$$u_f \to \frac{B}{\sqrt{2\lambda\,\mu_0\gamma_0}} = \frac{u_A}{\sqrt{2\lambda}}. \tag{9.30b}$$

In such fields the flow in short solenoids should be considered, taking into account of the actual equation of the medium state. The model of a noncompressible liquid in these conditions could introduce significant error.

The shock wave model is useful for calculations of processes in long solenoids. It is convenient to approximate the dependence determined by formula (9.29) by the power function. In the range of inductions $0 < B \le 4\bar{B}$ the following approximation [18] will be valid:

$$u_f \approx a_0 B^{3/2} \approx 0,43 \frac{C_1}{2\lambda} \left(\frac{B}{\bar{B}}\right)^{3/2} . \tag{9.31}$$

Here $a_0 = 0.43(\mu_0\gamma_0)^{-3/4}(2\lambda)^{-1/4}C_1^{-1/2}$. For copper $a_0 = 0.15$.

Further on we shall calculate the induction in the gap (see Figure 9.11b) and in a thick-wall long solenoid, switched on from an infinite power source. In the first of these tasks, equalizing the voltage of the source V_0 and the change of the flux $(d/dt)(ghB)$, using formula (9.29), for the stationary regime when $B = \text{const}$ and $dh/dt = u_f$, we obtain the expression:

$$U_0 = \frac{gBC_1}{\lambda}\left(\sqrt{1 + \frac{B^2}{\bar{B}^2}} - 1\right). \tag{9.32a}$$

In the range of fields less by approximately $4\bar{B}$, when approximation (9.31) is valid, we have

$$U_0 = 2g\,a_0\,B^{5/2}, \tag{9.32b}$$

whence

$$B = \left(\frac{U_0}{2ga_0}\right)^{5/2} . \tag{9.32c}$$

If $U_0 = 10^4$ V, $g = 10^{-2}$ m, $a_0 = 0.15$ (copper), $B = 390$ T, and $u_f \approx 4.5 \cdot 10^3$ m/s. In an asymptotic regime, when $B \gg \bar{B}$,

$$U_0 = \frac{2gB^2}{\sqrt{2\lambda\,\mu_0\gamma_0}}, \tag{9.33a}$$

$$B = \left(\frac{U_0}{2g}\right)^{1/2} (2\lambda\,\mu_0\gamma_0)^{1/4}. \tag{9.33b}$$

The latter relation is close to equation (9.7). From the given relationships it follows that, as in the regime of a two-dimensional flow of noncompressible liquid, in order to generate the field with an induction of few hundred Tesla, high voltage sources are necessary: the value $\langle E\rangle = U_0/(2g)$ should be of the order of tens of kV/cm. For instance, to obtain the magnetic induction field of a thousand Tesla it is necessary to provide $\langle E\rangle = 54$ kV/cm. This estimate determined from the analysis of the simplest slit model also remains applicable for the actual configuration of a single-turn magnet. If its length and wall thickness are large enough, then the flow becomes cylindrical instead of flat. Nevertheless, even in this case, for the calculations we will use the equations referring to a steady-state flat flow. One can ascertain, by comparison with more accurate numerical calculation, that the error is insignificant if the relative increase of the inner radius does not exceed 1.5–2. Considering a field with induction much more than $\bar{B}$, we use the formulas (9.33a) for the velocity of the boundary movement, and the equation $U_0 = (d/dt)(\pi R_1^2 B)$. The solution of this task differs from what was presented above in the calculations of the expansion of the solenoid with Alfven velocity only by the values of B_f and t_f in formulas (9.9a) and (9.9b). In the case of a shock wave,

$$B_f = (U_0/\pi R_0)^{1/2} (2\mu_0\gamma_0\lambda)^{1/4}, \quad t_f = R_0 (\pi R_0/U_0)^{1/2} (2\mu_0\gamma_0\lambda)^{1/4}. \tag{9.34}$$

In Table 9.3 we present the values of the average strength of the electric field $E_{AV} = U_0/2\pi R_0$ as applied to the task of generating fields with induction 500, 700, and 1000 T in copper and tantalum solenoids. These values are close to those that correspond to the motion of the boundary with Alfven velocity.

Thus, from the point of view of the requirements for the parameters of the power supply source, both the flow of noncompressible liquid and the flow behind the front of the shock wave differ insignificantly. The voltage of the source required for the generation of the field with the induction of the scale of 10^3 T should be of the order of hundreds of kV if the perimeter of the cavity of the single-turn magnet is of only a few centimeters. The megavolt small inductance-forming lines are sources of this kind. The experiments performed at SANDIA Lab (USA), which used such a line, displayed the capabiliy of reaching fields with induction of above 1000 T [34].

9.6 General information on the electric explosion of conductors

Experimental data shows that single-turn solenoids subjected to the action of the super strong magnetic field carry traces of not only plastic deformation, but also of the processes connected with the heating of conductors by the pulsed current. The inner coil surface is often fused and the mass of samples after the discharge is noticeably less than at the initial stage. As mentioned in Section 9.1, the ejection of metal from steel samples is significant, and in the case of fused metals this is the dominating factor determining the residual change of diameter after one discharge.

These facts can be attributed to the electric explosion resulting from the heating of the conductor by a pulsed current. Depending on the solenoid wall thickness the skin effect could be declared either sharp or weak. Therefore, it is meaningful to consider first the basic processes of the electric explosion with the absence of the skin effect, which takes place both at the explosion of the magnets, the wall thickness of which is less than the thickness of the skin layer, and at the explosion of wires and foils in the experiments on switching, acceleration of projectiles, etc.

The electric explosion of the conductor (EEC) is the process of its disruption as a result of electric current action. The phenomenon of EEC has been known since the late 18th century: however, its detailed study has begun only in the last decades, after the appearance of such experimental technology such as electron oscilloscopes, fast image convertors, X-ray radiography, etc. A large number of theoretical and experimental studies have been devoted to the detailed description of the processes developing during EEC. The results of these studies are summarized in [39, 51–53]. In a typical experiment on electric explosion, the wire or foil is placed in the break of the discharge circuit of the capacity bank. If the resistance of the circuit, including the initial resistance of the exploded conductor, is small enough, the current first varies according to the sinusoidal law, like in the conventional high quality *RLC* contour. The voltage in the wire is small. As the conductor is heated, its resistance rises, resulting in anomalies in the current curve and growth of the voltage on the conductor. The peculiarities of this process and the current–voltage characteristics could be quite different, depending on the circuit parameters, the storage energy, the material, the cross section and length of the conductor, and the properties of an ambient medium. Simultaneous oscilloscope study of the current in a onductor and the active voltage drop on it allows us to determine the dependence of the resistance on the energy input. If the conductor preserves its initial shape and size, then at relatively modest energy inputs its resistance increases approximately exponentially, depending on the current action integral (see Chapter 3) in correspondence with the tabulated dependence of conductivity on the energy input. The growth of the conductor's resistance may strongly affect the transitional process and lead to an increase of damping and transformation of the discharge to the aperiodic phenomenon. In the case of purely inductive circuit with resistance increasing according to the law, $R = R_0(1 + \beta\,\Delta q')$ the current with initial value i_0, is damping according to

$$i = i_0 Z\left(1 + \frac{Y_0^2}{2}\right)^{1/2}\left(1 + \frac{Y_0^2 Z^2}{2}\right)^{-1/2}, \tag{9.35a}$$

where $Y_0 = (i_0/S)(\rho_0 \beta L/R_0)^{1/2}$, $R_0 = \rho_0 l/S$, l is the length, S is the cross section of the conductor, and L is the inductance of the circuit [54]. In this case the voltage drop on the resistance is described by

$$U = i_0 R_0 \left(1 + \frac{Y_0^2}{2}\right)^{3/2} Z \left(1 + \frac{Y_0^2 Z^2}{2}\right)^{-3/2} . \tag{9.35b}$$

In these formulas $Z = \exp[-(tR_0/L)(1 + Y_0^2/2)]$. Under the condition $Y_0 > 1$ the voltage first increases, attaining the maximal value

$$U_m = \frac{U}{3\sqrt{3}\,Y_0} \left(2 + Y_0^2\right)^{3/2} , \tag{9.35c}$$

then drops. The processes in which resistance increases in correspondence with the tabulated value could be defined as the processes in circuits with a heated conductor. In contrast, at the electric explosion the dependence of resistance on the current action integral becomes more sharp than the exponent corresponding to the tabulated values of ρ. Such "over-exponential" regimes could be caused by the purely mechanical disruption of the conductor under the action of electromagnetic forces due to th edevelopment of MHD instabilities, as well as by its complete or partial evaporation.

With increasing resistance of the conductor during electric explosion in the circuit, the current more or less sharply decreases and, depending on experimental conditions, breaks down to zero or, after diminishing, begins to grow again. In the latter case this peculiarity appears on the current curve. Ata complete break of the current in a system with capacity storage, residual voltage on capacitors may persist. In this case a second breakdown of the gap is possible along with the recovery of its electroconductivity, after some time interval, referred to as the current pause. The current pause, under other equal conditions, is shorter the smaller the length of the conductor is. There is a "critical" length of wire or foil, at which the pause disappears, and the current does not totally terminate (Figure 9.15).

The voltage on an exploding conductor has, as a rule, the shape of a pulse with an increase and a following decay. Such a dependence of the voltage on time is inherent to the inductive circuit with growing resistance, as shown above in the example of resistance linearly rising with heating. If the resistance grows to infinity, the current in the circuit drops to zero and the voltage initially increases and then tends to zero.

In actual conditions of electric explosion, along with the disruption of the conductor and disappearance of metallic conductivity ionization of the metal vapor and ambient gas occurs. As a consequence, the resistance of the part of the circuit with the exploding conductor initially increases and then may begin to drop. In particularly, if the current break is not complete the processes of ionization lead to the formation of the electric arc before the current disappears in the circuit.

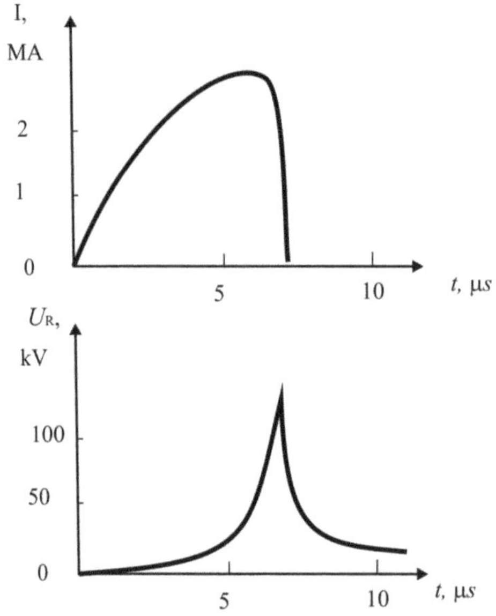

Fig. 9.15: The typical current-voltage oscillograms of an electrical foil explosion foil [51].

The hypervoltage arising at EEW strongly affects the ionization processes in the gap, and the recovery of the conductivity, in its turn, restricts hypervoltage, both in the regimes with an incomplete and with complete current break. The properties of an ambient medium exert an significant influence upon the conductivity recovery and hypervoltage amplitude. Thus, in the experiments described in [51, 52], the highest values of hypervoltage were obtained with a foil explosion in quartz dust. From the preceding it follows that EEW is quite a complex process, in which metal makes a transition from the solid phase to plasma. This transition is characterized by dramatic changes in the thermodynamic properties of the material, its electroconductivity, and, as a rule, by the appearance of strongly pronounced inhomogeneity.

The state equations describing the metal properties under local thermodynamic equilibrium show the quite complex relationship between pressure, density, and temperature. As an example we present the isotherms in Figure 9.16. Point ***A*** in this figure corresponds to the initial state of the conductor, and point S to the final state. For the final state the typical values are $\mathrm{T} = 10^4$–10^5 K, $\gamma \ll \gamma_0$. Transition from the initial state to the final one is marked by the line on the plane $(P, 1/\gamma)$, the appearance of which is determined by the process of the heating and expansion of the conductor and can be found by the solution of hydrodynamic equations, along with the state equation and the equation describing the dependence of electroconductivity on temperature and density. It should be taken into account that at the slow evaporation and $T_F < T_k$, where T_k is the critical tempera-

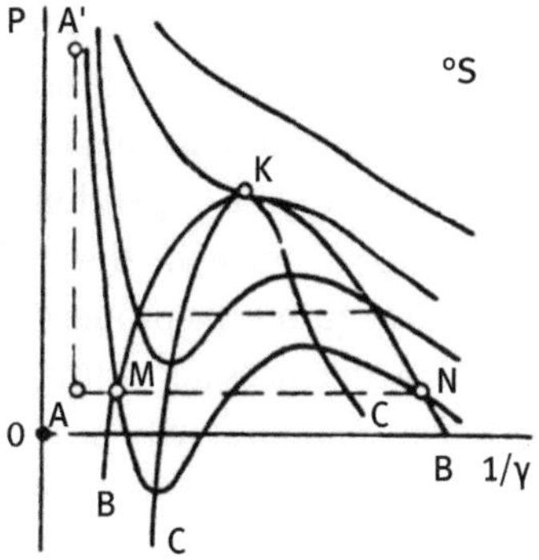

Fig. 9.16: Qualitative diagram of the state of metal. K is the critical point.

ture, T_F is the Fermi temperature (in a case of copper $T_k \approx 0.7$ eV or about $8 \cdot 10^3$ K, $T_F \approx 5$ ev or about $6 \cdot 10^4$ K), the isobaric expansion takes place, which described by the line AMN in Figure 9.16. When the processes are fast, states are possible corresponding to the region between the binodal BKB and spinodal CKC and the states with negative pressures (pulled out liquid) as well. The states corresponding to the region below spinodal cannot exist.

The most complete equations for a wide diversity of states of copper, aluminum and other metals were obtained by Fortov and colleagues [55]. The simplified equations obtained in [50, 56, 57] and others were taken up for specific conditions of wire explosions.

Each point in Figure 9.16, besides the region below the spinodal, corresponds to the determined value of conductivity. It is known that the high conductivity of metals in the solid and liquid phase is caused by the presence of degenerate "gases" of free electrons. These electrons populate the free-energy zone (conductivity zone) which is formed under the near approach of atoms. The lattice defects connected with the thermal oscillations of atoms cause the temperature dependence of the conductor resistance. Further, in Chapter 3 we considered the dependence of conductivity on the density in the range of comparatively low (subcritical) temperatures. In solid phases, the conductivity rises with the increase in density according to [2]:

$$\sigma = \sigma(\gamma_0)\left(\frac{\gamma}{\gamma_0}\right)^{2,7}. \tag{9.36}$$

In this formula $\sigma(\gamma_0) = 1/\rho(\gamma_0)$, where $\rho(\gamma_0)$ depends linearly on the temperature (see Chapter 3). Conductivity as a function of temperature and concentration in a broad range of their variations was studied in several works. The qualitative models of this dependence were considered in earlier publications [59–61]. To understand the character of this dependence in the regions of high density let us consider the first hypothetical case, when the density is close to the initial one

($\gamma = \gamma_0$), and the temperature increases (the line AA' in Figure 9.16). The similar situation could be realized when the conductor is heated so fast that it does not have enough time to expand. The average atomic concentration in this case remains constant. One would expect that at a density close to $\gamma_0$the dependence $\sigma(T)$ acquires a U-like character: at $T \to 0$, conductivity grows, since the thermal oscillations of ions in the lattice disappear; at $T > T_F$ metal acquires the properties of fully ionized nonideal plasma, the conductivity of which increases according to $T^{3/2}$ in accordance with the Spitzer formula. The increase in conductivity begins at $T \geq 10^5$ K, which takes place in a temperature range attainable only in very powerful discharges.

The second hypothetic case corresponds to a fast expansion of the medium and to a subsequent heating while retaining low concentration. Under expansion of the conductor (for example, when the process is described by the line AMN, Figure 9.16) electric fields of ions do not overlap any more, the configuration of energy levels drastically changes, and free electrons disappear. The so-called de-collectivization of free electrons occurs [59]. The conductivity sharply drops by a few orders with an increase of interatomic distance approximately twofold or with a decrease of the density by one order of magnitude. In spite of the fact that the electrons appear under conditions of the thermodynamic equilibrium due to thermal ionization, one can state that at $T \geq 10^5$ K the specific resistance is quite small.

The authors of [62–66] calculated the dependence of conductivity on temperature and concentration assuming thermodynamic equilibrium. The expected character of these dependences is confirmed by calculations carried out in [63, 66]. Their results are presented in Figure 9.17 from the review [53]. This figure shows the U-shaped behavior of the curve $\sigma(\gamma, T)$ at initial density, and the sharp

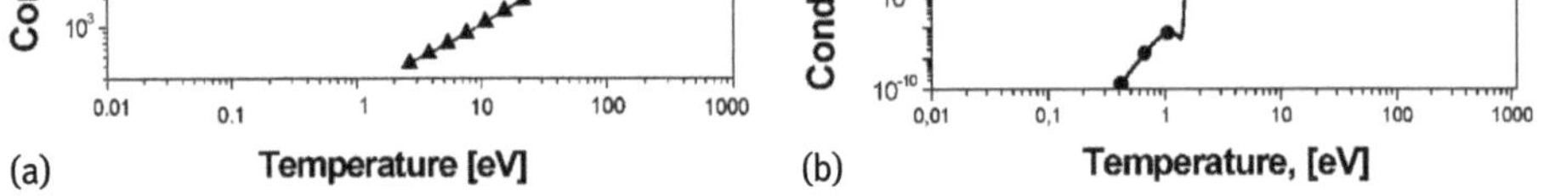

Fig. 9.17: Conductivity of copper with density 8.9 g/cm^3 (a) and 10^{-3} g/cm^3 (b). *Solid line* corresponds to the model given in [66]; (•) model [63]; (▲) Spitzer's model.

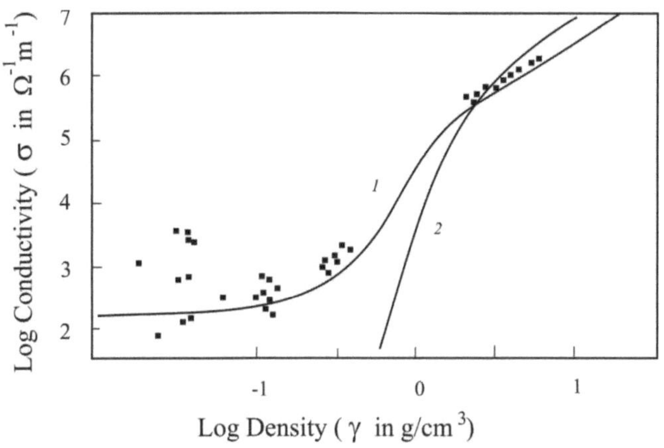

Fig. 9.18: Conductivity of copper as a function of density at temperature 6000 K. Curve 1 and experimental points are from Desjarlais's work [64]. Curve 2 is constructed by formulas given in the work of Bakulin and Luchinskiy [66].

drop of the conductivity for a medium of small density. At high temperature the conductivity is described by Spitzer's formula.

Figure 9.18 shows the conductivity as a function of density, according to calculations by Desjarlais [64] for copper at temperature 6 000 K, as compared to the experimental results. Note that the conductivity decreases approximately by a factor of 200 with a decrease in the density of copper from the initial value up to 1 g/cm^3.

Figure 9.19 shows the results of calculations according to the semiempirical model of Bakulin and Luchinskiy [66] for a wide range of parameters. In the range of density $(0.1-1)\gamma_0$ the results are close to those obtained in [64]. Curves 1 and 5 in Figure 9.19 show conventionally two dependences for the conductivity, corresponding to the mentioned hypothetical regimes. The U-shaped character $\sigma = f(\gamma, T)$ takes place for density higher than $0.3\gamma_0$. A minimum of conductivity corresponds to $\sigma \approx 10^{-2}\sigma_0$, where σ_0 is the conductivity of the metal at the initial state [65].

In the process of electrical explosion accompanied by heating and expansion of plasma, the conductivity of each element of the medium is described by the temperature dependence lying between the hypothetical ones. It should be noted that the presented data do not reflect some fine details of the metal electroconductivity, for instance, its disappearance near the critical point due to electron scattering during density fluctuations. Other processes that occur which are beyond the framework of the assumption on the local thermodynamic equilibrium may produce a noticeable influence on the electric properties of the medium. These

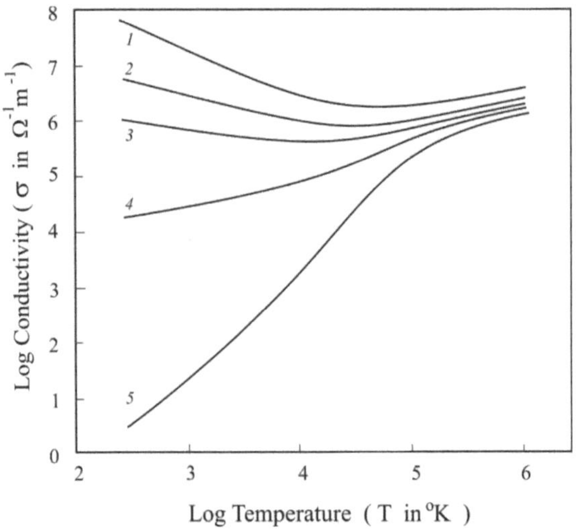

Fig. 9.19: Conductivity of copper as a function of temperature and density as calculated by the method of Bakulin and Luchinskiy [66]. Curves 1 to 5 correspond to the following values of the ratio γ/γ_0, respectively: 1, 0.8, 0.6, 0.4, 0.2.

are, first of all, the impact ionization of metal vapor that can cause a noticeable growth of conductivity compared to its equilibrium value at a low temperature. The ppearance of "run-away" electrons may also contribute to this. Along with that, at high current density, the excitation of the various kind of oscillations in metal plasma is possible, as well as the appearance of microturbulence and, as a consequence, the growth of the medium resistance. The nonequilibrium processes in the plasma produced at the EEW have not aet been practically studied.

Based on the local properties of the medium, it is possible to consider the phenomenon as a whole, if only the spatial distributions of the density and temperature are known. An electric explosion of conductors is an example of the process in which spatial distribution is sharply pronounced and plays an important role especially in relatively slow processes. Experiments made by Abramova, Zlatin, and Peregud [67] have shown that for pulses with duration of the order of 10^{-3} s the conductor, both in a solid and a liquid phase, acquires a s sprial shape, due to instability, and then breaks. At shorter pulses, before the full evaporation of the cylindrical conductor, the perturbations of the zero mode (flutes) appear, resulting in wire transformation into a sequence of cylindrical fragments with the metallic conductivity and breaks between them in which the discharge has the form of an arc in the metal vapor. The x-ray photographs show the alternate sequence of the dense opaque fragments and fragments filled with plasma of relatively low density which are transparent for soft x-rays. The onset and development of zero mode instability is affected by the initial perturbations of the conductor, forces of the surface tension, and electromagnetic forces. According to

[67] the maximal growth rate for the mode $m = 0$ is determined by the expression

$$\nu_m = \frac{V_A \cdot \Pi}{R_0}, \tag{9.37a}$$

where R_0 is radius of the conducting cylinder, the material of which is an ideal noncompressible fluid, V_A is Alfven velocity: $V_A = B_e/\sqrt{\mu_0 \gamma_0} = (i/2\pi R_0)(\mu_0/\gamma_0)^{1/2}$ (B_e is the induction on the conductor surface, and i is the current in it), and Π is the numerical term, which could be estimated as $\sqrt{2}$. The formula for characteristic time of the instability growth rate has the form

$$\tau_H = \frac{1}{\nu_m} = \frac{2\sqrt{\gamma_0}}{\Pi\delta\sqrt{\mu_0}}, \tag{9.37b}$$

where δ is the current density. For values δ in a range 10^{11}–10^{13} A/m^2, τ_H changes in the limits 10^{-8}–10^{-6} s. Thus, one could expect that only wth a rather small current increase time the conductor will expand uniformly along its length, and the "classic" EEW scheme take place, in which spatial inhomogeneity is not taken into account and attention is paid only to inhomogeneity along the radius.

The simplest criterion for the appearance of the "hyperexponential" growth of resistance can be attained, considering an electric explosion as the fast and uniform expansion of the conductor caused by energy input equal to the evaporation energy. If we assume that before this moment the linear dependence $\rho\ (\Delta q')$ is valid, then the criterion takes the form of the condition for the volume energy density

$$\Delta q' = \frac{1}{\beta}\ [\exp(\rho_0 \beta I) - 1] \geq Q_3, \tag{9.38}$$

where Q_3 is the volume density of the sublimation energy that is equal to the coupling energy of atoms in a lattice. This model, based upon the hypothesis of instantaneous loss of conductivity when achieving the threshold energy, can be considered as an idealized model. The experiments [68, 83], have shown that with short duration pulses (10^{-5} s and less) the approximate criterion of the onset of explosion (Anderson criterion) is valid. This has the form of the threshold relation for the current action integral. For copper, according to this criterion, an electric explosion occurs if $I \geq I_A \approx 2 \cdot 10^{17}$ A^2 m^{-4} s. For copper, the calculation using (9.38) gives $I \geq 4 \cdot 10^{17}$ A^{2}m^{-4} s, assuming $\beta = \beta_2 = 4.8 \cdot 10^{-10}$ m^3/J, $Q_3 = 5.5 \cdot 10^{10}$ J/m^3. This value by its order of magnitude is close to I_A. The Anderson criterion is rather approximate: it does not take into account that the initial point in time at which the resistance exhibits a sharp step depends on the increment of the energy input. With slow energy input, i.e., at relatively long discharges the "hyperexponential" growth of the total resistance begins before the Anderson criterion becomes applicable, due to the development of instability.

One can estimate the condition at which the sublimation energy will be received by the conductor before the development of instability occurs. For estimation we propose the current density to be constant. Then the above-mentioned condition acquires the form of inequality $\tau_H > \tau_A$, where ror the round-shaped conductor τ_H is determined by formula (9.33b) and can be found using the Anderson criterion $\int_0^{\tau_A} \delta^2 dt = \delta^2 \tau_A = I_A$. Thus, we have the inequality

$$\frac{2}{\Pi\delta}\sqrt{\frac{\gamma_0}{\mu_0}} > \frac{I_A}{\delta^2}, \tag{9.39}$$

or

$$\delta > \frac{I_A n}{2}\sqrt{\frac{\mu_0}{\gamma_0}} \approx 10^{12}\frac{\mathrm{A}}{\mathrm{m}^2} = 10^8\frac{\mathrm{A}}{\mathrm{cm}^2}. \tag{9.40a}$$

Such a current density is attained in the conductor of a diameter of about 0,1 mm, at current 10^4 A. But since under real conditions the current does not immediately increase, its increase time should be less than τ_H, which means that the current increase time should satisfy the condition

$$\frac{d\delta}{dt} \geq \frac{\delta}{\tau_H} = \frac{\delta^2 n}{2}\sqrt{\frac{\mu_0}{\gamma_0}} = \frac{I_A^2 n^3}{8}\left(\frac{\mu_0}{\gamma_0}\right)^{3/2} = 2 \cdot 10^{19}\frac{\mathrm{A}}{\mathrm{m}^2 \mathrm{s}}. \tag{9.40b}$$

For the conductor of cross section 10^{-8} m^2 the current increase velocity is $di/dt \geq 2 \cdot 10^{11}$ A/s.

Chase fnd Levine [39] have proposed to distinguish the types of explosion as "slow" and "fast". To the first type belong the disruption of the conductor caused by the development of the "flutes", while the second type includes those in which the explosion occurs uniformly along the length of the conductor. In a rather rough approximation one can assume that in events of the second type the Andersen criterion is fulfilled.

In a number of experiments made with conductors having a small cross section, at a high current increase time conditions have been achieved where the nonuniformity of the explosion on the length was not a crucial factor. Similar experiments were carried out both with cylindrical conductors and with foils. Interpretation of the experimental results with the help of numerical simulation containing the solution of the hydrodynamic equations has shown that for fast explosions the process is always nonuniform in thickness. Evaporation begins from the surface while the inner parts have no time to expand and preserve metallic conductivity, until the moment when the axis of the conductor is not approached by the expansion wave, propagating with the sound velocity. Before this moment the pressure sharply rises in the inner conductor area. In these conditions the sharp growth of wire resistance may begin when the values of the current action integral exceed the Andersen criterion. Therefore, for very fast energy inputs it is necessary to replace I_A by $K_A\ I_A$, where $K = 1$–3. Let us note that in recent years the electric explosion of superthin wires and foils has been intensively studied, using the most powerful energy sources, namely, the nanosecond forming lines with internal resistance of the order of 10^{-1}–1 Ohm, voltage of 10^5–10^6 V, and stored energy of 10^4–10^6 J. In these experiments a current interruption does not occur, and the conductor, evaporating at the very initial stage of the discharge, transforms to plasma formation with a particle energy up to hundreds of eV.

9.7 Electric explosion of the skin layer in superhigh magnetic fields. Ideal model

If the magnet wall is so thin that the skin effect is weak, then the current density is constant along the thickness. In this solenoid, along with the expansion under the action of electromagnetic forces, an electric explosion and consequent loss of conductivity are also possible, resulting in a current interruption in the circuit. Let us use an ideal model to estimate the induction attainable in such a device. In a given case the disruption criterion acquires the form

$$\int_0^t \delta^2 dt = \frac{1}{\mu_0^2 h^2} \int_0^t B^2 dt \geq I_A. \tag{9.41a}$$

Further on, we can find the induction amplitude for the "standard" pulse in the form of a half-wave of a sinusoid:

$$B_m \leq \mu_0 h \left(\frac{I_A}{t_m}\right)^{1/2}, \tag{9.41b}$$

where t_m is the current increase time up to the maximum. The formula (9.42b) gives rather high values of induction: for copper, for $t_m = 2 \cdot 10^{-6}$ s, $h = 1$ mm, we obtain $B_m \approx 400$ T. The analogous estimation for a circular conductor with radius R gives

$$B_m \leq \mu_0 \frac{R}{2} \left(\frac{I_A}{t_m}\right)^{1/2}. \tag{9.42c}$$

This is related to the fact that with increasing h or R we can theoretically obtain a field which is as strong as we want for a given current density defined by the Andersen criterion. Also, these estimates show that the explosion of thin conductors may occur when the induction amplitude on the boundary does not attain the characteristic value B_{S3} (about 370 T for copper).

In actual conditions, at the above-mentioned current increase time, the depth of the field penetration (the thickness of a skin layer) is of the order of fractions of a millimeter, and the condition of a sharply-pronounced skin effect is realized. Under these conditions, as shown in Chapter 3, the induction on the surface of a single-turn magnet and the volume energy density are coupled by a simple relation:

$$\Delta q'(0) = \frac{\vartheta B^2}{2\mu_0}, \tag{9.43}$$

where ϑ is close to unity.

Let us consider further, along with the explosion of wires, the ideal model of an explosion, according to which, the layer in which the volume energy density reaches Q_3 instantly losses its conductivity [69]. Using this extremely simplified model, one can calculate the velocity of the displacement of the field-conductor boundary and estimate the influence of this process on the induction value. Displacement of the boundary affects the current distribution: the surface layers lose their conductivity, and current passes into deeper layers. This should lead to a "dragging" of current and an increase in its density. The numerical solution of this problem, which is similar to the problem of the propagation of the phase separation (the Stephan problem, known in the theory of thermoconductivity) confirms the considerations on the expected "current drag" effect (Figure 9.20a).

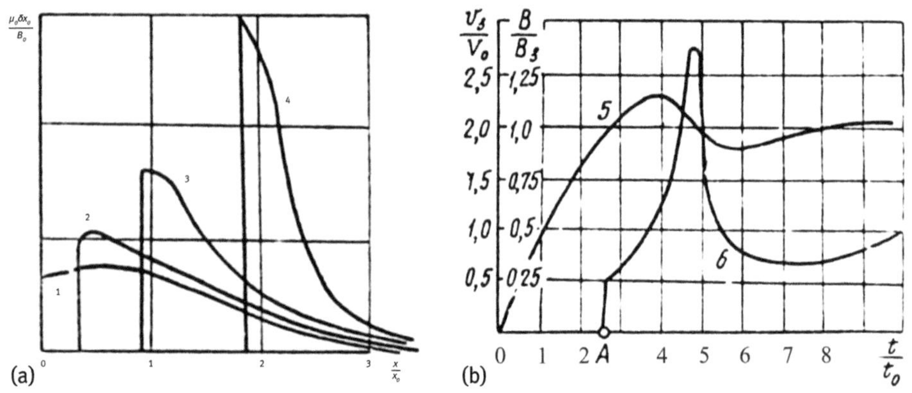

Fig. 9.20: The idealized model of an electric explosion. (a) The current density after the beginning of the explosion in the building-up field ($B_e = B_0 t/t_S$, $B_{S3} = \sqrt{2}B_0$, $x_0 = (\rho_0 t_S/\mu_0)^{1/2}$); (1) $t = 1.35t_S$; (2) $t = 1.5t_S$, (3) $t = 1.65\,t_S$, (4) $t = 1.5t_S$. (b) The induction in the slot and the velocity of the boundary movement; (5) $B/B_{S3} = f(t/t_1)$; (6) $u_s/V_0 = f(t/t_1)$. $t_1 = 4\rho_0 g^2 B_{S3}^2(1 + B_{S3}^2/B_0^2)/(\mu_0 U_0^2)$; $V_0 = U_0/2gB_{S3}$.

In this process the velocity of surface layer heating grows in the increasing external field, and each consequent layer losses its conductivity faster than the previous one. Thus, the model of the idealized explosion of the surface layer in an increasing field leads to the instability of the flat front of a sort: its velocity increases infinitly in a finite time. From here it can be seen that at the instant of conductivity loss by the heated layer, the electric explosion may become the main factor limiting the level of the attainable magnetic fields. To prove this, let us present the solution of the modeling problem on the source of the infinite power with voltage U_0, discharging through the inductance onto the slit, the edges of which (field-conductor boundaries) are shifted, since the conductivity disappears when explosion occurs. In the process, the boundary velocity changes peculiarly. It increases after the onset of the explosion (point A in Figure 9.20b), then stops growing becomes constant. The limitation of the velocity follows from the law of induction: the velocity of displacement cannot be arbitrarily high, since the sum of the drop of voltage on the slit and on the inductance should be equal to the source voltage U_0. (It should be noted that in increasing fields these components have the same sign). The solution is stabilized at $B = B_{S3} = (2\mu_0 Q_3)^{1/2}$, i.e., at the threshold value of the induction. Under this regime the current in the circuit does not change, and there is no voltage drop on the inductance of the circuit. Therefore, the velocity can be calculated using the equality $2u_S B_{S3} g = U_0$. From here,

$$u_S = \frac{U_0}{2gB_{S3}}. \tag{9.44}$$

For $U_0/2g = 10^7$ V/m and $B_{S3} = 370$ T we have $u_S = 2.5 \cdot 10^4$ m/s. It can be seen from this estimate that although the properties of the circuit limit the velocity of the conductor boundary, on an electric field of average strength needed to attain the fields with induction of about 10^3 T (see Section 9.5), this velocity significantly exceeds the Alfven velocity. It is necessary, however, to take into account the limited applicability of the model of the ideal explosion, within the framework of which the finite velocity of the expansion of the medium losing its conductivity is not taken into account.

9.8 The actual processes developing for "slow" and "fast" electric explosions of a conductor surface skin layer in a superhigh magnetic field

In actual practice the threshold values of energy (and, hence, of induction) at which electric explosion begins depend on the speed of the energy input and

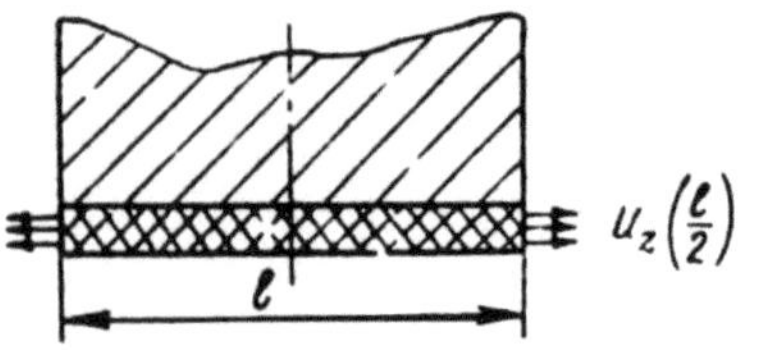

Fig. 9.21: A schematic drawing of a "slow explosion", which is the ejection of metal through the ends of the solenoid.

the geometry of the single-turn magnet. Similarly, with the electric explosion of wires and foils one can distinguish between "slow" and "fast" explosions (see Section 9.6). In the first case the determining process is the hydrodynamic flow, developing in the surface layer and resulting in the bursting of the conductor through the ends (Figure 9.21).

This process can occur in a hot softened layer, weakly coupled with the other part of the coil. However the most noticeable manifestation of this process is observed in conductors heated above melting point. The threshold induction at which the melting of the surface element occurs is close to the value $B_{S1} \approx \sqrt{2\mu_0 Q_1}$, where Q_1 is the melting energy related to the volume unit. According to Table 3.2, this induction for copper is about 120 T, and for lead only 30 T. Experiments showed that the modeling samples made of wood alloy practically did not have any traces of plastic deformation and expanded due to the ejection of the metal in a field with induction ≈ 50 T. The study of the sediment remaining after the explosion has shown that it mainly consisted of small droplets, produced by metal splashing. In these experiments a "slow" explosion was practically the only mechanism leading to the growth of the inner magnet diameter. In other experiments with fields below the limit, corresponding to the metal evaporation, the "slow" explosion was also the factor that determined the metal spray.

Quantitative evaluations of the processes occurring during a "slow" explosion could be derived for the simplest case of the steady-state process. One can find the velocity of the boundary field conductor, taking into consideration the mass balance and assuming that a displacement of the boundary by the distance ΔR_1 occurs due to metal ejection through the ends, with Alfven velocity [69]. This velocity corresponds to the stationary flow in the melted layer of thickness Δ_S. In fact, if we assume that the pressure in the layer of the metal in average is $P_M \approx B_e^2/2\mu_0$,[1] then for the steady state flow one can write down the following equation of motion of noncompressible fluid along the generatrix, i.e., in the direction of

1 In the preceding chapters we denoted the inductions on the internal and external sides of the winding with B_i, B_e. In the present chapter the symbol B_e stands for the induction of the field at the boundary of the conductor.

the axis:

$$\frac{1}{2}\gamma_0 \frac{\partial u_z^2}{\partial z} + \frac{\partial P}{\partial z} = 0,$$

or

$$\frac{1}{2}\gamma_0 u_z^2 + P = \text{const.} \tag{9.45}$$

In the middle plane of the coil, $u_z = 0$, $P = B_e^2/2\mu_0$, and on the cut of the magnet $u_z \approx u_z(l/2)$, $P \approx 0$. From here we find the velocity of the substance ejection through the plane of the face: $u_z(l/2) = V_A$. The velocity of the boundary is found from the mass balance condition $l(dR_1/dt) = 2\Delta_S V_A$. Further on we find

$$\frac{dR_1}{dt} = \Omega_S = 2V_A \frac{\Delta_S}{l}. \tag{9.4}$$

Assuming that thickness Δ_S is equal to the penetration depth of the magnetic field, we derive from the above formula that the boundary velocity is much less than the Alfven velocity if $\Delta_S \ll l$. In spite of this, the role of the "slow" explosion, as a factor leading to the growth of the inner radius before the induction reaches its maximal value, could be essential in those cases when the melting and ejection of the metal begins prior to overcoming the yielding level. There are grounds to believe that this process took place in the modeling experiments with samples from wood alloy [38], in which the induction amplitude did not reach its calculated value for the above-mentioned reason.

From the given estimates, it follows that in long single-turn coils the ejection of hot metal through the faces should not lead to growth of the inner radius at the stage of the current growth, which does not exclude radius growth at later stages of the discharge. Also, ejection of metal is possible not only along the force lines of the field, but also transversely to them. This process, may occur due to the effect of the reversal of the sign of the volume electromagnetic force in the skin layer after the current maximum.

One additional process which could lead to the expansion of the long thick-wall coils in the course of discharge, along with the shock wave considered in Section 9.5, is the evaporation of the surface layer, i.e., "fast" explosion. This process is not practically studied in experiments, since the threshold for its initiation can only be achieved fields with induction $B_{.S,3} \approx \sqrt{2\mu_0 Q_3}$ (Q_3 is the evaporation energy) of the order of 300–500 T (Table 3.2). In recent experiments made at SANDIA Laboratory [34] this threshold has been overcome. Therefore the evaluation of the role of the "fast explosion" as a factor affecting the displacement of the effective field conductor boundary becomes quite urgent. In these studies, likewise in the study by Shearer [18], regimes have been realized in which, due to the relatively long size of the single-turn magnet, the edge effects were insignificant. They could be ignored when describing the shock wave, the flow of the medium behind the shock wave front, field diffusion into the conductive medium, as well as the expansion of the medium accompanied by loss of conductivity. The model of an explosion of a skin layer which was considered in Section 9.7 describes the "catastrophic" scenario of this process based on the assumption of an instantaneous loss of conductivity by the elements of medium on the attainment of the sublimation energy. It was shown that even in this case a velocity of displacement of the conductor boundary remains finite at a given voltage. The additional limitation of this velocity is due to the fact that in order to provoke the conductivity loss, the density of the boundary elements of medium must become essentially less than its initial value. This limitation is considered in the present section in the frames of analytical model. Further (see Section 9.9), the results of the computer simulation are presented.

The qualitative outlay of the distribution of the main parameters, characterizing the steady flow behind the shock wave front is presented in Figure 9.22. The point s corresponds to the initial position of the conductor boundary. The current is concentrated in the skin layer (in the region $x > s$) where the heating of the medium and formation of the magnetic pressure occur. The combined action of the pressure of the heated medium and magnetic pressure results in formation of the shock wave.

Density and temperature exhibit a jump at the front of a shock wave initiated by the magnetic field. The velocity of elements of the medium before the front of the shock wave equals zero, and behind this front it equals u_f. Figure 9.22

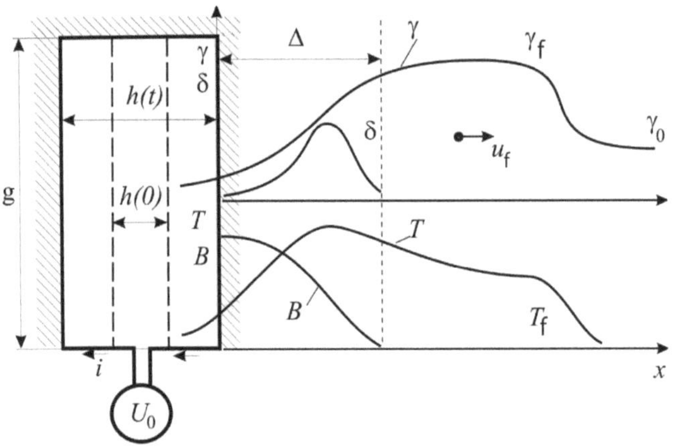

Fig. 9.22: The picture of one-dimensional flow occurring in an electric explosion of a skin layer: δ is the current density; γ_0 is the undisturbed density of a medium; γ_f is the density of a medium behind the wave front; $u_f$is the velocity of the flow behind the wave front; T is the temperature; B is the induction.

shows the qualitative current density distribution with the characteristic maximum, the origin of which was considered in Chapter 3. The faster the field grows, the stronger the heating of the medium in the skin layer, and the faster the field penetrates into the medium. It can be also retained in a steady regime, although the main part of the current is concentrated in the region between the boundary of the conducting medium and the shock wave front. In the limits of this region there exist the volume electromagnetic forces, the result of which (as calculated per unit of the surface) is $P_M = B_e^2/2\mu_0$. Thus, the given region plays the role of a piston producing the shock wave. In this region the additional heating of the medium by current takes place. Numerical calculations show [74] that in the field, the increase time of which exceeds 100 ns or less, the shock wave front does not separate for some time from the region where the current is flowing. In this case part of the current is concentrated before the front in the region where the metal still holds the initial temperature and has higher conductivity than the heated medium behind the front. In the frontal area the local maximum of the current density is formed. Such regimes are discussed in the next section.

It is natural to consider the point s as the conductor boundary. To the left of this point the external medium with low pressure is located. Hence, in the vicinity of the boundary a region is formed where the medium density and hydrodynamic pressure decrease. In this region the volume force caused by the pressure gradient is directed outward in the region of the external field. The conductivity here drops, due to the reduction of the electron density. Consequently the current density is reduced even to a greater extent than in the case of nonlinear field diffusion in a solid conductor. With the current density dropping, the volume of electromagnetic force is also reduced, and the resulting volume force is directed outward. As a result we obtain an effect similar to the one mentioned above when we discussed the explosion of thin-wall coils: the preconditions are created for slowing down the external layers due to expansion of the medium. The volume velocity of the boundary elements becomes negative in respect to the flow velocity behind the shock wave front. The flow velocity of the boundary elements in a laboratory coordinate system after the onset of the explosion can, for a while, even change its sign. In another words, it is reasonable to suppose that the "reverse" motion of the boundary may take place.

From a practical view point there are two problems of interest: namely, whether the ejection of the hot gas into an ambient medium is possible, and how the boundary motion and field diffusion affect the displacement of the effective field-conductor boundary. The first question is important for the cases where the conductor borders with the slit which provides for the vacuum insulation of the circuit. The second question is meaningful for the choice of the generator parameters producing the current in the magnet system.

The answers to both of these questions require the analysis of flows in the boundary region, taking into account the thermodynamic properties of a plasma and the dependence of its conductivity on the density and temperature.

One can perform this analysis using a numerical simulation. Some of the results will be presented below. Here we restrict ourselves to an estimation based on the simple assumption used in the description of a wire explosion [69]. Let us assume that the conductivity of the medium vanishes stepwise when the density in the boundary point s reduces to some value γ_S, $\gamma < \gamma_S$. It is evident that within the framework of this model the elements of the medium with density $\gamma < \gamma_s$ are not confined by the electromagnetic forces and freely move to the external medium, in another words, we have the ejection of the substance from the skin layer.

In a stationary regime the velocity of the displacement of point s is constant and equals u_S . Equalizing the mass flows on both sides of the plane passing through the point s, we obtain

$$\gamma_f \left(u_f - v_S\right) = \gamma_S \left(u_S - v_S\right), \tag{9.47a}$$

where u_S is the flow velocity at point s. Along with this, similar to the consideration of detonation waves, the Chapman–Zhuge condition should be fulfilled, which is the consequence of the assumption on the stationary character of the process

$$c_S + u_S = v_S, \tag{9.48b}$$

where c_S is the adiabatic sound velocity at point s. From equations (9.48a,b) we derive the expression for the displacement velocity of the conductive medium boundary in respect to the medium, the elements of which propagate in the positive direction with velocity of the flow behind the shock wave front:

$$\Omega_s = v_S - u_f = \frac{\gamma_S}{\gamma_f} c_S. \tag{9.49}$$

This parameter could be considered as a component of the boundary velocity, caused by ejection of the mass $m' = c_S \gamma_S = \Omega_s \gamma_f$ from the unit surface in a unit of time. One can evaluate Ω_s for hypercritical regimes, which is one of the main interests in the case of the generation of fields close to 10^3 T. The change of the internal energy due to the compression of the substance in the shock wave with pressure $P_f = P_M$, is (as calculated per unit mass) [48]

$$\varepsilon_f = \frac{1}{2}\frac{P_f}{\gamma_f}\left(\frac{\gamma_f}{\gamma_0} - 1\right) = \frac{B_e^2}{4\mu_0\gamma_f}\left(\frac{\gamma_f}{\gamma_0} - 1\right), \tag{9.50a}$$

while the Joule heating gives [Chapter 3]

$$\varepsilon_D \approx \frac{B_e^2}{2\mu_0\gamma_f}. \tag{9.50b}$$

The sum of these values is

$$\varepsilon = \varepsilon_f + \varepsilon_D = \frac{B_e^2}{4\mu_0\gamma_f}\left(1 + \frac{\gamma_f}{\gamma_0}\right). \tag{9.50c}$$

The increase in temperature is defined as

$$T_D \approx \frac{\alpha\ (\eta - 1)\ M B_e^2}{4\mu_0\,\gamma_f\,\vartheta_0\,k}\left(1 + \frac{\gamma_f}{\gamma_0}\right), \tag{9.51}$$

where M is the atomic mass, ϑ_0 is the coefficient accounting for the distinction of the heat capacity of the hot metal from that of the ideal gas (at $T \geq 3 \cdot 10^4$ K $\vartheta_0 \approx 1$), k is the Boltzmann constant,η is the effective index of the adiabat of the hot metal (more accurately, of its heavy component), and α is the factor accounting for the fraction of the energy possessing by the heavy plasma components (ions and atoms). At low temperatures the heat capacity of free electrons is small, and $\alpha \approx 1$. In a field with induction of about 10^3 T the metal behind the current front itself represents f the entirely ionized dense plasma of singly charged ions, the temperature of which is close to 10^5 K; in these conditions $\alpha = 0.5$. For an adiabatic expansion the density varies from γ_f to γ_S, and the temperature from T_D to $T_S = T_D(\gamma_S/\gamma_f)^{\eta-1}$. Assume that in the region of adiabatic expansion the metal is close in its properties to the ideal gas. Then the pressure at point s is $P_S = (\gamma_S/M)kT_S$. Then, let us find the adiabatic sound velocity at point s and the velocity Ω_s:

$$c_S = \sqrt{\frac{\eta\,P_S}{\gamma_S}} = \sqrt{\frac{\eta\,k\,T_S}{M}} \approx \left(\frac{\gamma_S}{\gamma_f}\right)^{\frac{\eta-1}{2}}\sqrt{\frac{\eta\ (\eta - 1)\ \alpha\ B_\alpha^2}{4\mu_0\gamma_f}\left(1 + \frac{\gamma_f}{\gamma_0}\right)}, \tag{9.52a}$$

$$\Omega_s = \frac{\gamma_S}{\gamma_f}c_S \approx V_A\left(\frac{\gamma_S}{\gamma_0}\right)^{\frac{\eta+1}{2}}\left(\frac{\gamma_0}{\gamma_f}\right)^{1+\frac{\eta}{2}}\sqrt{\frac{\eta\ (\eta - 1)\,\alpha}{2}\left(1 + \frac{\gamma_f}{\gamma_0}\right)}. \tag{9.52b}$$

Formula (9.52b) includes the Alfven velocity $V_A = B_e(\mu_0\gamma_0)^{-1/2}$ for a nonperturbed medium. For copper we have the relation $\gamma_f/\gamma_0 = 2.19$ if the induction is 10^3 T[2]. When estimating, assume $\eta = 5/3$ (the gas with three freedom degrees), $\alpha = 0.5$, then

$$\Omega_s = 0,23\ V_A\left(\frac{\gamma_S}{\gamma_0}\right)^{4/3}. \tag{9.52c}$$

In hypocritical regimes $\gamma_f/\gamma_0 \approx 0.1$. This choice is based on the representations on the “decollectivization” of electrons and the transformation of metal into

2 The limiting rate of copper compression behind the shock wave front is equal to 3.1, according to data from [50].

nonconductive vapor consisting of neutral molecules. In hypercritical regimes, at the temperatures of the order of 10^4–10^5 K, this approach is less justified, since the counterprocess takes place, namely, the thermal ionization and recovery of the conductivity. Nevertheless, the given evaluation has meaning, since the plasma conductivity is much less than that of the metal, extrapolated to the temperature T_S under the assumption that the metal maintains its initial density. If the condition of the sharp drop of conductivity with density reduction, at least up to one order of magnitude, is applicable, then, as follows from formula (9 .52c), the role of the ejection as a factor leading to the displacement of the boundary of the conductive medium is small, compared to the role of the substance compression behind the shock wave front.

9.9 Computer simulation of a skin layer explosion

When considering a shock wave and an explosion of a skin layer (Sections 9.5 and 9.7) we used simplified models. In particular, to calculate the velocity of the flow behind the shock front we used a formula describing the stationary flow. The effect of the diffusion on the formation of a shock wave was not taken into account. The computer-simulated models, based on the solution of equations of hydrodynamics provide a more complete analysis. The numerical calculations of a one-dimensional flow were first made in Shearer's work [18], and this problem was further considered in [71–76]. In the calculations the different equations of state have been used, as well as the dependences for conductivity on the concentration and density. In the experimental works [31, 49], a reverse motion of the boundary was reported, and the possibility of metal ejection from a skin layer was discussed. This effect was also observed in the numerical experiments described in [71, 72, 74]. In calculations carried out for copper [71] it was assumed that the induction at the boundary attained $B_\infty = 900$ T in a time 0.1 μs, and then remained constant. In these calculations the model of the conductivity developed by Polischuk and Bespalov [63] and the equation of state derived by Hachaturyants [56] were used. The authors note that at the beginning of the process motion of the boundary layers toward the field takes place, and that further the boundary changes the direction of the velocity and moves together with the remaining medium with a delay relative to the shock front. Figure 9.23. shows changes in the distribution of the current density, of the medium density and of the conductivity at instant of time $t = 1$ μs, as calculated in these numerical experiments. The detachment of the shock front from the region where the current density exhibits a maximum can be clearly seen. From the results of the calculations, the behavior

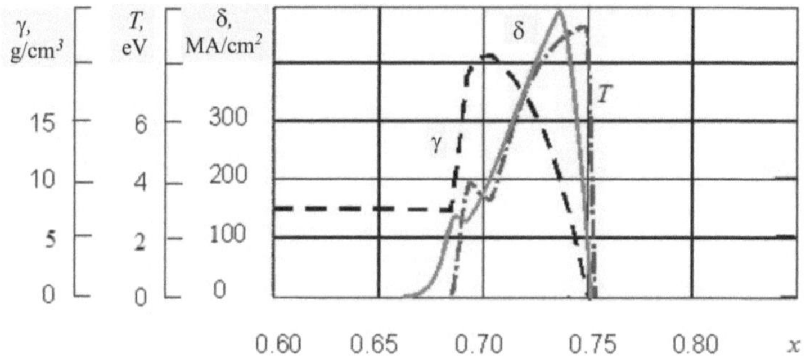

Fig. 9.23: The distribution of the current density, temperature and density of the medium in the skin layer at instant 180 ns for the case where the induction at the boundary attained B_∞ = 900 T in time 0.1 μs, and then remained constant [71].

of the velocity of the field-conductor boundary $\Omega_s$was found. This is described by dependence (9.52c), provided the coefficient 0.23 is replaced by 0.17.

The results of extensive numerical experiments are presented in [73]. They were obtained for stainless steel and tantalum and pertain to fields with induction up to 310^3 T.

In the examples considered above the front of a shock wave is not separated from the front of the magnetic pressure. Due to this fact, a significant displacement of the pressure jump, in relatiion to the peak current density, was not noticed.

The authors of [74] observed the additional specific features of an electric explosion of a skin layer in a building megagauss field. They calculated the diffusion of the magnetic field and the hydrodynamic flow in copper in accordance with the MAG program [77]. The medium was described by Garanin's equation of state [58], and the conductivity by the Bakulin–Luchinsky model [66]. Figures 9.24 and 9.25 show the results of the calculations when the field was assumed to increase linearly ($B_e = B't$) and to attain the stationary value at time t_1. The zero coordinate corresponds to an initial position of the separation boundary "vacuum conductor". The calculations were carried out for two values of a field increase rate: $B' = 2 \cdot 10^9 T/s$ and $B' = 2 \cdot 10^{10}\ T/s$, while attaining the stationary value at a time t_1, when the induction of the external field takes the value B_e = 500 T. Correspondingly, in the first case $t_1 = 500$ ns and in the second $t_1 = 50$ ns. In the external field increase stage the current distribution is essentially different for these regimes. When the induction increase rate $B' = 10^{10}$ T/s, the second maximum of the current density is observed, while for $B' = 10^9$ T/s it is absent.

The observed difference can be explained by the fact that in the case of a high rate of field increase a diffusion velocity can exceed the velocity of a shock wave.

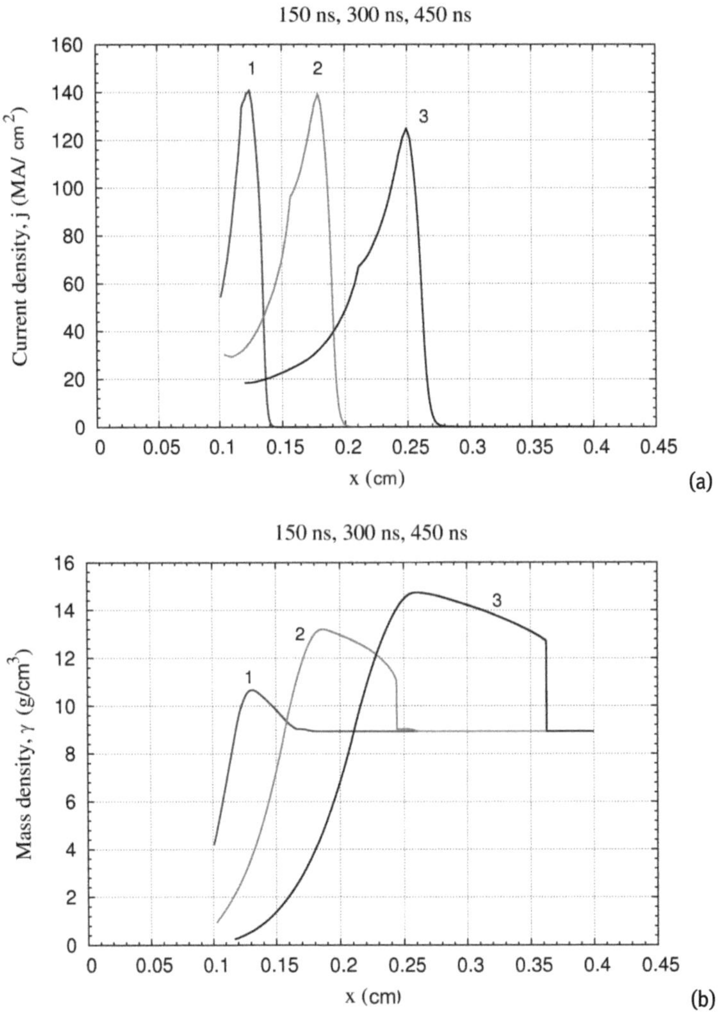

Fig. 9.24: The distribution of the current density (a) and of the density of the medium (b) in a space at different times: $B' = 2 \cdot 10^9$ T/s; (1) $t = 150$ ns ($B_e = 300$ T); (2) $t = 300$ ns ($B_e = 600$ T); (3) $t = 450$ ns ($B = 900$ T).

As a result, a part of the current may appear before the shock front in the region where the conductivity of a cold metal is higher than that of a heated metal in a skin layer. Due to this, the specific current distribution in the vicinity of the shock front appears, exhibiting two maximums. To confirm the this, a model problem was considered, using current distribution for the case of nonlinear diffusion described in Chapter 3. This distribution takes place when the conductivity depends

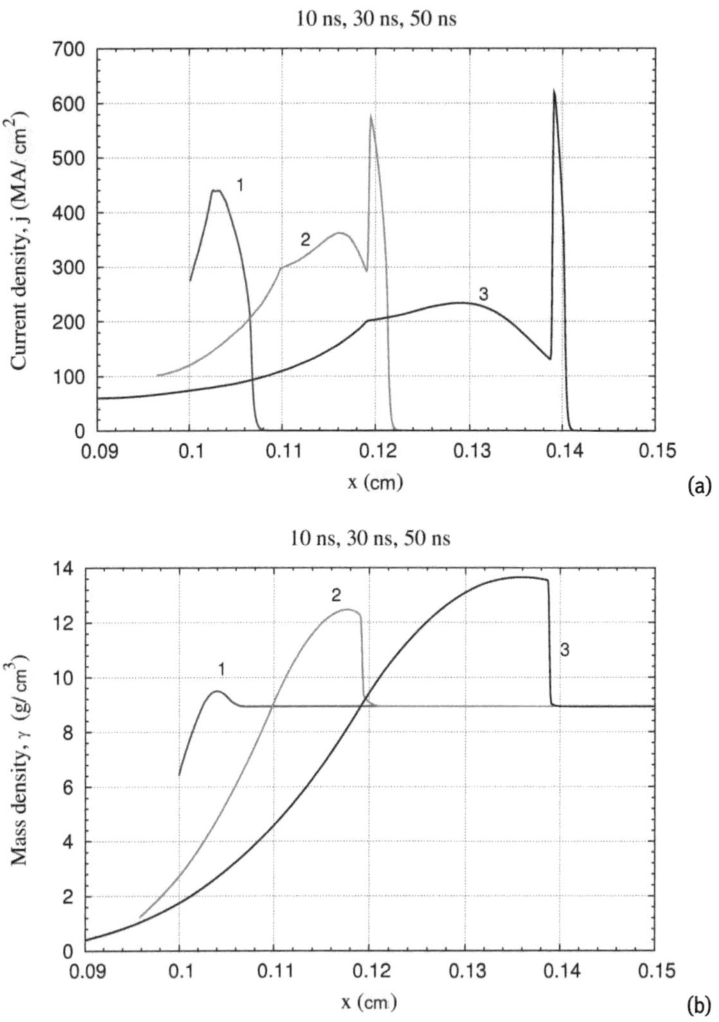

Fig. 9.25: The distribution of the current density (a) and of the density of the medium (b) in space at different times: $B' = 2 \cdot 10^{10}$ T/s: (1) $t = 10$ ns ($B = 200$ T); (2) $t = 30$ ns ($B = 600$ T); (3) $t = 50$ ns ($B = 1000$ T).

on the volume thermal energy $\Delta q'$as $\sigma = \sigma_0/(1 + 2\mu_0 \Delta q'/B_0^2)$, where B_0 is the characteristic induction (for copper $B_0 \approx 42$ T).

Under the condition $B_e \gg B_0$ there forms the specific non monotonic distribution of the current distribution with a characteristic maximum (Figure 9.25a). When estimated, the coordinate of the maximum

$$\Delta_j = \frac{B_e}{B_0}\sqrt{\frac{t}{\mu_0 \sigma_0}} \tag{9.53}$$

may be assumed to be the coordinate of the "magnetic piston" producing the pressure $B_e^2/(2\mu_0)$. The condition under which a part of the current will be concentrated before the shock front has the form

$$\Delta_j \geq \Delta_f, \tag{9.54}$$

where $\Delta_f \approx \int_0^t D dt$ is the distance the shock front travels. From formula (9.25a) we have the relation for D:

$$D = \frac{B_e^2}{2\mu_0 \gamma_0 u_f}. \tag{9.55a}$$

Using the Shearer's approximation ($u_f = a_0 B_e^{3/2}$), we have

$$D = \sqrt{B_e}/(2\mu_0 a_0 \gamma_0). \tag{9.55b}$$

When the field grows linearly, the distance Δ_f can be estimated by the formula

$$\Delta_f = \int_0^t D\,dt = \frac{\sqrt{B'_e}\,t^{3/2}}{3a_0\,\mu_0\gamma_0}. \tag{9.56}$$

From condition (9.54) we obtain the estimate of the threshold value of the increase rate of the induction B'_k, at which one can expect an appearance of the second maximum in the current density distribution:

$$B'_e \geq B'_k = \frac{B_0^2\sigma_0}{9\mu_0 a_0^2\,\gamma_0^2}. \tag{9.57}$$

For copper, $B_0 = 42$ T, $\gamma_0 = 8.9 \cdot 10^3$ kg $\cdot$ m^3, $\sigma_0 = 5.7 \cdot 10^7$ $(\Omega \cdot \text{m})^{-1}$, and then $B'_k = 5 \cdot 10^9$ T/s.

Thus, based on the obtained estimates, it is possible to separate two regimes of field diffusion: a hypocritical regime with an increase rate of less than $B'_k \approx 5 \cdot 10^9$ T/s, and a hypercritical rate, when the diffusion takes place with the formation of two maximums on the curve of the current density and with an induction increase rate exceeding B'_k. The results of some numerical calculations show that the threshold value of an external field increase rate is about $B'_k = (5\text{–}7) \cdot 10^9$ T/s, which approximately corresponds to the given estimate.

When an induction increase is stopped, the qualitative picture of the distribution of the current and density, for the case of a rather slow field increase ($B'_e < B'_k$), remains the same as at the shock front. Another picture takes place in the case of a fast field increase. With time the amplitude of the second maximum decreases. Part of the current concentrated before the shock front becomes progressively smaller (Figure 9.26). This can be explained by the fact that in a

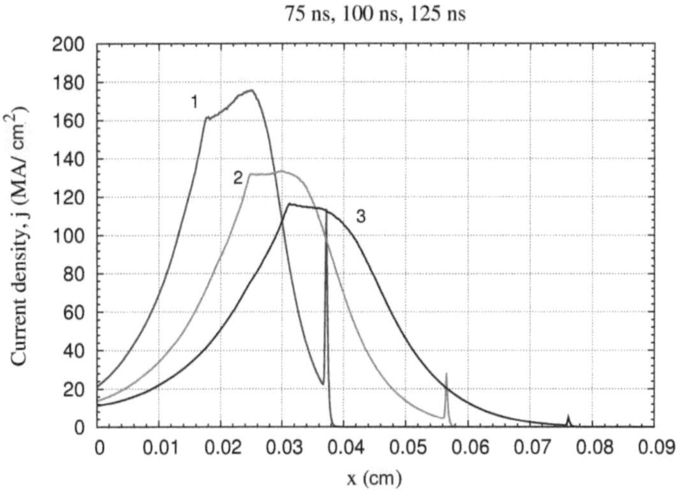

Fig. 9.26: The distribution of the current density in a space at different times after the external field attains the stationary value; $B = 500$ T in 50 ns (a field rise rate $B' = 10^{10}$ T/s); (1) $t = 75$ ns; (2) $t = 100$ ns; (3), $t = 125$ ns.

constant field the thickness of the skin layer increases as $\Delta_j \approx (B_e/B_0)\sqrt{\rho_0 t/\mu_0}$. At the same time, the velocity of the field penetration decreases as $t^{-1/2}$, while the velocity of the shock front is constant. Hence, the current front eventually lags behind the shock front. In the system of coordinates related to the front, the current diffusion takes place from the region of the second maximum through the plane of the front. Finally, the second maximum of the current will decrease.

Of some interest is the character of the flow of the elements of a medium near the boundary. The coordinate of the Lagrange point in the vicinity of the edge of the calculated region gives the appropriate indications. In advance of evaporation such an element is moved in the positive direction under the action of electromagnetic force. Similar to the studies mentioned above [71, 72, 74], when an explosion begins, the elements which lose conductivity change their direction of motion and acquire negative velocity. After a time, the velocity again changes the sign, and the flow becomes stationary in a positive direction. In a stationary regime the velocity of the boundary remains lower than the velocity of the flow behind the front (Figure 9.26). Differently, the boundary elements retain the negative velocity with relation to the elements of the medium near the front.

The “reverse motion” is caused by a change in the sign of the volume force $f_x = -\partial P/\partial x + jB(x)$. This is clear from the distribution of the volume force shown as an example in Figures 9.27 and 9.28.

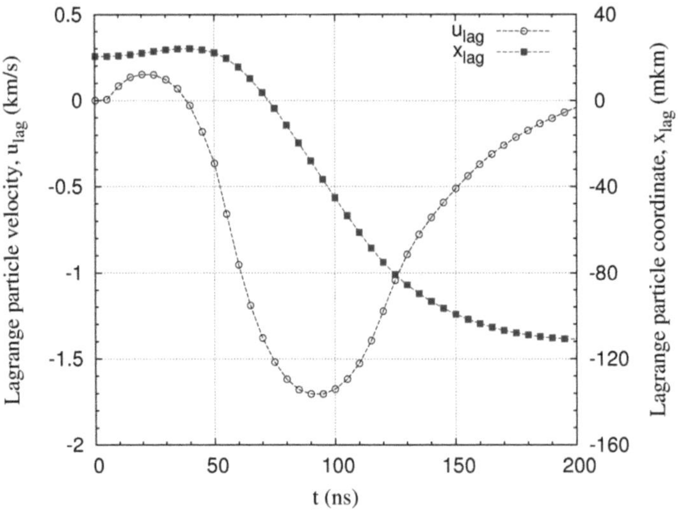

Fig. 9.27: The time dependences for the velocity and the coordinate of the Lagrange cell near the boundary of the medium.

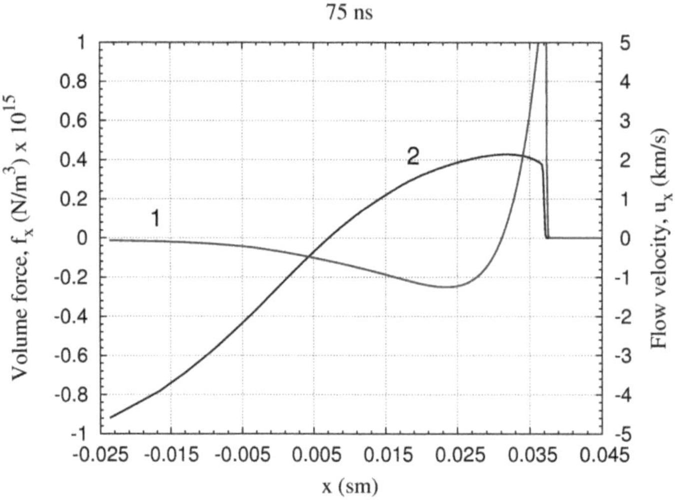

Fig. 9.28: The spatial distribution of the volume force, $f_x = -\partial P/\partial x + \delta B(x)$ (1), and of the velocity of the flow (2) at instant of time $t = 75$ ns.

An additional factor which affects the diffusion process of the field is the heating of media near the boundary, due to radiation generated by plasma from a skin layer. This effect was studied by Garanin, Ivanova, Karmishin, and Sofronov [76], and its role in wire explosion was discussed in [83]. According to the calculated data obtained in [83], in a wire with an initial radius of 1 mm, approximately half the current is concentrated in a narrow peripheral region where the density of the medium is approximately two orders lower then in the remaining part of the conductor. In the above example the induction at the boundary was 160 T, in the main part of the conductor the current density varied slightly, and the temperature was tens times lower than in the region of low density. Near the boundary the plasm temperature reached 50 eV, and the current density was several times larger than in the main part of the conductor.

In [76] it was shown that when an increasing field diffuses into a conductor with a plane boundary, the analogous effect occurs; however, a portion of the current which is concentrated in the layer of low density is relatively small. For instance, for the diffusion of a field increasing linearly ($dB/dt = 5 \cdot 10^8$ T/s), up to the moment of time $t = 1$ μs when the induction of an external field attains 500 T a hot region is formed near the boundary, where the temperature is as high as 12 eV, whereas in the remaining part of the skin layer it does not exceed 2 eV (Figure 9.29). The thickness of the hot region is about 0.01 cm, which is 10 times smaller than that of the skin layer. Only 4 % of the total current is concentrated in the region of low density, although plasma conductivity, defined by the Spitzer formula, is essentially higher than in the adjacent zone of the skin layer. This suggests that the occurrence of a hot external zone has little affects on the integral characteristics of the nonlinear diffusion process accompanied by a shock wave and a hydrodynamic flow behind the shock front.

The influence of field diffusion and hydrodynamic flow behind the shock front on the current in the circuit of a generator connected to a magnet was studied in several works [69, 71]. To estimate this effect it is convenient to use the notion of

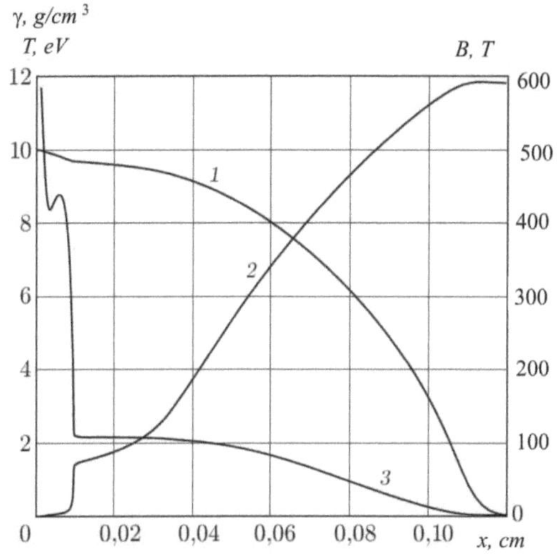

Fig. 9.29: The results of calculation of the diffusion field in the taking into account the heating of a media due to a radiation 76: (1) $B(x)$; (2) $\gamma(x)$; (3) $T(x)$.

an "effective thickness of a skin layer". It can be determined by the formula

$$\Delta'' = \left(\int_{S}^{\infty} B(x,t)dx \right) / B_e. \tag{9.58}$$

Here B_e is the induction of an external field, and the integral is taken from the boundary of the conductor over the whole region of current flow. Of some interest is the parameter $x_S + \Delta''$, which can be considered to be the coordinate of a conventional filed-conductor boundary. In the framework of the model, with a total break of the current, x_S is the coordinate of point s in which conductivity vanishes. In numerical calculations this point is the Lagrange coordinate of the calculated edge element of a medium. The velocity of the displacement of this conventional boundary can be written in the form

$$\Omega = dx_S/dt + d\Delta''/dt. \tag{9.58}$$

For estimation, most convenient is the configuration of a slot of length g, the edges of which are spaced at a distance h. This configuration is shown in Figure 9.22, and $x_s = \pm h/2$. The voltage at the input of the circuit is $U_0 = 2gd(B_e x_s)/dt + 2gd(B_e\Delta'')/dt$. Here the first term takes into account the change in the flux caused by the slot expansion due to the hydrodynamic flow, and the second term takes into account the change in the flux in the skin layer on either side of the slot.

The numerical calculations, using the known equations of state and the temperature and density dependences for the conductivity, yield results which are somewhat different from those obtained in Section 9.2 in the model of ideal conductivity. This is clear from the Figure 9.30, where the results of [75] are shown. The calculations were carried out using the one-dimensional program MAG [77] for Garanin's equation of state [58], and the Bakulin–Luchinsky model of conductivity [66]. Note that the discrepancy with the model of ideal conductivity takes place only in the initial state of a discharge, and then it decreases by up to several percent. This result corresponds well with the estimates given in Section 9.4 and obtained in numerical experiments. In these works it was noted that the velocity of displacement of the real boundary relative to the velocity of the flow is much less than the Alfven velocity.

Thus, the loss of conductivity experienced by the boundary elements of the skin layer makes a rather small contribution to the velocity of the effective field-conductor boundary as compared to the velocity of the flow of a medium behind the shock front. Hence, it is believed that in thick-wall single-turn magnets made from conventional conducting materials the role of the skin layer would not be decisive. Differently, the formulas for the velocity of an ideally conducting medium

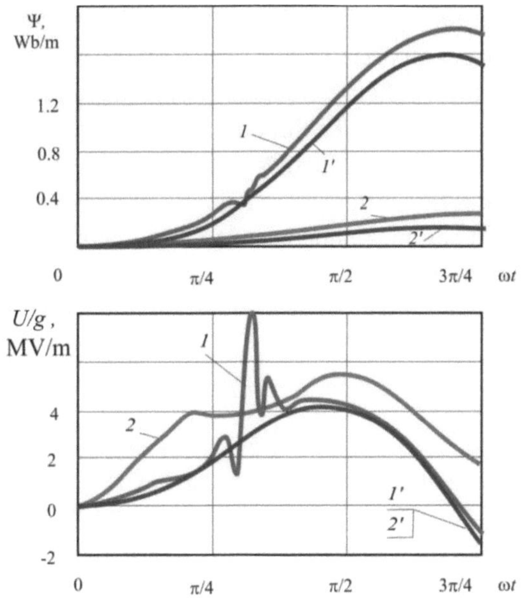

Fig. 9.30: The magnetic flux in the skin layer per unit length of the boundary (upper figure). The stress at the boundary per init length (lower figure): (1′) an ideal conductor, a current rise rate is $t_m = 500$ ns; (2′) ideal conductor, $t_m = 50$ ns; (1) the complete model, $t_m = 500$ ns; (2) the complete model, $t_m = 50$ ns.

can be used to hyperestimate the voltage drop at the boundary of the conductor subjected to an ultrahigh magnetic field. These estimates (see Section 9.5) show that to attain multimegagauss fields in magnets in which a one-dimensional flow is realized it is necessary to provide rather high voltage at the input of the magnetic system.

Studies of the skin layer are far from complete. Additional analysis of the flow in the boundary region is evidently needed, since there can be effects which would be beyond the model of the thermodynamic equilibrium to describe. They are the nonequilibrium processes intrinsic for to breakdown, and the development of “microturbulence” in a rather rarefied plasma with a high current density . These effects emerge in a one-dimensional flow. At the same time, in plasma there are the definite prerequisites for the formation of a more complicated flow, due to the appearance of the Rayleigh–Taylor instability. The above-mentioned reverse motion can also lead to the development of a Rayleigh–Tailor instability. The authors of [71] evaluated the development time of instabilities based on the simplest estimations of [78]. The analogous estimates can be obtained for the example considered above, shown in Figure 9.28. In this case the velocity of the counterflow is close to $5 \cdot 10^4$ cm/s . It exists in the region $d \approx 0.03$ cm and is formed in a time shorter than 100 ns. The acceleration in this region is directed toward the field, and its absolute value is equal to $g \approx 5 \cdot 10^{11}$ cm/s^2. The perturbation with a wavelength less than d is formed in a time $\tau \leq (2\pi\, g/d)^{-1/2} \approx 10^{-7}$s. In such a time the layer d may be destroyed, due to the development of perturbations.

Similar processes, which occur at the boundary of the shell subject to compression, were considered by Bud'ko and Liberman [79], and also in [80, 81].

Along with this, in the case of nonlinear field diffusion, thermal instabilities may develop [82]. This results in the appearance of a three-dimensional structure in which the elements with a sharply different conductivity are stirred. These processes may lead to the displacement of the current from region d and affect the character of the field diffusion, which can be analogous to that in the model of the "ideal" explosion considered above.

9.10 References

[1] H. P. Furth, M. A. Levine, and R. W. Waniek, *Rev. Sci. Instrum.* **28**(11) (1957), 949–958.
[2] H. Knoepfel, *Pulse high magnetic fields*, Amsterdam London, North-Holland, 1970.
[3] P. N. Dashuk, S. L. Zayents, V. S. Komelkov, G. S. Kuchinskiy, N. N. Nikolaevskaya, P. I. Shkuropat, and G. A. Shneerson, in: V. S. Komelkov, *Tekchnika bolshih impulsnih tokov i magnitnih poley*, Moscow, Atomizdat, 1970 (in Russian).
[4] F. Herlach (ed.), *Strong and Ultrastrong magnetic fields and Their Aplications*, Berlin New York Tokyo, Springer-Verlag, 1985.
[5] G. A. Shneerson, *Fields and Transients in Superhigh Pulse Current Devices*, New York, Nuova Science Publishers, Inc., 1999.
[6] A. S. Lagutin and V. I. Ozhogin, *Silniye impulsniye magnitniye polya v fizicheskom experimente*, Moscow. Energoatomizdat, 1988 (in Russian).
[7] S. I. Krivosheev, *Megagauss field and the strength of materials*, p. 152, Saarbrücken, Lap Lambert Academic Publishing. 2011.
[8] G. A. Shneerson, **MG-3** (1984), pp. 70–76.
[9] V. F. Demithev and G. A. Shneerson, **MG-4** (1987), pp. 49–63.
[10] N. Miura, Y. H. Matsuda, K. Uchida, F. Herlach, et al., **MG-IX** (2002), pp. 58–66.
[11] S. I. Krivosheev and G. A. Shneerson, **MG-X** (2004), pp. 29–38.
[12] O. Portugall, N. Puhlmann, H. U. Müller, et al., *J. Phys. D Appl. Phys.* **32** (1992), 2354–2369.
[13] F. Herlach, **MG-SF** (2006), pp. 1–12.
[14] S. I. Krivosheev and G. A. Shneerson, **MG-XII** (2008), pp. 107–134.
[15] G. A. Shneerson, *J. Tech. Phys.* **32** (1962), 1153–1158 (in Russian).
[16] D. W. ,J. C. Forster and *Les Champs magnetiques intenses* Martin, pp. 361–374, Paris, CNRC, 1967.
[17] J. W. Shearer, *Les Champs magnetiques intenses*, pp. 355–359, Paris, CNRC, 1967.
[18] J. W. Shearer, *J. Appl. Phys.* **40**(11) (1969), 4490–4497.
[19] A. M. Andrianov, V. F. Demichev, G. A. Eliseev, and P. A. Levit, *Pisma J. Eksp. Teor. Fiz.* **11** (1970), 582–585 (in Russian).
[20] H. Knoephel and R. Luppi, *J. Phys. E: Sci. Instr.* **5** (1972), 1133–1141.
[21] F. Herlach and R. McBroom, J. Appl. Phys. E., Sci.Instrum., 1973, V.6, pp.652–654.
[22] G. A. Shneerson, *J. Tech. Phys.* **44**(10) (1974), 2217–2228 (in Russian).
[23] Y. N. Bocharov, A. I. Kruchinin, S. I. Krivosheev, et al., **MG-II** (1980), pp. 485–496.
[24] N. N. Gennadiev, V. F. Demichev, and P. A. Levit, **MG-II** (1980), pp. 27–37.

[25] A. M. Andrianov, S. P. Anan'ev, V. F. Demichev, G. A. Eliseev, and P. A. Levit, *Pisma J. Tech. Phys.* **8**(4) (1982), 240–245.
[26] Y. N. Bocharov, S. I. Krivosheev, and G. A. Shneerson, *Pisma J. Techn. Phys.* **8**(4) (1982), 212–216.
[27] A. M. Andrianov, V. F. Demichev, EG. A. liseev, P. A. Levit, and V. I. Sinitsin, **MG-III** (1984), pp. 29–38.
[28] K. Nakao, F Herlach, T. Gote, et al., *J. Phys. E Sci. Instrum.* **18** (1985), 1018–1026.
[29] T. Goto, N. Miura, K. Nakao, et al., **MG-IV** (1987), pp. 149–158.
[30] M. Von Ortenberg, O. Portugall, N. Puhlmann, et al., in: L. Challis, J. Franse, F. Herlach, P. Wyder (eds.), Second European Workshop on Science in 100 T, Leuven, Belgiumm 30 September–1 October 1994, pp. 69–71.
[31] S. I. Krivosheev, V. V. Titkov, and G. A. Shneerson, *J. Tech. Phys.* **42**(4) (1997), pp. 352–366.
[32] N. Miura, Y. H. Matsuda, K. Uchida, and S. Todo, **MG-VIII** (1998), pp. 663–670.
[33] K. Mackay, M. Bonfin, and D. Gibord, **MG-VIII** (1998), pp. 175–178.
[34] R. W. Lemke, M. D. Knudsen, C. H. Harjes, et al., **MG-X** (2004), pp. 403–404.
[35] B. H. Novac, I. R. Smith, D. F. Runkin, and M. Hubbard, **MG-X**, (2004), pp. 39–45.
[36] A. S. Boriskin, A. Y. Brodsky, Y. V. Vlasov, et al., **MG-X**, (2004), pp. 49–53.
[37] T. J. Awe, B. S. Bauer, R. E. Siemon, S. Fuelling, V. Makhin, M. A. Angelova, et al., **MG-XII** (2008), pp. 515–521.
[38] V. P. Gordienko and G. A. Shneerson, *J. Tech. Phys.* **34** (1964), 376–378 (in Russian).
[39] W. G. Chace and M. A. Levine, *J. Appl. Phys.* **31**(7) (1960), 1298.
[40] V. P. Gordienko and G. A. Shneerson, *J. Tech. Phys.* **34**(2) (1964), 376–378 (in Russian).
[41] Y. N. Botcharov, S. I. Krivosheev, N. G. Lapin, and G. A. Shneerson, *Pribory i Technika Eksperimenta* **2** (199), 92–97 (in Russian).
[42] Y. N. Botcharov, S. I. Krivosheev, et al. Krutchinin,, **MG-1V** (1987), pp. 65–77.
[43] Y. A. Alekseev, N. N. Gennadiev, and V. F. Demichev, **MG-IV** (1987), pp. 79–88.
[44] J. W. Rayleigh, *Phil. Mag.* **34** (1917), 94–98.
[45] BY. N. otcharov, S. I. Krivosheev, A. I. Kruchinin., V. V. Titkov, and G. A. Shneerson, **MG-III**, (1984), pp. 77–84.
[46] C. H. Mielke and B. M. Novak, *IEEE Trans.on Plasma Science* **38**(8) (2010), 1739–1749.
[47] V. M. Mikhaylov, *Impulsniye elektromagnitniye polya*, p. 138, Kharkow, Vissha skola, 1979, (in Russian).
[48] Y. B. Zel'dovich and Y. P. Raizer, *Physics of Shock Waves and High-Temperature Hydrodynamic Phenomena*, New York. Academic Press, 1966–1967.
[49] N. Miura, Physica B, 201, 1994, pp. 40–48.
[50] Zharkov V.N, Kalinin V. A. *Uravneniya sostoyaniya tverdih tel pri visokih davleniyah i temperaturach*, Moscow. Nauka, 1968 (in Russian).
[51] V. A. Burtsev and N. V./ *Elektricheskiy vzriv provodnikov i ego primenenie v elektrofizicheskih ustanovkah* Kalinin, Moscow, Energoatomizdat, 1990 (in Russian).
[52] V. A. Burtsev, in: E. P. Velikhov, *Fizika i technika moschnih impulsnih system*, p. 211, Moscow, Energoatomizdat, 1990 (in Russian).
[53] G. V. Ivanenkov, S. A. Pikuz, T. A. Shelkovenko, V. M. Romanova, I. V. Glazirin, O. G. Kotova, and A. N. Slesareva, *Obzor literaturi po modelirivaniyu prozessov electricheskogo vzriva tonkih metallicheskih provolochek*, Chast 2. Preprint FIAN, Nr. 10, Moscow, 2004 (in Russian).
[54] V. G. Kuchinskiy, V. T. Mikhel'soo, and G. A. Shneerson, *Pribori i technika experimenta* **3** (1973), 108–112 (in Russian).

[55] E. N. Avrorin, B. K. Vodolaga, V. A. Simonenko, and V. E. Fortov, *Phys. Usp.* **36**(5) (1993), 337–364 [*Uspekhi Fizicheskih Nauk* **165**(5) (1993), 1–34] (in Russian).
[56] D. J. Steinberg, *Equation of State and Strength Properties of Selected Materials*, Livermore, CA, Lawrence Livermore National Laboratory, 1996.
[57] S. N. Kolgatin and A. V. Khachaturyanz, *Teplofisika Visokih Temperatur* **20** (1982), 447–451 (in Russian).
[58] A. M. Bujko, S. F. Garanin, V. A. Demidov, V. N. Kostjukov, A. A. Kuzjaev, V. I. Mamyshev, et al., **MG-V** (1990), pp. 743–748.
[59] E. David, *Zeitschrift für Physik* **150**(2) (1958), 162–171.
[60] G. Lehner, Springer Tracts in Modern Physics 47, pp. 67–110, Springer-Verlag, 1968.
[61] R. F. Kidder, **MG-I** (1966), p. 37.
[62] N. B. Volkov, *J. Tech. Phys.* **49**(9) (1979), 2000–2002 (in Russian).
[63] I. M. Bespalov and A. Y. Polischuk, *J. Tech Phys. Letters* **15**(2) (1989), 4–8 (in Russian).
[64] M. P. Desjarlais, *Contrib. Plasma Phys.* **41**(2–3) (2001), 267–270.
[65] V. A. Burtsev and N. V. Kalinin, in: V. E. Fortov, V. P. Efremov, and K. V. Khischetnko, et al. (eds.), *Fizika extremalnikh sostoyaniy*, Chernogolovka, IVTAN, 2005 (in Russian).
[66] J. D. Bakulin, V. F. Kuropatenko, and A. V. Luchinsky, *J. Tech. Phys.* **46**(9) (1976), 1963–1969.
[67] K. B. Abramova, N. A. Zlatin, and B. P. Peregud, *Zh. Eksp. Teor. Fiz.* **69**(11) (1975), 2007–2022 (1975).
[68] G. Anderson and F. Neilson, in: W. G. Chace and H. K. Moore (eds.), *Exploding Wires*, New York London, Plenum Press, 1959..
[69] G. A. Shneerson, *J. Tech. Phys.* **43**(2) (1973), 419–428 (in Russian).
[70] G. A. Shneerson, **MG-III** (1984), pp. 70–76.
[71] S. N. Kolgatin, A. Y. Polischuk, and G. A. Shneerson, *Teplofisika Visokih Temperatur* **31**(6) (1993), 447–451 (in Russian).
[72] S. I. Krivosheev, N. G. Karlykhanov, and G. A. Shneerson, **MG-IX** (2002), 529–536.
[73] R. B. Spielman, S. Chantrenne, and D. H. McDaniel, **MG-11** (2006), pp. 339–344.
[74] S. I. Krivisheev, V. S. Pomasov, and G. A. Shneerson, *J. Techn. Phys. Letters* **37**(18) (2011), 73–80 (in Russian).
[75] Y. E. Adamian, S. I. Krivosheev, and G. A. Schneerson, *Skin layer with an extreme current as an element of the electric circuit.* Report on European Pulsed Power Conference, Geneva, 2009.
[76] S. F. Garanin, G. G. Ivanova, D. V. Karmishin, and V. N. Sofronov, *IEEE Trans. on Plasma Science* **58**(8) (2010), 1815–1821.
[77] V. V. Rudenko and M. V. Shaburov, **MG-X** (2004), pp. 321–324.
[78] L. D. Landau and Lifshitz E. M., *Fluid Mechanics*, Moscow, Nauka, 1986 (Oxford, Pergamon, 1987).
[79] A. B. Bud'ko and Liderman M. A., **MG-VI** (1993), pp. 217–224
[80] V. V. Bychkov, S. M. Gol'berg, and M. A. Liberman, *JETP* **73**(4) (1991), 642–653 [*Zh. Eksp. Teor. Fiz.* **100**, 1162-1185].
[81] K. H. Almstrem, G. Bjarnholt, S. M. Golberg, and M. A. Liberman, **MG-7** (1996), pp. 146–153.
[82] V. I. Oreshkin and S. A. Chaikovsky, *Phys. Plasmas* **19** (2012), 022706.
[83] S. F. Garanin, S. D. Kuznetsov, W. L. Atchison, R. E. Reinovsky, T. J. Awe, B. S. Bauer, S. Fuelling, I. R. Lingemuth, and R. E. Siemon, *IEEE Trans on Plasma Science* **8** (2010), 1815–1821.
[84] E. C. Cnare, *J. Appl. Phys.* **37**(10) (1966), 3812–3817.

10 Magnetic cumulation

10.1 Initial idea. Brief history. Main trends in development and research

Principle of magnetic cumulation

The idea of magnetic cumulation (MC) was proposed by Sakharov in the early 1950s as one of viable solutions for the problem of controlled fusion: "Creation of a powerful gas discharge in a mixture of DT, transferring into the thermonuclear explosion" (i.e., creation of a discharge induced by the fast alternative ultrahigh magnetic field). In the succeeding years the development of the idea of MC was linked with the search for the solution of another problem, urgent at that time, namely, the transfer of small masses of an active substance into a hypercritical state (the nuclear charge with a modest energy release) [1]. In the device proposed by Sakharov [2, 3], known as the generator MC-1 (Figure 10.1), the hollow metal

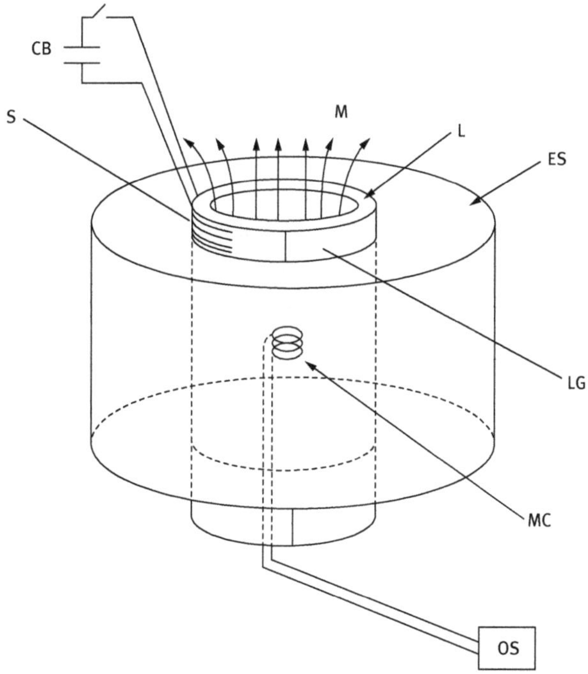

Fig. 10.1: The principal sketch (made by Sakharov himself) of the MC-1 generator: L = liner; LG = longitudinal gap; S = solenoid of initial field; ES = explosive substance; MC = measuring coil; OS = oscilloscope; CB = capacitor bank; M = magnetic lines.

cylinder is installed in a coil carrying current. It "embraces" the beam of magnetic force lines of the "initial" magnetic field produced by the coil. The charge with explosive substance (ES) is placed on the outside of the cylinder. At a certain moment it is exploded over the total external surface. The hollow cylinder is compressed by the products of the explosion, and, in turn, like a "giant fist" compresses the beam of magnetic force lines, increasing the intensity and energy of the magnetic field (A. D. Sakharov, the newspaper "Izvestiya", 1965).

Earlier in Los Alamos Fowler and colleagues were developing the powerful pulsed neutron source [4] and proposed generating the DT plasma in the initial magnetic field inside the cylindrical thin-wall liner, so that the subsequent explosive implosion would compress both the field and the plasma. A high neutron yield was expected in the case of the isentropical compression of plasma. Publication of the results at the Los Alamos laboratory in 1960 reported about attaining a magnetic field with induction of 1400 T [5] and initiated intensive investigation on magnetic cumulation in numerous laboratories over the world. In Fowler's view, the expectations for the creation of "pure fusion" were the major impetus to many projects on explosive-driven flux compression: "As said, it was Sakharov's dream, who is considered to be a founder of Russian research, and exactly the same dream was the foundation for studies in Frascatti" [4] (Euratom Laboratory of Ionized Gases). "The main incentive of these studies was the expectation of quick success in solving the problem of pulsed thermonuclear fusion" (A. I. Pavlovskii "Reminiscence of different years") [6, 7].

The alternative studies on MC were related to the possibility of constructing compact powerful explosive sources of current up to 10^9 A (energy up to 10^9 J, power above 10^{14} W). The constructive scheme of the MC-2 generator, proposed by Sakharov, referred to as the spiral-coaxial generator, is shown in Figure 10.2 [8]. The initial magnetic flux is induced by the discharge of a capacitor bank in the current circuit formed by the external cylindrical spiral (1), passing into the solid cylinder (2), and the switched-on central tube (3). Under the action of the explosion products (4) the shape of the tube is stretched into a cone, which flies up to the beginning of the spiral, and with the detonation wave through the HE in the tube, goes inside the spiral, diminishing its inductance. The magnetic flux in the region between the spiral and central tube is displaced first in the coaxial and then additionally compressed with further deformation of the circuit. An increase of magnetic energy occurs owing to the work done by the walls of the central tube the magnetic flux is compressed.

It is quite evident that the studies in this direction were begun in the defense-oriented laboratories of Russia (All-Russian Scientific Research Institute of Experimental Physics – RFNC-VNIIEF.) and of the US (LosAlamos National Laboratory – LANL): they possessed the well-developed techniques for an explosion-driven

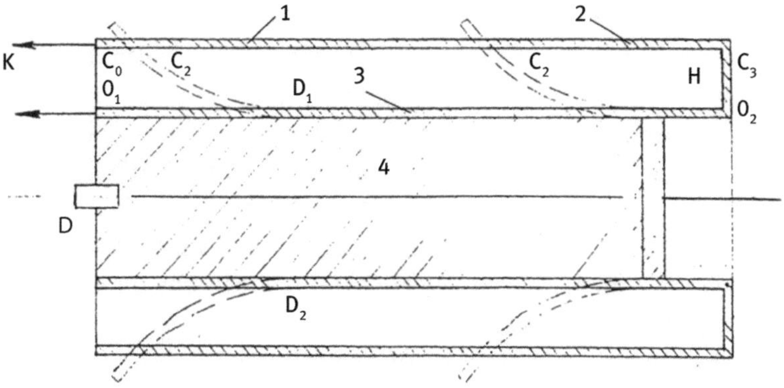

Fig. 10.2: The principal schematic drawing of the MK-2 generator: (1) the spiral coil with variable number of turns; (2) solid cylinder; (3) internal tube; (4) explosive substance; D = detonator. *Dashed line* marks the motion of cone along the spiral.

compression of shells, and for a long times the explosive substances were the basic source of the flux compression. Sakharov divided principally magnetocumulative devices into two types of MC-generators: the MC-1 generator of field (energy density) with a single imploding current shell ("liner"), and the MC-2 generator of energy with an immobile ("stator") and a moving ("armature") conductor. Nevertheless, both of these types have the same basic principle of operation: in both types of MC generators the explosion energy is transformed into kinetic energy of the conductor (liner or armature), which, in turn, is transferred into magnetic energy when the conductor is decelerated by the pressure of the magnetic field. Therefore references is often made to explosion-driven flux compression in general. Note that the magnetocumulative devices are the generators (amplifiers) of magnetic fields and, due to this feature, differ principally from the solenoids previously considered as sources of high magnetic fields, supplied by an external current source.

The whole long history of magnetic cumulutation as one of the branches of modern high energy density physics – at least the main stages of its development – were reflected at numerous international conferences. First of all, there are the megagauss conferences which integrate and assemble regularly the members of the international megagauss community, beginning in 1965 with the first conference in Frascatti and ending with the last, 14th conference in 2012 in Hawaii, USA [MG I–XI and MG–SF; see the Introduction]. Now, of course, most of these scientific works are only of historical interest. Very often, investigations which were at first very promising were never developed or applied later on. During the early stages of MC research it was sometimes difficult or even impossible to openly publish the study results. Some of the competent and experienced scientists working at that time in the fields of the physics and technology of high energy densities have published memoirs of their first steps in this field [4,8–11], as well as detailed reviews of the progress in research at the time of publication [12–22].

These papers and recently published reviews of the work done at the RFNC-VNIIEF on magnetic cumulation [23–25], along with the well-known monographs devoted to the problem of attaining high and ultrahigh magnetic fields, as well as their application in megagauss physics [21, 24, 25, 28], give us a good idea of the progress made in this area. The extremely useful bibliography on ultrahigh magnetic fields, compiled by Shvetsov and covering the period 1924–1985 [29], should be especially noted, as well as Knoepfel's monograph [18, 30], which is outstanding in its scope of the wide range of problems, and is especially useful as introduction in the technique of obtaining high magnetic fields, in general, and in explosion magnetic cumulation, in particular.

10.2 MC energy generators

The aim of this book is to describe the methods of obtaining ultrahigh magnetic fields. Therefore emphasis is placed on the generators of the MC-1 type. In this section explosion magnetic generators of energy and current are briefly discussed (MC-2, EMG). For a detailed description of their construction and applications the reader is referred to the monographs [19, 26, 27].

A special feature of MC-2 generators, as suggested by Sakharov, is the separation of the compressed volume and inductive load in circuit design and space. Mostly, the generator circuit of MCG is studied using the "engineering approach", when the processes in the magnetic system are described using a replacing circuit with variable inductance $L(t)$. In this case, assuming the conductance of the generator to be ideal and the flux constant, one can describe the current amplification by the relation $i(t) = i_0(L_0/L(t))$, where L_0 is the initial inductance. For practically used magnetic systems with the circuit resistance R, the concept of the inner impedance $|dL/dt|$ is introduced. The simplest analysis of the generator circuit with a load shows that the condition of the current amplification is $|dL/dt| > R$. The calculation of the conductor resistance takes the diffusion of magnetic field into account.

Following common practice, the moving conductors of generators are said to be armatures (or liners), and the fixed elements of devices are often referred to as stators. The terms of the EMG are usually related to the configuration of characteristic conductive units. For instance, there are "coaxial", "flat", "spiral" "disk" "strip" and other types of generators (see the simplified sketches in Figure 10.3 from the survey by Fowler and Altgilbers [12]).

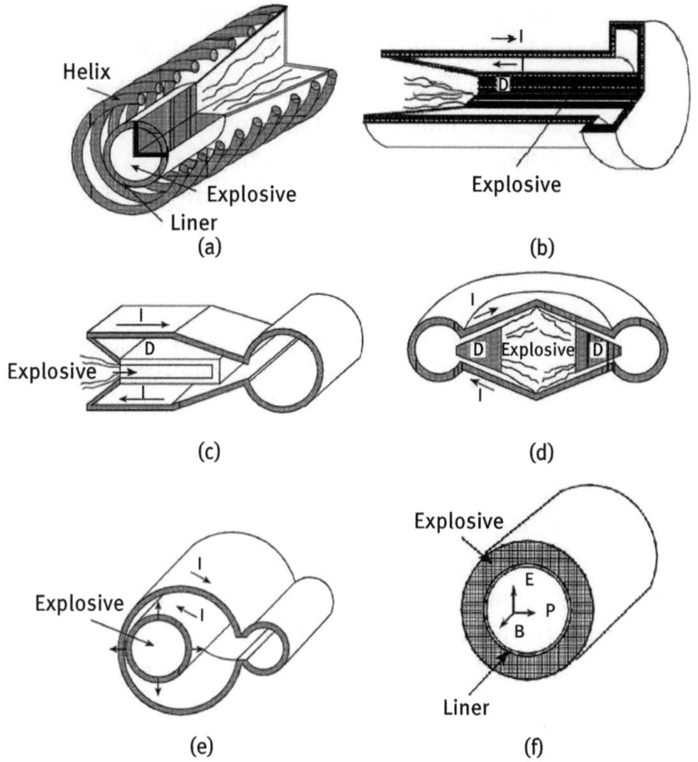

Fig. 10.3: Sketches of different devices for explosion-driven flux compression: (a) spiral generator; (b) coaxial generator; (c) flat generator; (d) toroidal generator; (e) turn generator; (f) (for comparison) MC-1 generator.

Often they are constructed to provide the power supply of the different powerful pulsed devices: sources of high frequency radiation, railguns, plasma focus installations, high power lasers, Z-pinch systems generating soft x-ray radiation, and so on. In such devices the outer size of the generator can reach a several meters and the mass of the explosive several kilograms. The spiral generator can be used to create an initial magnetic field in the MC-1 generator. It allows construction of autonomous (transportable) sources of the ultrahigh magnetic fields which can be used in a large variety of studies [31] without connection to the stationary sources of the initial magnetic field, such as large capacitor banks.

On the other hand, in the course of the adjustment of different units or for educational purposes, the small devices of cm scale with an explosive of a few grams are used. Systematical studies of the generators with spiral diameter less than 40 mm [19] included the supply from the permanent magnet and capacitor bank, construction materials of different type and shape, and various combinations of the straight and bent stators and armatures. For their estimation and comparison the quality coefficient was introduced, with regard to the initial and final currents, and the initial inductance of the load generator. In comparing the results obtained in different studies, it was established that the operation of small and large generators is dependent on the size of conductors. This fact was attributed to the decrease of current density in large generators and, correspondingly, the decrease of flux losses.

The most often studied and widely used were generators of moderate size of different design. In the late 1940s and later on at the LANL, different easily manufactured devices for flux compression were developed and used for generating fields of 100–250 T in loads of fixed volume. This resulted in the development of two systems of generators: the strip generator with a single-turn magnet as a load and maximum field about 100 T (Figure 10.4), and a two-cascade strip generator with final field about 200 T [32]. The load in the second one is also a single-turn magnet with a cavity with a diameter of 16 mm and length of 76 mm. Due to a large aspect ratio (ratio of length to diameter), the homogeneity of the field is quite high, calculated at about 0.0003, in a volume of 5 mm diameter in the center of a cavity, and this is still retained during the field pulse. The construction of these generators is particularly handy, allowing for an easy increase in the initial flux and the volume of the load or the field in the load of decreased size. To be sure, there are practical limits, but the generators of LANL are beyond them. During many years they were applied for investigations of substance at cryo temperatures, including magnetic properties, magnetooptical effects, and high temperature supersemiconductors as well.

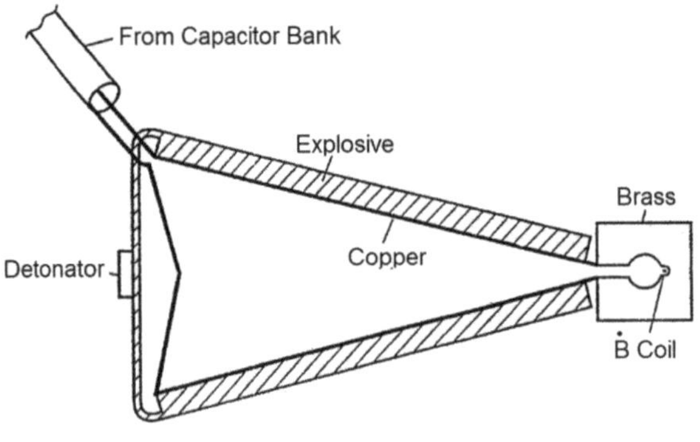

Fig. 10.4: Strip generator of magnetic fields up to 100 T (LANL).

Single-turn MC-generators with a reasonably simple construction have been also developed and applied in experiments [34, 35] at VNIIEF. A schematic drawing of a single-turn generator is shown in Figure 10.5. The current circuit of the cell of

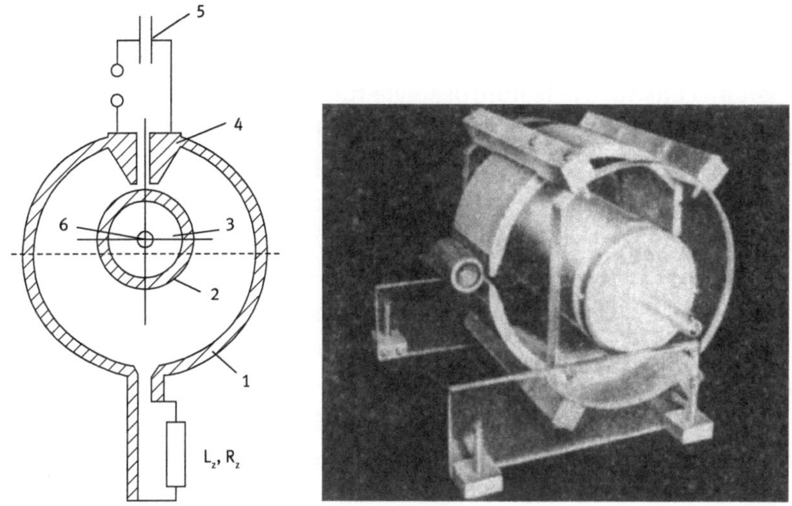

Fig. 10.5: Schematic drawing and external view of the turn generator used in WNIIEF. Notations are given in the text.

MCG has the shape of a wide turn (1) with two cuts for feeding with initial current and an outlet into the load for it. Inside the turn the cylindrical shell (2) filled with explosive (HE) (3) is installed. Initial magnetic flux is produced by discharge of a capacitor bank (5). The ES is initiated over the axis, at the same time on the whole length. On feeding the circuit with initial current, the HE is exploded and the cylindrical shell (2), expanding due to explosion, closes the input contacts (4) at the moment of maximum initial current. Then the trapped flux is quickly compressed by the whole surface of the expanding shell and is driven out to the load. With an optimal choice of magnetic field and current density, the deceleration of the shell and transformation of its kinetic into magnetic energy is most efficient. Simultaneous flux compression by the whole shell surface provides the small time interval of extraction of the shell energy.

To obtain uniform sliding of a shell contact over a turn and the discontinuous displacement of the magnetic flux into the load, the shell is installed with a shift of 4–10 mm relative to the turn axis toward the load. Experimental data shows that magnetic flux conservation strongly depends on the eccentricity. The flux conservation was sharply reduced with the decrease of the eccentricity below a critical value of 3.5 mm. The researchers explained this fact by the decrease of the speed of the sliding contact.

The single-turn generators have low initial inductance; therefore they do not give a large current amplification coefficient, e.g., it is increased by a factor of 7–10 while the time of the “e” fold-current amplification is 9–12 μs. The optimal value of the current linear density over the turn width is of the order 0.8 MA/cm. The specific energy output over the volume of generator is 200–225 MJ/m^3.

The important constructive feature of single-turn EMG is the convenient connection (both in series and parallel) of the turn cells in multielement systems. The magnetic flux is simultaneously compressed in all the cells, which conserves the short time of operation and provides the summation of the generated power of separate cells and, in the case of a parallel connection, the summation of currents as well.

A one-element EMG, with a turn of 30 cm diameter and 30 cm width with initial inductance 87 nH, induced in a load of 8 nH 50 MA of current, with voltage in the range 30–35 kV and energy of 10 MJ. An EMG of three cells (Figure 10.6), connected in series, with an initial inductance of 260 nH, induced in a load of 30 nH the current pulse of 46 MA, energy of 30 MJ and power of 4 TW.

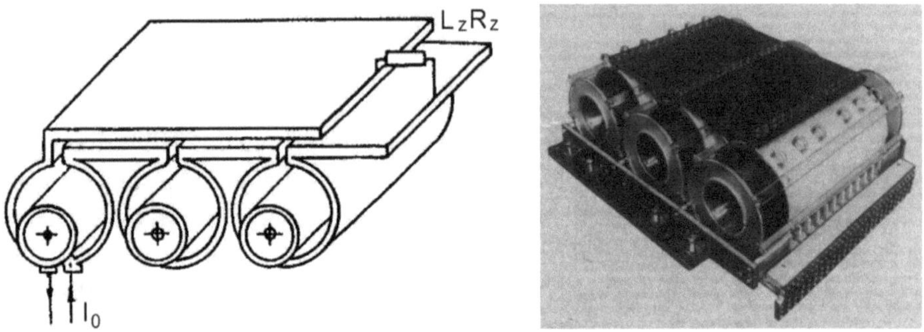

Fig. 10.6: Schematic drawing and external view of three single-turn generators connected in series.

Another important fact with regards to the application features of EMG single-turn generators is that the explosive charging procedure is convenient and can be made immediately prior to the completion of the experiment.

On the basis of a single-turn EMG a generator of pulsed ultrahigh magnetic fields has been built (Figure 10.5, to the right). This compact device consisted of a single-turn generator with an external diameter of 23 cm, width of 12 cm, and the load as a single-turn solenoid from tungsten alloy. The inner diameter of the solenoid cavity of 0.75 cm and a width of 2–3 cm was sufficient for placingf several samples in the homogeneous field. With a volume of 2.3 cm^3 a magnetic field up to 280 T was achieved.

The present devices of high density energy require not only a large amount of supplying energy, but also a high rate of energy input into the load, which called for the development of a quick-acting MCG and the use of current switchers and opening switches of different construction in the generator circuit.

A breakthrough in this direction was achieved in early 1970, when at VNIIEF tests of a multimodule disk generator (MMDG) were carried out. The MMDG consisted of several (5–15) equal generators with a profile surface of conductors compressing the magnetic flux. Some modules were connected in series in the course of the seed-feeding of the initial current, and in parallel, when working on the load. For seed feeding, quick-acting spiral explosive-driven magnetic generators (SEMG) were used [36]. As an example of recently achieved results, Figure 10.7 presents the scheme of the above-mentioned generator with an electroexplosive foil release. The load of the generator is the liner compressed by current in the Z-pinch configuration.

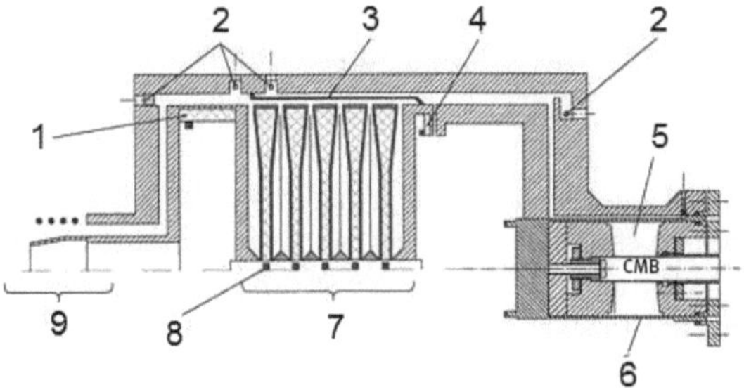

Fig. 10.7: Scheme of the DEMG (disc generator) with the electroexplosive foil release and the spiral EMG of the seed-feeding current: (1) an explosive switch-off unit; (2) detectors; (3), foil current release; (4) switching key; (5) loading unit; (6), liner; (7) DEMG; (8) detonators; (9) SEMG (spiral generator), CMB = central measuring block with speed detectors of the liner implosion. An explosive is shown with *dashed line*. It is not shown in the SEMG.

The device operates in the following way: the two-cascade EMG, consisting of spiral (9) and disk (7) generators, connected in series, is fed by initial current delivered by the discharge of a capacitor bank. On closing the circuit of the explosive system at the maximum initial current, the first cascade spiral generator (SEMG) begins to operate, creating an initial current in the disk generator. At the peak current of the disk generator the unit, releasing the SEMG and switching-on the DEMG, just operates. Simultaneously, the detonators on the axis of the disk elements are initiated and the process of current amplification begins. The DEMG is loaded on the foil current opening switch (3). The foil is heated by current and evaporates, but prior to its evaporation, the explosive switching key (4) connects the load to the DEMG, the load being the unit of acceleration of the thin-wall liner, containing the central measuring probe (CMP) on the axis. The main merit of this device is the low inductance resulting in the decrease of losses at transfer of energy from the DEMG to the load. In experiments accomplished with this facility record results have been obtained for explosive magnetic systems.

Over many years a similar device, functioning at various initial parameters, is under operation in joint VNIIEF-LANL studies of the high speed implosion of metal liners. In the experiment projected for the very near future the co-workers expect that at a current above 60 MA the speed of aluminum liner of 3 mm thickness will exceed 20 km/s [33]. These experiments make it possible to check the efficiency of the magnetic implosion of a cylinder liner and to carry out shock-wave measurements at pressure in aluminum of up to 1 TPa.

Recent advances of American laboratories are the double spiral-coaxial EMG with an output current of 98 MA and energy of 66 MJ, developed at the Livermore Laboratory and, as well, the tests of the half-model of this generator with 60 MA current and 10 MJ energy [37–38].

Finally, in his life report presented at the X Megagauss conference in 2004 Max Fowler summed up [39]:

> Over the years, our community has developed a wide range of high power energy sources and pulse shaping tools that allow construction of systems that can deliver pulses that drive a great variety of systems. The portability of these systems allows experiments to be performed in remote locations, including the upper atmosphere and beyond. As noted earlier, the output pulses of some large, costly machines have been simulated with our sources that have then been used as successful replacements in experiments. This capability should be exploited much more widely.

10.3 Physical processes in magnetic cumulation. Analytical estimates for the MC-1 system

It would be worthwhile to consider the physical processes involved in magnetic cumulation in the MC-1 system on the basis of the analytical solution of modeling problems. The estimates of characteristic parameters, such as induction amplitude, diameter of a volume of maximum field, and duration of a field pulse, are given below. These estimates enable us to define the main factors limiting the values of ultimate parameters attained in magnetic cumulation.

10.3.1 Induction amplitude and the radius of turnaround for flux compression by an ideal cylindrical shell

Let us consider a long closed cylindrical shell with inner radius r_1 that embraces the magnetic flux Φ_0 (Figure 10.8). Let at the initial instant when $r_1 = r_1(0)$, the liner wall be compressed with velocity $(dr_1/dt)_0 = -u_0$ and have kinetic energy W_0' (per unit length).

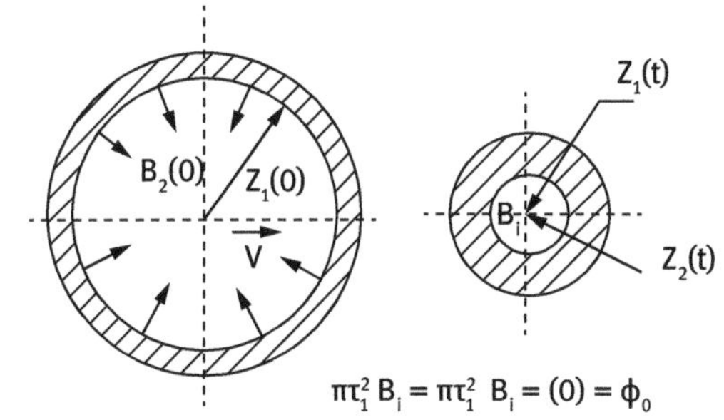

Fig. 10.8: Principle of magnetic cumulation: the magnetic flux Φ in a cavity remains constant in a time of compression by the shell.

At ideal conductance of the wall material, at any instant the following condition is valid:

$$\Phi_0 = \pi r_1^2 B_i = \text{const}, \tag{10.1}$$

where B_j is the induction in a cavity. In the time of compression the inner radius r_1 is decreased, while the induction $B_i = \Phi_0/\pi r_1^2$ is increased, as well as the energy density of the magnetic field

$$\frac{B_i^2}{2\mu_0} = \frac{\Phi_0^2}{2\pi^2\mu_0 r_1^4}, \tag{10.2a}$$

the total energy (per unit length)

$$W_M' = \frac{\Phi_0^2}{2\pi\mu_0 r_1^4}, \tag{10.2b}$$

and the magnetic pressure, decelerating the conductor, $P_M = B_i^2/2\mu_0$. If the shell is not subjected to the action of magnetic forces from the outside (but moves by inertia), and the wall material can be likened to the incompressible liquid, then at the moment t_M the whole energy of the system will be transferred to the magnetic field, and the shell will come to a turnaround. At this moment the induction attains the maximum B_M Further on the radius begins to increase under the action of electromagnetic forces. The minimum inner radius of the shell (the radius of the turnaround) r_M is related both to the B_M, and to the initial kinetic energy by an expression following from the law of energy conservation:

$$\frac{\pi r_M^2 B_M^2}{2\mu_0} = W_K' + \frac{\pi r_1^2(0) B_i^2(0)}{2\mu_0}, \tag{10.3}$$

where $B_i(0) = \Phi_0/\pi r_1^2(0)$ is the initial induction. Since $\pi r_M^2 B_M = \pi r_i^2(0) B_i(0)$, the equation for the induction amplitude can be derived from (10.3):

$$B_M = \frac{2\mu_0 W_K'}{\Phi} + B_i(0) = \frac{2\mu_0 W_K'}{\pi r_i^2(0) B_i(0)} + B_i(0). \tag{10.4}$$

Further we can find r_M:

$$r_M = \sqrt{\frac{\Phi_0}{\pi B_M}}. \tag{10.5}$$

Usually $B_M \gg B_1(0)$ and $r_M \ll r_1(0)$.

Relation (10.4) shows that the induction amplitude is higher when initial field is weaker. As an example we consider the case of the MK-1 device and calculate the initial induction and energy necessary for producing the field with induction of the order $B_M = 10^3$ T in the cylindrical volume with radius $r_M = 1$ cm. Magnetic energy in this volume (total magnetic energy of the system per unit length) is $W_K' \approx \pi r_M^2 B_M^2 / 2\mu_0 = 0.5 \cdot 10^8$ J/m. Assuming that the initial radius $r_1(0)$of the liner is 10^{-1} m, then the initial induction must be equal to $B_i(0) = B_M [r_M / r_i(0)]^2 = 10$ T, and the initial energy of the magnetic field $\pi r_i^2(0) B_i^2(0)/2\mu_0 = 2.5 \cdot 10^6$ J/m, which is much less comparable with the total energy. This example shows that the energetic scale of installations could be very large. This leads to the necessity of using explosive materials. The energy density stored by explosives is fairly high (in a range of 4–6 MJ/kg) [18], but calls for special chambers for carrying out the experiments at the explosion test field. Along with this, with the decrease of the radius of the turnaround by the order of magnitude, ($r_M = 10^{-3}$ m), W_K' decreases down to 25 kJ/cm, which is easily achieved under laboratory conditions when using capacitive or inductive energy storages. However, the experiments are troubled when in the mm volume.

10.3.2 Estimation of the pulse duration of a magnetic field in magnetic cumulation

For a number of applications, in particular, for the confinement of dense plasma, it is important to obtain not only the required induction amplitude, but also to provide for the necessary duration of the field existence at a given level. The pulse shape of induction can be determined considering the shell material as noncompressive ideally conductive liquid. From the continuity equation we have

$$\frac{\partial \gamma}{\partial t} + \frac{1}{r}\frac{\partial}{\partial r}(r\gamma u_r) = 0. \tag{10.7}$$

For the incompressible liquid ($\gamma = \gamma_0 = \text{const}$) we can write

$$r u_r = r_1 \dot{r}_1 = \text{const}, \tag{10.8}$$

where $\dot{r}_1 = dr_1/dt$. From here the following relation for kinetic energy of a moving conductor can be found:

$$W' = \int_{r_1}^{r_2} \left(\frac{1}{2}\gamma_0 u_r^2\right) 2\pi r dr = \frac{\pi}{2}\gamma_0 \left(r_1 r_1^2\right) \ln\left(1 + \frac{S_0}{\pi r_1^2}\right), \tag{10.9}$$

where $S_0 = \pi(r_2^2 - r_1^2)$ is a cross section of a conductor. At any time of the process the sum of kinetic and magnetic field energy is constant and equal to the field energy at the moment of the peak induction $W_K' + W_M' = W_M(t_M)$. Thus the equality

$$\frac{\pi}{2}\gamma_0 \left(r_1 \dot{r}_1\right)^2 \ln\left(1 + \frac{S_0}{\pi r_1^2}\right) + \frac{\pi r_1^2 B_i^2}{2\mu_0} = \frac{\pi r_M^2 B_M^2}{2\mu_0} \tag{10.10}$$

takes place.

Excluding B_i by means of the relation $B_i = B_M(r_M/r_1)$, we come to the equation for the shell inner radius:

$$\left(r_1 \dot{r}_1\right)^2 \ln\left(1 + \frac{S_0}{\pi r_1^2}\right) = \frac{r_M^2 B_M^2}{\mu_0 \gamma_0}\left(1 - \frac{r_M^2}{r_1^2}\right). \tag{10.11}$$

We are seeking an analytical solution of this equation, describing the radius change in the vicinity of the minimum point $r_1 \approx r_M$. For an approximate calculation we can change r_1 in the logarithm argument by r_M. We will characterize the induction pulse duration by the time interval τ_0, in the course of which the induction varies in the range $(1/\sqrt{2}) < B_i/B_M < 1$, which corresponds to the energy density changing from $(1/2)B_M^2/2\mu_0$to $B_M^2/2\mu_0$, and to the inner radius from $2^{1/4} r_M$ to r_M. Separating the variables we come to the equation

$$\int_{r_M}^{2^{1/4} r_M} \frac{r_1 dr_1}{\sqrt{1 - \left(\frac{r_M}{r_1}\right)^2}} \approx \frac{r_M B_M \tau_0}{2\sqrt{\mu_0 \gamma_0 \ln\left(1 + \frac{S_0}{\pi r_M^2}\right)}}. \tag{10.12}$$

On substituting $r_1/r_M = x$, we have

$$\int_1^{2^{1/4}} \frac{x^2 dx}{\sqrt{x^2 - 1}} = 0,68 \approx \frac{r_M B_M \tau_0}{2\sqrt{\mu_0 \gamma_0 \ln\left(1 + \frac{S_0}{\pi r_M^2}\right)}}.$$

From here we find

$$\tau_0 \approx 1,36\frac{r_M}{B_M}\sqrt{\mu_0\gamma_0 \ln\left(1+\frac{S_0}{\pi\, r_M^2}\right)}. \tag{10.13}$$

The numerical factor in this expression corresponds to the chosen time interval of the induction change. Independently on this choice, the parameter τ_0 by order of magnitude is the ratio of r_M to the Alfven velocity $V_A(t_M) = B_M(\mu_0\gamma_0)^{-1/2}$. At the moment the shell comes to a turnaround its acceleration has the maximum

$$\left.\frac{d^2 r_1}{dt^2}\right|_{r_1=r_M} \approx \frac{B_M^2}{\mu_0\gamma_0 r_M \ln\left(1+\frac{S_0}{\pi\, r_M^2}\right)}. \tag{10.14}$$

In order to show how this result could be used in practice, let us determine the parameters needed for dense plasma confinement in a time span corresponding to the parameter $n\tau_0 = 10^{20}\,\text{m}^{-3}\text{s}$ with the confining magnetic field of 300 T. If we assume that the average energies of ions and electrons in plasma are 10 keV, and the gas kinetic pressure is equal to the magnetic pressure, then the corresponding plasma concentration is $n = 10^{25}\,\text{m}^{-3}$, and hence $\tau_0 = 10^{-5}\text{s}$. Let us take $\ln(1 + S_0/\pi\, r_M^2) = \ln 10 = 2.3$. Then, according to (10.13), we have $r_M = 1.4\cdot 10^{-2}$ m. The energy of the magnetic field at the moment of maximum compression will be $W_M' \approx 2,2\cdot 10^7$ J/m.

10.3.3 The effect of field diffusion on the induction amplitude with magnetic cumulation

The effect of the diffusion on the flux compression was examined by Somon [68] and Knoepfel [18]. At the final stage of this process the field exists in the cylindrical cavity of radius r_1, where it is homogenous, and in the conductor, where it decays. Magnetic field diffusion into the shell wall leads to the displacement of the effective field-conductor boundary. As the coordinate of the boundary it is reasonable to take the radius r_{eff} such that the flux through the circle of this radius is Φ_0. The length $\Delta'' = r_{eff} - r_1$ is the effective depth of the flux penetration (conventionally, the skin layer in the conductor).

Let us first consider the case when the surface effect is sharply pronounced. In this case the condition $\Delta'' \ll r_M$ is valid, or

$$\Delta'' \approx a_0\frac{B_M}{B_0}\sqrt{\frac{\rho_0\tau_0}{2\mu_0}} \ll r_M, \tag{10.15}$$

where $\tau_0/2$ is the time characterizing the induction increase up to a maximum (estimated above), a_0 is of the order of the unity (see Chapter 3). Then as a first

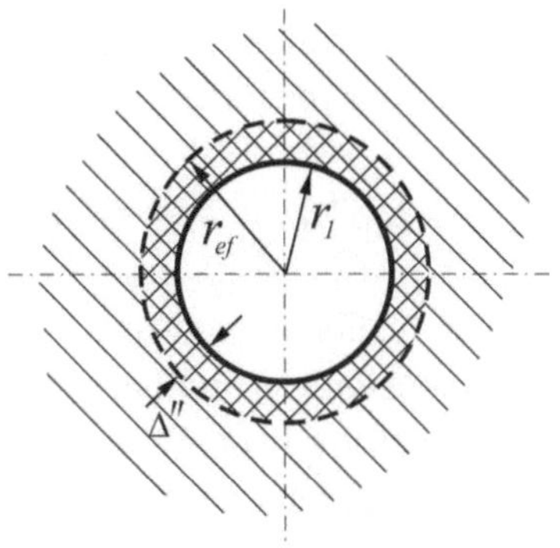

Fig. 10.9: Sketch for the estimation of the effect of the field diffusion on the induction amplitude at comparatively small penetration depth ($\Delta'' \ll r_i$).

approximation we can adopt the ideal conductance for a liner, but its effective inner radius is $r_{eff} = r_1 + \Delta''$ (Figure 10.9). We can conditionally suggest that on the liner inner surface with radius r_1 there is an interlayer of thickness Δ'', "transparent" for the field. In the framework of this model we can expect that the above relations, describing the ideal magnetic cumulation, remain valid, but kinetic energy W_K' should be decreased, since the energy stored in the layer of thickness Δ'' is not transmitted to the field. Since the layer thickness is small compared to the wall thickness, and the change of W_K' is insignificant, the induction amplitude is close to the calculated one. The calculated radius of the turnaround is now the radius of the field-conductor boundary, and consequently the true boundary of the shell is now $r_M - \Delta''$. with a sharply pronounced skin effect an additional compression of the shell takes place over the radius by the value Δ''. This resulted in compensation of the displacement of the field-conductor boundary and, due to this fact, there is no loss in the induction amplitude.

These considerations are confirmed by the following calculations. In the final stage of compression when the liner thickness essentially exceeds the field penetration depth, the condition of the total flux conservation is fulfilled:

$$\Phi_1 + \Phi_2 = \Phi_0 = \frac{2\mu_0 W_K'}{B_M}, \qquad (10.16a)$$

where $\Phi_1 = B_i(t)\pi r_i^2$ is the flux in the liner cavity, Φ_2, the flux in the skin-layer. The latter can be represented as $\Phi_2 \approx 2\pi r_i \Delta'' B_i(t)$, where $B_i(t)$ is the instant induction in the liner cavity, and Δ'' is the effective depth of the skin layer (see Section 3.11). Further, we assume the condition $\Delta'' \ll r_i$ to be valid.

Thus, the following relation takes place:

$$\Phi_0 \approx B_i(t)\pi r_i^2(1 + \Delta''/r_i). \qquad (10.16b)$$

At the moment of the liner turnaround $B_i(t) = B_{md}$. The equation of the energy balance of the shell, assuming an incompressible liquid, has the form

$$W'_K = \frac{B_{md}^2}{2\mu_0}(\pi r_{md}^2 + q2\pi \Delta'' r_{md}) = \frac{B_{md}^2}{2\mu_0}\pi r_{md}^2 (1 + 2q\Delta''/r_{md}). \tag{10.16c}$$

In this formula the second term is the sum of the magnetic field energy and thermal energy in the skin layer. In this expression the dimensionless number q is of the order of the unity and calculated in Supplement S10 at the end of this chapter. The joint solution of equations (10.16) leads to the following relations for the induction amplitude and inner radius of the liner at the moment of the turnaround:

$$B_{md} = \frac{1 + 2\Delta''/r_{md}}{1 + 2q\Delta''/r_{md}} B_M, \tag{10.17a}$$

$$r_{md} = ((\Delta'')^2 + r_M^2 B_M/B_{md})^{1/2} - \Delta''. \tag{10.17b}$$

The formula (10.17a) gives the somewhat underestimated value of the induction amplitude.

Another situation takes place if $\Delta'' > r_M$. In this case the total implosion of the shell is possible, since the region of the surface layer is "transparent" for the field and only slightly decelerated. At complete implosion the induction amplitude remains finite. It can be estimated, assuming that the flux Φ_0 threads the circle of radius Δ''. This estimate is given in [18]. The time span for filling the cavity of radius Δ'' with a shell material can be reasonably taken as the characteristic induction rising time. This time is of the order of Δ''/u_0, where u_0 is the velocity of the inner boundary, which we assume to be constant. Then, with an accuracy up to the order of the magnitude, the equality is valid where the numerical factors are dropped, and instead of the calculated induction amplitude B_M the practically-obtained induction B_m is introduced:

$$B_m \approx \frac{\Phi_0}{\pi \Delta''^2}. \tag{10.18a}$$

In the estimating formula (10.15) we can take $\tau_0 \approx \Delta''/u_0$ and replace B_M by B_m Then the expression for the effective thickness of the skin layer takes the form $\Delta'' \approx (B_m/B_0)(\rho_0\Delta''/\mu_0 u_0)^{1/2}$, or $\Delta'' \approx (B_m/B_0)^2(\rho_0/\mu_0 u_0)$; also, we use the equality $\Phi_0 = \pi r_1^2(0) B_i(0)$. Further on, after substitution of the expression for Δ''in (10.16) and simple transformations, we have

$$B_m \approx B_0^{4/5} B_i^{1/5}(0)\,(\mu_0 v_0 r_1(0)/\rho_0)^{2/5}. \tag{10.18b}$$

Numerical calculations give a close result: it differs only by an additional factor 1.1 [18]. Let us introduce in the ultimate expression the Reynolds magnetic number $Re_M = \mu_0 u_0 r_1(0)/\rho_0$. As a result we have the estimate for the induction in the completely compressed liner:

$$B_m = 1.1 Re_M^{2/5} B_0^{4/5} B_i(0)^{1/5}. \tag{10.18c}$$

Numerical calculations for the shell from the incompressible material show that the induction peak is achieved not at the complete implosion of the shell, but somewhat earlier, when the inner radius is

$$r_m \approx 0.25 r_1(0) B_i(0)^{2/5} B_0^{-2/5} Re_M^{-1/5}. \tag{10.18d}$$

In this case the ratio of the flux in the cavity with a cross section πr_m^2 to the initial one (i.e., the compression coefficient) is

$$\lambda = \frac{B_m r_m^2}{B_i(0) r_1^2(0)} \approx 0.07. \tag{10.19}$$

The given estimates of the field diffusion influence show that, unlike the model of ideal magnetic cumulation, the induction amplitude does not increase unlimitedly at $B_i(0) \to 0$, but tends to zero as $B_i^{1/5}(0)$. As an example, we estimate the induction amplitude, induced in a copper shell with the following initial parameters: the radius is $r_1(0) = 3 \cdot 10^{-2}$ m, wall thickness $h_0 = 2.5$ mm, velocity $u_0 = 2 \cdot 10^3$ m/s, $B_i(0) = 5$ T. The Reynolds magnetic number is $Re_M = 3.8 \cdot 10^3$. The induction amplitude, accordingly to (10.18c), is $B_m = 720$ T, if $B_0 = 38$ T. The initial kinetic energy of the shell is $W_K' = 2\pi r_1(0) h \gamma_0 u_0^2 = 1.7 \cdot 10^7$ J/m, and the initial flux $\Phi_0 = 1.4 \cdot 10^{-2}$ Wb. At ideal cumulation the induction amplitude is $B_M = 1500$ T. So, the induction here is almost halved because of the field diffusion into the conductor.

10.3.4 Restrictions on the induction amplitude conditioned by the compressibility of a medium

A role of compressibility as a factor restricting induction increase at magnetic cumulation appears most clearly when considering, for example, the flux compression in a gap of width 2 g between two conductors with a flat boundary (Figure 10.10) (the model of the flat piston [18]). As the gap is compressed, the induction in it increases, as does simultaneously the magnetic pressure, and the shock wave is driven in both conductors.

For an observer associated with one of the conductors, say, with the left one (Figure 10.10), and located before the shock-wave front at point A, the boundary "runs" towards him with the velocity of the flow behind the shock wave front

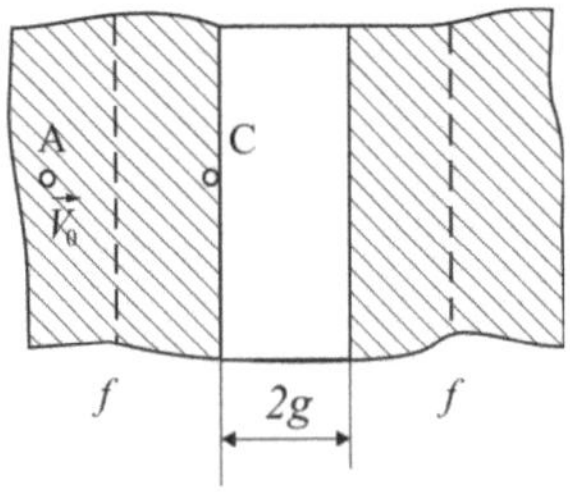

Fig. 10.10: Flux compression in the gap between two conductors with a flat boundary. Φ is the plane of the shock-wave front.

$(-u_f)$. In the laboratory the frame of reference for the velocity of point C is $u_C = V_0 - u_f$. The maximum induction amplitude is achieved at the moment when the boundary reaches turnaround, i. e., $u_C = 0$. At this moment $V_0 = u_f$. The piston velocity u_f can be related to the magnetic pressure by means of equation (9.27) but this leads to an inaccurate solution, since the given equation describes a stationary shock wave. In conditions where magnetic pressure increases, the process is nonstationary. However, comparison with the numerically calculated values shows that using the Rankine–Hugoniot relation does not lead to a significant error in the value of the field initiating the shock wave with a flow velocity of the medium behind the front V_0. The induction of this field is determined from the equation

$$B_\Pi = V_0\sqrt{2\mu_0\gamma_0\left(\lambda + \frac{C_1}{V_0}\right)}, \tag{10.20}$$

where C_1 is close to the sound velocity, and λ is the numerical parameter (for copper $C_1 = 4 \cdot 10^3$ m/s, $\lambda \approx 1.5$). Using the approximation of Shearer [9.18] $V_0 = a_0 B_\Pi^{3/2}$, the induction B can be related to the wall velocity in a following way:

$$B_\Pi = \left(\frac{V_0}{a_0}\right)^{2/3}. \tag{10.21}$$

As noted in Section 9.5, this approximation and hence formula (10.21) are valid for copper ($a_0 = 0{,}15$) up to fields with induction of 10^3 T. Let $V_0 = 2 \cdot 10^3$ m/s; then $B_m = 560$ T. The thickness of the conductors compressing the flux does not appear in (10.21). With an unlimitedly large thickness of conductors, i.e., at the indefinitely large kinetic energy of the colliding bodies, the induction and hence the energy transferred to the field remains constant and depends only on the velocity of the bodies. Only the kinetic energy of the layer, which will loose its velocity at the moment of the peak induction, is transferred to the field energy. The thickness of this layer d_0 can be derived from the equality

$$\frac{B_\Pi^2 g_\Pi}{2\mu_0} = \frac{\gamma_0 V_0^2 d_0}{2}, \tag{10.22a}$$

where on the left side of the equation one can see the magnetic field energy and the value of the half width of the gap at the moment of peak induction, and on the right side the kinetic energy of the layer of thickness d_0. The induction amplitude determined in this example can be compared with the case when the wall material is incompressible and the kinetic energy is completely transferred to the field when the wall is at the turnaround. If the thickness of each conductor is h, then for incompressible conductors the following relation is valid:

$$\frac{B_M^2 g_M}{2\mu_0} = \frac{\gamma_0 V_0^2 h}{2}, \tag{10.22b}$$

where B_M and g_M are the calculated values of the induction and half-width of the gap, respectively, at the moment of a boundary turnaround in the case of incompressible medium. Since at the ideal conductance the flux in the gap is conserved, the condition $B_\Pi g_\Pi = B_M g_M$ takes place. After the term-by-term division of (10.22a) by (10.22b), accounting for the above mentioned condition, we have

$$\frac{B_\Pi}{B_M} = \frac{d_0}{h}. \tag{10.23}$$

This equality is valid as soon as $d_0 < h$. If d_0 becomes equal or exceeds h, then, in the framework of the considered model, the induction amplitude will be equal to the calculated value B_M. More accurate calculations permit determination of the induction amplitude B_m and its comparing with the evaluated values. The dependence $B_m/B_\Pi = f(B_M/B_\Pi)$ is given in Figure 10.11.

At the ideal cumulation $B_m = B_M$ (straight line 1). This is a hypothetical case: when the material behaves like an incompressible liquid, the kinetic energy is completely transferred into the field energy at the moment of the liner turnaround, and formula (10.22b) is applicable. As the flux is compressed between the flat sheets, the field, calculated with regard to the compressibility of the medium, will be close to the one determined by relation (10.22b) if the induction amplitude is less than B_Π. Namely, in this case the conductors transfer the energy to the field and behave like an incompressible liquid. At $B_M > B_\Pi$ the estimated value is not achieved, and the induction amplitude is determined by the compressibility and equal to $B_{\Pi?}$ (curve 3). Numerical calculations give a result close to the one discussed above (see Figure 10.11). This confirms the applicability of the relation (10.20), despite the fact that the process is nonstationary.

In the case of a cylinder the dependence $B_m/B_\Pi = f(B_M/B_\Pi)$ is more complicated, because the velocity of the inner layers of the hollow cylindrical conductor increases when compressed. Therefore the induction rises, although insignificantly, even at the condition $B_M > B_\Pi$. Numerical calculations [60] show that in

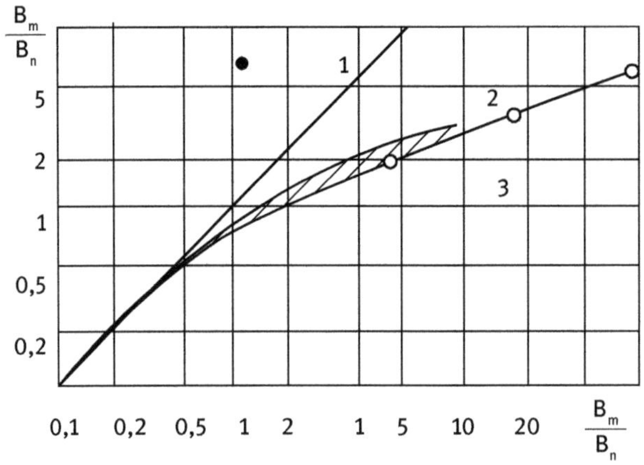

Fig. 10.11: Dependences characterizing the effect of compressibility in magnetic cumulation (based on results of Somon [60] and Volkov et al. [69]); (1) the ideal cumulation ($B_m = B_M$); (2) compressed cylindrical shell (calculated values are marked with points, the others are in the *dashed* region); (3) flat case.

the region, where $B_M > B_\Pi$, the following approximation is justified:

$$\frac{B_m}{B_\Pi} \approx \left(\frac{B_M}{B_\Pi}\right)^{1/3} . \tag{10.24a}$$

From the above it follows that the compressibility leads to a significant limitation on the induction amplitude. According to (10.24a),

$$B_m \approx B_\Pi^{2/3} B_M^{1/3} , \tag{10.24b}$$

or

$$B_m = \left[2\mu_0 \gamma_0 V_0 \left(C_1 + \lambda V_0\right)\right]^{1/3} \left[\frac{2\mu_0 h_0 \gamma_0 V_0^2}{r_1(0)\, B_i(0)}\right]^{1/3} , \tag{10.25}$$

where h_0 is the initial thickness, and $r_1(0)$ is the initial radius of the shell. In the limit case $\lambda V_0 \gg C$ (or $B_\Pi \gg \bar{B}$) we have $B_m = \text{const} \cdot \gamma_0^{2/3} V_0^{4/3} = \text{const} \cdot (W_K')^{1/3}$.

10.3.5 Violation of the stability of a liner at flux compression

There are two stages in the process of flux compression by the shell when a loss of stability of the shell shape is possible. In the initial stage of the compression of the thin-wall cylinder from the elastic material, when the field is relatively weak and the material retains elastic properties the violation of the stability of the thin shell can be described in the framework of elasticity theory. At the sudden stressing with pressure P from the outside, violations with wavelength

$$l = \frac{2\pi\, r_1\,(0)}{n} \tag{10.26}$$

arise in the shell wall; in the formula above n is the number of longitudinal wrinkles (gofers) on the cylinder surface. According to the theory worked out by Ishlinsky and Lavrent'ev [57], the violations with the number of gofers

$$n = n_M = \sqrt{\frac{1}{2}\left[\frac{P r_1^3\,(0)}{E h_0^3}\cdot 12\,(1-\nu^2) + 1\right]} \tag{10.27}$$

have the greatest velocity. Here E is the elasticity module, $\nu \approx 0.32$ is the Poisson coefficient, and h_0 is the wall thickness. In experiments with thin-wall cylinders made from copper, brass, and aluminum alloys ($h_0/r_1(0) = 0.025 - 0.1$) [58], it was shown that at relatively slow compression the cylinder wall is not increased in thickness, but rather gofers are produced, which leads to a decrease of the mid-radius: the cylinder wall is folded into a "bellows". In these regimes large plastic deformations are developed in the liner, which cannot be described in the framework of the model of small perturbations of the cylinder of elastic material. As was established, at a relatively low velocity of the radius change (less than 500–600 m/s), the initial number of gofers by the order of magnitude is close to n_M, and in the course of compression their number is decreased in such a way that the perturbation wavelength is not practically changed as the radius decreases down to about $0.2\, r_1(0)$. This fact was also pointed out also in a paper by Belan, Durmanov, Ivanov, Levashov, and Podkovirov [59]. The authors mentioned that towards the end of compression the maximum amplitude is reached by perturbation modes of numbers 8–10. The violation of the axial symmetry of the external pressure may lead to the appearance of initial perturbations of symmetry. This effect is sharply pronounced in magnetic cumulation, and its significance will be discussed in Sections 10.4 and10.11.

In the final stage of magnetic cumulation the inner surface of the conductor is subjected to magnetic pressure, and the whole conductor turns out to be in the field of the volume forces of the inertia $f_r = -\gamma du_r/dt$, where u_r is the velocity of the medium element. Since $u_r < 0$ and decreases by absolute value, the f_r will be directed toward the field, and Rayleigh–Taylor instabilities can be developed in the conductor. The magnetohydrodynamic theory of this process enables us to estimate in the framework of linear theory the time τ_1, after a lapse of which the amplitude of the sine perturbation increases by **e** times:

$$\tau_1 = \left(\frac{2\pi}{l}\,\frac{du_r}{dt}\right)^{-1/2}, \tag{10.28a}$$

where l is the wavelength of the perturbation. Here account of the forces of the surface tension was not taken. On the inner boundary of the shell we have $du_r/dt = d^2r_1/dt^2$.

More comprehensive analysis of the stability loss of the liner can be found in the works of Somon [60] and Almstrem, Bjarnholt, Goldberg, and Liberman [61]. Numerical calculations give the more complete pattern of the instability development with regards both the radial character of the flow and nonlinear effects. According to [60], the sine perturbations, as their amplitude is increased, change shape: their "humps" become sharper and valleys more smooth. Although the theory of small perturbations does not give the whole pattern of the development of instability, it permits the time τ_1 to be expressed in terms of the characteristic parameters of the magnetic field pulse, using the relation (10.14) describing the acceleration of the inner boundary of the shell. For the perturbation with the mode $n = 2\pi r_M/l$ we obtain

$$\tau_1 \approx \frac{r_M}{\sqrt{n}B_M}\sqrt{\mu_0\gamma_0 \ln\left(1 + \frac{S_0}{\pi r_M^2}\right)} = \frac{\theta\,\tau_0}{\sqrt{n}}, \tag{10.28b}$$

where θ is the numerical factor close to the unity. Since the perturbation amplitude increases as $\Delta r_n \exp(t/\tau_1)$, where $\Delta r_n$is the initial perturbation of the given mode, then, at the condition $t \gg \tau_1$ this amplitude, even with a rather small initial value, may increase up to large values. However, in magnetic cumulation the time interval of the field confinement and that of the development of the instability are close, if, as follows from (10.29), the number n is of the order of the unity. Therefore, over a time close to τ_0 , there is the possibility that amplitudes of the most dangerous long wave perturbations do not manage to increase up to values commensurate with r_M, if the initial perturbations are small. This makes it possible to conserve the symmetry of the liner in the final stage of compression. The methods for decreasing the initial perturbations are discussed below when considering some experiments.

In the models of magnetic cumulation discussed so far, we took into account the limitations imposed on the induction amplitude by one of the involved processes, whereas in practice they act in combination. Analysis of the joint action of all computable factors and, in particular, of their effect on the induction amplitude can be made using numerical methods. Below we dwell on the comparison of our estimated results with numerical calculations and experimental data. For relatively weak fields (about a few hundred Tesla) it is possible to use the known dependences for the conductance and the equations of state. Extrapolation of these dependences to the range of parameters corresponding to the final stage of cumulation in the fields close to the record ones (of order of 1000 T and above) does not give the reliable initial data for describing of heating, hydrodynamic flow, and field diffusion. This fact makes questionable the applicability of the estimates given above, but the main qualitative conclusion remains acceptable: in order to increase the final magnetic field of the MC-1 generator, it is necessary to increase the speed of the liner implosion and/or the density of the liner material. As for the finite diameter, the given quantitative estimates are not very suitable for practice, if it concerns the fields close to the record ones. However, the general relation for the induction in the cavity liner retains its meaning:

$$B(t) = B(0)\left(\frac{r_0}{r(t)}\right)^2 \phi\,(t), \tag{10.29}$$

where ϕ (t) is the conservation coefficient of the flux. The effect of the processes actually involved in the operation of the MC-1 generator is concealed in the dependence ϕ (t). In practice, if the induction of the magnetic field is less than approximately 100 T, the conservation coefficient is close to 1, then with field increase it drops down to 0.1–0.3. In this case, as follows from (10.4), with decreasing flux it is possible to increase the field, owing to the higher compression of the liner.

10.4 Flux compression systems not using the explosion energy for liner acceleration

Not only chemical explosion can be applied for the acceleration of the liner up to a large speed and for flux compression, but the other methods as well.

Among the not numerous methods for nonexplosive acceleration of a liner, it is necessary to mention the study of Velikhov, Vedenov, Bogdanetz, et al. [40]. They proposed using the acceleration of a thin-wall copper liner with the help of the pulsed gas puffing under high pressure (up to 2000 atm). This method has not been commonly applied, unlike the magnetodynamic cumulation based on the compression of the liner by electromagnetic forces.

Magnetodynamic cumulation is the method for obtaining the fields of megagauss range by means of electromagnetically-driven flux compression without using explosive substances. It makes the experiment more safe and less expensive. In the experiments considered below, the compression of a metal liner has been produced in two types of systems. In both cases the compression of the poloidal field occurs. In the system of the first type (Θ-pinch) the liners were accelerated by the magnetic pressure of the poloidal field, and in the second (Z/Θ-pinch), by pressure of the azimuth field.

The compression of the thin cylindrical shell in this system occurs when the current increases in the solenoid, coaxially to which the liner is installed. Pioneering experiments on magnetodynamical cumulation have been done in the USA (work of Cnare in the Sandia Laboratories [41]) and in Russia at the Institute of Nuclear Physics (works of Alikhanov et al. [46–48]). Studies of the MDC method were carried out in many works [42–45, 49–56]. Table 10.1 presents the results of certain experiments carried out with the initial field and without it.

Table 10.1: Some experiments on magnetodynamical cumulation.

First author Type of experiment	Ref., date	Liner parameters				Capacitor bank energy	Flux density		Probe external diameter
		Material	Radius, mm	Length, mm	Thickness, mm	kJ	$B_i(0)$, T	B_m, T	mm
Cnare θ	10.41 1966	Al	70	17	0,4	136	0	210	2
Freeman θ	10.42 1967	Al	39,5	20	0,8	136	–	120	–
Alikhanov θ	10.46 1967	Al	54	145	2	700	≈ 1	140	3.5–7
Alikhanov Z-θ	10.47 1968	Cu	30	150	2	570	3–5	310	2.7
Guillot θ	10.43 1969	Al	16	15	0.2	27	0	160	1.5
Kachilla θ	10.44 1970	Al	15,8	14,3	0,3	27	0	134	1.8
Michkelsoo θ	10.45 1974	Cu	22,5	20	0,4	92	0	340	1.6
Miura θ	10.49 1979	Al	33	20	1	280	2,8	280	–
Alikhanov Z/θ	10.48 1984	Al	28	85	1	–	1.5	350	3
Miura θ	10.51 1989	Cu	75	70	1,5	4000	3,0	350	–
Miura θ	10.52 1994	Al	75	55	1,5	4000	2,1	545	4.5
Matsuda θ	10.53 2002	Al	58	45	1,8	5000	3,2	622	5.8
Novac θ	10.56 2004	Al	25	20	0,54	63	0	230	2.45
	10.55 2010, a	Cu	57.5	50	2	5000	4.4	730	≈ 1
Takeyama θ	10.55 2010 b	Cu	58	50	1.5	5000	4.4	710	≈ 1
	10.55 2010 c	Cu	62	53	2	5000	4.4	640	≈ 1

10.4.1 MDC systems with azimuth current in a liner

In the first experiments by Cnare [41] and in certain other experiments [43–56] the flux, compressed by a shell, was not initiated a priori ($B_i(0) = 0$). The flux penetrates in the cavity of the shell in the initial stage when the current increases in the discharge circuit, while in the gap between the solenoid (1) and the liner (2) (Figure 10.12a,b) the generated field compresses the liner.

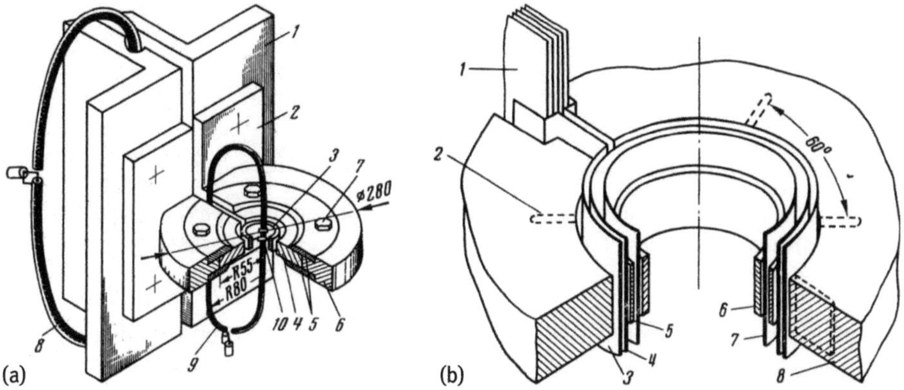

Fig. 10.12: (a) Magnetic system used for flux compression in an experiment without the initial field [45]: (1) buses of the step-down transformer; (2) contact of the single-turn coil; (3) liner; (4) inner replaceable part of the magnet; (5) washer for binding the replaceable part of the magnet; (6) external part of the magnet; (7) bolts; (8) Rogowski belt for measuring currents of the magnet and shell; (10) detectors of magnetic field. **(b)** Displacement of the liner and insulation: (1) insulation of the transformer; (2) placement of supplemented cuts; (3) basic insulation (fluoroplastic); (4) and (7) polyethylene insulation; (5) copper screen; (6) liner; (8) replaceable part of the magnet.

Typical temporal dependences of characteristic parameters were demonstrated in [45], Figure 10.13. In these experiments the energy from the capacitor bank, used for producing the current in a single-turn solenoid, did not exceed 100 kJ. The field was generated with the compression of copper and aluminum shells with diameter 40 mm and wall thickness from 0.2 to 0.5 mm. With the rising current in the solenoid, the opposing azimuth current is induced in the shell, but part of the field penetrates into the cavity of the liner. Simultaneously with compression of the shell, trapping and compression of the flux occur in its cavity. The liner accelerates as far as the magnetic pressure outside the wall exceeds the pressure from the inner side of the wall, i.e., while the condition $|B_e| > |B_i|$ is fulfilled. Let us note the characteristic moment t_0, at which the modules of induction on both sides of the wall are equalized and the deceleration of the liner begins. At this

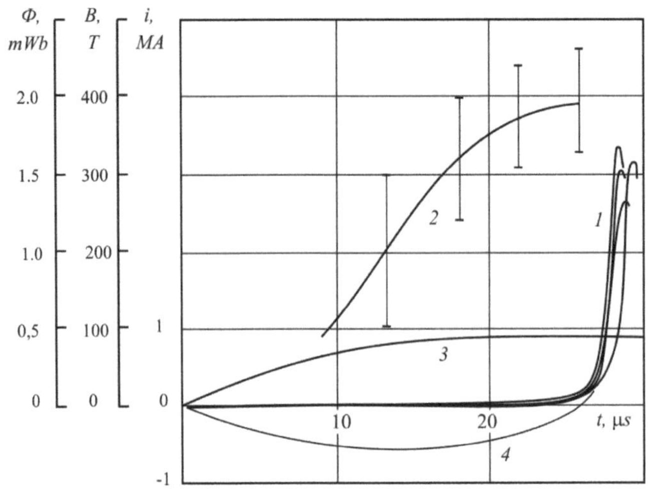

Fig. 10.13: Temporal variations of induction, flux in the liner, current of the liner and of the magnet: (1) induction; (2) flux; (3) current of the magnet; (4) current of the liner.

moment the current in the shell passes through zero and then increases by the absolute value with the increasing induction of the "trapped" field.

We emphasize that there are two stages of magnetodynamic cumulation. The first one, the stage of acceleration, is finished at the moment t_0. Next is the final stage when the kinetic energy of the liner is transferred magnetic field energy, as occurs in other devices for flux compression. Note that the field in the final stage of the process can be much stronger than in the gap. This is seen, for example, from the temporal dependence of the liner current, shown in Figure 10.13. In this experiment the peak field attains 340 T at a turnaround radius about 1 mm.

The drawback to this simple experiment is the difficult control over the field parameters, namely, over the induction amplitude and turnaround radius of a shell. This is due to the fact that the trapped flux depends on characteristics of the liner and on the rate of the current rising in the gap. There is a more adaptable system, in which the initial field is induced by an added magnet in the form of Helmholtz coil, as seen in Figure 10.14. In experiments without an initial field or with a small one the first stage of the process terminates at the moment when the current in the liner equals zero. In contrast, in experiments with a fairly strong negative initial field, at the moment t_0 the linear density of the current in the liner has the absolute value $2|B_e(t_0)|/\mu_0$. The final stage of the process is the same as in the experiments with and without an initial field.

In recent years Japanese researchers used the capacitor bank with megajoule energy as the source of power supply in experiments on acceleration of the liner.

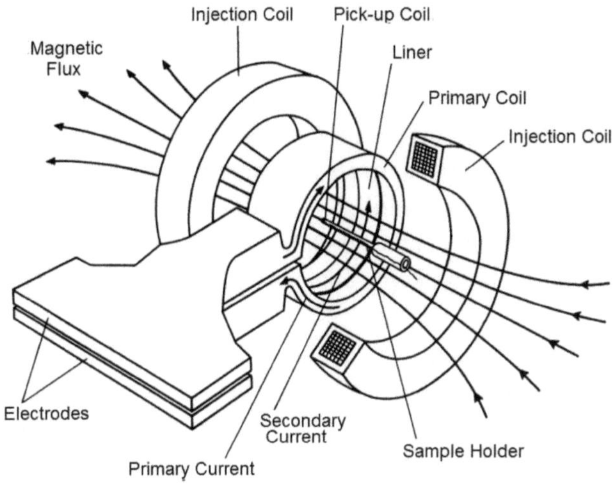

Fig. 10.14: Schematic drawing of the device for generation of ultra high magnetic field by means of magnetodynamical cumulation with the initial field [50].

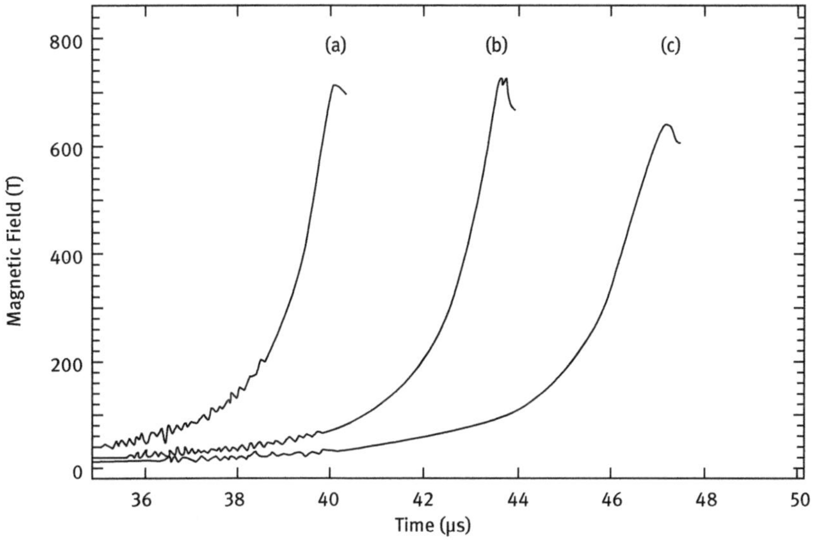

Fig. 10.15: Oscillograms obtained in the experiments on MDC with the record values of induction [55]. The oscillograms correspond to the three experiments mentioned in Table 10.1.

This permitted not only achieving higher induction amplitude but inducing the field in a volume with a fairly large diameter. Thus, in experiments with the field of 620 T the single-turn magnet has been supplied by a capacitor bank of 6.5 MJ, the turnaround radius being about 3 mm [53, 54]. The strongest fields have been attained in studies by Japanese physicists (Institute for Solid State Physics, University of Tokyo – ISSP).

In the MDC method, similar to the explosion-driven flux compression of the shell, the problem of liner stability acquires paramount significance. As we shall see in Section 10.5, compression symmetry is conserved over the field rising time, if the initial disturbances are reduced to a minimum. To do this, it is necessary, first of all, to achieve the high uniformity of the shell. For, example, in the experiments [46] the thickness of the shell was kept constant with an margin of error not exceeding 1 %.

Also of great importance is the homogeneity of magnetic fields in the gap between the accelerating magnet and liner. Even the weak disturbances produced by the insulated slot of a single-turn magnet can provoke the disturbance in the symmetry. The winding "remembers" the initial disturbances, which can be insignificant in the initial stage of acceleration, but sharply pronounced at the end. This effect has been directly demonstrated in the preliminary stage of the experiments described in [45]. In the wall of the single-turn magnet with a diameter of 40 mm the five thin radial cuts with a length of 1 cm and width of about 1 mm have been added to the insulation slot. They produced a negligible disturbance of the field, but at the end of compression the shell exhibited the shape of a hexahedron as seen in x-ray patterns.

The problem of stability loss is not fatal. Techniques have been developed and studied which allow for a decrease of the slot effect. In studies [49, 50, 113] for this purpose cuts in the body of a single-turn magnet were added[1]. To a certain degree they change the space distribution of the current and correct the field in the working gap. A more effective correction method is the use of insulated conductive spacers (compensators) placed between the liner and magnet, which results in an acceptable compression symmetry. For example, in [45] a thin-insulated compensator was used (Figure 10.12b). Due to this, as seen from the oscillograms in Figure 10.13, a reasonable reproducibility of results has been obtained. As in Figure 10.13, in a series of six experiments the amplitude of induction ranged 260–340 T. In the experiments done in the ISSP, a field of 620 T was achieved using compensators, which are shown in an assembly of the magnetic system in Figure 10.16 [53, 54].

In this magnetic system the symmetry of the liner was conserved, as seen in Figure 10.17. The application of compensators makes it possible reliably obtain record magnetic fields using the MDC method. This is confirmed by the ocsillogram shown in Figure 10.18, derived from [53]. In experiments A and B compen-

1 The displacement of the cuts is shown in Figure 10.12b.

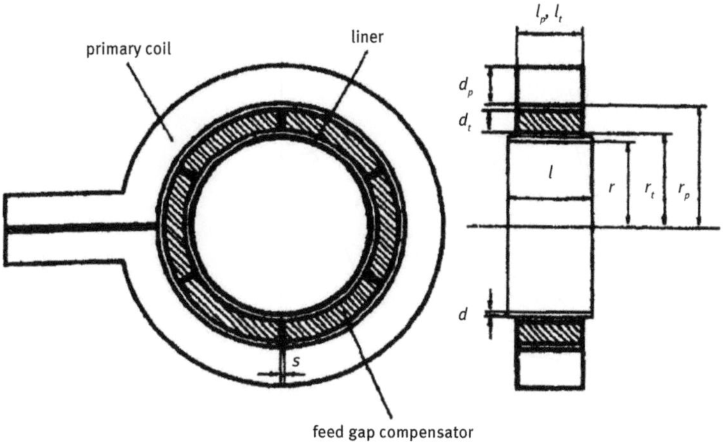

Fig. 10.16: Assembly of the primary coil and liner without (left) and with (right) a feed gap compensator [53].

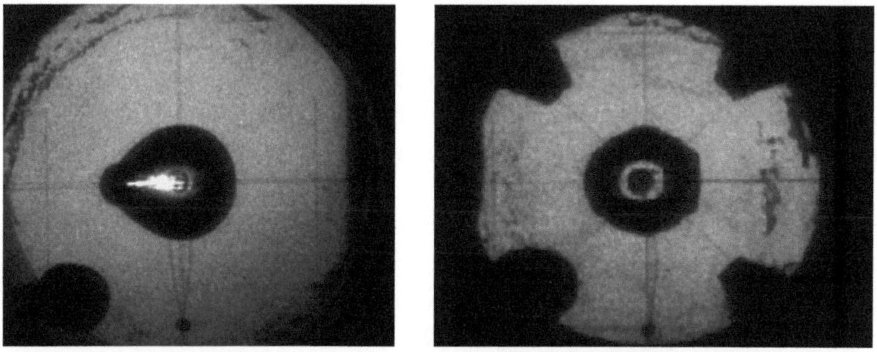

Fig. 10.17: Effect of insulated compensators on the symmetry of compression of the liner. *Left:* the experiment without compensators. *Right:* experiment with compensators [54].

sators were not used, and the induction amplitude was noticeably lower in two groups of experiments with the compensators. These groups differ in the values of the initial field, and slightly differ in size and energy inside each group. With regard to this fact, note that the results in experiments using compensators are quite reproducible. The described technology made it possible to carry out the above-mentioned experiments with a record field.

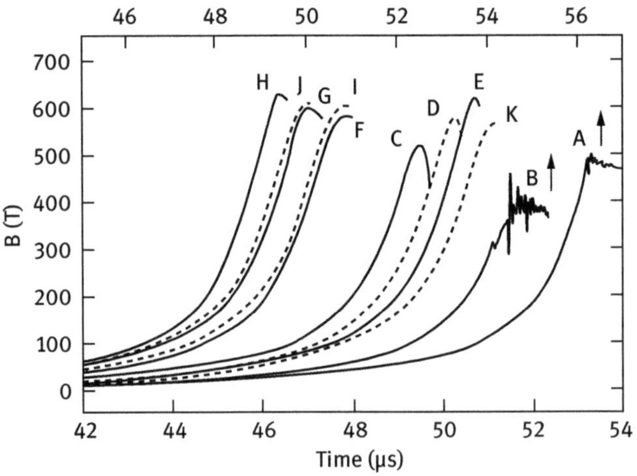

Fig. 10.18: Oscillograms of induction in [53].

10.4.2 Magnetodynamic cumulation in a *Z-Θ* pinch system

The first experiments on flux compression by the walls of a cylinder, accelerated by the magnetic pressure of an azimuth field, were carried out at the Institute of Nuclear Physics (Novosibirsk) [47], where a field with induction up to 310 T was obtained. Figure 10.19 shows the experimental set-up.

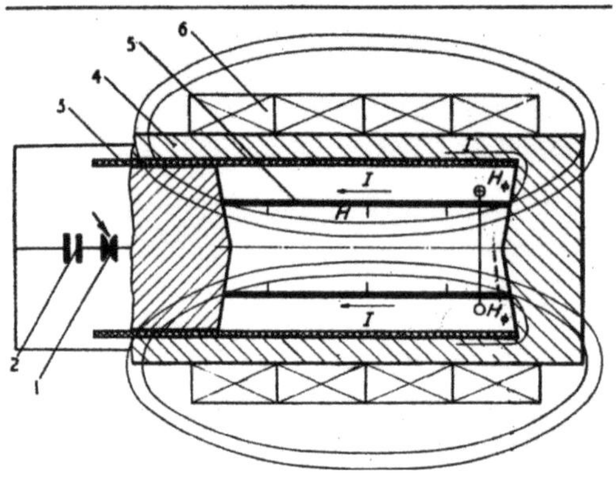

Fig. 10.19: *Z-Θ* magnetic system [47]: (1) spark gap; (2) capacitor bank; (3) insulation; (4) steel cylinder; (5) liner; (6) coil of initial magnetic field.

The liner (5) is compressed by the field of the axial current which is led through current-carrying disks (4). The initial field is induced by a coil (6). The liners are fabricated with the accuracy, producing a the spread in wall thickness less than 1 %. Only for this condition could repeatable results be obtained and the "turnaround motion" of the liner reliably detected by the decrease of induction after a maximum. Later on these experiments were developed in other studies [48]. In these experiments a field with induction of about 350 T was attained. The authors pointed out the effect of the conicity (taper) of the contact surface of the end electrodes (regions ***ab*** in Figure 10.19). From the analysis of the motion of the contact surface it was established that the liner edge is "blurred" over the ends when compressed. On simulation, the optimal tilt angle of the end surface to the axis was found to be 84°. Based on their experience, the authors of [48] conclude that "only with the careful fabrication, preparation, and installation of liners is it possible to obtain stable results on flux compression in the megagauss range."

The idea of field compression with the help of a plasma liner has been regularly discussed in the literature. Let us recall some types of such generators which are at least of scientific or educational interest if not of practical use [13, 62–65]. Few decades ago Linhart [62], evolving the subject of plasma compression by a magnetic field, constructed a device in which the initial magnetic field was compressed by a surrounding layer of plasma produced by the electrical explosion of the cylinder which was made from thin metal foil. He obtained a field with the induction amplitude of 6 T at initial value 0.2 T. Later, Felber et al. measured 160 T in a system of the plasma Z-Θ-pinch [63, 117, 118]. The authors of [61] state that in such a system the fields with induction up to thousands of tesla can be achieved. However, in actual experiments the measured induction rarely exceeds several hundred tesla in the field volumes of submillimeter dimensions.

The use of large-scale installations of pulsed high-power engineering stimulates interest for applications with Z-Θ-configurations in inertial controlled fusion.

As an example, an approximate schematic drawing of such a device designed for study of magnetic liner inertial fusion (MagLIF) at the ZR Installation of Sandia National Laboratories, USA, is presented in a general view on the left side of Figure 10.20. On the right of the figure is the external view of this device, in which the pair of coils of the initial magnetic field is clearly seen. The main goal of these proposals is, with a help of a multimegagauss field (tens and hundreds of MG) attained by compression of the liner with the field and plasma inside, to approach the conditions of thermonuclear ignition of plasma inside the liner [65–67].

One further source of initial energy of the liner worth mentioning is laser radiation. A field with induction of the order of thousands of Tesla can be achieved with compression of a liner by pressure arising from ablation of the liner surface subjected to a laser pulse of high intensity (> 10^{14} W/cm^2) [70–72]. Gotchev et al. proposed to use this field to decrease the losses associated with plasm thermal conductivity, compressed by a liner. The pulse of the powerful laser, operating at Rochester University, OMEGA (60 rays, 30 kJ), is used for the ablation compression of the cylinder target, filled with a gas (deuterium or compound D^3 He). In the cavity of the target the seed field with induction above 5 T [71] is induced with the help of a separate pulsed generator (Figure 10.21). This magnetic field is frozen

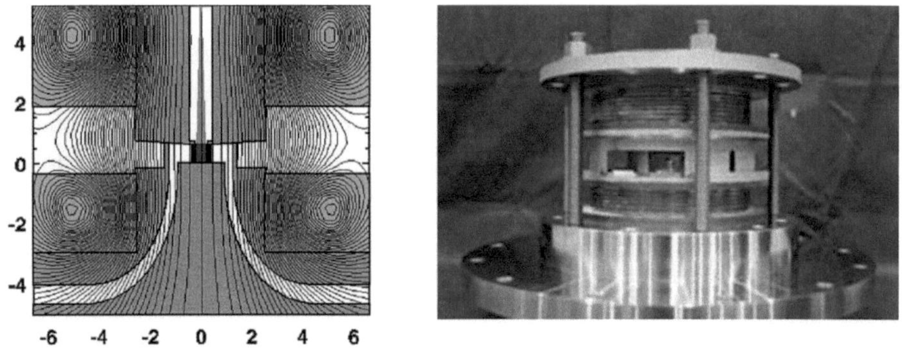

Fig. 10.20: Schematic drawing and external view of the MagLIF device for compression of magnetic flux inside the liner by current of the *Z*-pinch at the ZR installation in the SANDRIA Laboratory, USA. Field lines of initial magnetic field, generated by two coils (on the left), are shown. Sizes are in cm.

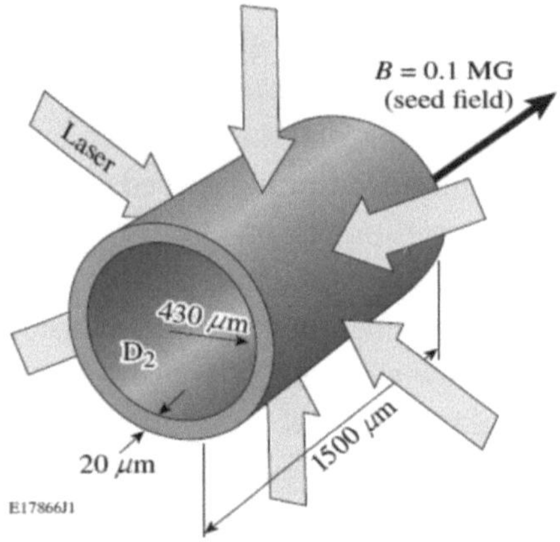

Fig. 10.21: Schematic drawing of magnetic flux compression by evaporation of the external surface of the liner under the action of a powerful laser pulse.

in plasma and compressed by a shell imploding with high velocity with minimal resistive losses.

The magnetic field in the liner cavity was not directly measured in the experiments. In order to estimate the maximum induction in the volume under compression, the method of proton deflectometry was applied. The protons were of the energy 14.7 MeV, and produced as the products of thermonuclear fusion, due to the ablation imploding of the D^3He target. The linearly averaged value of the magnetic induction was found to be in the limits of $(3–4) \cdot 10^3$ T (theoretically, the maximum induction should be about 10^4 T [71]).

10.5 Analytical estimations and simulation of magnetodynamic cumulation

In the majority of experiments on MDC described in the literature, the combination of coaxial liners was not used (cascade systems, see below), except in [83], as mentioned in Section 10.9. The simplest MDC system includes the liner, the material of which (copper, aluminum, and other metals) is adequately described by known characteristics. This fact simplifies to a certain degree the simulation of liner acceleration and flux compression, produced by a liner (as compared with the simulation of the multicascade systems of the MC-1 type with a metal-composite liner, described below). The motion of the liner can be simply described in the framework of a model of one-dimension motion of an ideally conductive incompressible liquid. With this assumption the dependence of the velocity on the radius is described by (10.8), and the equation of motion takes the form

$$\gamma_0 \left[\frac{1}{r} \frac{\partial}{\partial t} \left(r_i \frac{\partial r_i}{\partial t} \right) + \frac{1}{2r^2} \left(r_i \frac{\partial r_i}{\partial t} \right)^2 \right] = -\frac{\partial p}{\partial r} - \frac{\partial}{\partial r} \left(\frac{B^2}{2\mu_0} \right), \tag{10.30}$$

where γ_0 is the medium density, p pressure.

The variations of the inner radius r_i is given by

$$\frac{d^2 r_i}{dt^2} = \left(\left(\frac{B_e^2 - B_i^2}{2\mu_0 \gamma_0} + \left(\frac{dr_i}{dt} \right)^2 \frac{S_0}{S_0 + \pi r_i^2} \right) \frac{1}{\ln \frac{S_0 + \pi r_i^2}{\pi r_i^2}} - \left(\frac{dr_i}{dt} \right)^2 \right) \frac{1}{r_i} \tag{10.31}$$

Equation (10.31) is the result of the integration of (10.30) in limits from the inner radius r_i to the external one r_e. Here $S_0 = \pi [r_e(0)^2 - r_{i1}(0)^2]$ is a cross section area of a liner with initial inner radius $r_i(0)$ and external radius $r_e(0)$, and B_e, B_i are the induction values on the external and inner boundaries, respectively, of a shell.

As seen from (10.30), the acceleration of a liner occurs as long as the magnetic pressure outside of the wall exceeds the pressure from the inside, i.e., while the condition $|B_e| > |B_i|$ is valid. In the system of a Z-Θ-pinch the external azimuth field is generated by an axial current i_z. This field does not penetrate the cavity of the liner. The magnetic pressure on its external boundary is

$$P_e = \frac{B_e^2}{2\mu_0} = \frac{\mu_0 \, i_z^2}{8\pi^2 \, r_e(t)^2}. \tag{10.32}$$

The current i_z can be calculated by solving the equation describing the discharge of the capacity bank to which the accelerated cylinder is connected (see Chapter 4). The inner field is determined by the trapped flux, which, assuming ideal conductivity, remains equal to its initial value $\Phi = \pi \, r_i(0)^2 \, B_i(0)$. Therefore, in the course of compression the magnetic pressure on the inner boundary of the liner is given by

$$P_i = \frac{B_i(0)^2}{2\mu_0} \cdot \left[\frac{r_i(0)}{r_i(t)}\right]^4 . \tag{10.33}$$

The numerical calculations carried out in [47, 48], with the assumption of ideal conductivity and ignoring compressibility, are compared with the experiment. The experimental results agree with the calculations for the initial stage of acceleration, but disagree with the data on the induction amplitude. Thus, in [47] the calculated value of the amplitude was predicted to be 320 T, while the measured value was 280 T. In [48] the calculated and measured values of induction were 520 T and 350 T, respectively. The authors believe that this discrepancy is caused by the effect of diffusion, which was not taken into account.

A more accurate model considering the field diffusion and heating was used in [112]. In these calculations an equivalent circuit was used instead of a liner, which was a set of magneto-connected cylindrical layers. The motion of the medium was considered, assuming an ideal incompressible liquid. Figure 10.22 shows the calculated dependences characterizing the experiment in [47]. The induction amplitude in this numerical experiment was 306 T, which is close to what was calculated under the assumption of ideal conductivity.

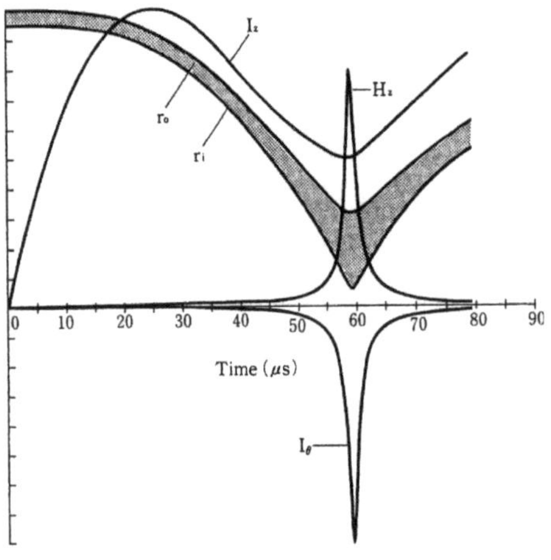

Fig. 10.22: Calculated parameters of the final stage of MDC in the system of z-θ pinch [112] for conditions in Alikhanov's et al. [47]. The maximum induction is 306 T, the minimum inner radius is 2.18 mm.

The distribution of the azimuth current at the peak induction (Figure 10.23) demonstrates that the region carrying current (thickness of a skin layer) is less than the radius of the inner boundary of the liner. There are grounds to think that what we are seeing is the "compensation effect", which reduces the diffusion

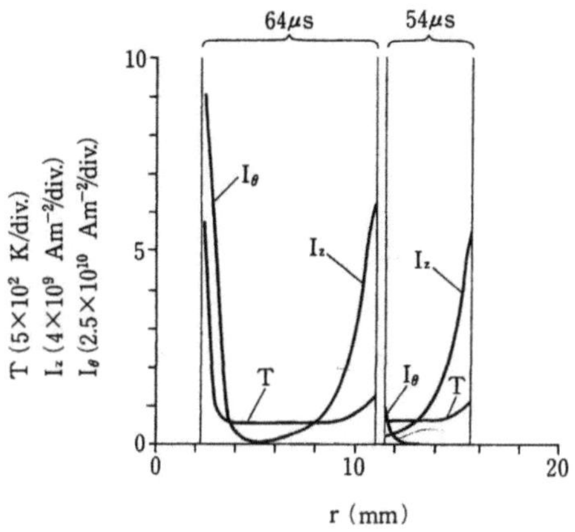

Fig. 10.23: Distributions of temperature and current density, calculated in [112] with the incompressible liquid approximation for the experimental conditions in [47].

influence on the induction amplitude. The compressibility of the medium may have the more pronounced effect, but it was not taken into consideration.

This factor may have an even more pronounced impact in a case of an experiment with s higher field, as described in [48]. As in a previous case, the numerical calculations for diffusion but not for compressibility gave a result close to the one calculated on the assumption of ideal conductivity: the induction amplitude is 513 T, noticeably exceeding the experimental one (350 T). Along with this, the results of calculations (Figure 10.23) show a temperature increase of the liner inner surface of up to approximately 2700 K in a field with induction close to 300 T. Alhough in this experiment the field is weaker than in the latter experiment, nonetheless, the phase transition resulting in the change of conductivity may affect the induction amplitude.

A more detailed pattern of liner compression in the experiments done by Alikhanov et al. [47] was given in the work of Sheppard et al. [102]. The authors of [102] used the model of MHD flow of an actual medium while taking the tabulated equation of state and of the conductivity into consideration. The results of the calculations agreed closely with the experimental data: for five experiments at the induction amplitude of about 300 T the discrepancy between the calculated and experimental data did not exceed 5 %.

The processes in the initial stage of the motion of a θ-pinch-type shell are more complicated than those in the Z-Θ-pinch. In this case the diffusion of the field into the liner cavity is of importance, since, namely, the diffusion determines the value of the trapped flux, especially in experiments without external injection of field. The calculations of this stage are given, for example, in [50, 69, 103–107, 109, 112].

In the study by Latal [103] an approximate theory ofthe Cnare-effect was constructed, but the effect of the liner motion on the penetration of the external field in the liner cavity was not taken into consideration.

The motion of the thin shell under compression and the flux variations are described by the following relations:

$$m'\frac{d^2r}{dt^2} = -\frac{B_e^2 - B_i^2}{\mu_0}\pi r, \tag{10.34a}$$

$$\frac{d\Phi}{dt} = \frac{2\pi r\rho}{\mu_0 h}(B_e - B_i) = \frac{2\pi r^2\rho}{\mu_0 h_0 r_0}(B_e - B_i). \tag{10.34b}$$

Here r is the mid-radius of a shell, h is the thickness, $m' = 2\pi r_0 h_0 \gamma_0$.

For most of the first phase the condition $|B_e| \gg |B_i|$ is fulfilled. Therefore the term B_i in equations (10.34a) and (10.35b) can be neglected. In equation (10.34b) the change in shell thickness $h = h_0 r_0/r$ caused by compression is taken into account. This relation is valid if the shell thickness is small compared to the radius. Analytical calculations will be carried out on the assumption that the skin effect in the shell is lacking and the specific resistance constant.

Ignoring B_i, we come to the approximate relations. It is reasonable to write them in a dimensionless form. To do this, we introduce the new variables $\tau = t/t_0$ and $x = r/r_0$. In a system of finite length the average induction value B_e can be presented in the form

$$B_e = \frac{\mu_0 i}{l} f(x) = \frac{\mu_0 i_m}{l} f_1(x) f_2(t), \tag{10.35}$$

where i_m is the current amplitude, $f_1(x)$ is the function, characterizing the dependence of the average induction of external field on an instant value of a shell radius at the given instant value of current in a solenoid, $f_2(t) = i(t)/i_m$. If we choose $t_0 = (m'/\mu_0)^{1/2}(l/i_m)$, as the basic time, we obtain in new variables the following equations:

$$\frac{d^2x}{d\tau^2} = x f_1^2(x) f_2^2(\tau), \tag{10.36a}$$

$$\frac{d\Phi}{d\tau} = \frac{2\pi\rho}{h_0}\sqrt{\frac{m'}{\mu_0}} \cdot x^2 f_1(x) f_2(\tau). \tag{10.36b}$$

Equations (10.34a,b) and (10.36a,b) ignore the processes at the end of compression. At this stage one cannot consider a liner as being thin and use the condition $B_i \ll B_e$. However, as seen from equation (10.34b), the flux increment slows down sharply at this stage, when the radius is small.

Therefore the trapped flux Φ_0 can be estimated by integrating equation (10.36b) from 0 to τ_1, where τ_1 is a time interval over which the value $x = 0$ is achieved:

$$\Phi_0 = \frac{2\pi\rho}{h_0}\sqrt{\frac{m'}{\mu_0}}\int_0^{\tau_1} x^2 f_1^2(x) f_2^2(\tau)\,d\tau. \tag{10.37}$$

Using equation (10.37), at given dependences $f_1(x)$ and $f_2(\tau)$ one can calculate the time interval of the total compression τ_1. This dimensionless integral depends on the specific form of the functions $f_1(x)$ and $f_2(\tau)$. The obtained formula is interesting because it includes the characteristic flux

$$\Phi' = \frac{2\pi\rho}{h_0}\sqrt{\frac{m'}{\mu_0}} \tag{10.38}$$

and the numerical factor, which is determined by specific conditions of the experiment.

Further we consider a magnetic system of large length, such that we can assume $f_1 = 1$. In our first example we also assume that the current is increasing linearly. This is true, if the basic acceleration of a liner occurs on the front of the current pulse. On these assumptions the induction B_e grows linearly, and equation (10.34a) takes the form

$$\frac{d^2r}{dt^2} + \frac{\pi B'^2 t^2 r}{\mu_0 m'} = 0, \tag{10.39a}$$

and the equation for flux is

$$\frac{d\Phi_0}{dt} = \frac{2\pi\rho\, r^2 B' t}{h_0 r_0 \mu_0}, \tag{10.39b}$$

where $B' = dB_e/dt$.

The solution of equation (10.39a) was given above:

$$r = r_0\frac{\Gamma(3/4)}{\sqrt{2}}\sqrt{\xi}\cdot J_{-1/4}\left(\frac{\xi^2}{2}\right); \tag{10.40a}$$

$$\frac{dr}{dt} = \frac{\Gamma(3/4)}{\sqrt{2}}\xi^{3/2}\cdot J_{3/4}\left(\frac{\xi^2}{2}\right)\cdot r_0\left(\frac{\pi B'^2}{\mu_0 m'}\right)^{1/4}, \tag{10.40b}$$

where $\xi = t(\frac{\pi B'^2}{\mu_0 m'})^{1/4}$, $J_{-1/4}, J_{3/4}$ are Bessel functions.

Dependences for a dimensionless radius and speed are shown in Figure 10.24. It is of specific interest that the basic acceleration occurs at the initial length of the trajectory: a liner gains approximately half its energy over the compression time of the order $0.6r_0$. Total compression takes place at the moment t_1, when

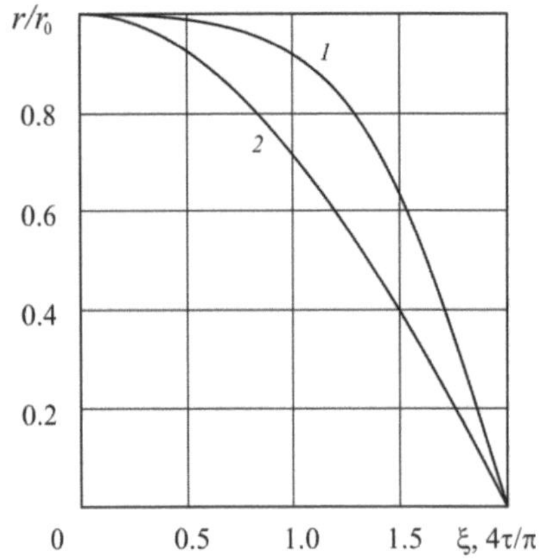

Fig. 10.24: Calculated dependences for the radius in a magnetic system without an initial field (the length of dthe magnetic system essentially exceeds the initial diameter of the liner): (1) $r/r_0 = f(\xi)$, external field is building-up linearly; (2) $r/r_0 = f(4\tau/\pi)$, external field is building-up instantly, and then remains constant.

$\xi = \xi_1 = \sqrt{2\lambda_{-1/4}} \approx 2$, where $\lambda_{-1/4}$ is the first root of Bessel function $J_{-1/4}(x)$. The speed at this moment is

$$(dr/dt)_{\xi_1} \approx 1.4 r_0 \left(\frac{\pi B'^2}{\mu_0 m'}\right)^{1/4}. \tag{10.40c}$$

The deceleration by pressure of the "trapped" magnetic field begins to affect the liner motion only at the end of acceleration, and therefore the obtained value can be used as the estimate for the maximum speed of the liner. With the obtained formula we can find the magnetic field in the shell cavity:

$$\Phi(t) = \frac{2\pi\rho}{h_0 r_0} \int_0^t r^2 B' t \cdot dt\,. \tag{10.41a}$$

We can estimate the trapped flux, if we let the upper limit be equal to the calculated time interval of the total compression $t_1 \approx \xi_1 (\mu_0 m'/\pi B'^2)^{1/4}$:

$$\begin{aligned}\Phi_0 &= \frac{\rho r_0}{h_0}\left(\Gamma\left(\frac{3}{4}\right)\right)^2 \left(\frac{m'\pi}{\mu_0}\right)^{1/2} \int_0^{\tau_1} \tau^3 \left[J_{-1/4}\left(\frac{\tau^2}{4}\right)\right]^2 d\tau \\ &= 3.46\frac{\rho r_0}{h_0}\left(\frac{m'}{\mu_0}\right)^{1/2} = 0.55\,\Phi'. \end{aligned} \tag{10.41b}$$

In another example we assume that the induction of the external field builds abruptly up to, B_E and then remains constant. In this case equation (10.34) takes the form

$$m'\frac{d^2 r}{dt^2} = -\frac{\pi r B_e^2}{\mu_0} \tag{10.42a}$$

Its solution is

$$r = r_0 \cos\tau\,, \tag{10.42b}$$

where $\tau = B_E t (2\mu_0 m')^{-1/2}$ [18]. The trapped flux is determined by the relation

$$\Phi_0 = \frac{\pi^2}{\sqrt{2}} \frac{\rho_0 r_0}{h_0} \sqrt{\frac{m'}{\mu_0}} = 1.1\,\Phi'. \tag{10.42c}$$

It is interesting to note that, in contrast to the results of Latal [103], according to the obtained formula the trapped flux in the first case does not depend on the increment of an accelerating field. This fact can be explained: with increasing B' the speed of flux changing also increases, but the time of field trapping decreases, since the compression is going faster. The influence of both factors is mutually compensated, and the flux Φ_0 turns out to be only dependent on the characteristic transversal dimensions, liner conductivity, and its mass per unit length. In a similar manner, in a second case there is no dependence on the induction of an accelerating field. According to the general estimation given above, the trapped flux depends, to a degree, on the shape of the external field, but the order of magnitude is determined by the parameter Φ', and does not depend on the intensity of the external field.

It is of interest to compare the results of calculations based on the derived formula with the data of actual and simulated experiments. The given formula refers to a system of large length, which does not take into consideration the skin effect and change of resistance due to the heating of the shell, whereas these factors are taken into account in the simulation. As expected, for short systems the calculation using equation (10.41b) gives the hyperestimated value of speed in comparison with the experiment. For example, for the conditions of the experiment in [45], the calculated maximal value of speed is 2.5 km/s, while the experimental value is about 1.7 km/s. The value Φ_0, determined using (10.41b), is 3.3 mWb, while the measured value is (2 ± 0.3) mWb at the end of first phase of the process. Thus, the given simple relations, describing expanded magnetic systems with the permanent conductivity and linearly growing external field, are not only illustrative, but can also be used for estimating the parameters characterizing MDC in magnetic systems whose length is commensurate with the initial radius of a liner.

Numerical calculations enable us to describe a motion of a liner for nonlinear diffusion of a field and the actual configuration of a magnetic system.

In calculations, as applied to the experiments of [45], a shell was replaced by a system of coaxial cylindrical layers. Each layers was a circuit, associated with others through mutual inductance. The current in each circuit i_k was determined by solving the system of equations

$$\frac{d\Psi_k}{dt} = -R_k i_k, \tag{10.43}$$

where R_k is a resistance of each layer, and Ψ_k is the flux coupling of each layer. The heating of conductors was taken into consideration when calculating a resistance. Magnetic induction from inside and outside the shell was calculated with regard to the actual configuration of the system of the single-turn magnet – the shell. Here linear dependence was used, which links the middle induction inside and outside of the shell with currents circling in the shell and the magnet. Constant coefficients in this dependence had been found by measuring the high frequency field on the model from [69, 111]. One-dimensional motion of a shell was described by equation (10.31) on the assumption of incompressible liquid. This approach permitted the description of the process, with acceptable accuracy, up to the moment of passage of the shell current through zero. In Figure 10.25 from [69], the results of the calculations are presented. The current distribution has a complicated character: in the external layers a current is in opposition to the current in a solenoid, while on the inner side it us in parallel. At the moment $t_0 \approx 25$ μs the current of the shell is zero. A similar pattern of the current radial distribution is mentioned by Miura and Nakao [54].

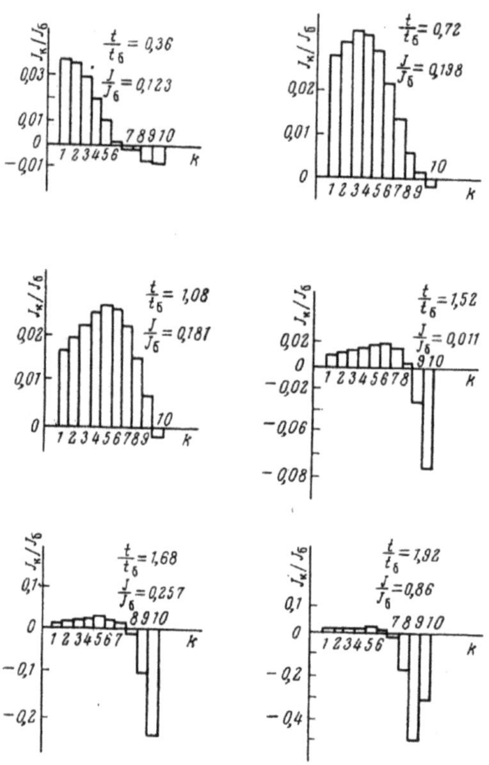

Fig. 10.25: Distribution of current density in a liner wall as calculated for conditions in experiments in [69]. One-dimensional model.

It is possible to describe with more accuracy the field diffusion and acceleration taking into consideration the edge and skin-effect, if splitting the shell into a large number of magneto-coupled annular circuits (method of filamentary currents). This method is widely used for problems dealing with diffusion of axially symmetric field into conductors having the form of a body of revolution. An example is the calculation of single-turn magnets, considered in Chapter 8. In some studies by Japan and British authors, the circuit method turned out to be applicable for describing two-dimensional diffusion into the liner, accelerated by a magnetic field [105–109, 112]. There are grounds to believe that with a great many units the method of coupled circuits gives a sufficiently comprehensive description of field diffusion, heating, and motion of the medium without taking its compressibility under consideration. Some eculiarities of the current distribution in a short liner are shown in Figure 10.26, derived from [106]. The current distribution exhibits the edge effect, as well as a noticeable skin effect and the heating of edges. This method adequately describes the first phase of the process, but its applicability is doubful for a description of the final stage of magnetic cumulation, where the compressibility of the medium should be taken into consideration.

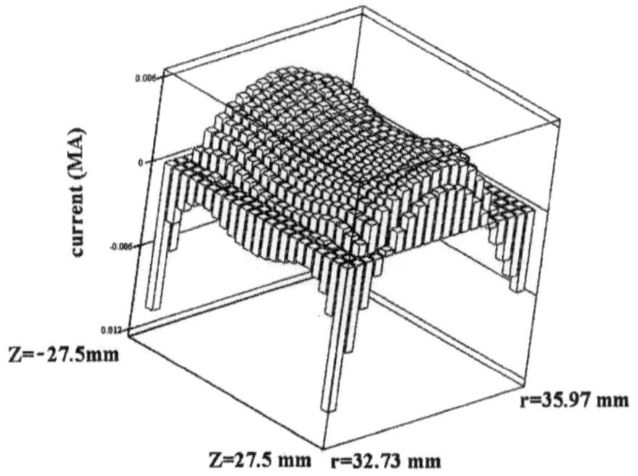

Fig. 10.26: Distribution of the current in the thickness of a short liner [106].

As was shown in calculations of the initial stage of liner compression, if the thickness of the shell is small, the field penetrates inside the cavity so fast that the induction is hardly different on each side of the wall. This results in a drastic decrease in compression efficiency. Therefore, The wall thickness should exceed a certain value: for the experiments described in [45], wall thickness exceeded 0.3 mm. In the specific conditions of the experiment [45], a further increase in thickness leads to an increase of the conversion efficiency of the capacity bank energy into kinetic energy of a shell, as well as to a decrease in the trapped flux and increase of the induction amplitude together with a simultaneous decrease of the radius of a turnaround. From the preceding it follows that the optimization of the MDC in a system with Θ-pinch requires the rational choice of the initial dimensions of a shell and the characteristics of the generator [109].

The edge effects result in a distortion of the cylindrical form of the liner. Calculations of the liner motion, on the assumption of ideal incompressible liquid without regard for the diffusion of the field into a shell, give the qualitative pattern of this process [50, 112] As seen in Figure 10.27 derived from the paper by Miura and Nakao [112], under compression not only the liner thickness increases, but its cross section also deforms. Along with this, the given data show that at the end of compression the shell holds the shape of a regular circular cylinder for the most part of its length.

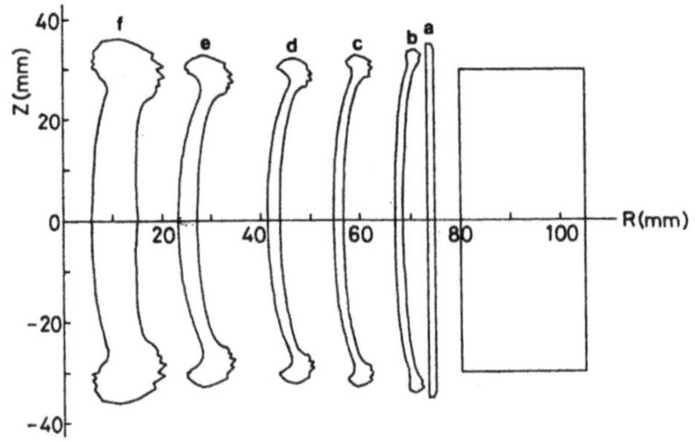

Fig. 10.27: Configuration of a shell at different times of compression in thefield of a single-turn magnet, as calculated in [112], assuming ideal conductivity.

The processes involved in the final stage of MDC are similar to those occurring in the single-stage generators of the MC-1 type (see below). Note that these processes, by their physical nature, are the same ones that occur in one-turn magnets in an extremely high field. We are dealing with hydrodynamic flow, accompanied by shock-wave generation, onlinear diffusion, and, lastly, the explosion of a skin layer. An explosion starts in fields above 400 T, when the medium is evaporating and its conductivity drastically changes. Simulation of these processes with and excluding certain factors has been done in [69, 104]. The system of equations of magneto-hydrodynamics, as applied to the experiments described in [45], as well as to certain experiments on the liner compression by explosive products, has been solved using the extended-state equation and the model dependence of the conductivity on the temperature and concentration, described in [110].

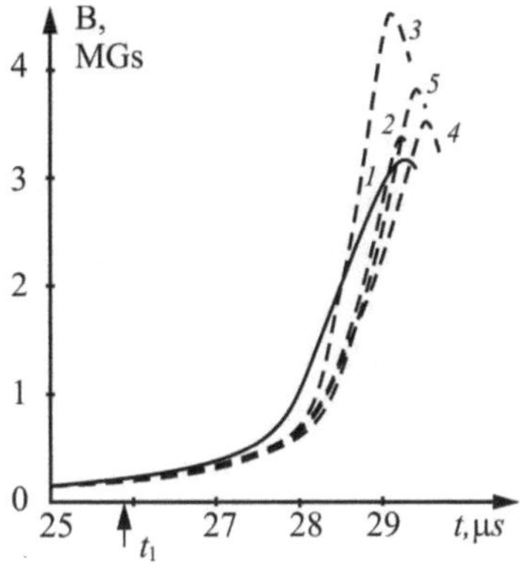

Fig. 10.28: Oscillograms of induction compared to results of simulation. The experiment [45]. (1) Calculations on the total model. (2) Calculations on the assumption of incompressible ideally conductive medium. (3) Calculations with regard for compressibility, but without regard for field diffusion. (4) Calculations on the model of incompressible liquid with regard for field diffusion.

Figure 10.28 shows the temporal behavior of the induction on the axis for the experiment with the induction amplitude $B_{m,0} = 320$ T (curve 0, [45]), and the calculated. results for the final stage of this experiment. The induction amplitude, calculated on the full model ($B_{m,1} \approx 335$ T, curve 1) is sufficiently close to the measured one. Variations of conditions in calculations reveal the influence of certain factors in a final stage of magnetic cumulation. The simplest model, in which the liner material is considered to be the medium with constant density and ideal conductivity, gives the hyperestimated value for the induction amplitude ($B_{m,2} \approx 430$ T, curve 2). The inclusion of the compressibility with retention of the ideal conductivity leads to the value $B_{m,3} \approx 350$ T, (curve 3). The effect of the field diffusion is of secondary importance, since the thickness of the skin layer is small compared with the turnaround radius. As was mentioned in Section 10.4, at this condition the "compensation effect" takes place: simultaneously with diffusion the additional compression of a liner occurs. This effect is revealed in the numerical experiments. With diffusion taken into consideration at the constant medium density the calculated amplitude decreases from $B_{m,1}$ to $B_{m,4} \approx 370$ T. The diffusion effect is exhibited to less extent in the actual compressible medium. It is seen from the comparison of calculated amplitudes $B_{m,3}$ and $B_{m,1}$, as well as from the estimate given below, based on modification of the relation (10.17a). It was derived for a medium with constant density, but, as was shown in [69, 104], in the case of a skin-layer with small thickness it can be used for a compressible medium as well. For this purpose, we should take the value of the turnaround radius, calculated with regard for compressibility. As a result, we come to the relation for induction amplitude:

$$B_{m,3} \approx B_{m,1} (1 + 2\Delta/r_{m,3})/(1 + 2.5\Delta/r_{m,3}), \tag{10.44}$$

where $r_{m,3}$ is the turnaround radius, calculated for an ideally conductive medium with regarding for compressibility, and Δ is the thickness of the skin layer. A correction factor is close to unity even at condition $\Delta/r_{m,3} \approx 1$.

One can estimate the amplitude $B_{m,3}$ using the relations given in Section 10.4. The estimation for the above-considered example gives $B_{m,3} = \breve{B}_{m,3} \approx 334$ T. Though it somewhat differs from the result obtained by numerical calculation, the simple estimate by the mentioned relation can be used for taking the medium compressibility into consideration when analyzing the experimental data. In the example under consideration the discrepancy between the estimated amplitude and the experimental one is less than expected (about 4 %).

Other example is the experiment described in [50]. The researchers, instead of the expected peak field of 590 T, calculated without regard for compressibility, obtained 323 T. For comparison one we can use relations (10.24a), in which the based value of induction B_{Π}, can be estimated by the formula (10.20). At the same time, the experimental data should be taken for the flux and kinetic energy. For example, for the instant t_1 (Figure 10.29) we have $\Phi = 4.2 \cdot 10^{-2}$ Wb, $B(t_1) =$ 33 T, $u = 1.54$ km/s. Using this initial data, we find $B = 440$ T and, according to (10.24a), arrive at $B_M \approx 300$ T.

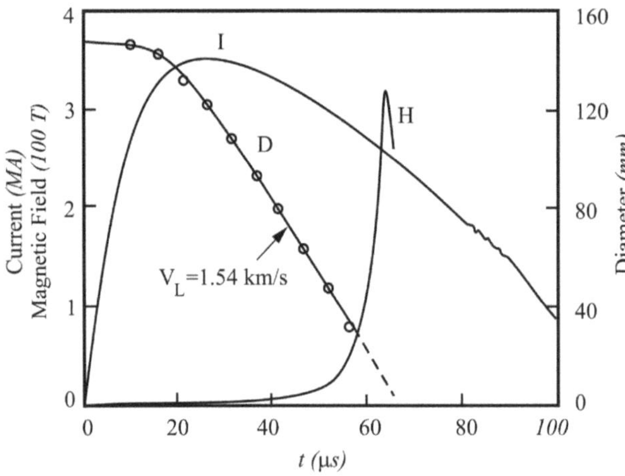

Fig. 10.29: Calculated and experimental characteristics of MDC in [50].

The essential role of compressibility appears in experiments with flux compression, described in the study [53]. This is confirmed by comparison of the diffusion speed with the speed of the flow. In the field with induction of about 600 T the speed of the medium flow behind the shock front u_f, calculated by formula (9.29), is about 2 km/s, while the diffusion speed was much less, close to 0.15 km/s, as estimated in [53]. The maximum compression speed correlates both with the calculated induction value (formula (9.29)) and the experimentally obtained one. Thus, it is reasonable to believe that the model is confirmed, according to which the turnaround of the liner occurs at the instant when the compression speed becomes close to the speed of the flow behind the shock front.

An electrical explosion of a skin layer in the final stage can have a pronounced effect on magnetic cumulation, if the induction exceeds 400 T (for copper). In such a field, as in single-turn coils, phase transition occurs. The processes considered in Chapter 9 take place in a magnetic cumulation as well. The pattern of the MHD radial flow essentially differs from the pattern which occurs when a skin layer is exploded in a conductor with a flat boundary. However, the calculations, based on certain models [104], show that in this case the loss of conductivity due to medium heating is also possible. As a result, the induction amplitude appears to be somewhat decreased. Nevertheless, based on the numerical analysis of many experiments, the authors suggested that at an axially symmetric magnetic cumulation, when copper and aluminum liners are used in fields of the order of more than hundreds of Tesla, compressibility is the main factor leading to the decrease of the induction amplitude in comparison with the calculated value $B_{m,1}$. Along with this, the calculations show that in the case of liners from poorly conductive materials, the effect of diffusion on the induction amplitude can be decisive. In order to confirm these suggestions, calculations made with up-to-date methods, are required. The extensive experimental data, partially presented in Table 10.1, can provide the basis for the calculations.

In some papers the prospects of generating extremely high fields using the MDC method are discussed. Already in his first publication Cnare [41] discussed the effect of Joule heating of the shell and considered the conditions at which the Anderson's criteria are attained, i.e., the evaporation of a liner occurs. An action integral can be compared to the speed obtained by liner when accelerated. Without regard for the pressure of a trapped field, the absolute value of the compression speed of a thin liner is described by the relation

$$u_0 = \frac{\pi}{m'\mu_0} \int B_e^2 r dt. \tag{10.45}$$

Ignoring the skin effect we obtain

$$u_0 = \frac{\mu_0 \pi}{m'} \int r h^2 \delta^2 dt = \frac{\mu_0 h_0}{2\gamma_0} \int \frac{r_0}{r} \delta^2 dt > \frac{\mu_0 h_0}{2\gamma_0} S, \tag{10.46}$$

where $S = \int_0^t \delta^2 dt$. This relation enables us to find the upper limit of the acceptable value of the action integral at which the given speed can be attained. Here the increment of temperature of a liner will be restricted by a given value. In all the experiments presented in Table 10.1, the heating of the main part of the liner was not more than a few hundred degrees. Although the calculations in [112, 106] have shown that in a field of 500–600 T the heating of the angular points and inner liner surface can exceed 2000 K, these factors appeared in the final stage of compression. As known from the literature, during the acceleration stage the liner kept a sufficiently high conductivity in the whole mass, which ensured the acceptable conservation of flux.

In principle, it is possible to compress a liner, heating it up to temperatures at which the metal, when evaporated, turns into plasma. However, in order to obtain the plasma with conductivity of the order 10^7 $(\Omega\ \mathrm{m})^{-1}$, intrinsic to good metallic conductors, it is necessary to heat the plasma up to temperature of the order 10^7 °K. Such conditions are reached at the compression of a light multiwire liner in a Z-θ pinch configuration in the experiments on inertial thermonuclear fusion. Here, along with Cnare, we focus on the possibility of the liner acceleration in systems of aθ-pinch not involving electrical explosion.

Assume for further estimations that the temperature increment of a copper liner over the time of acceleration is 300 °K. Then the specific conductivity rises from the initial value of $1.810^{-8}\Omega \cdot \mathrm{m}$ up to $5.110^{-8}\Omega \cdot \mathrm{m}$. Here the action integral is 310^{16} $\mathrm{A^2cm^{-4}s}$. The relation for the acceptable speed (10.46) can be used for the estimation of parameters of the hypothetical system with MHD cumulation. Using the relations from Section 10.4, and taking the compressibility into consideration, one can calculate that at a flow speed of 3 km/s it is possible to generate a field with induction $B_{m3} \cong 1200$ T, if the value of the induction amplitude for the ideally conductive incompressible medium was estimated to be $B_{m1} = 2000$ T. When heating up to 300 °K, the acceptable value of the action integral for the given speed value corresponds to the relation (10.46), permitting us to find the thickness of a liner wall:

$$h_0 = \frac{2\gamma_0 u_0}{\mu_0 S} = 1.4 \cdot 10^{-3}\text{м}. \tag{10.47}$$

For a liner of the given initial radius and length it is possible to calculate the kinetic energy:

$$W_k = \pi\, r_0 \gamma_0 h_0 l u_0^2. \tag{10.48}$$

At the indicated speed of a liner of length 0.2 m and initial radius 0.03 m we have $W_k \approx 2.1$ MJ. Further calculations enable us to estimate the time interval of compression ($t \approx r_0/u_0 = 10^{-5}$ s), and the current density ($\delta_{eff} \approx (S/t_0)^{1/2} \cong 5.5 \cdot 10^{10}\ \mathrm{Am^{-2}}$). Further we can estimate the induction of the accelerating field, which is $B \approx \mu_0 h_0 \delta_{eff} \approx 100$ T. The pulsed field with such induction is quite possible to be achieved in single-turn magnets, and the construction of a generator energizing a system of this kind is not beyond the possibilities of current pulse power technology.

Theis illustrative example testifies that neither the known physical processes responsible for magnetic cumulation nor technological restrictions are the obstacles for attaining fields of the order of 10^3 T using magnetic cumulation.

It is of interest to compare two methods for obtaining extremely high magnetic fields: the direct discharge of a capacitor bank into a single-turn magnet and magnetodynamic cumulation. As shown in Chapter 9, the first method requires a low-inductance capacity bank, providing a fast rising current, and for this goal the simplest single-turn magnet can be used. In experiments on pulsed magnetic cumulation the requirements for the generators are essentially simplified, if they are used for acceleration of a liner over time of the order of and above 10^{-5} s. In this case it is not necessary to construct a capacity bank which provides a fast rising current. Along with this, as mentioned above, the MDC method requires specific efforts ensuring the symmetrical compression of a shell. Also, in contrast to single-turn magnets, in magnetic cumulation the object under study is installed in the cavity of the liner and destroyed when compressed.

10.6 Explosion devices and solenoids of an initial field used in magnetic cumulation

10.6.1 Detonation of the explosive charge

The chemical explosive charge in ring form is a commonly applied source of energy for an explosion magnetic cumulation. In order to give a general idea on the explosives used in magnetic cumulation, we consider several important characteristics of one of the most powerful explosives (in American notations PBX 9501). Its mass density is 1.84 g/cm^3, detonation speed 8.8 km/s, energy detonation 11 MJ/cm^3, density of the energy flux (the detonation speed multiplied by the unit-volume energy) 9.7 GW/cm^2. The pressure behind the detonation front is 37 GPa (370 kbar). The performance of the explosion devices calls for a developed technique, adequate technology, and experience with the safe operating of explosives. The energy density stored by explosives is fairly high (in a range of 4–6 MJ/kg) [18], and calls for special chambers, or for the carrying outof the experiments at an explosion testing ground.

From the technical view the most complicated problem is the production of the converging detonation wave, satisfying the strong requirement of uniformity along a circle. This requirement is rather essential, since it determines the azimuth symmetry of the compressed liner. At the same time, the requirement on the uniformity of the detonation wave along the cylinder axis is not so important: it only determines the longitudinal dimension of the region where the magnetic field is

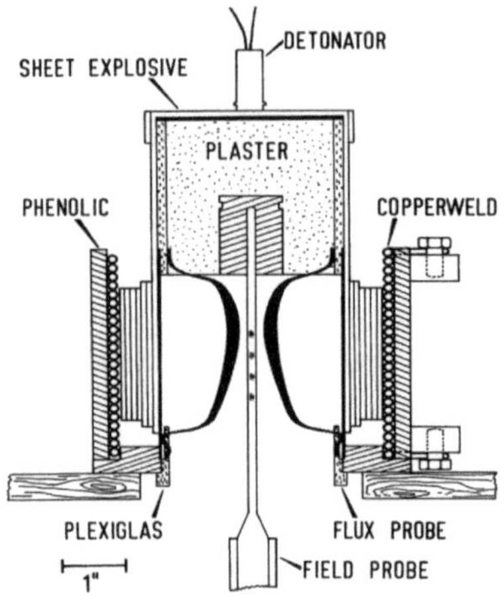

Fig. 10.30: The "top-hat" MC generator with one detonator, sheets of plaster explosive and stainless steel [114].

amplified. The reduction of the requirements for the axial uniformity of the explosion have been used in one of the most simple and inexpensive constructions of MC generators, developed as early as the 1960s at the Illinois Institute of Technology [1.114] (Figure 10.30). This arrangement used a stainless steel cylinder, on the external surface of which a few sheets of plaster explosive were coiled. The first turn of the explosive was much longer compared to the liner, and connected to the disk from the similar explosive at its far end. A single detonator was mounted in the center. The detonation wave moves away from the center of the disk to its edge and is swept through the first turn to the main charge, initiating it around the circle. The liner was compressed symmetrically over the circle, but irregularly along its length. Therefore, as seen in Figure 10.30, the volume of a strong magnetic field, generated due to liner compression, was small in these experiments. Another disadvantage of this arrangement was one-sided access to the field volume.

In his monograph [18] Knoepfel described several possible techniques for initiating a ring explosive charge (see Figure 10.31).

Experiments have shown that the shape of the detonation wave differs slightly from a regular cylinder in the case of a multipoint initiation of the charge external surface. This can be done, e.g., with the help of numerous point detonators, evenly located over the charge surface. The main requirement for such an initiating system is as small a spread of the actuation times of detonators as possible. One recently constructed device of this kind involving 240 so-called "slapper" detonators was presented in the report of Tasker, Goforth, and Oona at the 12th Megagauss conference [73].

Figure 10.32 shows the external view of this device, which consists of four parts incorporating 60 detonators each, located on the cylindrical surface. The

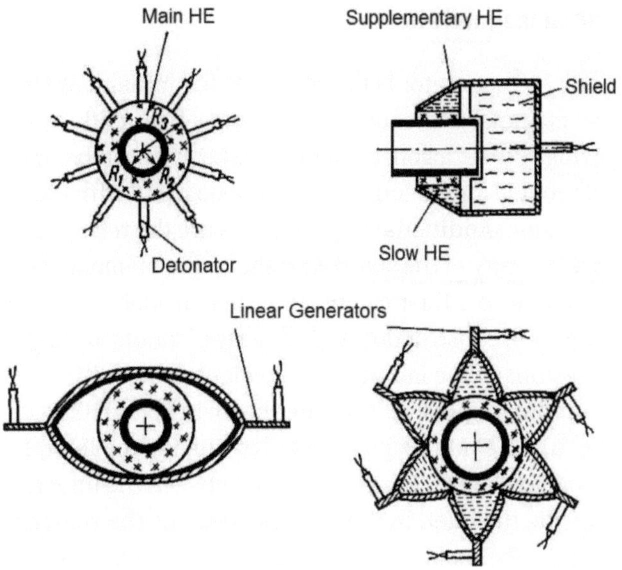

Fig. 10.31: Several possible techniques for initiating of the explosive charge for cylindrical flux compression.

Fig. 10.32: 240 “slapper” detonators evenly spread over the cylinder surface, initiating the cylindrical detonation wave [73].

explosion of each part is initiated by a capacitor of 1 μF, charged to 9.5 kV. The film with detonators can be stacked on the surface of arbitrary curvature.

10.6.2 Generation of the initial magnetic field

A necessary component of any MC generator is the source of the initial magnetic flux. In the majority of cases it is a pulsed solenoid supplied most frequently by the capacity bank. As the supplying source, a generator of the MC-2 type can be used. The magnetic system should provide a sufficiently homogeneous field with the required induction in a given volume. Additional requirements are the retention of the mechanical and electrical integrity of the solenoid right up the moment of the peak of the initial magnetic field in the liner cavity, and the reasonable requirement of simplicity and cheapness of construction and the ease of mounting on the test site. There are ony few versions of the mutual arrangement of the initial field solenoid and liner with surrounding explosive. They are shown in Figure 10.33, derived from [18]. These can be (a) a Helmholtz pair, installed outside the liner, (b) a coil outside both the liner and explosive charge, (c) a coil between the liner and explosive charge, (d) a coil inside the liner. In the first two versions the source of the initial field does not affect the dynamics of liner compression and field amplification, but the use of the upplying energy is not efficient. Therefore, considering the natural restrictions on the value of this energy, a sufficiently high initial field cannot be obtained in such magnetic system.

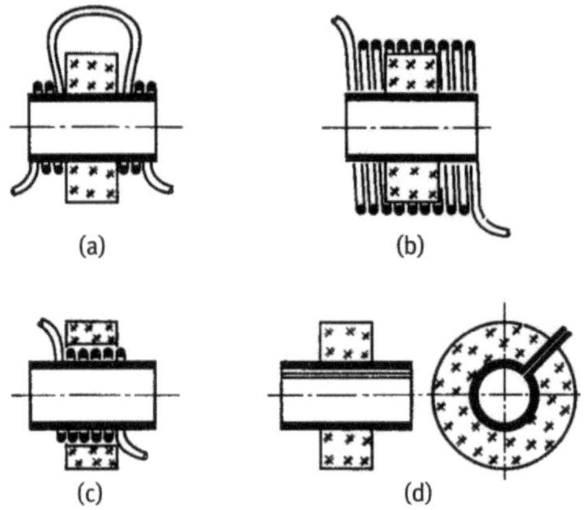

Fig. 10.33: Versions of devices for generating initial magnetic flux in the MC-1 generator.

The last two versions use the supplying energy in a most efficient way, but the symmetry of the liner shape is essentially affected by the operating units of the solenoid: steep variations in the material density in the turns and interturn insulation lead to distortions of the liner shape with the formation of jets that are directed inward the coil and result in the destruction of the liner. Note that in the last version the coil of the initial field occupies the useful volume with the amplified field and restricts the level of magnetic flux compression.

The numerous tests of varied devices for the generation of initial fields carried out in the All-Russia Scientific Research Institute of Experimental Physics (now Russian Federal Nuclear Center – VNIIEF), from a coil, turned from a thick copper tube, to the pair of coils of the "Bitter" kind, flooded with epoxy compound, have shown that the destruction of coils occurs most frequently due to the detachment of the edge turns aside from the current collector and to the interturn breakdown, mainly, of the same edge turns. The axial compressing force leads to the displacement of the edge turns resulting in the destruction of the interturn insulation and breakdown.

In the 1970s researchers used a solenoid in the form of a multistart multilayer coil made from insulated thin copper wire to solve the problem of generating a relatively high initial field [74]. The supplying energy is efficiently used in this device, and the symmetry of the liner compression is not degraded. The appropriate solution was the application of the multistart winding. One layer of such a winding contains 500 conductors, connected in parallel, each of them forming two turns. Standard enameled copper wire with a diameter of 0.3 mm is used. The insulation has adequate electrical strength (more than 1 kV) and is convenient for application, since it can be destroyed by boiling in an acid medium. A winding is soaked with epoxy compound in vacuum conditions. The photo in Figure 10.34 shows that the far ends of the solenoid make up the "beard" consisting of hundreds of insulated wires. The technology of stripping and of welding of wire ends to the current collectors was developed. The single-layer magnet produced by this strategy was not destroyed in a field with induction 4.5 T.

The next versions of wire solenoids already consisted of five or more layers. Such a solenoid, having been soaked with compound, is the cylindrical shell consisting of the glued tightly-packed small wires entirely and evenly filling the shell volume.

As seen in Figure 10.34, the supplying cable surrounds from the outside the explosive charge, which complicated the mounting of the generator and impaired

Fig. 10.34: Single-layer multiwire solenoid at various stages of fabrication.

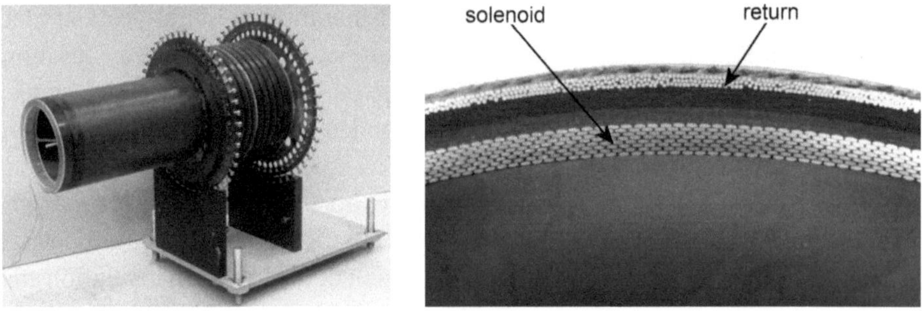

Fig. 10.35: Single-layer multiwire solenoid at various stages of fabrication.

the efficiency of the power supply. The next sufficiently important change of the solenoid construction was the displacement of the solenoid reverse withdrawal inside the explosive ring, thus minimizing the volume between the solenoid and the reverse withdrawal, and consequently, the energy consumption on the generation of the field. For this purpose the wire extension is fixed on the surrounding insulated layer and laid in parallel to the axis of the solenoid.

An external view of the assembled wired solenoid and a fragment of its cross section are shown in Figure 10.35. Supply cables are connected on one side of the generator, and the explosive charge is fitted over the solenoid once the device is mounted on the testing grounds, so that, if the occasion requires, it can be easy removed, as well as the generator itself. In one of the versions, a solenoid with a diameter about 15 cm and length of 30 cm (Figure 10.35) has the initial inductance below 0.25 μH and active resistance $\sim 210^{-4}$ Ohm. It can be powered both by the capacity bank and an EMG of the MC-2 type. With the current at about 2 MA the induction of the magnetic field in the solenoid center is 16 ± 0.05 T, in a more intensive regime with a current of 3.5 MA, an induction of 25 T and an initial magnetic flux of 0.37 Wb can be achieved without any signs of solenoid destruction up to the first field maximum.

10.7 Liners of MC-1 generators

10.7.1 Commonly used metal liners

The amplification of the magnetic field in a MC-1 generator occurs due to the compression of magnetic flux by the cylindrical conductor under the action of external forces. Over the course of development, various liners have been used in magnetic cumulation. The simplest of them are solid metal cylinders. If they are fabricated from highly conductive metal, e.g., copper, then the pulse of the initial magnetic field has to be long enough for the initial field manages to diffuse into the cavity of the liner. Another version is the solid liner made from a poorly conductive metal, commonly, from stainless steel of the type X18H9T (chrome-nickel-titan) with a specific resistance 40 times larger than copper. In this case diffusion occurs faster. The liners, made of highly conductive metal, provided with a longitudinal tangent slot and filled with an insulation material, are also used, but in this case a problem associated with the symmetry of implosion appears. Solid stainless steel liners have been also applied, lined with a thin electrolyzed layer of copper. In generators with a similar liner, a field induction of 2500 T [8] has been achieved some time ago.

10.7.2 Metal composite liners

At the RFNC-VNIIEF, an additional liner for the MC-1 generator was rejected in favor of the multilayer multiwire solenoid described above. Thus, on the one hand, the idea of the most efficient location of the initial field source (between the liner and the explosive) was realized, and, on the other hand, the idea of generating the electrical conductance in an azimuth direction in a composite material consisting of tightly packed wires and epoxy binder at the required time moment was also implemented. The regularity of the structure made possible the use of solenoid as a liner due to the switching of wires under the action of a shock wave. This decreased the power consumption for the generation of the initial magnetic field, while simplifying the design of the generator itself, including its mounting on at testing grounds and the adaptation of diagnostics to rigid safety rules.

When the shock wave transits through the solenoid cross section, first the system of parallel conductors forms reverse withdrawal closes. Then the insulation layer between these wires and the winding is compressed. The part of the flux, inverse relative to the flux in the coil cavity, travels in this layer just prior to compression. This flux penetrates in the cavity and somewhat decreases the field (the negative derivative of induction appears, as seen in the oscillogram of the inductive probe signal shown in Figure 10.36). Further, the wave of closure travels through the cross section of the solenoid, winding with a small amplification of a field in a cavity, and finally the coil as a whole begins to move to the center, and now it is the liner. In the implosion process the derivative of induction first steeply increases (by one order and more) and then continues to rise smoothly during 12–14 μs. The density of the material of the solenoid-shell in the course of acceleration increases from an initial value close to 6 g/cm^3 up to value close to the density of copper (8.9 g/cm^3). In the process of implosion, directed to the center, the complicated structure of the solenoid-shell is not obvious: the shell behaves like a cylinder made from a solid conductive material.

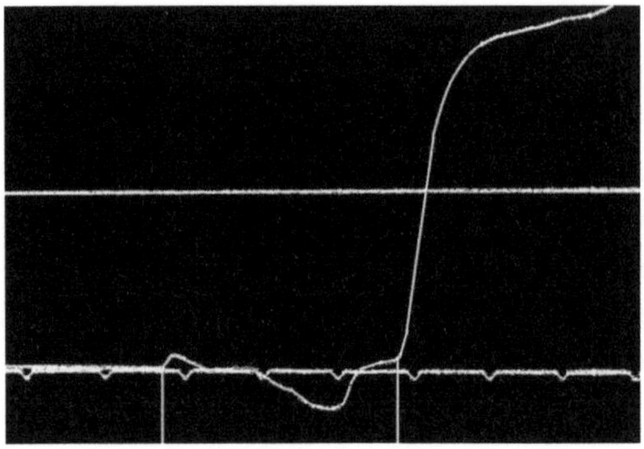

Fig. 10.36: Oscillogram of the inductive probe signal at the instant of the closure and of the beginning of implosion of the metal composite liner. Time markers are 0.5 μs.

In the stage of the explosive amplification of the field (up to 150 T), the conservation coefficient of magnetic flux in the wired shell is not lower than 0.9, testifying that flux capture and compression are rather effective.

Thus, the essential new construction of an MC-1 generator has been elaborated in which the functions of the device generating the initial magnetic field and of the liner are combined in the same device and acquire high operating characteristics. In the described system with a multiwired composite liner, ideas have been realized which supplemented and evolved the initial idea of magnetic cumulation: a sufficiently high initial field is generated, it is additionally amplified at the beginning of the process with a rather small (easily induced) compression flux, the solenoid is converted into the highly conductive liner, which is further accelerated by detonation products and compresses the magnetic flux. The technology for the short-run production of solenoids of three sizes (with inner diameter 139, 175, and 330 mm) have now been developed, with record values of initial magnetic flux in the simplest MC-1 generator, consisting of the solenoid-liner and explosive charge.

10.7.3 Shock-wave liners with phase transitions

In the history of magnetic cumulation, different types of liners (or techniques of flux compression) are known, which distinguishes the different groups of generators of ultrahigh magnetic fields, and which transform by one or another method the chemical energy of the explosive.

In the 1950s–1960s a certain amount of attention was given to generators compressing the magnetic flux due to the phase transition in a compressed media under the action of an explosion shock wave: these are the ferroelectric and the piezoelectric generators, with a switchin on or off of the polarization. Ceramics on the base of lead and barium titanate and of lead zirconate have been applied as a working substance. Between the back side and the shock wave there occurs the stress pulse serving as a current source in the resistive load [75]. Up to now the problem has been the production of large uniform polarized sheets from these materials. Currently it has been shown that a number of processes occurring at relatively low pressure without the forming of shock waves can be applicable. These are the devices based on the shock-wave demagnetization of ferromagnetic materials: at demagnetization the magnetic flux changes, and the e.m.f. is induced in the output coil, so that sometimes such generators are used for generating the initial current in the other EMG. As an example we mention a study with the participation of Shkuratov, Talantsev, Baird, Altgilbers, and Stults [76], in which a fully autonomous compact pulsed power source is described (Figure 10.37). It consists of the initial power source with the shock-wave demagnetization of the ferromagnetic $Nd_2Fe_{14}B$ (FMG) and of the spiral generator of the compression flux (FCG). The armature of the FCG was loaded with 197 g of explosive material, and the current, generated by the system, attained the peak of 33.13 kA at the amplification coefficient 114.

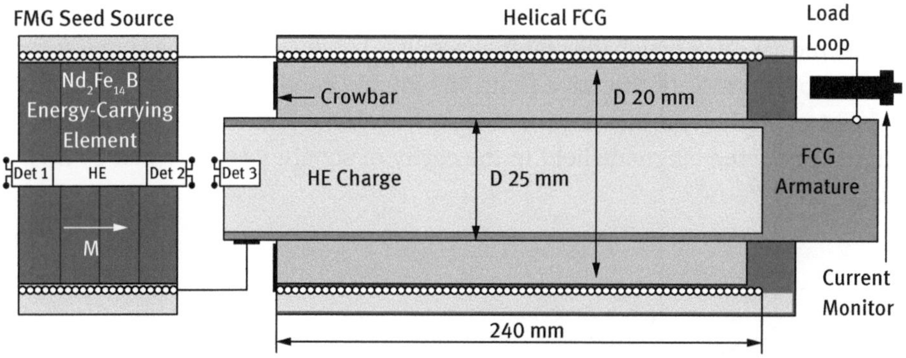

Fig. 10.37: Schematic drawing of an autonomous current source, FMG-FCG.

Shock-wave (semiconducting) generators are worthy of more detailed consideration. They are based on variations of the conductance in the detonation shock wave. Some dielectrics or semiconductors with very small specific conductivity in the regular state (silicon, germanium, grey tin, silicon oxide, cesium iodide, germanium iodide, powder of aluminum oxide) transfer in the conductive state at high pressure. Bichenkov, Gilev, and Trubachev [77] and Nagayama [78, 79] were the first to report that the magnetic flux can be compressed by means of the wave of the phase transition in the semiconductor. They also discussed the possibility of the conductivity being switched by the wave in nonconductive metallic powder [80, 81]. Over the next years generators of this type were studied in different countries [82, 83]. Prischepenko, Barmin, Markov, and Melnik studied both cylindrical and spherical versions of them [84]. Many researches were involved in this subject, because of the lack, as was initially expected, of instabilities of the substance-field boundary, which restrict, as we know, the magnetic field amplitude.

An extremely profound and interesting theoretical and experimental study of the physical processes in the shock-wave generators was carried out by Bichenkov, Gilev,and Trubachov in Novosibirsk. The content and results, along with the comprehensive literature, are formed in the Gilev's Ph.D. thesis [85] and in [121]. As was experimentally shown in this work, the compression waves in different materials generate the sharp changes in conductivity. The zones of high or low conductivity, transported by the shock or detonation wave, can be used for the control of electromagnetic fluxes so as to generate electromagnetic energy of high density and for the current commutation. Briefly, the physical pattern of the processes occurring in the shock wave generator is as follows. In the initially nonconductive material with the transverse magnetic field B_0, a closed configuration of the converging shock wave (circular or, in the general case, of any shape) is produced, under the action of which the material acquires conductivity. In the simplest case (the zero thickness of the wave front, the infinite conductivity of the compressible material, the constant ratio of the mass velocity u to the shock wave speed D; Figure 10.38) the magnetic field in the cavity of square S is described by $B(t) = B_0(S_0/S(t))^{u/D}$.

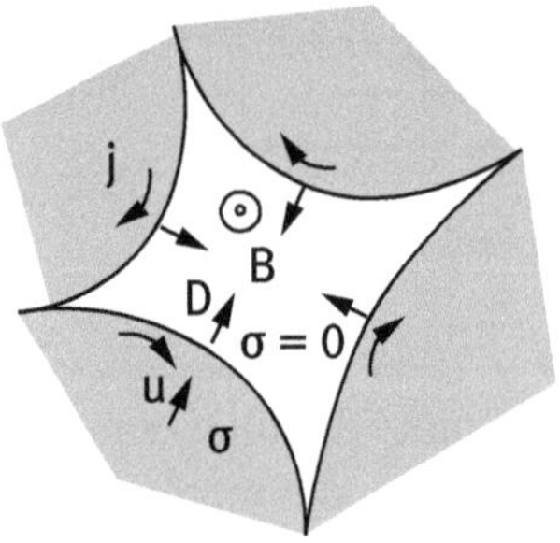

Fig. 10.38: Magnetic flux compression by means of the converging shock wave with the formation of the conductive phase.

In the meantime the authors of a fairly large number of studies in this line of investigation have pointed out the following merits of the discussed method for the generation of ultrahigh magnetic fields [85]: (a) the practically instantaneous input of initial magnetic flux into the cavity; (b) the substance on the conductor-field boundary is constantly renewed, and therefore the growth of the MHD instabilities is suppressed; (c) the feed system is simplified, and the mass of explosive charge is decreased; (d) the working substance protects the central region against the jets. The physical meaning of field amplification is that the boundary of the substance subjected to the metallization travels with the shock wave speed D, and the work against the forces of magnetic pressure is done by a substance moving with the speed u. Due to the difference in speeds, part of the magnetic flux is frozen into the substance and lost with regards to magnetic cumulation. The rest part is displaced to the region before the shock front. According to the above mentioned relation, in order to increase the ultimate field it is necessary to apply a strongly compressed substances with a large ratio u/D.

The practical results of the long-standing studies in the physics and technique of magnetic field generation by means of shock wave are as follows. Several versions of the construction of shock wave generators have been tested, the serviceability of the method was proved, and fields of 800 T have been registered [121]. With a help of a computer package named MAG (a product of RFNC-VNIIEF, [9.77]) the numerical MHD model of cumulation in metallic powders was developed. The theoretical dependence describing the field in the generator coincides well with the experimental one up to the arrival of the shock wave at the detector. As was understood with a help of simulation, when approaching the axis the shock wave is decelerated by the field pressure and stops, and the magnetic field attains a maximum and, further on, decreases: the concept of the residual field still remains as it takes place in the standard MC-1 generator. This has somewhat reduced the initial optimism of the developers concerning the potentials of this method, and has enabled Gilev to indicate on the “paradox” of magnetic cumulation: (1) the increase of the initial magnetic field or sizes of the generator resulted in a decrease of the ultimate field; (2) the increase of conductivity of the compressible material resulted in a decrease of the ultimate field. Thus we face the evolution from the initial considerations on the absence of restrictions on the achievable maximum field by means of the shock-wave flux compression, to the conclusions, intrinsic to the liner-method, on the existence of the limiting field, depending on initial parameters of the device.

The amplification of the magnetic field was detected in experiments with the longitudinal and transverse compression of the liner, described in reports by Prokop’ev, Anisimov, Matrosov, and Shvetsov at the XIth and XIIIth Megagauss conferences [86, 120]. They reported on the generation of an ultrahigh magnetic field by conical MHD flows in a device based on longitudinal and transverse flux compression. The device is schematically shown in Figure 10.39. This device consisted of eight copper tubes (2) with a wall thickness of 1.5 mm, a diameter of 9 mm, forming the cone: the lower edges of tubes are closed on the surface of the dielectric rod with the induction detector, while the upper diverge outward, similar to a fan. The initial field is generated by the discharge onto the coil (1).

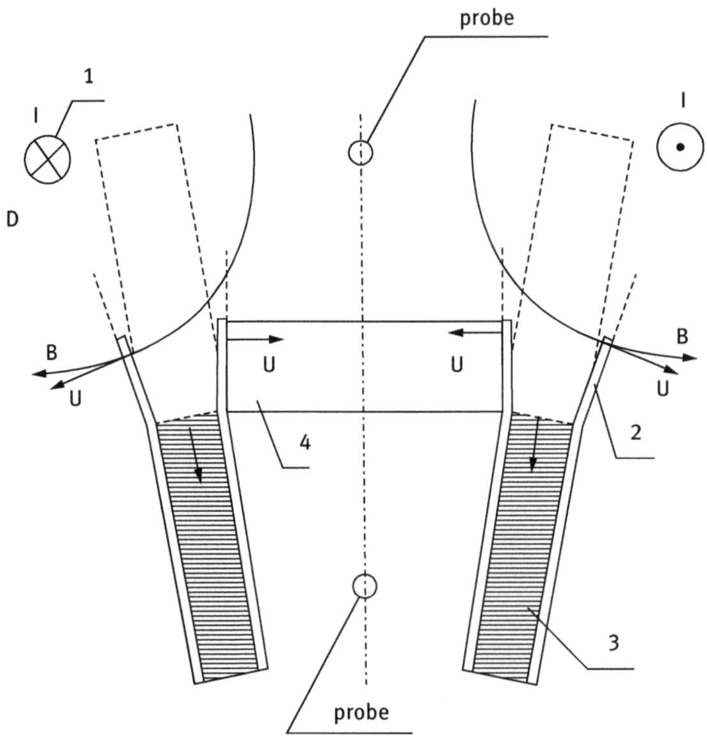

Fig. 10.39: Schematic drawing of the experimental device with the longitudinal-transverse flux compression: (1) coil; (2) copper tubes; (3) explosive material; (4) dynamic liner; (D) the speed of detonation.

The tubes are insulated and filled with an explosive material, which is exploded instantly on one side. With expansion the walls of the tubes are close in and form the dynamic liner which traps the magnetic flux induced by the coil. The liner is moving along the cylinder axis and simultaneously compresses over the radius. The results of the measurements showed that in the center of the small radius of the "cone" the field increased by several hundred times.

10.8 Violation of liner stability in flux compression

In experiments with a simple MC-generator (see above), consisting only of a wired solenoid and the explosive charge, under conditions of the insufficient protection of probes, the maximum field did not exceed 300 T. The typical "life-time" of the probes is ~ 13 μs, during which time the liner cannot closely approach the measuring unit. The successive increase of the strength of the protection system enabled prolonging the measurement and thus obtaining an increased maximum value of the field. Only the strongest and most complex protection system, occupying a volume of not less than 13 mm in diameter after it was shocked by the imploded liner, provided a repeatable measurement of the field levels up to ~ 600 T with probes of up to 1 mm diameter. Since this is noticeably less than the expected level of 1000 T, and, in addition, as was also known in other laboratories, the peak field achieved was less than the expected one (as a rule, below 300 T), it was important to study the conditions of the generator in the vicinity of the field peak.

In the case of a generator with a wired solenoid and high initial magnetic field, this problem is alleviated, due to the fact that the processes associated with heating and decelerating of liner substances manifest themselves at relatively large diameters of the inner surface of the shell, so that the evolution of the substance-field boundary could be distinctly observed. Using a specially-developed technique of pulsed x-ray radiography in combination with the numerical one-dimensional model of the actual generator, it has been possible to obtain reliable information on the dynamics of the shape and state of a shell compressing the ultrahigh magnetic field.

Of primary interest were the initial perturbations of the compression symmetry, caused by an initiating system consisting of ten identical elements (multipoint initiating), evenly used on the circle of the explosive substance, and, as well, the possible deterioration of the compression symmetry, associated with the construction of the composite wired solenoid.

A comparison between the x-ray photographs of a cross section of the steel cylinder with a linear mass equal to a linear mass of the wired solenoid-shell and that of the wired solenoid-shell (Figure 10.40) shows that there is no essential difference between the solid metal and the composite: the shape of the inner boundary of the composite wired shell as well as that of the pure steel shell (without the magnetic field in the cavity) equally repeats the basic features of the detonation wave (the number of the maximal harmonic is 5), and the amplitude of deviations from a regular circle does not exceed 1 mm.

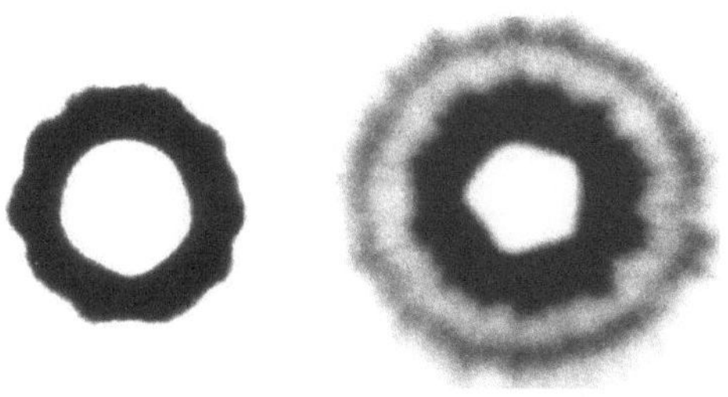

Fig. 10.40: X-ray photographs of a cross section of the solid steel shell (*left*) and composite wired shell (*right*) without the magnetic field inside.

The shape of the inner boundary of both shells is practically identical. More noticeable is the difference in the symmetry of the external boundary: in the case of a steel cylinder it is soft, with a fundamental harmonic 10, and in the case of the composite layers it exhibits more pronounced small-scale jets near the first solenoid layer and near the second longitudinal reverse withdrawal with the insulated layer of lower dense between them.

Several x-ray photographs (from the set, enclosing the whole compression process) of the generator with the field while operating in the regime with the initial magnetic field 0.16 T are shown in Figure 10.41. In the first photograph (from the left) the magnetic field in the cavity is above 250 T; the shell is noticeably compressed, and the temperature of the inner surface achieved the boiling point for a metal, but the shell remained of a sufficiently regular shape with maximal deviations below 1 mm.

When the magnetic field attains 300 T and above, the shell substance intensively evaporates, and in a short time the configuration of the inner surface is subject to distortion, as shown in the next photograph. Small- and large-scale asperities arise with jets of substance on their ends, remaining behind the main mass of the shell. The correlation of the shape of a shell with that of the detonation wave becomes less evident, and, every so often, almost disappears, which is confirmed by the quantitative examination of the inner surface of shells presented in the x-rays photographs in Figure 10.41.

On the next (third and forth) photos, taken at ~ 0.5 μs prior and at the instant of peak field, one can see the catastrophic consequences of the development of the instabilities which have happened previously. The highly dense and, obvi-

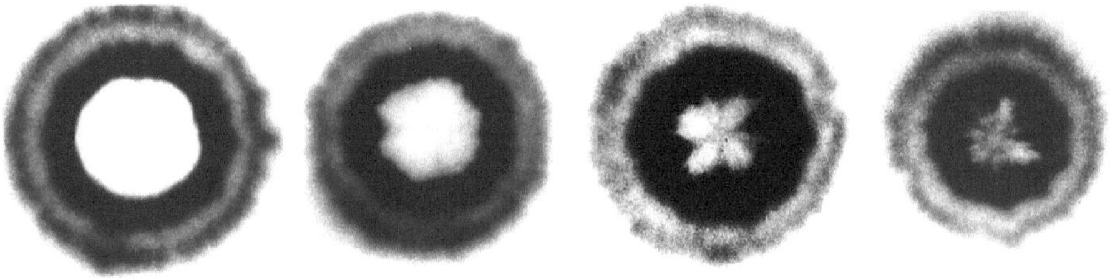

Fig. 10.41: Sequence of x-ray photographs of the wired shell, compressing the magnetic field. Magnetic induction in the cavity is (from left to right) about 250 T, 350 T, 500 T, and 600 T.[2]

2 In Figure 10.41 magnetic field values above 400 T were measured in generators with enhanced protection of probes, which, obviously, affected the shape and sizes of the liner impacting with the protection system, so that the shape and dimensions of the liner only approximately correspond to the magnetic field value at the moment of exposure.

ously, current-carrying part of the inner layer was decelerated by magnetic counterpressure attaining 1.5 Mbar, and its shape differs greatly from the initial one. Although the mid-radius of the inner boundary has achieved the minimum value, we cannot speak in a traditional way of the shell is stopping. The inner part of the shell broke up into fragments, some of which move with great speed to the center, others remain static or slowly move outside. A mixing of the evaporated substance and magnetic field takes place. In the center one can observe the annular layer of the substance, which is the result of the reflection of the jets from the cumulation axis. Evident is the mechanism of faster transformation of kinetic energy of a shell into Joule heating, compared to that in the process of magnetic cumulation.

A description of the phenomena occurring in the cavity of the MC-1 generator, measurements of field induction and its derivative are essentially supplemented with numerical calculations based on the full-scale one-dimension MHD model of the actual construction of a generator (the model will be considered in more detail below). The results obtained due to a combination of actual and numerical experiments are given in Figure 10.42 in the form of graphs showing changes in: the boundaries between the layers of the wired liner, $D(t)$; the behavior of magnetic induction $B(t)$ and its derivative $dB(t)/dt$; and the conservation coefficient of the magnetic flux in the cavity, limited by inner boundary of the liner, $\Phi(t)$.

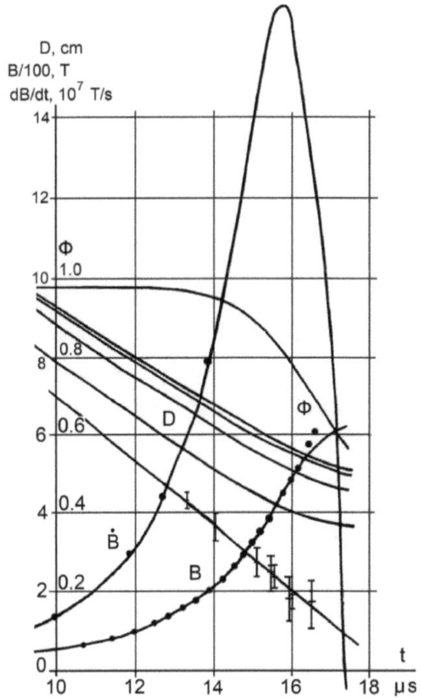

Fig. 10.42: Magnetic cumulation in the MC-1 generator with a wired liner, calculations and experiments (points). Behavior of diameters of boundaries between layers of the wired liner = $D(t)$; magnetic induction = $B(t)$ and its derivative = $dB(t)/dt$; Conservation coefficient of the magnetic flux in the cavity limited by inner boundary of the liner = Φ(t).

Experimental data, obtained using inductive and optical (based on the Faraday effect) methods, are marked with points, as in the case of x-ray radiography measurements of the inner diameter of the liner. The high level of agreement between the magnetic and x-ray measurements at large diameters, when symmetry distortion is little, shows that the numerical model provides a reasonably fair pattern for magnetic cumulation as a whole.

Among other results of the simulation, let us note that the kinetic energy of the shell is almost entirely converted, as was established with regards to the energy balance: at the moment of peak field about 40 % was transformed into magnetic field energy, contained in the shell cavity, where the field is constant; the rest was distributed approximately evenly between the magnetic field energy, diffused into the shell, and the internal energy of the shell substance.

At the moment of the measured peak field of 620 T, the mid-inner diameter of the shell is ≥ 15 mm, and the conservation coefficient of the magnetic flux ~ 0.5 (the initial magnetic field is 1.6 T).

Thus, due to the possibility realized in wired-solenoid to essentially increase the initial magnetic field which, consequently, leads to a certain restriction of the compression degree of the liner (to an increase of the final field volume diameter) and then results in the possibility of determining the basic parameters of the MC generator in the final stage of the process [87]. The experimental results confirm that with the slowing-down of the shell a Rayleigh–Taylor instability of the field-matter boundary is developed. Due to this, the matter is ejected into the cavity of the shell, which prevents obtaining the field of the calculated amplitude and carrying out the studies in the magnetic field of the generator. In this regard the instability of the inner boundary of the shell compressing the ultrahigh magnetic field may be considered to be a main factor which restricts the usable magnitude of the magnetic field of the MC-1 generator.

The measuring unit of the generator involving the detectors and samples, located on the shell axis, is destroyed prior to the moment when the amplified field achieves a maximum, and the cavity with the field is filled with a substance. The unstable, accidental character of this phenomenon explains why the magnetic fields below 400 T, (the level above which the instabilities just begin to evolve) obtained are sufficiently reproducible, whereas the reliable measurement of the magnetic field in the immediate vicinity of a maximum (when the shell is almost or fully decelerated) meets with significant difficulties.

10.9 Principle of cascading in MC generators of ultrahigh magnetic fields

In addition to the idea of use a multiwired solenoid as a liner, the authors [88] advanced a proposition, possibly more important, developing the idea of magnetic cumulation. The case in point is the method of stabilization of the magnetic cumulation of energy. The essence of the proposal is to compress the magnetic flux not only by a shell but by a set of coaxial shells made from specific material. Every time when the loss of stability of the inner boundary of a shell becomes rather probable, a shell is changed by another, which continues to compress the flux. Thus it is possible to produce stable compression at large changes of the radius of the volume occupied by the magnetic field, and, consequently to increase the induction amplitude of the amplified field.

For this purpose it is required that the conductivity of the shell change during the performance. The inner shell consists of a large number of thin insulated wires, and therefore in the initial state (prior to the impact) it freely lets pass the axial magnetic flux inside. At the impact of an external shell on a stationary one its insulation is destroyed and it becomes conductive. From this moment, part of the kinetic energy is transferred from the first to the second shell, and the last begins to compress the flux.

Along with this, at impact the inner boundary of a skin layer passes to the inner surface of the shell switched to flux compression, whereas its external boundary remains in the body of the previous shell. In other words, after impact the current is circulating in both shells, and as a result the energy exerted in the substance of the inner layer of each shell reduces. The heating of the substance of the inner layer of a shell, compressing the flux, and the counter-pressure of the magnetic field are determined by the induction difference in the cavity and outside the layer. In contrast to the flux compression produced by a single shell, this difference, with an increase of the amplified field, increases more weakly than the field itself. In this case the density of the thermal energy in the shell substance and the counterpressure appear to be sufficiently less compared to the density of magnetic energy of the generated field. It is worth noting that every time when a new shell is switched to flux compression, on the substance-field boundary the heated substance is changed by another, cold one, and the full time of operation of each shell reduces, as, consequently, does the time interval of the instabilities development.

Magnetic cumulation, produced by a system of coaxial shells, is a version of the cascade construction of amplifying devices. Therefore each of these shells is

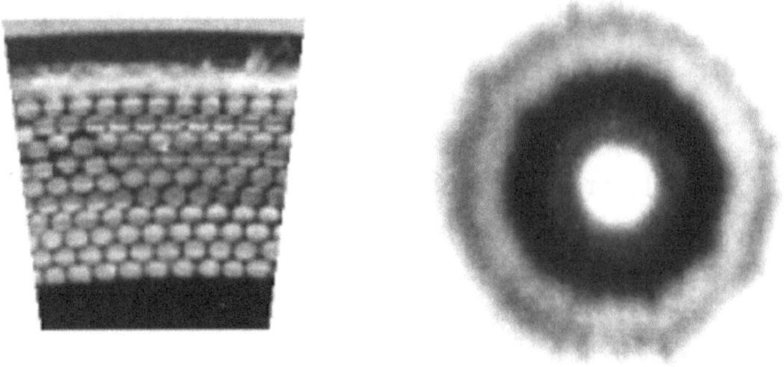

Fig. 10.43: Fragment of a cross section of the wired cascade and the x-ray photograph of the two-cascade generator in the absence of a magnetic field.

referred to as a cascade of the MC-1 generator. The first cascade is the solenoid-shell.

The construction and performance of the rest of the cascades are based on the same principle of the controlled conductivity of a substance as the solenoid-shell. Initially this is a cylinder, made from closely-packed insulated thin copper wires, packed in parallel to the axis and glued with epoxy compound (Figure 10.43). If a diameter of wires is sufficiently small, then such a cylinder freely passes inside the magnetic flux and does not affect field amplification until it is involved in the process of magnetic cumulation. This occurs when the moving cascade collides with the stationary one, and the shock wave travels in its substance, switching the azimuth conductivity. Later both cascades are imploded to the center, and the magnetic flux is compressed by the second (inner) cascade. The location of each sequent cascade is chosen in such a way that on the inner surface of the previous one the noticeable instabilities did not manage to develop.

The influence of cascades on the shape dynamics of the inner surface of the shell, compressing the ultrahigh magnetic field, is clearly seen in results of x-ray investigations.

The efficiency of instability suppression is demonstrated in an experiment without a magnetic field. The x-ray photograph of a device consisting of a solenoid-shell and additional wired cascade with initial inner and external diameter, respectively, 35 and 28 mm, is given in Figure 10.43 on the right. In contrast to the experiment with a single liner (Figure 10.42), the compression of the second cascade occurs without the symmetry distortion of the inner boundary of the liner.

In Figure 10.44 we see photographs of a two-cascade generator with an initial magnetic field of 1.6 T. On the left, the field induction in the cavity of the second cascade is about 400 T (compare with the corresponding picture of a one-cascade generator!), and the inner boundary of the shell is almost regular, with

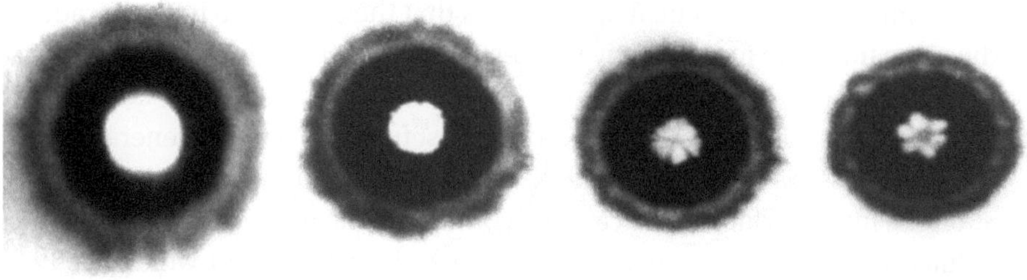

Fig. 10.44: Successive (1 μs apart) x-ray photographs of a two-cascade generator. The third one corresponds to the field peak.

small in amplitude, long-wave, smooth deflections. The next three pictures were made during a time of about 1 μs, close to the instant of field peak, i.e., prior and immediately after. One can see the different phases of the development of disturbances: the formation of cumulation jets, their collision in the center, involving more and more of the shell substance of the second cascade.

Due to the third cascade it was possible to completely solve the problem of stability of the shell surface compressing the magnetic field of induction 900–1000 T. Figure 10.45 shows the three-cascade generator at the peak field: the substance-field boundary is smooth, and its minimum diameter is about 11 mm. When the fourth cascade is installed on a deeper radius, the third cascade does not fly to it and does not compress the magnetic flux, but additionally mechanically protects the measuring probe. Note that between the second and third cascade one can see the gaps, in which part of the magnetic flux is lost, and the stable state of the shell does not exist long: in the subsequent picture, made approximately in 0.5 μs, it is seen that the boundary still suffers shape loss, and the cavity is filled with the substance of the shell.

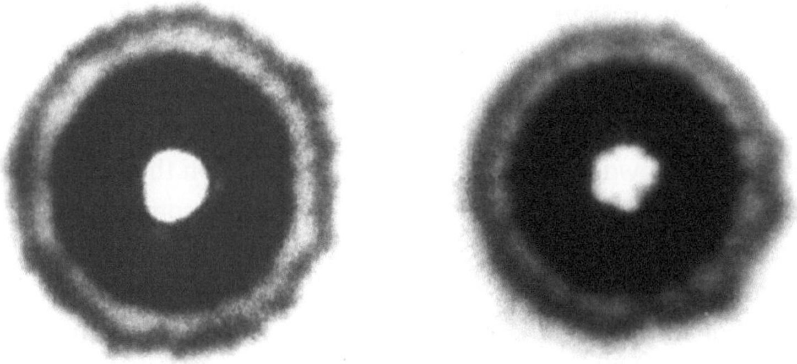

Fig. 10.45: X-ray photographs of a three-cascade generator taken at the moment of the field peak and at intervals of about 0.5 μs.

It is important for practical applications that the composite liner cascade is, at the same time, the most effective protection of the measuring unit. In contrast to various dielectric mechanical obstacles, the liner cascade is the “active” protection and does not lead to a decrease in the useful volume of the generator. This is due to the passage of the current layer, responsible for the existence of magnetic flux, on the inner surface of the cascade continuing the flux compression.

Numerous experiments using a three-cascade generator in which the field induction was registered to pass over a maximum testify that, at first, the liner, compressing the magnetic flux, has stopped, and its kinetic energy is exhausted, and, then the measuring unit in the liner cavity is not damaged. Such a generator can be used for studying fields in the 10 megagauss range.

Besides the composite material for cascades with copper conductors described above, there are also other types of materials, the use of which is dictated by two factors. Firstly, the induction of the final field is increased with an increase in density of the composite material, since the kinetic energy (and its density) transferred to a liner is also increased. Secondly, since the device is destroyed by detonation products, the cost and adaptability to production of the generator cascades is important. In order to increase the density of the composite material, it was proposed to use tungsten instead of copper.

A first version of the tungsten composite from tungsten isolated wires instead og copper ones was described in [89]. This resulted in a rise of the cost and complicated the technology due to the rigidity and elasticity of tungsten wires. In the second version the researchers applied the composite, pressed up to desired density, from the metal powder and polymeric binder, evenly mixed in the blank [115] in the proportion providing a small initial conductivity. Present-day technology of the production of powder composites is relatively cheap; in the composite may be practically any metal can be used, and the dimensions of the cascade are arbitrary and can be easy varied by simple turning from blanks.

The initial material density of tungsten cascades, as compared to cascades with copper conductors, was increased approximately from 6 g/cm^3 to 10 g/cm^3. Experiments [116] demonstrated that the wired (copper and tungsten) composites produce flux compression with almost equal efficiency, whereas the flux conservation in the liners from powder composite material depends on the initial liner sizes: it is much lower in liners with diameters of 5–15 cm and is comparable with liners with diameters of 1–3 cm. The poor conductivity of powder composite in the liners of large diameters becomes apparent by comparison of the enlarged images of cross sections of both composites. During compression of the wired composite, in order to over-switch the wires they should be shifted in a radial direction relative to the adjacent layer by a very small distance equal to the thickness of enamel insulation. This is easy attained even with a weak shock wave travelling in a liner of any sizes. In order to properly switch on the metal particles in the powder composite, compression in both, radial and azimuth, directions should be produced. Such two-dimensional compression of the composite occurs only with small diameters and highly regular cylindrical liner shape.

At one time Novac, Smith et al. from the University of Loughborough, Great Britain, used "the cascade based on the phase insulator-metal junction" at the electromagnetically-driven flux compression in the θ-pinch configuration [56, 117]. In one version the cavity between the copper liner with an external diameter of 51 mm, thickness of 0.4 mm, and length of 20 mm, and the ceramic tube on the axis with probes was filled with aluminum powder. In two identical experiments fields up to 300 T were reliably measured, in spite of the development of essential instabilities, as the x-ray photographs of the liner have shown. In the experiments using the thoroughly fabricated aluminum liners the field level exceeded 350 T. They assumed that the application of the powder cascade results not only in the correction of the liner shape but also in the protection of the measuring unit.

10.10 MC-1 cascade generator. Numerical simulation and experiment

The history of magnetic cumulation is filled with the many attempts to achieve the most adequate theoretical description of an MC generator. This description is necessary for the comprehensive understanding of the complicated physical processes involved in cumulation and for the development of actual constructions of generators with the practically required ultimate parameters, at the same time solfing the problem of economically viable physical experiments. Initially there were attempts at analytical description with one or another degree of idealization. Descriptions of these can be found in Section 10.4. This is also presented in the voluminous literature on the subject, an important part of which belongs to one of the VNIIEF founders of the MC method Prof. Lyudaev [90]. Then along with the development of the numerical simulation methods and the appearance of adequate software, numerical models of generators in the format of the software packages were also developed.

The complexity of the problem (even in the simplest one-dimensional approach) is determined by the physics of the process and by the complexity of design of modern MC generators of superhigh magnetic fields, as well as by the necessary knowledge and correct description of the properties of teh construction material. This description is difficult, since the materials are used in extreme regimes of high energy density associated with the generation of multimegagauss fields.

The design of the MC-1 generator possesses a number of peculiarities requiring additional modeling efforts. First of all, there is the multilayer structure of the generator shells with one or few layers made of composite metal-dielectric material. The main property of such a material is controlled electroconductivity. At the initial stage this material does not conduct, and conductivity appears only when a compression wave passes through it.

The processes occurring in the central part of MC-1 (inside the ring with the explosive) are described in terms of MHD (i.e., ignoring the shift currents). The motion of the compressible viscous medium is described by the balance equation system for the density ρ, position r, velocity u, specific internal energy ε, and magnetic induction B for points on a continuous medium. To make the system closed, it is supplemented with equations connecting the pressure, conductivity, shear modulus, and yield point with the density and specific internal energy, as well as with corresponding initial and boundary conditions.

For the solution of the one-dimensional MHD problems using the Lagrangian regular grid, the program SMOG-DISK [91] was developed at VNIIEF and then modernized, which also includes frequent upgrading and corrected library of wide-range basic relations such as equations of state, conductivity laws for different substances, thermal (caloric) equations, elasticity modulus, etc. The program SMOG-DISK was written in FORTRAN and adapted for the PC with high precision. The program SMOG-DISK permits calculation without delay of the main characteristics of the processes occurring in the central elements of MC-1 generators. At present, for the majority of the solved problems the calculation time on the PC is within 0.5–5 min.

For the solution of any problems it is necessary to introduce the proper initial and boundary conditions. The position of points of a continuous medium at the zero moment of time is determined by the arrangement of knots of the calculation grid prior to calculation. The initial values of the deviator of tension tensor, magnetic field, and velocity for all points of a continuous medium are zero in most problems. The density and specific internal energy at the zero time moment are chosen in such a way that the initial temperature and initial pressure would be equal to the given values. (Usually these are room temperature and atmospheric pressure). And finally, in order to solve the equation of motion on the internal r_{int} and external r_{ext} boundaries of region of calculation, the pressure or velocity should be given, while for the equation of the field diffusion on these boundaries whether the magnetic field or its spatial derivative as functions of time should be defined. In all cases of interest the internal boundary coincides with the coordinate axis ($r_{int} = 0$). Since on the axis of the device the radial motion and azimuth currents are absent, the radial velocity and radial derivative of magnetic induction will be equal to zero on the internal boundary. Namely, $u_r(0, t) = 0$ and

$dB_z/dr(0,t) = 0$. The external boundary of the region of calculation is assumed to be free: the pressure on it at all moments of time equals the atmospheric value. As for the second boundary condition for a magnetic field, it is necessary to outline the following: from the zero time on and in the course of inducing the initial magnetic field of MC-1, the radius of the region for which the field diffusion is calculated can be less than the radius of the whole region of calculation. During the entire time interval, on the surface confining the region of calculation, the magnetic field as some function of time should be specified (for instance, as varying harmonically). In the residual part of the region of calculation the field is introduced in the form of a spatial-temporal function, found from the known current distribution. In other words, beyond the limits of the region of calculation, we ignore the induced field but not the field from external sources. (Let us note that this omission is not always possible). When the region of the field diffusion coincides with the whole region of calculation, then on its external boundary the field is also specified as a certain time-dependent function, but in most problems of interest this function identically equals zero.

The whole region where the solution is sought may consist of few physical domains (layers), each of which is characterized by its own set of determining relations. For instance, one or several domains could be related to the explosive substances (ES) and the detonation products (D. P.). For each of them a time of initiation of ES may be specified.

It is assumed that at the moment of the onset of explosion uniform pressure is applied over the whole external surface of the ES, which is sufficient for the formation of the detonation wave.

In conclusion, we outline that the time of initiation of the ES charge is chosen in such a way that the shock waves induced by the detonation waves reach one or another surface at fixed moments of time.

For the cascade generator MC-1 the cylinder-symmetrical region of calculation includes the external layer of ES and the solenoid, consisting of two composite layers and two layers of epoxy compound. It also includes internal cascades, the composition of which could vary, depending on the character and the purpose of calculation.

The calculation begins from the stage of formation of the initial magnetic field by layers according to a given algorithm. At a given time the initiation of detonation “occurs” on the external boundary of ES and the calculation of the gas-dynamic processes in ES and adjacent solenoid starts. With the passing of the shock wave through the different regions of calculation, the conductivity is turned on in them, the field diffusion through them is sharply decreased, and, as a result, they are subjected to the pressure of the magnetic field. As a rule, the problem is under calculation up to the time of peak field in the central region.

At the initial moment the azimuth conductivity of the solenoid composite material and cascades is assumed to be zero. When the compression of the composite exceeds some definite value δ_1 larger than unity, it becomes a conductive. In this case its specific conductivity is determined by the formula $\sigma = \sigma_m \lambda k$. In this formula σ_m is the conductivity of the metal, λ its volume fraction in the composite, and k the correcting coefficient ($0 < k < 1$). This coefficient takes into account the deviation of the conductivity from the tabulated values and is matched to the results of the comparison between calculated dependences of $B(t)$ and the experimental data. If in further performance of the generator the compression degree of the composite becomes less than δ_2 ($0 < \delta_2 < 1$), the conductivity is assumed to be equal to zero.

A proper description of the performance of MC generator requires that both the compressibility of substances and their thermal expansion be taken into consideration. Therefore the construction of the equations of the state including the whole range of densities and energies attained in the process of magnetic cumulation is of paramount importance for the modeling problem. It is also necessary to keep in mind that the current-carrying shells of MC generators experience strong heating.

Hence, the corresponding equation should take into account one way or another the opportunities of shell melting and evaporation. It means that the equation of state should be determined not only in the energy-density region, corresponding to the homogeneous phases, but also in the regions corresponding to the phase mixing.

The second program MAG used for calculations was developed initially as an educational research-user complex MASTER [92]; therefore this program, in comparison with the SMOG-DISK, has a more elaborate interface and is better adapted to the requirements of an individual user. This is the reason why the large volume of model testing and the matching with the experimental data is often done using the MAG program.

The general picture of the performance of the simplest one-cascade MC generator is shown in Figure 10.46 as the temporal dependences of the main parameters characterizing the magnetic cumulation. They describe the processes developing in the cavity of the generator and in a liner itself, namely, the temporal changes of liner dimensions, magnetic induction and its derivative, as well as the state of the liner material. For plotting the graphs a sufficiently detailed numerical model of the actual generator was used, and the points on the curves indicate the results of measurements in actual explosion conditions. The graphs are intentionally separated into two parts: on the left, there is the initial stage of the generator performance, namely, the runaway of a liner and a rise of magnetic induction from initial level of 16 T to approximately 150 T.

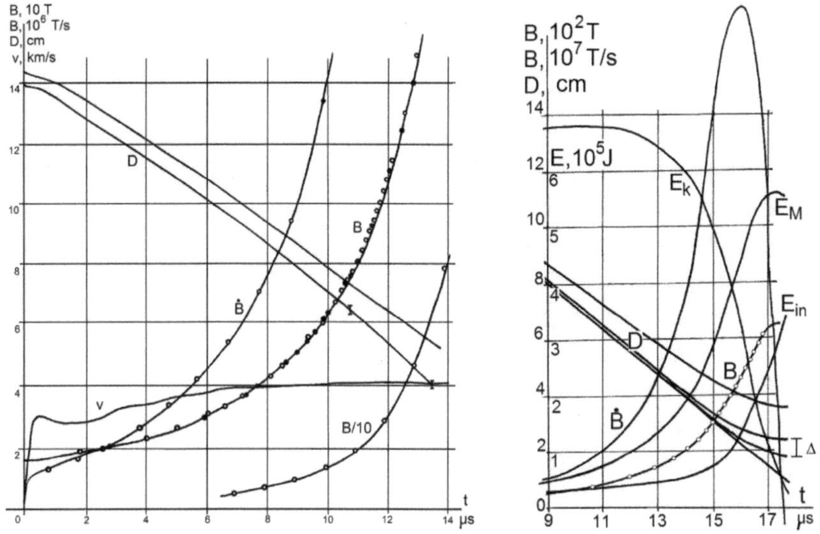

Fig. 10.46: The qualitative behavior of the internal and external radius *D* of a liner, induction *B* and its derivative dB/dt, the linear kinetic energy E_k; Δ indicates the position of the internal boundaries of the skin layer in a liner. The points mark the experimental data.

In the first stage flux compression occurs almost without losses, the state of the liner material is close to the initial one, and the numerical model gives the results in a good accordance with experimental data. The speed of compression increases on a length of about 3 cm up to the value close to 4 km/s. This result is typical for generators with an initial liner diameter of about 15 cm, linear mass up to 100 g/cm, and linear mass of the HE charge of order of 1 kg/cm. In this case about 20 % of the HE energy is transformed to kinetic energy of the liner.

On the right, it can be seen how during the second stage of the process the effective deceleration of the liner by the counter-pressure of the magnetic field begins and develops. The kinetic energy of the liner achieves a maximum and then quickly drops, while the internal (thermal) energy of the liner substance increases at the same rate. In the beginning liner material is melted from the inside, then it evaporates. During this process a part of the vapor loses its conductivity and does not slow down at the magnetic field (the internal boundary of the skin layer in the liner enters into the liner wall). During this final stage of flux compression the main role passes to other factors, which differ from those featured in an ideal scheme of magnetic cumulation: the finite conductivity of the liner material and its growth depending on absorbed energy are becoming fairly well pronounced as a result of nonlinear diffusion due to heating of the liner; the compressibility of the substance and stability of the internal liner boundary should be taken also into account. They ultimately determine the maximal value of the generator field and its volume.

The most convenient tool for study and practical application of reproducible magnetic fields in the 10 megagauss range [93] turns out to be the three-cascade generator developed in VNIIEF at the the beginning of the 1980s, which features an explosive charge of moderate size and mass. The basic parameters of this generator are the following: the explosive substance is a 50/50 trotyl-hexogen alloy, internal and external charge diameters are 152 and 300 mm, respectively, the length 181 mm, mass of HE 16 kg. Internal and external solenoid-shell diameters are 139 and 151.7 mm, the ength of the winding is 300 mm, the linear mass is at about 100g/cm, and the maximum implosion velocity is 4 km/s.

The internal diameter of the second cascade is 28 mm, thickness (including the layer of epoxy compound above the winding) is 4 mm, linear mass is 19 g/cm. The corresponding data for the third cascade are 12 mm, 2.5 mm, and 5 g/cm, respectively.

The development of the MC-1 cascade generator marked the beginning of "megagauss physics" with real opportunities to carry out studies of extremely high magnetic fields [94].

Figure 10.47 presents the temporal dependences of the main parameters of a three-cascade generator in the "standard" regime of performance with the final field of 950 ± 50 T. They were plotted on the basis of experimental data.

The generator may be displaced on an explosion test site in horizontal or vertical positions. Figure 10.48 exhibits the photograph of the MC-1 three-cascade generator in a horizontal position, prepared for an explosion and equipped with

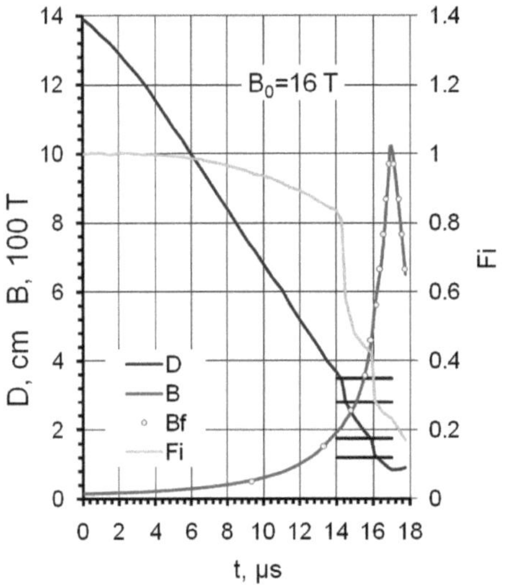

Fig. 10.47: Temporal behavior of the internal liner diameter $D(t)$, magnetic field $B(t)$, and the coefficient of conservation of the magnetic flux $Fi(t)$ of an MC-1 three-cascade generator (points B_f are the data obtained by optical methods).

Fig. 10.48: Photograph of an MC-1 three-cascade generator prepared for an ignition in experiments with x-ray and optical diagnostics.

x-ray, optical and inductive diagnostics. On the detached table the cassette with x-ray film is placed. On the front plane the cables and light guides are seen. The generator and ES charge are mounted on the supports with adjusting clamps.

In the specific experiments the vertical displacement of the generator is preferred, for instance, if necessary to cool the samples. In this case the Dewar vessel with cryoliquid is mounted under the table, the cryopipeline being made of coaxial tubes with vacuum insulation between them (made of magneto-transparent glass or superthin stainless steel tubes). The cooling liquid or gas is fed through a cryo pipeline upward to the generator center and cools the samples. The vapor given off then goes up outside the generator. Outlets of the diagnostic probes are also placed on the top of the device. The photograph of the experimental device of this kind with MC-1 generator is seen in Figure 10.49.

The results of numerical modeling compared to the corresponding explosion experiment using the three-cascade MC-1 generator gives us an impression of how the cascade generator operates. The dependences shown in Figures 10.44 and 10.45 describe the integral characteristics of magnetic cumulation. Along with it, the physical picture of the MHD flow at the moment of peak induction is also of great interest. To obtain this picture, it is necessary to consider the distribution of the characteristic parameters in the final stage, calculated with a help of characteristics of a medium, formed at the transition of the composite liner to the conductive state.

Figure 10.50 shows the radial cross section of the generator at the instant of the impact of the second on the third cascade. It can be seen how at the instant of the impact a compression wave is produced, propagating in the substance of the

Fig. 10.49: Experimental setup with a vertically positioned MC-1 generator and with a system of cooling samples.

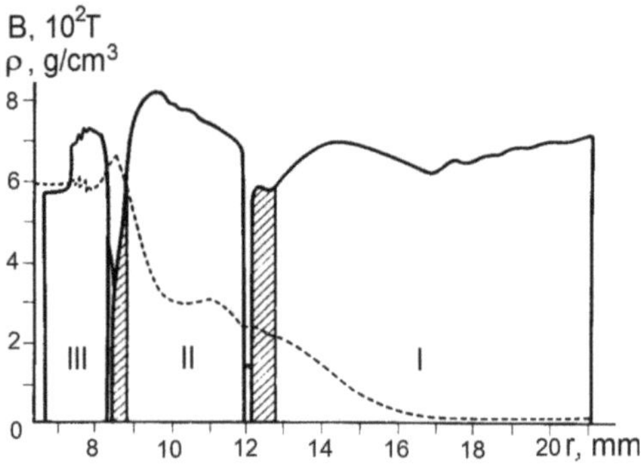

Fig. 10.50: The radial distribution of the substance density ρ and magnetic field induction B (*dotted line*) on the cross section of a three-cascade generator at the instant of the impact ofthe second cascade on the third one. Only the internal layer of the first cascade is shown, namely, solenoid-shell MC-1. The evaporated part of the cascade is *dashed*.

third cascade and trapping the magnetic flux. The magnetic field is constant in the cavity and body of the third cascade prior to the crest of the compression wave. Then it somewhat increases, when compressed in the gap between the second and third cascades. The external boundary of the skin layer in shells is in the body of the third cascade. In other words, the region of the current flow in a three-cascade generator spans the third, then second cascade, and terminates in the body of the first cascade. The fraction of the magnetic flux, therefore, remains frozen in the cascades. This explains the jumps of the coefficient of magnetic flux at the switching of each following cascade, even at ideal compression symmetry. The dashed layers on the internal surface of the first and second cascades define the evaporated fraction of the shell substance, while the material of the third cascade, according to the calculations, does not manage to melt.

10.11 Ways of increasing the induction amplitude. Methods for control over the pulse shape. Capabilities of the MC-1 generator

10.11.1 Methods for control over the shape and amplitude of an induction pulse

In order to extend the use of the cascade generator as a physical tool generating extremely high magnetic fields, and for a profound understanding of the physical processes involved in the magnetic cumulation, certain methods have been advanced and studied which permit the control over the pulse shape and an increase of the pulse amplitude.

The simplest way is to decrease the initial magnetic flux. In this case the effective deceleration of the liner occurs later, the maximum speed increases, and, correspondingly, the maximum induction derivative and final induction also increase. A series of experiments with various initial fields were carried out, which made it possible to obtain statistically reliable data. As a result, it was shown that the induction can be increased by the factor 1.5, but in this case the final diameter of a liner will be decreased approximately to 5 mm, and the probability of obtaining of the maximum final induction sinks to approximately 50% [95]. Figure 10.51 shows the dependences of the final induction B_{fin} and the liner final radius R_{fin} on the induction of initial field B_0, derived in numerical experiments (for the three-

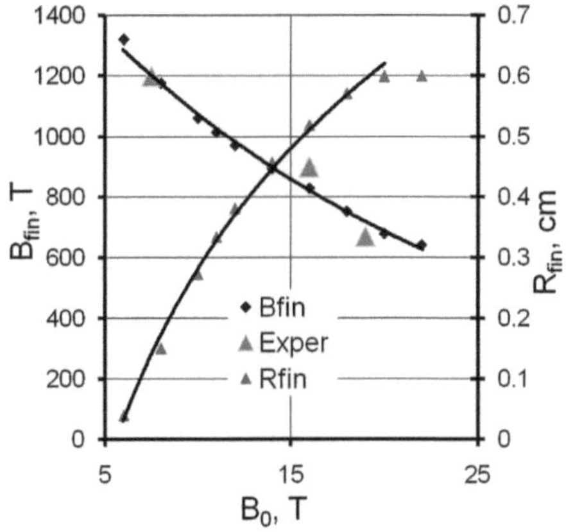

Fig. 10.51: Final parameters of the three-cascade MC-1 generator: induction of the final magnetic field B_{fin} and the final radius of the cavity R_{fin} as functions of the induction of the initial magnetic field B_0. As mentioned above, it is possible to increase the energy of a liner as a whole and its density on the inner surface using, in particular, material of higher density, for instance, tungsten instead of copper in the wired or powder composite [89].

cascade MC-1 generator), together with the experimental points of the dependence $B_{fin} = f(B_0)$.

The numerical model of the MC-1 generator, adjusted according to the data of experiments with tungsten composite, makes it possible to find out the optimal construction and operating regime for the generator with cascades made from the most dense composites, including the heavy inner layer of the first cascade made from insulated tungsten wires. The calculation has shown that in such generators the final magnetic field is higher by 15–20 %, compared to the standard generator with cascades from insulated copper wires. It was also determined that the amplification of the final field was achieved owing to the increase of kinetic energy transferred to the third cascade.

It is obvious that the ultimate parameters of the MC-generator, such as the induction amplitude and dimensions of the final field volume, depend on the type of the explosive substance (high explosive HE), mainly, on its energy capacity. Figure 10.52 shows the results of the modeling calculation of the MC-1 generator with HE of various type. The given induction dependences, as calculated for charges of different energy capacity, show that the greater the charge energy is, the greater is the final magnetic field and the less the final diameter of a liner.

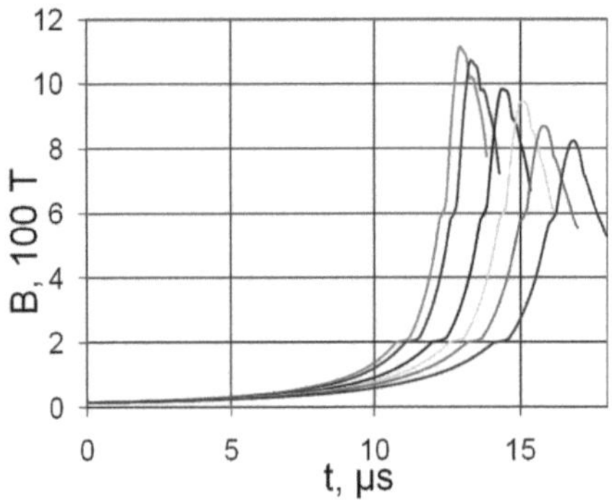

Fig. 10.52: Induction of MC-1 generator for different charge power. Energy capacity increases from right to left.

The first international experiments with a cascade generator, carried out in the laboratory of one of the founders of magnetic cumulation Max Fowler (Los Alamos National Laboratory) played an important role in the advantageous use of cascade generators. Several solenoids with inner cascades and mechanical blocks of a multipointed initiating system have been delivered there, while charges have been made of two explosive substances (types of HE: Comp.B and PBX 9501) in the laboratory, and just there the initiating elements have been provided with HE. Note that the charges were more powerful than those applied in Russian VNIIEF (composition TH 50/50). The experiments with generators were carried out with the participation of VNIIEF researchers [96].

The test results of the three-cascade generator with an initial magnetic field of 1.6 T and different HE are presented in Figure 10.53 as the experimental dependences of magnetic induction *B*(*t*) (points obtained using optical diagnostics), its derivative *dB*(*t*)/*dt* (inductance diagnostics), and trajectories of the liner *D*(*t*), (build up according to marks on derivative curve, featuring the impacts of cascades). This data confirms the above-discussed regularity: the increase in energetics of a liner (in a given case, the implosion speed) at all other equal conditions leads to a rise of the peak field and to the essential decrease of a diameter of the volume with this field.

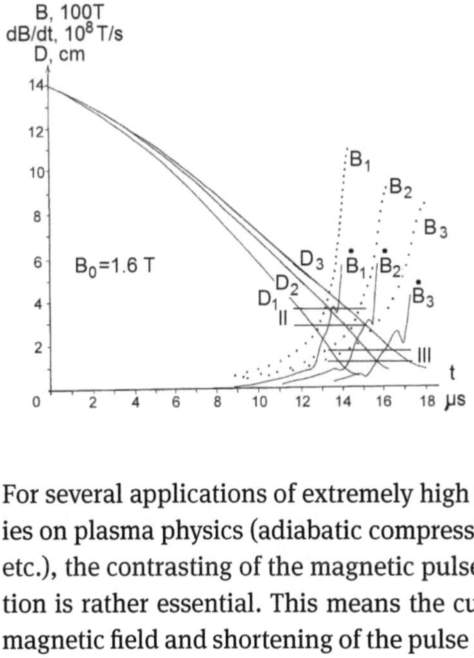

Fig. 10.53: Temporal dependences of the magnetic field, *B*(*t*), its derivative, *dB*/*dt*, and the inner diameters of liners of a three-cascade generator, *D*(*t*), with different *ES*: (1) PBX-9501; (2) Comp B; (3) *TG* 50/50.

For several applications of extremely high magnetic fields, in particular, in studies on plasma physics (adiabatic compression, acceleration of the plasma shells, etc.), the contrasting of the magnetic pulse acting on the object under investigation is rather essential. This means the cutoff of the slow stage of an amplified magnetic field and shortening of the pulse pedestal. It was shown in experiments that the rise time of the ultrahigh magnetic field in an MC-generator of the MC-1 type can been reduced with the help of a simple key such as the thin-wall cylinder installed on the outer surface of the second cascade. The cylinder key is electrically exploded when the field diffuses through its wall. Varying the cylinder thickness, one can control the pulse shape of the magnetic field of the MC-1 generator. The measurements of the magnetic field on the axis of the generator, clearly testified that the rise time of magnetic field was successfully reduced from the value close to zero up to 800 T over time less than 4 μs [97] at duration of flux compression exceeding 16 μs.

10.11.2 Capabilities of MC-1 generators of ultrahigh magnetic fields

The level of understanding of physics of magnetic cumulation along with the experimental and theoretical treatment of actual constructions of MC-1 generators permitted estimation of the capabilities and restrictions in the generation of ultrahigh magnetic fields based on the explosion-driven flux compression with the help of a conventional (chemical) explosive substance.

A set of cascading effects resulted in the overcoming of the essential factor restricting the cumulation, which is the influence of the instabilities of the substance-field boundary, and gave prominence to the main restriction, an energetic one, associated with the fact that the limiting level of the magnetic field, at the given dimensions of the cavity, is fully determined by the storage of chemical energy in the explosive charge. In experiments with the cascade MC-1 generator at a sufficiently high initial field, the passage of the field induction through a maximum was repeatedly registered. This indicates that the initial storage of kinetic energy of the liner is exhausted.

As mentioned above, the simplest method to increase the liner speed is the increase in the amount of explosive substance: in the limiting case of the infinite diameter of the explosive charge, the implosion speed of a liner far from the cumulation axis attains the detonation speed, which is close to 4 km/s and nearly twice as large as the usually attained implosion speed. Removal of the restriction on the outer diameter of the explosive charge made it possible to study the different possibilities for efficiency increase. Besides the increase of the density of the liner material, this can be also done by applying the gas dynamic cumulation of kinetic energy. It is based on successive impacts of the cylinder shells with a decreasing mass.

In order to obtain higher magnetic fields, later on a so-called large explosive charge of type TH 50/50 was used: the inner and outer radius was, respectively, 18 and 32.5 cm, length 36 cm, mass about 140 kg. It made it possible to build a variety of MC-1 generators for specific applications. Thus far, two constructions of MC-1 generators of different dimensions have been built with this type of charges. The schematic drawing and outer view of these generators together with the a "small" generator considered above are shown in Figures 10.54 and 10.55. The basis of the first construction is the solenoid-shell with an inner and outer radius of 16.25 cm and 16.7 cm, respectively, and the basis of the second is the solenoid-shell with radii 8.75 cm and 10 cm. In the second generator the intermediate explosive charge providing a steel inpactor is applied, i.e., the one-cascade system of gas-dynamic acceleration is used.

In experiments where the MC-1 generator had a "large" explosive charge and nearly doubled dimensions and a mass eight times larger, the expected high levels

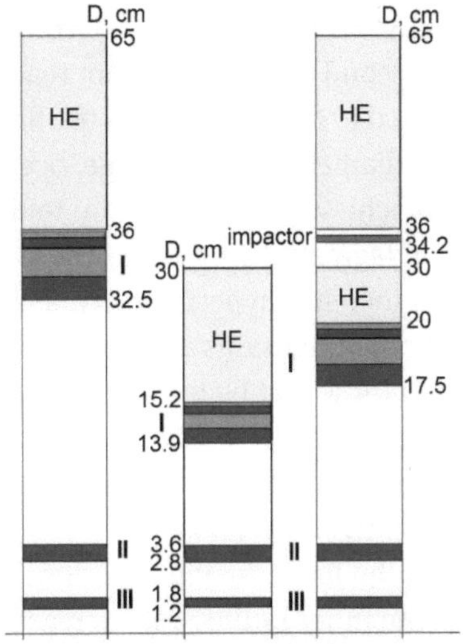

Fig. 10.54: Schematic sectional view of the three cascade MC-1 generators of different dimensions.

Fig. 10.55: Photograph of the three solenoid-shells with an outer radius of 17.8 cm (left), 10 cm (right) and 7.6 cm (center); the last one contains a model of a small explosive charge.

of the final magnetic field have not been achieved, and certain results could at first not be explained. Particularly, in the experiment with a three-cascade generator at sufficiently high initial field (induction of 13 T), a field in excess of 800 T was detected using optical diagnostics, but after the first maximum the field began to decrease, then, in approximately 1.5 µs began to increase again. Only after that was the recording signal terminated. The second unexpected fact was that the impact of the second cascade on the third one was not detected in the recording signals of the inductive probes.

At present these results have been confirmed and explained with a help of a numerical model of a large generator. According to the calculations, a simple increase of dimensions of the explosive charge together with the increase of the liner dimensions does not lead to an increase of the implosion speed, and the magnetic induction remains very nearly the same as in a small MC-1 generator. The total kinetic energy of the liner increases, but the energy density does not, whereas the diameter of the volume with the final field is also increased. In the case when the initial field is sufficiently high, the second cascade does not reach the third one, but stops, and then it is again subjected to compression by the first cascade, which still holds kinetic energy. The large diameter of the volume, occupied by the final field, shows that it is possible to achieve high induction: to do this requires attenuating the initial field to such a degree that a liner would not manage to decelerate before its velocity begins to increase hyperbolically, as in the case of a cylinder. However in this regime the final dimensions are very small, and the probability that the functioning of a generator would be terminated due to the disturbance of the liner symmetry sharply increases.

In order to essentially increase the speed of the liner, the liner of the small generator should be compressed by a large explosive charge. A version of a mediate generator has been developed and tested in which the first cascade is of dimensions somewhat larger than the small generator: the inner radius was increased from 6.95 cm up to 8.75 cm (Figure 10.54, right). Along with this, the intermediate charge is added to the large explosive charge, and it is initiated by a steel impactor (inner radius 17.1 cm, thickness 0.4 cm) driven with the help of a large charge.

As was shown in the numerical calculation of the medium generator, the speed of the inner boundary of the liner increases from cascade to cascade and achieves approximately 10 km/s. Figure 10.56 presents the temporal behavior of the magnetic induction $B(t)$, inner diameter $D(t)$, and speed of the inner boundary of cascades for the three-cascade medium MC-1 generator with the initial field 12 T.

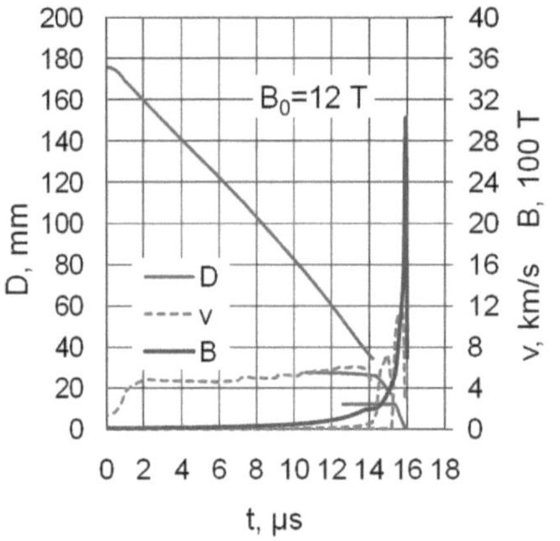

Fig. 10.56: Calculated temporal dependences of the magnetic induction$B(t)$, inner diameter $D(t)$ of the liner cascades, and the speed of the inner boundary of the cascades (*dotted*) for the three-cascade medium MC-1 generator with an initial field of 12 T.

It can be seen that the maximum induction attains 3000 T. Note that the low density nonconducting layer of the vaporized substance, after leaving behind the conducting part of the liner, almost reaches the generator axis (see in Figure 10.57

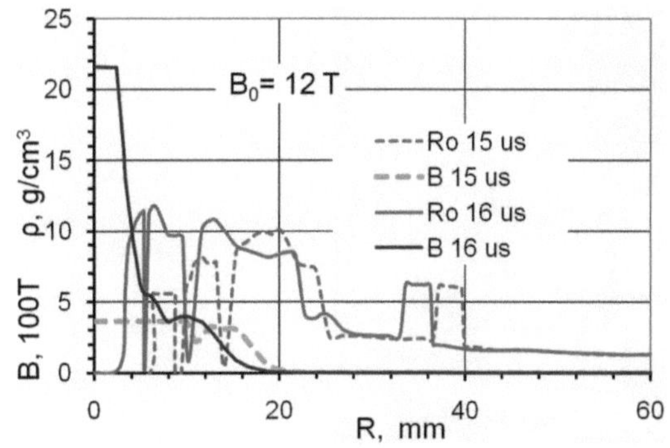

Fig. 10.57: Near-axis fragment of the radial distribution of the substance density R_0 and magnetic induction $B(r)$ of a three-cascade medium MC-1 generator with an initial field of 12 T for two instants of time near the peak field (as obtained in the calculation).

graphs of the radial distribution of the substance density $\rho(r)$ and magnetic induction $B(r)$ for two instants of time: prior to the hit on the third cascade and near the peak field).

Figure 10.58 shows the schematic drawing of the medium generator MC-1 with a large explosive charge installed on the support, and Figure 10.59 shows an external view of the generator being prepared for the experiment. In contrast to the experiments with the small 10-megagauss MC-1 generator, the preparations and experiments with a large generator are much more complicated, long-lasting, and expensive. In particular, the feed supply from the used earlier capacity bank used earlier is not sufficient in a case of the medium or large generator, and the initial

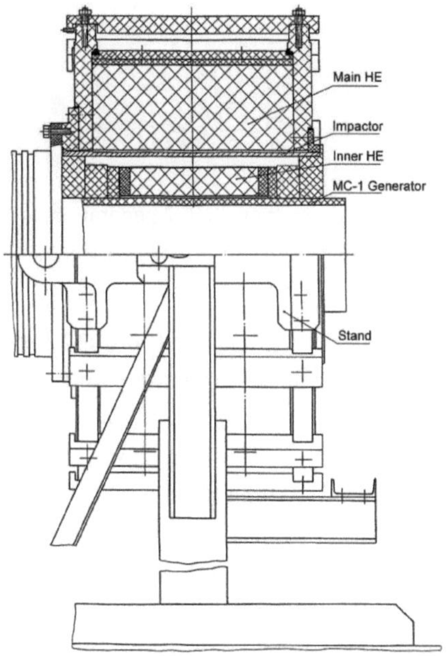

Fig. 10.58: Schematic drawing of the construction of the cascade generator in the 20-megagauss range (medium generator with a large explosive charge).

Fig. 10.59: External view of a medium MC-1 generator prepared for explosion.

Fig. 10.60: External view of a MC-1 generator in the 20-megagauss range (MC-2 generator is a power supply of the initial magnetic field of the MC-1 generator).

field had to be induced by a current of the spiral EMG (MC-2) of the C320 type. Figure 10.60 demonstrates the external view of the similar complicated generator.

In some experiments with this version of MC-1 generator fields of 2000 T have been achieved, and in one of these experiments a record field close to 2800 T (28 MGs, the energy density of the field 30 MJ/cm^3) [98] was detected. The main parameters of this generator MC-1 are shown in Figure 10.61; the maximal value of the derivative of magnetic induction dB/dt achieved $1.5 \cdot 10^{11}$ T/s. It is reasonable to believe that the record field, detected with the most reliable optical diagnostics, was achieved due to the rare combination of the necessary regime and improved

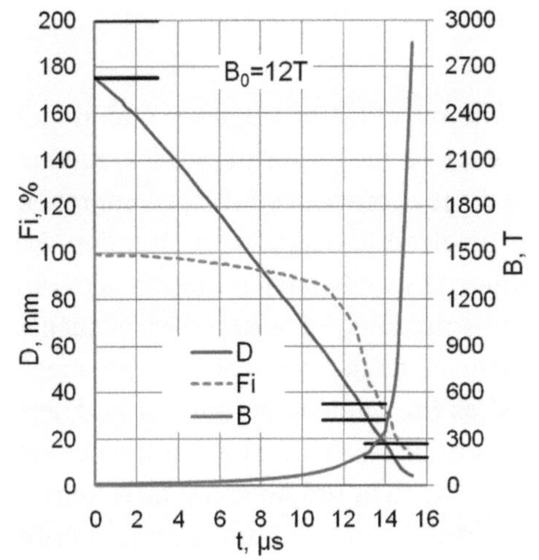

Fig. 10.61: Temporal dependences of the magnetic induction $B(t)$, inner diameter $D(t)$ of a liner, and the conservation coefficient $Fi(t)$ of magnetic flux energy in an experiment with a record magnetic field in excess of 2 800 T. The positions of the cascades are marked with horizontal lines.

symmetry of the compression of cascades. Therefore the generation of extremely strong fields casts some doubt on their practical applications.

The field level in the range of 2000 T is about the limit achievable in the experiments of the acceptable scale and complexity, providing the reliable generation of fields in the region of mm dimension.

Numerical calculations [99] have shown that in the volume approximately of the same transverse dimension it is possible to obtain a field of about 3000 T, if the diameter of explosive charge can be brought to one meter (in this case the HE mass would be less than 500 kg). In such a generator it is reasonable to use the heavy tungsten layer inside the solenoid and cascades. The experiments of the given scale could still be safe and convenient in operation. The advisability of pursuing research in this direction can be proved by the valuable results obtained in experiments with extremely strong fields.

To date the most powerful source of energy is the atomic explosion. It is known that an atomic explosion provides the acceleration of the shells up to 100–500 km/s, which results in an increase of the field level that can be achieved. Realistic evaluations show that the energy of the atomic explosion of the equivalent power of order 100 t TNT is sufficient to be converted with efficiency of about 0.5 % into the energy of a magnetic field with induction of the order 10^5 T in a volume with a diameter 1 mm and length 10 cm [64].

10.12 Conclusion

As a result of long-standing research, Sakharov's idea of magnetic cumulation was intensively developed in VNIIEF and became the powerful physical method for generating superstrong magnetic fields. In the course of studies a comprehensive understanding of physical processes involved in magnetic cumulation was achieved, and the interrelations between the initial and final parameters of the applied devices have been established. The constructions of physical devices such as the cascade generators of ultrahigh magnetic fields in the ten and twenty megagauss range have been developed, as well as the implementations of magnetic cumulation methods into the practice of scientific research. This has opened opportunities for carrying out systematic study in such fundamental directions of physics as solid state physics (studies of optical, magnetic, transport properties of the substance), physics of properties of substances in the extreme conditions (the iso-entropical compression of mega bar range).

To be sure, experiments with MC-1 generators are sufficiently complicated and expensive, with drawbacks intrinsic to the single-action explosion device, but the achieved level of the development of magnetic cumulation makes it possible to obtaing the expected conditions with a probability of about 100%, provided that the experiment is carefully prepared and correctly conducted. The value of these results, and the fact that they could not be obtained in any other way, justify the cost of the experiment. In this case, as has been pointed out many times by the founders of magnetic cumulation, the usefulness of international cooperation is becoming of paramount importance. The possibility of such cooperation is proved by the successful results obtained in "Dirac" experiments, earlier done in the USA in LANL [100], and in "Kapitsa" experiments regularly conducted in Russia at VNIIEF [101].

Supplement S10. Calculation of the skin layer thickness and of the parameter *q*, characterizing the energy in the skin layer

In order to calculate the effective thickness of the skin layer $\Delta'' = \Phi_2/(2\pi r_i B_i)$, one can use the equation describing the diffusion of the field into the shell, the motion of which is described with the help of the model of noncompressible liquid:

$$\frac{1}{r}\frac{\partial}{\partial r}\left(r\rho\frac{\partial B}{\partial r}\right) = \mu_0\left(\frac{\partial B}{\partial t} + u_r\frac{\partial B}{\partial r}\right), \tag{S10.1}$$

where $u_r = (r_i/r)(dr_i/dt)$. After that one can multiply both parts of the equation by $2\pi r \cdot dr$, then integrate from r_i to ∞ and use the relations

$$2\pi\mu_0 \int_{r_i}^{\infty} \frac{\partial B}{\partial t} r \cdot dr = \frac{d\Phi_2}{dt} + 2\pi\mu_0 B_i r_i \frac{dr_i}{dt}, \tag{S10.2}$$

$$2\pi\mu_0 \int_{r_i}^{\infty} u_r \frac{\partial B}{\partial r} r \cdot dr = -2\pi\mu_0 B_i r_i \frac{dr_i}{dt}. \tag{S10.3}$$

The relation (S10.3) contains the value Φ_2, which is the magnetic flux in the skin layer. Then we arrive at the equation

$$\frac{1}{r_i}\frac{d}{dt}\left(r_i B_i \Delta''\right) = \rho_i \left(\frac{\partial B}{\partial r}\right)_{r_i} = \frac{\rho_i B_i}{\mu_0 \Delta'} \approx \frac{\rho_0 B_i^3 \vartheta\, p'}{\mu_0 B_0^2 \Delta''}. \tag{S10.4}$$

Here we used the relations from Section 3.9, appearing there in the model of the skin layer: $\rho_i \approx \vartheta\,(B_i^2/B_0^2)\rho_0$, $\Delta' = \Delta''/p'$.

Then we take into consideration the approximate equality $\vartheta p' \approx 0.3$. Multiply both parts of equation (S10.4) by $\Delta'' B_i r_i^2$ and integrate over from $B_i(0)$ to $B_i(t)$. Since $B_i(0) \ll B_i(t)$, the lower limit of the integration may be set equal to zero. As a result we obtain

$$\left(r_i \Delta'' B_i\right)^2 \approx \frac{2\rho_0 \vartheta\, p'}{\mu_0 B_0^2} \int_0^{B_i} r_i^2 B_i^4 \frac{dt}{dB_i} dB_i, \tag{S10.5}$$

where, according to $r_i = r_M\,(B_M/B_i)^{1/2}$, we have

$$\frac{dt}{dB_i} = \frac{dt}{dr_i}\frac{dr_i}{dB_i} = \frac{dt}{dr_i}\sqrt{B_M}\left(-\frac{1}{2}\right) r_M B_i^{-3/2}. \tag{S10.6}$$

Expressing dr_i/dt with a help of relation (10.10), we find the penetration depth Δ'':

$$\begin{aligned}\frac{\Delta''(t)}{r_M} &= \frac{1}{B_0}\sqrt{\frac{B_M}{B_i}}\left(\frac{\Lambda\gamma_0}{\mu_0}\right)^{1/4}\left(\frac{\rho_0\vartheta\, p' B_M}{r_M}\right)^{1/2}\left(\int_0^{B_i/B_M}\frac{x\cdot dx}{\sqrt{1-x}}\right)^{1/2} \\ &= \frac{1}{B_0}\left(\frac{\Lambda\gamma_0}{\mu_0}\right)^{1/2}\left(\frac{\rho_0\vartheta\, p' B_M}{r_M}\right)^{1/2}\left\{\frac{B_M}{B_i}\left[\frac{4}{3}-\frac{2}{3}\left(\frac{B_i}{B_M}+2\right)\sqrt{1-\frac{B_i}{B_M}}\right]\right\}^{1/2}.\end{aligned} \tag{S10.7}$$

At the instant when the induction calculated at the approximation of the ideal conductivity achieves a maximum, we have $B_i = B_M$. Hence,

$$\frac{\Delta''(t_M)}{r_M} = \frac{2}{\sqrt{3}B_0}\left(\frac{\Lambda\gamma_0}{\mu_0}\right)^{1/4}\left(\frac{\rho_0\vartheta\, p' B_M}{r_M}\right)^{1/2}. \tag{S10.8}$$

Now it is possible to calculate the energy delivered in the skin layer prior to the instant t_M:

$$W_1'(t_M) = \int_0^{t_M} \frac{2\pi r_i E^*(r_i) B_i}{\mu_0} dt = \int_0^{t_M} \frac{2\pi r_i E^*(r_i) B_i}{\mu_0} \frac{dt}{dB_i} dB_i, \tag{S10.9}$$

where $E^* = \rho_i \delta_i = \frac{\rho_0 \vartheta p' B_i^3}{\mu_0 \Delta'' B_0^2}$ is the electric field intensity in the system of reference related to the element of the moving medium.

Further, using the relations (S10.7)–(S10.9), on rearrangement, we obtain

$$W_1'(t_M) = \sqrt{\frac{3}{2}\frac{\pi}{\mu_0}} \left(\frac{\Lambda \gamma_0}{\mu_0}\right)^{1/2} \frac{B_M^{5/2}}{B_0} r_M^{3/2} \left(\vartheta p' \rho_0\right)^{1/2} I,$$

where

$$I = \int_0^1 \frac{x^2 dx}{\sqrt{(1-x)\left[2-(2+x)\sqrt{1-x}\right]}} = 1.18.$$

In conclusion we have

$$q' = \frac{\mu_0 W_1'(t_M)}{\pi r_{M0} \Delta''(t_{M0}) B_M^2} = 1.25.$$

10.13 References

[1] A. D. Sakharov, *Magneto-Explosive Systems and Some Possibilities of their Application*, pp. 40–43, Atom, 2001 (in Russian).

[2] A. D. Sakharov, R. Z. Lyudaev, et al., *Doklady Academii Nauk* **165**(1) (1965), 65–68 [*Soviet Phys. Doklady*].

[3] A. D. Sakharov, *Uspekhi Fiz. Nauk* **88**(4) (1966), 725–734 [*Soviet Physics Uspekhi* **9**(2) (1966), 294–299],

[4] C. M. Fowler, Veiwpoint of an Aging Physicist on Achievements and Prospects of Explosive Sources of Energy and Magnetic Fields, in: G. A. Shvetsov (ed.), *High Energy Density Hydrodynamics*, pp. 37–44, Novosibirsk, Hydrodynamics Institute of SD RAN, 2004.

[5] C. M. Fowler, W. B. Garn, and R. S. Caird, *J. Appl. Phys.* **31**(3) (1960), 588–594.

[6] A. I. Pavlovskii, Memoirs of Various Years, in: B. L. Al'tshuler, B. M. Bolotosky, et al., *He lived with Us … Memoirs for Sakharov*, Moscow, RAN Institute of Physics, Praktika, 1996.

[7] A. I. Pavlovskii, Memoirs of Various Years, *Soviet Phys. Uspekhi* **161**(5) (1991), 137–152.

[8] A. I. Pavlovskii and R. Z. Lyudaev, Magnetic Cumulation of Energy (History and state-of-the-art), *Voprosy Atomnoi Nauki i Tekhniki, seriya: modelirovanie* **4** (1992), 3–18 (in Russian).

[9] C. M. Fowler, **MG-VI** (1994), pp. 3–8.

[10] C. Fowler, D. Thomson, and W. Garn, Explosive Flux Compression: 50 Years of Los Alamos Activities, **MG-VIII** (1998), pp. 21–28.

[11] A. I. Pavlovskii, Magnetic Cumulation – a Memoir for Andrey Sakharov, **MG-VI** (1994), pp. 9–24.
[12] C. M. Fowler and L. L. Altgilbers, Magnetic Flux Compression Generators: a Tutorial and Survey, *Electromagnetic Phenomena* **3**(11) (2003), 306–357.
[13] L. L. Altgilbers, Recent Advances in Explosive Pulsed Power, *Electromagnetic Phenomena* **3**(12) (2003), 497–520.
[14] N. Miura, Y. H. Matsuda, K. Uchida, S. Ikeda, T. Sekitani and F. Herlach, **MG-IX** (2004), pp. 58–67.
[15] S. Takeyama, **MG-XIII** (2010), pp. 163–169.
[16] A. I. Pavlovskii and R. Z. Lyudaev, Magnetic Cumulation, in: A. P. Alexandrov (ed.), *Voprosi sovremennoy experimentalnoy I teoreticheskoy fisiki*, p. 255, Moscow, Nauka, 1984 (in Russian).
[17] R. S. Caird, C. M. Fowler, D. J. Erickson, B. L. Freeman, and W. B. Garn, A Survey of Recent Work on Explosive-Driven Magnetic Flux Compression Generators, in: V. Nardi, H. Sahlin, and W. H. Bostick (eds.), *Energy Storage, Compression, and Switching*, vol. 2., pp. 1–18 Plenum Publishing Corp., 1983.
[18] H. Knoepfel, *Pulsed High Magnetic Fields*, Amsterdam London, North-Holland, 1970.
[19] B. M. Novac and I. R. Smith, Brief History and Classification of Magnetic Flux Compression Generators, *Electromagnetic Phenomena* **3**(11) (2003), 358–365.
[20] F. Herlach, Megagauss Magnetic Fields, *Reps. Prog. in Phys.* **X**(1) (1968), 341–417.
[21] F. Herlach, K. Dransfeld, and J. Hajdu (eds.), *Strong and Ultrastrong Magnetic Fields and Their Applications*, Berlin Heidelberg New York Tokyo, Springer-Verlag, 1985.
[22] E. I. Bichenkov and G. A. Shvetsov, Megagauss Magnetic Fields. Physics. Technology. Applications, *Zh. Prikl. Mech. Tech. Fiz.* **5** (1997), 90–102 [*J. Appl. Mech. Techn. Phys.* **38**(4) (1997), 578–589].
[23] A. I. Pavlovskii, A. I. Bykov, M. I. Dolotenko, A. A. Karpikov, N. P. Kolokolchikov, and O. M. Tatsenko, Ultrahigh Magnetic Fields Cumulation, in: *High Energy Densities: Collection of Scientific Papers*, pp. 446–468, RFNC-VNIIEF, Sarov, 1997.
[24] B. G. Grinevich, V. A. Demidov, A. V. Ivanovski, and V. D. Selemir, Magnetoexplosive Energy Generators and their Application in Scientific Research, in: *VNIIEF 65 Anniversary. High Energy Density Physics and Technology*, Issue 1, pp. 266–311, Sarov, RFNC-VNIIEF, 2011.
[25] G. V. Boriskov, A. I. Bykov, M. I. Dolotenko, N. I. Egorov, Y. B. Kudasov, V. V. Platonov, V. D. Selemir, and O. M. Tatsenko, *Uspekhi Fiz. Nauk* **54**(4) (2011), 441–447 [*Physics – Uspekhi* **54**(4) (2011), 421–427].
[26] L. L. Altgilbers, M. D. J. Brown, I. Grishnaev, B. M. Novac, I. R. Smith, Y. Tkach, and Y. Tkach, *Magnetocumulative Generators*, New York: Springer-Verlag, 2000.
[27] V. E. Portov (ed.), *Explosive Generators of Powerful Pulses of an Electrical Current*, Moscow, Nauka, 2002 (in Russian).
[28] F. Herlach and N. Miura (eds.), *High Magnetic Fields. Science and Technology*, World Scientific Publishing Co., 2003.
[29] G. A. and L. D. Vakulenko (compilers) Shvetsov and *Ultrahigh Magnetic Fields. 1924–1985 Years Bibliographic index*, Novosibirsk, Hydrodynamics Institute of SB RAN, 1986.
[30] H. Knoepfel, *Magnetic Fields, A Comprehensive Theoretical Treatise for Practical Use*, New York, JohnWiley & Sons, Inc., 2000.
[31] D. Rickel, C. M. Fowler, and B. Freeman, **MG-VI** (1994), part 1, pp. 157–164.
[32] V. A. Demidov, A. I. Bykov, L. N. Plyashkevich, A. M.,.S. A. Shuvalov Kazakov, I. K. Fetisov, E. V. Shapovalov, S. N. Golosov, and M. I. Dolotenko, **MG-VII** (1997), pp. 137–139.

[33] A. M. Buyko, S. F. Garanin, and V. V. Zmushko, Powerful Pulsed Electrophysics, in: *Intern. Conf. XIV Khariton Scientific Readings*, Sarov, 2012, Abstracts, p. 21.
[34] A. I. Pavlovskii, R. Z. Lyudaev, V. A. Vasyukov, A. S. Russkov, V. N. Kataev, G. F. Makarcev, and V. F. Bucharov, **MG-III** (1984), pp. 292–297.
[35] V. A. Vasyukov, **MG-VII** (1997), vol. 1, pp. 302–303.
[36] V. K. Chernyshev, V. A. Demidov, S. A. Kazakov, I. K. Fetisov, and V. A. Shvetsov, *MG-VI* (1994), pp. 519–524.
[37] D. B. Reisman, Mini-G: the Development of an Optimized FCG Device, in: 18th IEEE Intern. Pulsed Power Conf. Record – Abstracts, June 19–23, p. 66, 2011, Chicago, IL, USA.
[38] D. A. Goerz, R. A. Anderson, A. D. White, J. B. Javedani, D. B. Reisman, T. J. Ferriera, E. V. Baluyot, R. D. Speer, and D. P. Milhous, **MG-XIII** (2010)
[39] C. M. Fowler, Megagauss X: A Conference Milestone, **MG-X** (2005), pp. 1–7.
[40] E. P. Velikhov, A. A. Vedenov, A. D. Bogdanets, V. S. Golubev, E. G. Kasharsky, F. G. Kiselev A. A., Rutberg, and V. V. Chernukha, *Zh. Tech. Phys.* [*Sov. Phys. Tech. Phys.*] **18**(2) (1973), 274–279.
[41] E. C. Cnare, *J. Appl. Phys.* **37**(10) (1966), 3812–3816.
[42] J. R. Freeman, E. C. Cnare, and R. C. Waag, *Appl. Phys. Let.* **10**(4) (1967), 111–113.
[43] M. M. Guillot, *Comptes Rendus Acad. Seances. Serie B* **268**(12) (1969), 830–832.
[44] D. Kachilla, F. Herlach, and T. Erber., *Rev. of Sci. Instrum.* **41**(1) (1970), 1–7.
[45] V. T. Mikhkel'soo, G. A. Shneerson, and A. P. Shcherbakov, *Pribory I Technika Experimenta* **2** (1974), 212–215. [*Instrum. Exp. Techn.* **17** (1974), 551–554].
[46] S. G. Alikhanov, V. G.,G. I. Belan Budker, A. I. Ivanchenko, and G. N. Kichigin, *Atomnaya Energija* **23**(6) (1967), 536–541.
[47] S. G. Alikhanov, A. I. Ivanchenko, V. N. Karasjuk, and G. N. Kichigin, *J. Sci. Instrum. (J. Phys. E) Ser. 2* **1**(5) (1968), 543–545.
[48] S. G. Alikhanov, V. P. Bachtin, and D. A. Toporkov, **MG-III** (1984), pp. 213–217.
[49] N. Miura, KG. ido, A. Akihiro, and S. Chikazumi, *J. Magnetism and Magnetic Matetials* **11** (1979), 275–283.
[50] T. Goto, N. Miura, K. Nakao, S. Takeyama, and ST. akakibara, **MG-IV** (1987), pp. 149–158.
[51] N. Miura, T. Goto, K. Nakao, S. Takejama, T. Sakakibara, T. Naruyama, and T. Kikuchi, *Phys. B* **155** (1989), 23–32.
[52] N.,*Physica B Miura* **201** (1994), 40–48.
[53] Y. H. Matsuda, F. Herlach, S. Ikeda, and N. Miura, *Rev. of Sci. Instr.* **73**(12) (2002), 4288–4294.
[54] N. Miura and K. Nakao, **MG-IV** (1987), pp. 137–147.
[55] S. Takeyama, **MG-XIII** (2010), pp. 163–169.
[56] B. M. Novac, I. R. Smith, D. F. Rankin, and M. Hubbard, *J. Phys. D. Appl. Phys.* **37** (2004), 3041–3055.
[57] A. Y. Ishlinskiy and M. A. Lavrentyev, *Dokl. Acad. Nauk USSR* **64**(6) (1949), 779–782 (in Russian).
[58] V. T. Mikhkel'soo and G. A. Shneerson, *Zh. Tech. Phys.* **40**(10) (1970), 2198–2208 [*Sov. Phys. Tech. Phys.*].
[59] V. G. Belan, S. T. Durmanov, I. A. Ivanov, V. F. Levashov, and V. L. Podkovirov, **MG-III** (1984), pp. 218–220.
[60] J. P. Somon, *J. Fluid. Mech.* **38**(part 4) (1969), 769–791.

[61] K. H. Almstrem, G. Bjarnholt, S. Y. Goldberg, and M. A.. **MG-VII** (1997) Liberman, pp. 146–153
[62] J. G. Linhart, **MG-1** (1966), pp. 387–396.
[63] F. S. Felber, F. J. Wessel, N. G. Wild, H. U.,A. Rahman Fisher, C. M. Fowler, M. A. Liberman, and A. L. Velikovich, **MG-IV** (1986), pp. 117–124.
[64] A. I. Pavlovskii, S. A. Kholin, A. I. Bykov, O. M. Tatsenko, M. I. Dolotenko, N. P. Kolokol'chikov, et al., **MG-VI** (1994), pp. 149–156.
[65] S. A. Slutz, M. C. Herrmann, R. A. Vesey, A. B. Sefkow, D. B. Sinars, D. C. Rovang, and M. E. Peterson K. J., Cuneo, *Phys. Plasmas* **17** (2010), 056303, 15 pages.
[66] 1st MagLIF Workshop, SNL, Albuquerque, NM, USA, Feb. 5–8, 2012, http://www.sandia.gov/pulsedpower/Workshop.html.
[67] A. L. Velikovich, Some Comments on Physics of Magnetic Field Compression in Plasmas, 1st MagLIF Workshop, SNL, Albuquerque, NM, USA, Feb. 5–8, 2012 / http://www.sandia.gov/pulsedpower/Workshop/Velikovich_DiscussionSlides.pdf.
[68] J. P. Somon, **MG-I** (1966), p.235.
[69] N. B. Volkov, V. T. Mikhkel'soo, E. N. Nagel, and G. A. Shneerson, *Izvestija Acad. Nauk* **1976** (Energetica i Transport), 6, 146–153 (in Russian).
[70] O. V. Gotchev, P. Y. Chang, J. P. Knauer, D. D. Meyerhofer, O. Polomarov, J. Frenje, et al., *Phys. Rev. Lett.* **103** (2009), 215004 [4 pages].
[71] O. V. Gotchev, J. P. Knauer, P. Y. Chang, N. W. Jang, M. J. Shoup, D. D. Meyerhofer, and R. Betti, *Rev. Sci. Instrum.* **80**(4) (2009), 3504.
[72] J. P. Knauer, O. V. Gotchev, P. Y. Chang, D. D. Meyerhofer, O. Polomarov, R. Betti, et al., *Phys. Plasmas* **17** (2010), 056318 [8 pages].
[73] D. Tasker, J. Goforth, and H. Oona, **MG-XII** (2008), pp. 706–714.
[74] A. I. Pavlovskii, N. P. Kolokolchikov, M. I. Dolotenko, and A. I. Bykov, *Pribory I Technika Experimenta* **5** (1979), 195–197 [*Instrum. Exp. Techn.*]
[75] J. E. Besancon, J. David, and J. Vedel, **MG-1** (1966), p. 315.
[76] S. I. Shkuratov, E. F. Talantsev, J. Baird, L. L. Altgilbers, and A. H. Stults, **MG-SF** (2006), pp. 337–342.
[77] E. I. Bichenkov, S. D. Gilev, and L. M. Trubachev, *Zh. Prikl. Mech. Tech. Fiz.* **5** (1980), 125–129 [*J. Appl. Mech. Techn. Phys.* **21** (1980), 678–682].
[78] K. Nagayama, *Appl. Phys. Lett.* **38**(2) (1981), 109–110.
[79] K. Nagayama, T. Oka, and T. Mashimo, *J. Appl. Phys.* **53**(4) (1982), 3029–3037.
[80] S. D. Gilev and A. M. Trubachev, Hydrodynamics Nonsteady Problems, in: Continuous Medium Dynamics 48, pp. 39–342, Novosibirsk, 1980.
[81] K. Nagayama and T. Mashimo, *J. Appl. Phys.* **61**(10) (1987), 4730–4735.
[82] A. B. Prishchepenko, A. A. Barmin, V. V. Markov, and O. E. Mel'nik, **MG-VII** (1997), pp. 154–158.
[83] B. M. Novac, I. R. Smith, D. F. Rankin, Z. Pu, and M. Hubbard, *Proceedings of the IEE Pulsed Power Symposium 2003*, Loughborough University, 23 September 2003.
[84] A. A. Barmin and Prishchepenko A. B., **MG-VI** (1994), pp. 35–40.
[85] S. D. Gilev, *ElectroDynamic Processes at Condensed Matter Shock Compression*, D. Phil, Hydrodynamics Institute, SB RAN, 2009.
[86] V. S. Prokop'ev, A. G. Anisimov, A. D. Matrosov, and G. A. Shvetsov, **MG-XI** (2006), pp. 231–235.
[87] A. I. Pavlovskii, M. I. Dolotenko, N. P. Kolokolchikov, A. I.,N. I. Bykov Egorov, and O. M. Tatsenko, *Pisma v Zh. Tech. Phys.* **9**(22) (1983), 1360–1364 [*Tech. Phys. Lett.*],

[88] A. I. Pavlovskii, M. I. Dolotenko, N. P. Kolokolchikov, A. I. Bykov, N. I. Egorov, and O. M. Tatsenko, *JETPh Lett* **38**(9) (1983), 437–439.
[89] G. V. Boriskov, A. I. Bykov, M. I. Dolotenko, N. I. Egorov, D. I. Zenkov, M. B. Kozlov, N. P. Kolokolchikov, I. M. Narkevtsev, A. N. Moiseenko, and O. M. Tatsenko, **MG-IX** (2004), pp. 44–48.
[90] R. Z. Liudaev, **MG-VII** (1997), pp. 87–114.
[91] A. I. Pavlovskii, A. A. Karpikov, M. I. Dolotenko, and V. I. Mamyshev, **MG-V** (1989), pp. 21–28.
[92] V. V. Aseeva, G. V. Boriskov, M. I. Dolotenko, M. B. Kozlov, V. V. Rudenko, and M. V. Shaburov, The 28th IEEE International Conference on Plasma Science and The 13th IEEE International Pulsed Power Conference, Digest of Technical Papers, R. Reinovsky R. and M. Newton (eds.), Las Vegas, USA, 17–23 June 2001, pp. 1185–1188.
[93] A. I. Pavlovskii, N. P. Kolokolchikov, M. I. Dolotenko, A. I. Bykov, and O. M.. **MG-II** (1980) Tatsenko, pp. 627–639.
[94] A. I. Bykov, M. I. Dolotenko, KN. P. olokol'chikov, A. I. Pavlovskii, and Tatsenko O. M., *Physica B.* **216** (1996), 215–217.
[95] A. I. Pavlovskii, A. I. Bykov, M. I. Dolotenko, A. A. Karpikov, M. I. Dolotenko, N. P. Kolokol'chikov, V. I. Mamyshev, and O. M. Tatsenko, **MG-IV** (1987), pp. 159–166.
[96] M. G. Sheppard, J. H. Brownell, J. M. Christian, C. M. Fowler, B. L. Freeman, J. D. Goettee, J. C. King, C. L. Lund, B. J. Papatheofanis, P. J. Rodriguez, Veeser L. R., Zerwekh W. D., Bykov A. I., Dolotenko M. I., Kolokol'chikov N. P., Y. B. Kudasov, V. V. Platonov, O. M. Tatsenko, W. Lewis, and B. R. Marshall, MC-1 Generator Performance with Higher-Energy Explosives, LANL Report, 1995, LA-UR-95–02.
[97] A. I. Pavlovskii, M. I. Dolotenko, N. P. Kolokol'chikov, A. I. Bykov, and O. M. Tatsenko, **MG-VI** (1994), pp. 751–756.
[98] B .A. Boyko, A. I. Bykov, M. I. Dolotenko, N. P. Kolokol'chikov, I. M. Markevtsev, and A. M. Tatsenko O. M., Shuvalov, 12th IEEE International Pulsed Power Conference, Digest of Technical Papers, C. Stallings C. and H. Kirbie (eds.) IEEE, 1999, pp. 746–749.
[99] M. I. Dolotenko, M. B. Kozlov, and V. D. Selemir, Analysis of Possibilities for Magnetocumulative MC-1 Generator Final Magnetic Field Increasing, in: G. A. Shvetsov (ed.), *High Energy Density Hydrodynamics*, pp. 24–36, Novosibirsk, Hydrodynamics Institute SB RAN, 2004.
[100] J. C. Solem, C. M. Fowler, J. D. Goettee, D. Rickel, L. J. Campbell, et al., **MG-VII** (1997), pp. 756–760.
[101] V. D. Selemir, O. M. Tatsenko, and V. V. Platonov, **MG-X** (2004), pp. 219–226.
[102] M. G. Shappard, P. B. Barch, D. Skudder, J. S. Salem, and L. R. Veeser, **MG-VII** (1997), pp. 205–212.
[103] H. G. Latal, Theory of the Cnare Effect, *Annals of Physics* **42**(2) (1967), 352–364.
[104] N. B. Volkov, V. T. Mikhkel'soo, and G. A. Shneerson, *Zh. Prikl. Mech. Tech. Fiz.* **5** (1982), 15–26. [*J. Appl. Mech. Tech. Phys.* **23**(5) (1982), 607–616].
[105] B. M. Novac, I. R. Smith, and M. Hubbard, *IEEE Trans. on Plasma Sci.* **32**(5) (2004), 1896–1901.
[106] B. M. Novac and I. R. Smith, *Jap. J. Appl. Phys.* **45**(4A) (2006), 2807–2811,
[107] D. F. Rankin, B. M. Novac, and I. R. Smith, *IEEE Proc.-Sci. Meas. Technol.* **153**(3) (2006), 130–138.
[108] C. H. Mielke and B. M. Novac, *IEEE Trans. on Plasma Sci.* **38** (2010), 8, 1739–1749.
[109] N. Miura and H. Nojiry, *Physica B* **216** (1996), 153–157.

[110] N. B. Volkov, *Zh. Tech. Phys.* **49**(9) (1979), 2198–2208 [*Sov. Phys. Tech. Phys.*].
[111] V. T. Mikhkel'soo, in: *Electromagnitniye Processi w Energeticheskih Ustroystvah*, Leningrad, Polytehnichheskiy Institut, 1971, pp. 56–64.
[112] N. Miura and K. Nakao, *Jap. J. Appl. Phys.* **29**(8) (1990), 1580–1599.
[113] N. Miura, G. Kido, T. Goto, H. Miyajima, K. Nakao, and S. Chikasumi, **MG-III** (1984), pp. 116–129.
[114] F. Herlach and J. E. Kennedy, *J. Phys. D: Appl. Phys.* **6** (1973), 661–676.
[115] A. A. Aseeva, B. A. Boyko, A. I. Bykov, M. I. Dolotenko, N. P. Kolokol'chikov, and V. N. Tatsenko O. M., Troitsky, in: 11th IEEE International Pulsed Power Conference. Digest of Technical Papers, G. Cooperstein and I. Vitkovitsky (eds.), IEEE, 1997, pp. 1491–1496.
[116] V. V. Aseeva, B. A. Boyko, A. I. Bykov, M. I. Dolotenko, N. P. Kolokol'chikov, and V. I. Tatsenko O. M., Timareva, **MG-VIII** (1998), pp. 80–84.
[117] B. M. Novac, I. R. Smith, D. F. Rankin, Z. Pu, and M. Hubbard, in: Proceedings of the IEE Pulsed Power Symposium 2003, Loughborough University, 23 September 2003.
[118] H. U. Rahman, P. Ney, F. J. Wessel, A. Fisher, and N. Rostoker, in: Proc. 2nd Int. Conf. Dense Z-pinches, Laguna Beach, 1989, p. 351.
[119] F. J. Wessel, N .C. Wild, A Fisher, H. U. Rahman, A. Ron, and F. S. Felber, *Rev. Sci. Instrum.* **57** (1986), 2246.

Index

Ablation compression, 363
Accelerating system, 100
Acceleration, 259
Accelerator, 102, 260
Action integral, 46, 71
Adiabatic heating, 147
Alfven velocity, 346
Alikhanov, 356
All-Russian Scientific Research Institute of Experimental Physics – RFNC-VNIIEF, 334
Alloy MP35N, 157
Altgilbers, 336
Anderson criterion, 309
Angle, 22, 68
Angle rim, 57
Apex, 22
Artsimovich, 100
Aspect ratio, 133, 169, 203, 214
Asymptotical values, 57
Axial compressing force, 383

Bakulin–Luchinsky model, 321
Beryllium bronze, 270, 273
Bessel functions, 12, 184, 245
Bichenkov, 387
Bifilar elements, 178
Bitter coils, 148
Bitter's solenoid, 120, 175
Body, 266
Boltzmann constant, 319
Boundary, 256, 298, 312, 321
Boundary condition, 14, 52, 62
Bronze, 149, 150, 175
Buses, 239, 240

Calculation, 23
Capacitors, 239
Capacity, 282
Capacity energy storage, 92
Carbon fiber, 157
Cascade generator as a physical tool, 407
Catastrophic failure, 169
Chapman–Zhuge condition, 318
Characteristic magnetic field, 81
Chemical explosive charge, 379
Circuit, 98, 101
Closely-packed insulated thin copper wires, 396
Cnare, 356
Cnare-effect, 368
Coaxial modules, 170
Cockroft, 26
Coil, 16, 35, 39, 94, 99, 171, 227, 279
Complete elliptic integrals, 10, 76
Compound, 125, 150, 152
Compressibility, 349
Compressing, 176
Compression symmetry, 391
Computer simulation, 159
Concentrator, 244, 249
Condition, 201
Conditions of the generator in the vicinity of the field peak, 391
Conductivity, 14, 47, 249, 261, 290, 305, 320
Conductor, 122, 140, 219, 283
Configurations, 183, 186, 201
Conformal mapping, 202
Conformal mapping method, 42
Conservation coefficient of the flux, 355
Consideration of dimensions, 58
Constructions of solenoids, 149
Contacts, 149
Contrasting of the magnetic pulse, 409
Converging detonation wave, 379
Copper, 272
Copper matrix, 154
Correction coefficient, 41
Cracks, 233, 234, 258
Cross section, 57
Cu–Ag conductors, 155
Cu–Nb composite, 164
Cu–Nb microcomposite, 154
Cu–Nb wire, 171
Current, 8, 11, 38, 42, 46, 96, 99, 117, 121, 122, 137, 188–190, 195, 199, 254
Current amplification coefficient, 339

Current amplitude, 92
Current density, 13, 32, 54, 79, 84, 118
Current opening switch, 341
Current pause, 303
Cut, 29
Cutoff of external part, 133
Cylinder, 35, 41, 108, 117, 118, 121, 157, 160, 184
Cylindrical coordinates, 10

De-collectivization, 306
Deformation, 141, 143, 256, 257, 275, 276, 281, 283, 290
δ-function, 62
Density, 238, 257, 259, 262, 293, 296, 304, 311, 317
Density of the liner material, 355
Destruction, 270
Dewar, 163
Dielectric materials, 157
Diffusion, 57, 61, 76, 78, 81, 281, 292, 326
"Dirac" experiments, 416
Disk, 106, 175
Distribution, 308
Distribution of current density in a liner, 372
Dresden Magnetic Field Laboratory (DMFL), 173
Duhamel integral, 61
Duration, 238
Dynamic yield, 224
Dynamics of the liner shape and state, 391

Edge, 11, 22, 31, 54
Edge effect, 21, 236, 373
Educational research-user complex MASTER, 402
Effect of the slots, 244
Effect of training, 165
Effective depth of the flux penetration, 346
Efficiency of instability suppression, 396
Ejection, 281, 317
Elastic, 113
Elasticity, 113, 117
Elasticity theory, 353
Electric arc, 303
Electroconductivity, 304
Electromagnetically-driven flux compression (MDC), 355
Electron, 308
"Electrotechnical" model of the accelerator, 98
Enameled copper wire, 383
Ends, 119
Energy, 91, 92, 98, 103, 109, 110, 294, 309, 316, 323
energy balance at peak field, 394
Energy capacity storage, 100
Energy losses, 52, 53
Energy storage, 99
Energy transfer, 239, 260
Energy transferred to the field, 350
Equation of state, 367
Equilibrium, 114, 121, 123, 201, 202, 304, 306, 329
Equivalent stress, 124
Erosion, 231, 233
Estimate, 213
Euler constant, 42
Euler's method, 111
Evaporation, 279, 310
Expansion, 270
Experiments of the acceptable scale and complexity, 415
Explosion, 259, 277, 297, 298, 301, 307, 310, 314, 315
Explosion test field, 344
Explosive substance (ES), 334
Exponentially damped sinusoid, 68
Expressions for functions of the flux, 211

Fabri, 94, 95
Factor, 272, 280, 281, 290
Factor R/h, 115
Failure, 237
Felber, 363
Fermi temperature, 305
Field, 8, 91, 133, 250, 252, 262, 290, 294, 295, 301, 313, 321
Field diffusion, 50, 346
Field penetration, 325
Field-conductor boundary, 347
Filling factor, 125
Flat helical solenoids, 175

Flow, 278, 283, 284, 289, 290, 294, 295, 298, 314, 320, 327, 328
Flow function, 8
Fluid, 275, 309
Flux, 20, 188
Flux compression, 334
Flux concentrator, 242
Foner and Kolm, 176
Force-free field, 182
Force-free winding, 182
Forces, 20, 113, 114, 117, 130, 247, 254, 261, 318
Fourier–Bessel integral, 12
Fourier–Bessel transformation, 32
Fowler, 336
Free-elastic oscillations, 226
Free-energy zone, 305
Frenel integrals, 64
Frequency, 249
Front, 320, 327

Gap, 41, 42
Garanin's equation, 321
Gas dynamic cumulation of kinetic energy, 410
Generator MC-1, 333
Geometric factors, 261
Gersford, Müller, and Roeland, 152
Gilev, 387
Glass cloth, 151
Glid Cop material, 155

Heating, 46, 65, 70, 73, 76, 162, 230, 247, 281
Heating of conductors, 45
Heating-cooling cycle, 230
Helmholtz pair, 382
High and ultrahigh magnetic fields, 336
High energy density physics, 335
High rate, 236
High rate of energy input, 340
High-voltage impulse technology, 291
Hooke law, 141
Hydraulic press, 176
Hypervoltage, 304
Hypocritical regime, 319
Hysteresis loop, 143

Ideal conductivity, 54
Incorrect problem, 191
Independent supply, 170
Inductance, 42, 56, 99, 107, 166, 168, 248
Induction, 10, 29, 32, 71, 94, 122, 137
Inductive storages, 109
Initial current, 339
Initial magnetic flux, 334
Instability, 257, 279, 308, 329
Insulated conductive spacers, 360
Integral, 302
Integral equation, 36
Integral equation of the first kind, 16
Intensity, 286
Interaction, 254
International Annealed Copper Standard, 155
International megagauss community, 335
Ionization, 306, 308

Joule heating, 60

"Kapitsa" experiments, 416
Kapitsa's experiments, 149
Kapton/zylon material, 173
Kelvin's problem, 96
Kirenskii Institute of Physics, 151
Knoepfel, 336
Kondorskiy and Susov, 176
Kristoffel–Schwarz's formula, 70

L–R–C-circuit, 93
Lagrange method, 97
Lagutin, Rossel, Herlach, and van Bruynseraede, 163
Laplace equations, 40
Laplace transformation, 51
Large MC-1 generator, 411
Laser interferometer, 216
Law, 62, 127, 284
Layers, 75, 126, 128–131, 136, 148, 160, 183, 193, 194, 198, 199
Legendre polynomials, 36
Lifetime, 251
Liner, 334
Liners of MC-1 generators, 384
Linhart, 100, 363
Liquid, 283, 290

Liquid nitrogen, 163, 169
Load, 240
Longitudinal and transverse compression of the liner, 389
LosAlamos National Laboratory – LANL, 334
Losses, 66
Lyudaev, 399

Macrocomposites, 156
Magnet, 29, 35, 37, 106, 131, 133, 136, 147, 150, 213, 219, 240, 242, 246, 248, 270, 273, 278, 292, 295, 300, 312
Magnetic constant, 293
Magnetic counterpressure, 393
Magnetic cumulation, 342
Magnetic cumulation (MC), 333
Magnetic energy, 136
Magnetic field, 8, 60, 70, 114, 147, 255, 270, 273
Magnetic field diffusion, 50
Magnetic field energy, 99
Magnetic flux conservation, 339
Magnetic limit, 276
Magnetic liner inertial fusion (MagLIF), 363
Magnetic momentum, 10
"Magnetic piston", 324
Magnetic pressure, 203, 235, 255, 350
Magnetic system, 8, 109, 196, 259, 265
Magnetocumulative devices, 335
Magnetodynamic cumulation, 358
Manson and Koffin empirical formula, 144
Material state, 259
Materials, 154
Matter ejection, 277
Maximum, 65
Maxwell equation, 253
Maxwell model, 257
MC-2 generator, 334
McDonald function, 78
Measuring unit of the generator, 394
Mechanical properties, 157
Mechanical stresses, 220
Mediate MC-1 generator, 412
Medium, 78, 80, 81, 289, 298
Megagauss conferences, 335
Megagauss fields, 219
Megagauss physics, 336
Method of coupled circuits, 373
MHD radial flow, 377
"Microturbulence", 329
Minimization, 76
Miura, 356
Mixing of the evaporated substance and magnetic field, 393
Modules, 173
Modulus, 238
Monitoring the winding state, 168
Most effective protection of the measuring unit, 398
Multilayer systems, 159
Multimodule disk generator (MMDG), 341
Multistart multilayer coil, 383
Multistart winding, 383

Nagayama, 387
Nb-nanowhiskers, 155
New construction of an MC-1 generator, 386
Niobium-titanium alloy, 154
Noncompressive ideally conductive liquid, 345
Nondimensional variables, 102
Nonequilibrium, 308
Nonexplosive acceleration of a liner, 355
Nonferromagnetic, 265
Number of gofers, 353
Number of turns, 140
Numerical estimation, 59
Numerical modeling, 102
Numerical one-dimensional model of the actual generator, 391

One-layer magnet, 216
One-modular configurations, 202
Optimal number of turns, 93
Optimization, 94, 113, 163
Optimization problem, 111
Osaka University magnets, 161
Oscillations, 51
Outer zone, 210
Outlets, 153

Parceval's formula, 66
Parseval formula, 72
Partial compensation, 197

Penetration, 249
Penetration depth, 13, 51
Periphery, 192
Phase transition, 377
Photos, 278
Plane, 18, 41, 130, 175
Plane-wave, 250
Plasma, 239, 270, 304
Plasma compression, 247
Plasma liner, 363
Plasma Z-Θ-pinch, 363
Plastic deformation, 140, 257
Point, 284
Poisson coefficient, 117
Potential, 8, 40, 253
Powder composite, 399
Power, 98, 282, 313
Power loss, 56
Power supply source, 91
Powerful physical method, 416
Pressure, 255, 258, 292, 294, 299
Principle of cascading in MC-1 generators, 395
Principle of the controlled conductivity of a substance, 396
Problem of streams, 201
Procedure of iterations, 202
Process, 165, 166
Program SMOG-DISK, 400
Project, developed at NHMFL and LANL, 171
Projectile, 99, 100, 105
Properties, 304
Prototype method, 216
Pulse, 61
Pulsed x-ray radiography, 391

Quasi-force-free magnets, 210, 214
Quasi-force-free winding, 182, 194, 213
Quick-acting MCG, 341

Radial slots, 236
Rankin–Hugonio conditions, 299
Ratio, 88
Rayleigh equation, 289
Rayleigh–Taylor instabilities, 353
Record fields, 169
Regime, 66, 101, 102, 105, 147, 165
Region, 187, 191, 210
Reheat cracking, 234
Reinforcement, 133, 136, 137, 149, 150, 152, 160, 161, 164, 173, 174, 176, 188, 203, 210
Relative deformation, 225
Residual stresses and sizes, 213
Resistance, 33, 47, 52
Resistivity, 48, 293
Resource, 140, 144, 228, 231
Reverse motion, 325
Reynolds magnetic number, 349
Rigid sealing, 119
Ring, 29
Rise time, 236, 240
Root, 185
Rosenbluth, Furth, and Case, 76
Rounding, 26
Rupture, 222
Russian Federal Nuclear Center – VNIIEF (RFNC-VNIIEF), 383

Sakharov, 333
Sandia National Laboratories, 363
Schneider-Muntau and Simson, 174
Screen, 20, 203, 204
Self-similar solutions, 85
Service life, 166
Shell, 150, 252
Sheppard, 367
Shvetsov, 336
Silicon-organic polymer, 151
Single-turn MC-generator, 338
Skin effect, 13, 72, 147, 261, 311
Skin layer, 52, 227, 234, 328, 346
Skin layer method, 85
Slot, 31, 41, 236, 275
Slot edges, 231
Small (standard) MC-1 generator, 412
Small deformations, 216
Solenoid, 8, 10, 18, 20, 96, 113, 116, 154, 246, 260, 283, 288, 289, 300
Solenoid reverse withdrawal, 384
Solid phase, 304
Source, 300
Source of initial magnetic flux, 382
Spatial inhomogeneity, 309
Specific resistance, 78, 80

Speed of the liner implosion, 355
Spinodal, 305
Spiral, 20, 128
Spitzer formula, 306
Stability of the shell shape, 352
Static regime, 222
Steel, 150, 161
Steel 40, 224
Storage, 111
Strength, 116, 219, 270, 301
Strength parameter, 203, 209
Stress B_M, 115
Stress peaks, 221
Stresses, 113, 114, 121, 123, 126, 129, 141, 150, 152, 169, 198, 221, 238
Strip transmission line, 24
Strips, 177
Surface, 37, 52, 53
Switches, 239
Switching, 111
Synchronous generator, 173
System, 171, 207
System of solenoids, 92

Takeyama, 356
Tantalum, 270
Temperature, 46, 75, 152, 154, 304
Temperature dependence, 224
Temperature increment, 227, 378
Ten and twenty megagauss range, 416
Tensile strength, 223
Tensor, 230
Tests, 164
The processes involved in the final stage of MDC, 374
Thermal processes, 45
Thermoconductivity, 312
Thermoelastic stresses, 227, 229
Three-cascade generator, 397
Threshold, 277
Threshold induction, 228, 231
Threshold value, 223
"top-hat" MC generator, 380
Toroid, 32
Transformer, 241, 242, 249
Tungsten composite, 398
Turn, 114, 279, 286, 287, 290, 292, 297
Turnaround, 343
Turns, 272
Two boundary conditions, 190
Two-cascade generator, 396
Two-conductor line, 56
Two-module magnet with a field of 100 T, 171
Types, 147

University of Oxford, 150

Van Bockstal, Heremans, and Herlach, 160
Variations of resistance, 166
Vector diagram, 196
Velocity, 50, 87, 274, 276, 283, 287, 289, 290, 299, 301, 309, 310, 313, 314, 319, 328
Versions of the mutual arrangement of solenoid, liner, and explosive, 382
Virial theorem, 186
Voltage, 102, 139, 140
Volume density of thermal energy, 47
Von Mises criteria, 119
Von Mises formula, 200
Von Mises stress, 221

Wall, 16, 37
Wave, 82, 276, 300, 317
Ways of increasing the induction amplitude, 407
Welding, 250
Wiener–Hopf method, 37
Winding, 20, 116, 117, 122, 125, 128, 131, 147, 149, 150, 159, 161, 190, 192, 196, 201
Wire, 150
Wittaker formula, 12
Wood alloy, 247

X-ray, 278, 308

Yield point, 141

Z- and θ-pinches, 108
Zero thickness, 24
Zylon/epoxy composite, 157, 173

Bei Fragen zur Produktsicherheit wenden Sie sich bitte an:
If you have any questions regarding product safety,
please contact:

Walter de Gruyter GmbH
Genthiner Straße 13
10785 Berlin
productsafety@degruyterbrill.com